T0203595

# Pile Design and Construction Practice

## Sixth Edition

# Pile Design and Construction Practice

## Sixth Edition

Michael Tomlinson and John Woodward

CRC Press
Taylor & Francis Group
Boca Raton  London  New York

CRC Press is an imprint of the
Taylor & Francis Group, an **informa** business

A SPON PRESS BOOK

The publishers and authors disclaim any liability in whole or part, arising from information contained in this book. The reader is advised to consult with an appropriate licensed professional before taking action or making any interpretation of the material in this book.

CRC Press
Taylor & Francis Group
6000 Broken Sound Parkway NW, Suite 300
Boca Raton, FL 33487-2742

First issued in paperback 2020

© 2015 by The estate of Michael J Tomlinson and John C Woodward
CRC Press is an imprint of Taylor & Francis Group, an Informa business

No claim to original U.S. Government works

ISBN-13: 978-1-4665-9263-6 (hbk)
ISBN-13: 978-0-367-65901-1 (pbk)

This book contains information obtained from authentic and highly regarded sources. Reasonable efforts have been made to publish reliable data and information, but the author and publisher cannot assume responsibility for the validity of all materials or the consequences of their use. The authors and publishers have attempted to trace the copyright holders of all material reproduced in this publication and apologize to copyright holders if permission to publish in this form has not been obtained. If any copyright material has not been acknowledged please write and let us know so we may rectify in any future reprint.

Except as permitted under U.S. Copyright Law, no part of this book may be reprinted, reproduced, transmitted, or utilized in any form by any electronic, mechanical, or other means, now known or hereafter invented, including photocopying, microfilming, and recording, or in any information storage or retrieval system, without written permission from the publishers.

**Trademark Notice:** Product or corporate names may be trademarks or registered trademarks, and are used only for identification and explanation without intent to infringe.

---

**Library of Congress Cataloging-in-Publication Data**

---

Tomlinson, M. J. (Michael John)
  Pile design and construction practice / Michael Tomlinson, John Woodward. -- Sixth edition.
    pages cm
  Includes bibliographical references and index.
  ISBN 978-1-4665-9263-6 (hardback)
  1. Piling (Civil engineering) I. Woodward, John, 1936- II. Title.

TA780.T65 2014
624.1'54--dc23                                                                2014024302

---

**Visit the Taylor & Francis Web site at**
**http://www.taylorandfrancis.com**

**and the CRC Press Web site at**
**http://www.crcpress.com**

# Contents

# Preface to the sixth edition

Two factors are driving the development of modern pile design and construction–the growth in demand for high-rise buildings and the subsequent requirement for ever-larger piles, frequently in areas with poor subsoils. New piling techniques and powerful piling rigs have effectively addressed the problems of producing piles to cope with the larger structural loads, and significant improvements have taken place in understanding the behaviour of piles. However, despite the advances in analytical and numerical methods using sophisticated computer software which allow theoretical soil mechanics solutions to be applied to aspects of pile design, much reliance still has to be placed on empirical correlations. The late Michael Tomlinson was an empiricist committed to the scientific method with extensive practical knowledge, and these principles and applications are still the backbone of practical pile design.

A guiding precept in this edition was therefore to keep to the spirit of MJT's work, retaining a substantial amount of his writings on the technicalities of pile design, particularly the demonstration of the basic principles using his hand calculation methods and the reviews of the extensive case studies. However, there are new codified design procedures which have to be addressed. For example, the formal adoption in Europe of the Eurocodes for structural design (and 'load and resistance factor design' more generally elsewhere) has led to new ways of assessing design parameters and safety factors. One of the main objectives in this edition has been to give an overview of the current Eurocode requirements combined with the practicalities of applying the new suite of British Standards which relate to construction materials and installation procedures. However, compliance with the more systemised Eurocode rules has not necessitated any significant changes to the well-established procedures for determining ultimate geotechnical values for routine pile design. For more complex structures, such as offshore structures and monopiles, the new design methods for driven piles in clays and sands, developed from the extensive laboratory research and field testing by Imperial College for example, represent an important practical advance in producing economical foundations.

The author wishes to thank David Beadman and Matina Sougle of Byrne Looby Partners for a review of the reworked examples, Chris Raison of Raison Foster Associates for comments on current Eurocode 7 pile design; Paul Cresswell of Abbey Pynford for his contribution on micropiles; Colin O'Donnell for comments on contractual matters; and Tony Bracegirdle, David Hight, Hugh St John, Philip Smith and Marina Sideri of Geotechnical Consulting Group for their reviews, contributions and inputs on many of the topics. Any remaining errors are the authors.

Many specialist piling companies and manufacturers of piling equipment have kindly supplied technical information and illustrations of their processes and products. Where appropriate, the source of this information is given in the text. Thanks are due to the following for the supply of and permission to use photographs and illustrations from technical publications and brochures.

| | |
|---|---|
| Abbey Pynford Foundation Systems Ltd | Figure 2.14 |
| ABI GmbH | Figures 3.1 and 3.2 |
| American Society of Civil Engineers | Figures 4.6, 4.11, 4.12, 4.13, 4.39, 5.15, 5.28, 6.29, 9.29 and 9.30 |
| Bachy Soletanche | Figures 2.28a and b, 3.15 and 3.35 |
| Ballast Nedam Groep N.V. | Figure 9.23 |
| Bauer Maschinen GmbH | Figure 3.27 |
| David Beadman | Figure 4.43 |
| BSP International Foundations Limited | Figures 3.10 and 8.17 |
| Building Research Establishment | Figure 10.2a and b |
| Roger Bullivant Limited | Figure 7.18 |
| Canadian Geotechnical Journal | Figures 4.34, 4.36, 4.37, 5.18, 5.26 5.36 and 6.9 |
| A. Carter | Figure 9.24 |
| Cement and Concrete Association | Figure 7.14 |
| Cementation Skanska Limited | Figures 3.30, 3.34, and 11.11 |
| Construction Industry Research and Information Association (CIRIA) | Figures 4.8 and 10.4 |
| Danish Geotechnical Institute | Figures 5.6 through 5.10, |
| Dar-al-Handasah Consultants | Figure 9.14 |
| Dawson Construction Plant Limited | Figures 3.5, 3.19 and 3.20 |
| Department of the Environment | Figure 10.1 |
| DFP Foundation Products | Figure 2.27 |
| Frank's Casing Crew and Rental Inc | Figure 2.16 |
| Fugro Engineering Services Ltd | Figure 11.3b |
| Fugro Loadtest | Figures 11.12, 11.13 and 11.19 |
| GeoSea and DEME | Figure 8.16 |
| Gregg Marine Inc | Figure 11.4 |
| Highways Agency | Figure 9.18 |
| International Construction Equipment | Figure 3.3 |
| International Society for Soil Mechanics and Foundation Engineering | Figures 3.40, 5.22, 6.30 |
| Institution of Civil Engineers/Thomas Telford Ltd | Figures 4.21, 4.28, 5.24, 5.25, 5.31, 5.32, 5.33, 5.39, 5.40, 9.22, 9.24, 9.26 and 9.27 |
| Keller Geotechnique and Tata Steel Projects | Figures 9.4 and 9.5 |
| Large Diameter Drilling Ltd | Figure 3.31 |
| Liebherr Great Britain Limited | Figure 3.4 |
| Macro Enterprises Ltd | Figure 9.1e |
| Malcolm Drilling Company | Figures 2.29, 2.34 and 3.32 |
| Maxx Piling | Figure 2.15 |
| MENCK GmbH | Figure 3.11 |
| Moscow ISSMGE | Figure 5.23 |
| National Coal Board | Figures 4.26 and 8.2 |
| Numa Hammers | Figure 3.33 |
| Palgrave MacMillan | Figures 7.12 and 7.13 |

(continued)

| | |
|---|---|
| Pearson Education | Figure 4.22 |
| Oasys Ltd | Figure 9.19 |
| Offshore Technology Conference | Figures 4.16, 5.27 and 8.19 |
| Seacore Limited | Figures 3.7, 3.12 and 3.37 |
| Sezai-Turkes-Feyzi-Akkaya Construction Company | Figure 4.23 |
| Sound Transit, Seattle | Figure 3.38 |
| Spanish Society for Soil Mechanics and Foundations | Figure 9.21 |
| Steel Pile Installations Ltd | Figure 3.9 |
| Stent Foundations Limited | Figure 2.32 |
| Swedish Geotechnical Society | Figures 5.20 and 5.26 |
| Test Consult Limited | Figure 11.14 |
| TRL | Figures 9.17 and 9.20 |
| Vibro Ménard (Bachy Soletanche Group) | Figure 3.15 |
| John Wiley and Sons Incorporated | Figure 4.10 |

The cover photograph shows two vertical travel box leads, 60 m long, as supplied by Bermingham Foundation Solutions company to Gulf Intracoastal Constructors, being erected to drive the 48 m long by 760 mm diameter steel piles for the pumping station at Belle Chasse, Louisiana. Pile driving was by the B32 diesel hammer (see Table 3.4) for vertical and 3:1 batter piles. With permission of Bermingham Foundation Solutions of Hamilton, Ontario.

Figure 4.42 is after Figure 4.47 on page 136 of 'Piling Engineering' 3rd edition 2009, by Fleming, Weltman, Randolph and Elson, published by Taylor & Francis, with permission.

Figure 9.25 is published with the permission of the Deep Foundations Institute as originally published in the DFI 2005 Marine Foundations Speciality Seminar proceedings. Copies of the full proceedings are available through Deep Foundations Institute, Hawthorne, NJ; Tel: 973-423-4030; E-mail: dfihq@dfi.org.

Permission to reproduce extracts from British Standards is granted by BSI. British Standards can be obtained in PDF or hard copy formats from the BSI online shop: www.bsigroup.com/shop or by contacting BSI Customer Services for hard copies only: Tel: +44 (0)20 8996 9001, E-mail: cservices@bsigroup.com.

Extracts from AASHTO LRFD Bridge Design Specification, copyright 2010 by the American Association of State Highway and Transport Officials, Washington, DC, are used by permission.

Extracts from Australian Standard AS 2159-2009, Piling – Design and installation, are reproduced with permission from SAI Global Ltd under licence 1311-c073. The standard may be purchased online at http://www.saiglobal.com.

John C. Woodward
*Princes Risborough, United Kingdom*

**LIST OF TABLES**

# Preface to the first edition

Piling is both an art and a science. The art lies in selecting the most suitable type of pile and method of installation for the ground conditions and the form of the loading. Science enables the engineer to predict the behaviour of the piles once they are in the ground and subject to loading. This behaviour is influenced profoundly by the method used to install the piles, and it cannot be predicted solely from the physical properties of the pile and of the undisturbed soil. A knowledge of the available types of piling and methods of constructing piled foundations is essential for a thorough understanding of the science of their behaviour. For this reason, the author has preceded the chapters dealing with the calculation of allowable loads on piles and deformation behaviour by descriptions of the many types of proprietary and non-proprietary piles and the equipment used to install them.

In recent years, substantial progress has been made in developing methods of predicting the behaviour of piles under lateral loading. This is important in the design of foundations for deep-water terminals for oil tankers and oil carriers and for offshore platforms for gas and petroleum production. The problems concerning the lateral loading of piles have therefore been given detailed treatment in this book.

The author has been fortunate in being able to draw on the worldwide experience of George Wimpey and Company Limited, his employers for nearly 30 years, in the design and construction of piled foundations. He is grateful to the management of Wimpey Laboratories Ltd. and their parent company for permission to include many examples of their work. In particular, thanks are due to P. F. Winfield, FIStructE, for his assistance with the calculations and his help in checking the text and worked examples.

Michael J. Tomlinson
*Burton-on-Stather, United Kingdom*
*1977*

# Chapter 1

# General principles and practices

## 1.1 FUNCTION OF PILES

Piles are columnar elements in a foundation which have the function of transferring load from the superstructure through weak compressible strata or through water onto stiffer or more compact and less-compressible soils or onto rocks. They may be required to carry uplift loads when used to support tall structures subjected to overturning forces – from winds or waves. Piles used in marine structures are subjected to lateral loads from the impact of berthing ships and from waves. Combinations of vertical and horizontal loads are carried where piles are used to support retaining walls, bridge piers and abutments and machinery foundations.

## 1.2 HISTORY

The driving of bearing piles to support structures is one of the earliest examples of the art and science of the civil engineer. In Britain, there are numerous examples of timber piling in bridgeworks and riverside settlements constructed by the Romans. In mediaeval times, piles of oak and alder were used in the foundations of the great monasteries constructed in the fenlands of East Anglia. In China, timber piling was used by the bridge builders of the Han Dynasty (200 BC to AD 200). The carrying capacity of timber piles is limited by the girth of the natural timbers and the ability of the material to withstand driving by hammer without suffering damage due to splitting or splintering. Thus, primitive rules must have been established in the earliest days of piling by which the allowable load on a pile was determined from its resistance to driving by a hammer of known weight and with a known height of drop. Knowledge was also accumulated regarding the durability of piles of different species of wood, and measures were taken to prevent decay by charring the timber or by building masonry rafts on pile heads cut off below water level.

Timber, because of its strength combined with lightness, durability and ease of cutting and handling, remained the only material used for piling until comparatively recent times. It was replaced by concrete and steel only because these newer materials could be fabricated into units that were capable of sustaining compressive, bending and tensile forces far beyond the capacity of a timber pile of like dimensions. Concrete, in particular, was adaptable to in situ forms of construction which facilitated the installation of piled foundations in drilled holes in situations where noise, vibration and ground heave had to be avoided.

Reinforced concrete, which was developed as a structural medium in the late nineteenth and early twentieth centuries, largely replaced timber for high-capacity piling for works on land. It could be precast in various structural forms to suit the imposed loading and ground conditions, and its durability was satisfactory for most soil and immersion conditions. The partial replacement of driven precast concrete piles by numerous forms of cast-in-place piles has been due more to the development of highly efficient machines for drilling pile boreholes of large diameter and great depth in a wide range of soil and rock conditions, than to any deficiency in the performance of the precast concrete element.

Steel has been used to an increasing extent for piling due to its ease of fabrication and handling and its ability to withstand hard driving. Problems of corrosion in marine structures have been overcome by the introduction of durable coatings and cathodic protection.

## 1.3 CALCULATIONS OF LOAD-CARRYING CAPACITY

While materials for piles can be precisely specified, and their fabrication and installation can be controlled to conform to strict specification and code of practice requirements, the calculation of their load-carrying capacity is a complex matter which at the present time is based partly on theoretical concepts derived from the sciences of soil and rock mechanics but mainly on empirical methods based on experience. Practice in calculating the ultimate resistance of piles based on the principles of soil mechanics differs greatly from the application of these principles to shallow spread foundations. In the latter case, the entire area of soil supporting the foundation is exposed and can be inspected and sampled to ensure that its bearing characteristics conform to those deduced from the results of exploratory boreholes and soil tests. Provided that the correct constructional techniques are used, the disturbance to the soil is limited to a depth of only a few centimetres below the excavation level for a spread foundation. Virtually, the whole mass of soil influenced by the bearing pressure remains undisturbed and unaffected by the constructional operations (Figure 1.1a). Thus, the safety factor against general shear failure of the spread foundation and its settlement under the design *applied load* (also referred to as the *working load*) can be predicted from knowledge of the physical characteristics of the 'undisturbed' soil with a degree of certainty which depends only on the complexity of the soil stratification.

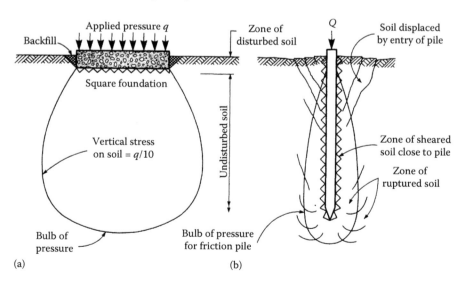

Figure 1.1 Comparison of pressure distribution and soil disturbance beneath spread and piled foundations: (a) spread foundation; (b) single pile.

The conditions which govern the supporting capacity of the piled foundation are quite different. No matter whether the pile is installed by driving with a hammer, jetting, vibration, jacking, screwing or drilling, the soil in contact with the pile face, from which the pile derives its support by shaft friction and its resistance to lateral loads, is completely disturbed by the method of installation. Similarly, the soil or rock beneath the toe of a pile is compressed (or sometimes loosened) to an extent which may affect significantly its end-gearing resistance (Figure 1.1b). Changes take place in the conditions at the pile–soil interface over periods of days, months or years which materially affect the shaft friction resistance of a pile. These changes may be due to the dissipation of excess pore pressure set up by installing the pile, to the relative effects of friction and cohesion which in turn depend on the relative pile–soil movement, and to chemical or electrochemical effects caused by the hardening of the concrete or the corrosion of the steel in contact with the soil. Where piles are installed in groups to carry heavy foundation loads, the operation of driving or drilling for adjacent piles can cause changes in the carrying capacity and load/settlement characteristics of the piles in the group that have already been driven.

Considerable research has been, and is being, carried out into the application of soil and rock mechanics theory to practical pile design. However, the effects of the various methods of pile installation on the carrying capacity and deformation characteristics of the pile and ground cannot be allowed for in a strict theoretical approach. The application of simple empirical factors to the strength, density and compressibility properties of the undisturbed soil or rock remains the general design procedure to determine the relevant resistances to the applied loads. The various factors which can be used depend on the particular method of installation and have been developed over many years of experience and successful field testing.

The basis of the *soil mechanics approach* to calculating the carrying capacity of piles is that the total resistance of the pile to compression loads is the sum of two components, namely, shaft friction and base resistance. A pile in which the shaft-frictional component predominates is known as a friction pile (Figure 1.2a), while a pile bearing on rock or some other hard incompressible material is known as an end-bearing pile (Figure 1.2b). The need for adopting adequate safety factors in conjunction with calculations to determine the design resistance of these components is emphasised by the statement by Randolph[1.1] 'that we may never be able to estimate axial pile capacity in many soil types more accurately than about ±30%'. However, even if it is possible to make a reliable estimate of total pile resistance, a further difficulty arises in predicting the problems involved in installing the piles to the depths indicated by the empirical or semi-empirical calculations. It is one

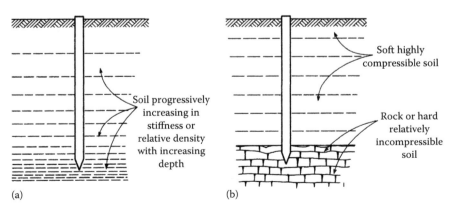

(a)                                              (b)

*Figure 1.2* Types of bearing pile: (a) friction pile; (b) end-bearing pile.

problem to calculate that a precast concrete pile must be driven to a depth of, say, 20 m to carry safely a certain applied load, but quite another problem to decide on the energy of the hammer required to drive the pile to this depth, and yet another problem to decide whether or not the pile will be irredeemably shattered while driving it to the required depth. In the case of driven and cast-in-place piles, the ability to drive the piling tube to the required depth and then to extract it within the pulling capacity of the piling rig must be correctly predicted.

Time effects are important in calculating the resistance of a pile in clay; the effects include the rate of applying load to a pile and the time interval between installing and testing a pile. The shaft-frictional resistance of a pile in clay loaded very slowly may only be one-half of that which is measured under the rate at which load is normally applied during a pile loading test. The slow rate of loading may correspond to that of a building under construction, yet the ability of a pile to carry its load is judged on its behaviour under a comparatively rapid loading test made only a few days after installation. Because of the importance of such time effects both in fine- and coarse-grained soils, the only practicable way of determining the load-carrying capacity of a piled foundation is to confirm the design calculations by short-term tests on isolated single piles and then to allow in the safety factor for any reduction in the carrying capacity with time. The effects of grouping piles can be taken into account by considering the pile group to act as a block foundation, as described in Chapter 5.

## 1.4 DYNAMIC PILING FORMULAE

The method of calculating the load-carrying capacity of piles mentioned earlier is based on a soil mechanics approach to determine the resistance of the ground to static loads applied at the test-loading stage or during the working life of the structure. Historically, all piles were driven with a simple falling ram or drop hammer and the pile capacity was based on the measurement of the ground resistance encountered when driving a pile. The downward movement of the pile under a given energy blow is related to its ultimate resistance to static loading. Based on the considerable body of experience built up in the field, simple empirical formulae were derived, from which the ultimate resistance of the pile could be calculated from the *set* of the pile due to each hammer blow at the final stages of driving. However, there are drawbacks to the use of these formulae when using diesel hammers due to the increase in energy delivered as the ground resistance increases and changes in hammer performance related to the mechanical condition and operating temperature. Driving tests on preliminary piles instrumented to measure the energy transferred to the pile head together with a pile driving analyser (PDA) can provide a means of applying dynamic formula for site control of working piles.

The more consistent hydraulic hammers overcome many of the problems of energy transfer and the availability of a large database of hammer performance and improvements in the application of PDAs has meant that under the right conditions, dynamic formulae can be reliable (see Section 7.3). Hence, the Eurocode for geotechnical design (EC7-1 Clause 7.6.2.5; see Section 1.5) allows the use of pile driving formulae to assess the ultimate compressive resistance of piles where the ground conditions are known. Also, the formula has to have been validated by previous experience of acceptable performance in similar ground conditions as verified by static loading tests on the same type of pile.

While the dynamic formula approach may now be more reliable, it can only be applied to driven piles and is being replaced by the use of pile driveability and stress wave principles. The basic soil mechanics design approach, and the associated development of analytical and numerical methods, can be applied to all forms of piling in all ground conditions.

## 1.5 INTRODUCTION OF EUROCODES AND OTHER STANDARDS

The Eurocodes[1.2], formulated by the transnational technical committees of the European Committee for Standardisation (CEN), are the Europe-wide means of designing works to produce identical, harmonised specifications for safe buildings, structures and civil engineering works. The United Kingdom, which adopted the European Public Procurement Directive of 2004 (2004/17/EC) through the Public Contracts Regulations of 2006, must ensure that all public projects in England, Wales and Northern Ireland are specified in terms of Eurocodes. Although there is no current legal requirement for structural design for private sector works to comply with Eurocodes, this is likely to change in the future under European trade directives.

The Eurocodes make a fundamental change to traditional UK design practice. They are not based on allowable stress and allowable capacity of materials calculated using overall (*global*) factors of safety, but on limit state design principles and partial factors applied to separate elements of the design, depending on the reliability which can be placed on the parameters or calculations. There are 10 structural Eurocodes made up of 58 parts which supersede the previous UK design standards, largely withdrawn by the British Standards Institute (BSI) in 2010. The main Codes of Practice, BS 8002 and BS 8004 dealing with foundation design and construction, are therefore no longer available. The concrete design standard, BS 8110 which was based on limit state principles, has also been withdrawn.

The BSI adopts and publishes, on behalf of CEN, the following *normative* standards for geotechnical design (with the prefix BS EN and the commonly used abbreviations):

EC7-1   BS EN 1997-1:2004 Eurocode 7: Geotechnical design, Part 1 General rules
EC7-2   BS EN 1997-2:2007 Eurocode 7: Geotechnical design, Part 2 Ground investigation and testing

EC7, which deals with the variable nature of soils and rock, differs in some respects from other structural codes where materials are more consistent in strength and performance. EC7 has to be read in conjunction with the following structural Eurocodes referenced in this text which bear on foundation design:

EC1-1   BS EN 1991-1-1:2002 Eurocode 1: Part 1-1 Actions on structures. General actions – Densities, self-weight, imposed loads for buildings
EC2-1   BS EN 1992-1-1:2004 Eurocode 2: Design of concrete structures, Part 1-1 General rules and rules for buildings
EC3-1   BS EN 1993-1-1:2005 Eurocode 3: Design of steel structures, Part 1-1 General rules and rules for buildings
EC3-5   BS EN 1993-5:2007 Eurocode 3: Design of steel structures, Part 5 Piling
EC4-1   BS EN 1994-1:2005 Eurocode 4: Design of composite steel and concrete structures, Part 1 General rules
EC5-1   BS EN 1995-1-1:2004 Eurocode 5: Design of timber structures, Part 1-1 General rules
EC6-1   BS EN 1996-1:2005 Eurocode 6: Design of masonry structures, Part 1 General rules
EC8-1   BS EN 1998-1:2004 Eurocode 8: Design of structures for earthquake resistance, Part 1 General rules
EC8-5   BS EN 1998-1:2004 Eurocode 8: Design of structures for earthquake resistance, Part 5 Foundations, retaining walls and geotechnical aspects

The objectives of the suite of Eurocodes are set out in BS EN 1990:2002, Basis of structural design, namely, to demonstrate structural resistance, durability and serviceability for the structure's designed working life. The clauses designated *principles* (*P*) in all Eurocodes are mandatory (i.e. *shall* clauses); the *informative* clauses indicate the means by which the principles may be fulfilled.

Each part of the Eurocode has to be read in conjunction with its corresponding *National Annex* (an *informative* document referred to here as the NA) which provides, within prescribed Eurocode limits, nationally determined parameters, partial factors and design approach to meet a country's particular conditions and practices for the control of its design process. The NA factors, published separately from the Eurocodes, are to be distinguished from those in *Annex A* (*normative*) in the Eurocode. The NA also sets out the procedures to be used where alternatives to the Eurocode are deemed necessary or desirable. Not all countries have produced NAs, but the UK Annexes for both parts of EC7 (and most of the other Eurocodes) are now applicable and importantly modify the parameters and factors published in Annex A. Designers therefore must be aware of the many variations to EC7 which exist in Europe when designing piles in one country for execution in another. Designers will be free to apply higher standards than given in the Eurocodes if considered appropriate and may use unique design factors provided they can be shown to meet the prime objectives of the Eurocodes. Such alternatives will have to be supported by relevant testing and experience.

Eurocodes introduce terms not familiar to many UK designers, for example *load* becomes *action* and *imposed load* becomes *variable action*. *Effect* is an internal force which results from application of an action, for example settlement. These and other new load conditions, *permanent unfavourable* and *permanent favourable*, require the application of different load factors depending on which of the *design approaches* and factor *combinations* are being used. The structural engineer is required to assess which actions give the critical effects and special care is needed when deciding on which actions are to be considered as separate variable actions; actions include temperature effects and swelling and shrinkage.

The United Kingdom has modified the EC7 partial factors in its NA to reflect established practice and has adopted Design Approach 1 (DA1) for foundations using partial factor combinations 1 and 2 in which the factors are applied at source to actions and ground strength parameters, requiring reliable and technically advanced soils testing laboratories. However, for *pile design*, the partial factors must be applied to the ground *resistance* calculations. This is inconsistent with the rest of EC7.

Clause 7 of EC7-1 deals with piled foundations from the aspects of actions on piles from superimposed loading or ground movements, design methods for piles subjected to compression, tension and lateral loading, pile-loading tests, structural design and supervision of construction. In using Clause 7, the designer is required to demonstrate that the sum of the ultimate limit state (ULS) components of bearing capacity of the pile or pile group (ground *resistances R*) exceeds the ultimate limit state design loading (*actions F*) and that the serviceability limit state (SLS) is not reached. New definitions of *characteristic* values (cautious estimate based on engineering judgement) and *representative* values (tending towards the limit of the credible values) of material strengths and actions are now given in BS EN 1990 and BS EN 1991 which must be considered when examining the various limit states (see Section 4.1.4). The use of cautious estimates for parameters can be important in view of the limitations imposed by the partial factors for resistance, especially for values of undrained shear strength at the base of piles. The representative actions provided by the structural engineer to the foundation designer should state what factors have been included so that duplication of factors is avoided.

EC7-1 does not make specific recommendations on calculations for pile design; rather, emphasis is placed on preliminary load testing to govern the design. Essentially, EC7-1 prescribes the succession of stages in the design process using conventional methods to calculate end-bearing resistance, frictional resistance and displacement and may be seen as the means for checking (*verifying*) that a design is satisfactory. This edition exclusively applies DA1 and the UK NA, and the reader who needs to consider DA 2 and 3 is referred to examples in Bond and Harris[1.3] which show the differences in design outcomes using the specified parameters from EC7-1. CIRIA Report C641 (Driscoll et al.[1.4]) highlights the important features of the Eurocodes applicable to geotechnical design using DA1 and the NA factors. The guide by Frank et al.[1.5] outlines the development of the code and gives a clause-by-clause commentary. The limit state and partial factor approach in EC7 should result in more economic pile foundations – particularly in the case of steel piles where the material properties are well defined.

The current EC7 procedures are not very amenable to the application of sophisticated computational developments in theoretical analyses, which in due course may produce further savings. In order to capitalise on these advances, two factors will have to be addressed: firstly, significant improvements in determining in situ soil parameters are required and, secondly, designers must have gained specialist expertise and competence to undertake the necessary modelling and be aware of the limitations. In any event, it is considered that a good understanding of the proven empirical geotechnical approach will be essential for future economic pile design, with continued validation by observations and publication of relevant case studies.

EC7 is to undergo a significant evolution over the next few years which should avoid the anomalies and difficulties in interpreting some of the current procedures; a new version will be published sometime after 2020.

New European standards (EN) have also been published dealing with the 'execution of special geotechnical works' (bored piling, displacement piles, sheet piles, micropiles, etc.) which have the status of current British Standards (and also designated BS EN). These, together with new material standards, are more prescriptive than the withdrawn codes and are extensively cross-referenced in this text. Selection of the design and installation methods used and the choice of material parameters remain within the judgement and responsibility of the designer and depend on the structure and the problems to be solved. Generally, where reference is made in Eurocodes to other BS, the requirements of the corresponding BS EN should take precedence. However, parts of existing standards, for example amended BS 5930: 1999 and BS 1377: 1997, are referred to in EC7-2 in respect of ground investigation and laboratory testing.

Where there is a need for guidance on a subject not covered by a Eurocode or in order to introduce new technology not in the ENs, BSI is producing 'noncontradictory' documents entitled 'Published Documents' with the prefix PD. Examples are PD 6694 which is complementary to EC7-1 for bridge design and PD 6698 which gives recommendations for design of structures for earthquake resistance; all come with the rider that 'This publication is not to be regarded as a British Standard'.

Geotechnical standards are also prepared by the International Standards Organisation (ISO) in cooperation with CEN. When an ISO standard is adopted by BSI as a European *norm*, it is given the prefix BS EN ISO. It is currently dealing with the classification of soil and rock and ground investigations generally and, when completed, the new set of ISO documents will supersede all parts of BS 5930 and BS 1377.

The UK Building Regulations 2010[1.6] set out the statutory requirements for design and construction to ensure public health and safety for all types of building; the complementary 'Approved Documents' give guidance on complying with the regulations. Approved Document A now refers exclusively to British Standards based on Eurocodes.

As noted earlier, some aspects of withdrawn standards are still referred to in the new BS ENs but designers should be aware of the risks of inappropriately mixing designs based on the new standards with withdrawn BS codes[1.7]. Designers should also be aware that compliance with a BS or BS EN does not confer immunity from the relevant statutory and legal requirements and that compliance with Eurocodes may be mandatory.

Working to code rules is only part of the design process. An understanding of the soil mechanics and mathematics behind the codes is essential, and designs and procedures should always be checked against comparable experience and practice. It is also important to avoid over-specification of design and construction as a result of applying new structural Eurocodes and the associated execution codes[1.8].

Alternative forms of limit state design, usually referred to as *load and resistance factor design* (LRFD), are being adopted and codified in many jurisdictions (see Section 4.10). Here, the factored load should not exceed the factored resistance, whereas the EC7-1 principle is that factored load should not exceed the resistance as determined by factored shear strength parameters (but note the previous comment for pile design).

A list of current and pending British Standards relating to geotechnical design is given in Appendix B.

## 1.6 RESPONSIBILITIES OF EMPLOYER AND CONTRACTOR

Contract conditions and procurement methods for construction in Britain for both main contracts and specialist work have changed significantly in recent years to meet new legal obligations and to implement the Eurocodes. These changes, which are considered in more detail in Section 11.2.1, have altered the relative responsibilities of the parties to a contract and the delegation of responsibilities to the parties' advisors and designers. Under the traditional piling contract arrangements, the employer's engineer is responsible for the overall design and supervision of construction. In this case, the engineer is not a party to the contract between the employer and contractor and must act impartially when carrying out duties as stated in the contract. With regard to the foundations, the engineer will have prepared, possibly with a geotechnical advisor[1.9], the mandatory Geotechnical Design Report and determined the geotechnical categories as required in EC7-1 and EC7-2 (see Section 11.1). The responsibility for the detailed design of the piles may then lie with the engineer or the piling contractor.

The New Engineering and Construction Contract (NEC3)[1.10], which is increasingly being used on major projects, does not provide for the employer to delegate authority to an engineer. A project manager is appointed under a contract with the employer to employ designers and contractors and to supervise the whole works, in accordance with the employer's requirements and instructions. The piles may be designed by the project manager's team or by the contractor.

The engineer/project manager has a duty to the employer to check the specialist contractor's designs, as far as practically possible, before approval can be given for inclusion in the permanent works. This will include determining that proper provision has been made by the piling specialist to cope with any difficult ground conditions noted in the ground investigation, such as obstructions or groundwater flow. Checks will also be made on pile dimensions, stresses in the pile shaft, concrete strengths, steel grades, etc. in accordance with specifications, relevant standards and best practice. However, the risks and liabilities of the piling contractor for his designs will not normally be reduced by prior approval. If the employer through the project manager provides the design, the risk for a fault in the design will generally fall to the employer.

The basic methods of undertaking the works either by employer-provided design or contractor design are outlined in Section 11.2.1. In all cases, the piling contractor is responsible for ensuring that reasonable skill and care has been and will be exercised in undertaking the piling works, usually confirmed in a form of warranty from the specialist.

The Eurocodes do not comment specifically on responsibility for checks, but require that *execution* is carried out by 'personnel having the appropriate skill and experience'; also that 'adequate supervision and quality control is provided during execution of the work, i.e. in design offices…and on site'. Here, 'execution' must be taken to mean both the design and construction of the piles. 'Adequate supervision' is not defined, but under the auspices of the Ground Forum of the Institution of Civil Engineers, a *Register of Ground Engineering Professionals*[1.9] has been developed to meet the European requirement to identify suitably qualified and competent personnel to address the issue.

The liability for dealing with unforeseen ground conditions should be explicitly addressed in the contract conditions. Similarly, the party liable for providing any additional piles or extra lengths compared with the contract quantities should be identified. If the piling contractor had no opportunity to contribute to the ground investigation, it would be reasonable for the contract to include rates for extra work and for payment to be authorised. Payment would not be appropriate if the piling contractor is shown to have been overcautious, but a decision should not be made without test pile observations or previous knowledge of the performance of piles in similar soil conditions. Contractor-designed piling has promoted the development of highly efficient and reliable piling systems, which means a contractor is less able to claim for extra payments.

Whichever form of contract is used, it is the structural designer's responsibility to state the limit for settlement of the foundation at the applied loads based on the tolerance of the structure to total and differential settlement (the serviceability). He must specify the maximum permissible settlement at the representative load and at some multiple in a pile load test, say, 1.5 times, as this is the only means that the engineer/project manager has of checking that the design assumptions and the piles as installed will fulfil their function in supporting the structure. It frequently happens that the maximum settlements specified are so unrealistically small that they will be exceeded by the inevitable elastic compression of the pile shaft, irrespective of any elastic compression or yielding of the soil or rock supporting the pile. However, the specified settlement should not be so large that the limit states are compromised (Section 4.1.4). It is unrealistic to specify the maximum movement of a pile under lateral loading, since this can be determined only by field trials.

The piling contractor's warranty is usually limited to that of the load/settlement characteristics of a single pile and for soundness of workmanship, but responsibilities regarding effects due to installation could extend to the complete structure and to any nearby existing buildings or services; for example, liability for damage caused by vibrations or ground heave when driving a group of piles or by any loss of ground when drilling for groups of bored and cast-in-place piles. The position may be different if a building were to suffer damage due to the settlement of a group of piles as a result of consolidation of a layer of weak compressible soil beneath the zone of disturbance caused by pile driving (Figure 1.3). In the case of an employer-designed project, the designer should have considered this risk in the investigations and overall design and specified a minimum pile length to take account of such compressible layer. The rights of third parties in respect of damage due to construction are now covered by statute (see Section 11.2.1).

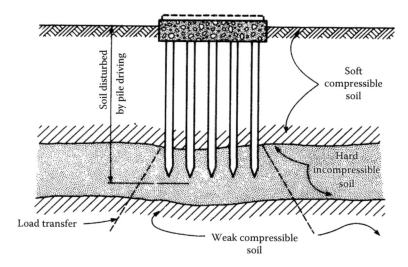

*Figure 1.3* Pile group terminating in hard incompressible soil layer underlain by weak compressible soil.

## REFERENCES

1.1 Randolph, M.F. Science and empiricism in pile foundation design, *Geotechnique*, 53 (10), 2003, 847–875.
1.2 British Standards Institution. BS EN 1990 to BS EN 1999. Eurocodes 0–10: BSI, London, UK.
1.3 Bond, A. and Harris, A. *Decoding Eurocode 7*. Taylor & Francis, London, UK, 2008.
1.4 Driscoll, R., Scott, P. and Powell, J. EC7 – Implications for UK practice. Eurocode 7 Geotechnical design, Construction Industry Research and Information Association, Report No C641, 2008.
1.5 Frank, R., Bauduin, C., Driscoll, R., Kavvadas, M., Krebs Ovesen, N., Orr, T. and Schuppener, B. *Designer's Guide to EN 1997-1 Eurocode 7: Geotechnical Design – General Rules*. Thomas Telford, London, UK, 2004.
1.6 *The Building Regulations 2010*. Department for Communities and Local Government, London, UK, 2010.
1.7 Department for Communities and Local Government, Circular to Local Government executives and heads of Building Control dated 29 January 2010.
1.8 Infrastructure Cost Review. Industry Standards Group of HM Treasury. *Specifying Successful Standards*. Institution of Civil Engineers, London, UK, 2012.
1.9 ICE 3009(4). *UK Register of Ground Engineering Professionals*. Institution of Civil Engineers, London, UK, 2011.
1.10 NEC3. *Engineering and Construction Contract*. Thomas Telford, London, UK, September 2011.

# Chapter 2

# Types of pile

## 2.1 CLASSIFICATION OF PILES

The traditional classification of the three basic categories of bearing piles is as follows:

1. *Large-displacement piles* comprise solid-section piles or hollow-section piles with a closed end, which are driven or jacked into the ground and thus displace the soil. All types of driven and cast-in-place piles come into this category. Large-diameter screw piles and rotary displacement auger piles are increasingly used for piling in contaminated land and soft soils.
2. *Small-displacement piles* are also driven or jacked into the ground but have a relatively small cross-sectional area. They include rolled steel H- or I-sections and pipe or box sections driven with an open end such that the soil enters the hollow section. Where these pile types plug with soil during driving, they become large-displacement types.
3. *Replacement piles* are formed by first removing the soil by boring using a wide range of drilling techniques. Concrete may be placed into an unlined or lined hole, or the lining may be withdrawn as the concrete is placed. Preformed elements of timber, concrete or steel may be placed in drilled holes. Continuous flight auger (CFA) piles have become the dominant type of pile in the United Kingdom for structures on land.

Eurocode 7 Part 1[1.2] (EC7-1, all Eurocodes are referenced in Section 1.5 and Appendix B) does not categorise piles, but Clause 7 applies to the design of all types of load-bearing piles. When piles are used to reduce settlement of a raft or spread foundation (e.g. Love[2.1]), as opposed to supporting the full load from a structure, then the provisions of EC7 may not apply directly.

Examples of the types of piles in each of the basic categories are as follows:

### 2.1.1 Large-displacement piles (driven types)

1. Timber (round or square section, jointed or continuous)
2. Precast concrete (solid or tubular section in continuous or jointed units)
3. Prestressed concrete (solid or tubular section)
4. Steel tube (driven with closed end)
5. Steel box (driven with closed end)
6. Fluted and tapered steel tube
7. Jacked-down steel tube with closed end
8. Jacked-down solid concrete cylinder

11

### 2.1.2 Large-displacement piles (driven and cast-in-place types)

1. Steel tube driven and withdrawn after placing concrete
2. Steel tube driven with closed end, left in place and filled with reinforced concrete
3. Precast concrete shell filled with concrete
4. Thin-walled steel shell driven by withdrawable mandrel and then filled with concrete
5. Rotary displacement auger and screw piles
6. Expander body

### 2.1.3 Small-displacement piles

1. Precast concrete (tubular section driven with open end)
2. Prestressed concrete (tubular section driven with open end)
3. Steel H-section
4. Steel tube section (driven with open end and soil removed as required)
5. Steel box section (driven with open end and soil removed as required)
6. Steel sheet piles used as combined retaining wall and vertical load bearing

### 2.1.4 Replacement piles

1. Concrete placed in hole drilled by rotary auger, baling, grabbing, airlift or reverse-circulation methods (bored and cast-in-place or in American terminology *drilled shafts*)
2. Tubes placed in hole drilled as earlier and filled with concrete as necessary
3. Precast concrete units placed in drilled hole
4. Cement mortar or concrete injected into drilled hole
5. Steel sections placed in drilled hole
6. Steel tube drilled down

### 2.1.5 Composite piles

Numerous types of piles of composite construction may be formed by combining units in each of the preceding categories or by adopting combinations of piles in more than one category. For example, composite piles of a displacement type can be formed by jointing a timber section to a precast concrete section, or a precast concrete pile can have an H-section jointed to its lower extremity. Tubular steel casing with a spun concrete core combines the advantages of both materials, and fibreglass tubes with concrete or steel tube cores are useful for light marine structures.

### 2.1.6 Minipiles and micropiles

Both replacement piles and small-displacement piles may be formed as mini-/micropiles.

### 2.1.7 Selection of pile type

The selection of the appropriate type of pile from any of the above-mentioned categories depends on the following three principal factors:

1. The location and type of structure
2. The ground conditions
3. Durability

Considering the first of these factors, some form of displacement pile is the first choice for a *marine structure*. A solid precast or prestressed concrete pile can be used in fairly shallow water, but in deep water, a solid pile becomes too heavy to handle, and either a steel tubular pile or a tubular precast concrete pile is used. Steel tubular piles are preferred to H-sections for exposed marine conditions because of the smaller drag forces from waves and currents. Large-diameter steel tubes are also an economical solution to the problem of dealing with impact forces from waves and berthing ships. Timber piles are used for permanent and temporary works in fairly shallow water. Bored and cast-in-place piles would not be considered for any marine or river structure unless used in a composite form of construction, say as a means of extending the penetration depth of a tubular pile driven through water and soft soil to a firm stratum.

Piling for a structure on *land* is open to a wide choice in any of the three categories. Bored and cast-in-place piles are the cheapest type where unlined or only partly lined holes can be drilled by rotary auger. These piles can be drilled in very large diameters and provided with enlarged or grout-injected bases and thus are suitable to withstand high applied loads. Augered piles are also suitable where it is desired to avoid ground heave, noise and vibration, that is, for piling in urban areas, particularly where stringent noise regulations are enforced. Driven and cast-in-place piles are economical for land structures where light or moderate loads are to be carried, but the ground heave, noise and vibration associated with these types may make them unsuitable for some environments.

Timber piles are suitable for light to moderate loadings in countries where timber is easily obtainable. Steel or precast concrete driven piles are not as economical as driven or bored and cast-in-place piles for land structures. Jacked-down steel tubes or concrete units are used for underpinning work.

For the design of foundations in *seismic situations*, reference can be made to criteria in EC8-5 which complement the information on soil–structure interaction given in EC7-1. However, the codes and the recommendations in the British Standard Institute document PD 6698:2009 give only limited data on the design of piles to resist earthquakes. The paper by Raison[2.2] refers to the checks required under EC8-1 rules for piles susceptible to seismic liquefaction at a site in Barrow (see Section 9.8).

The second factor, *ground conditions*, influences both the material forming the pile and the method of installation. Firm to stiff fine-grained soils (silts and clays) favour the augered bored pile, but augering without support of the borehole by a bentonite slurry cannot be performed in very soft clays or in loose or water-bearing granular soils, for which driven or driven and cast-in-place piles would be suitable. Piles with enlarged bases formed by auger drilling can be installed only in firm to stiff or hard fine-grained soils or in weak rocks. Driven and driven and cast-in-place piles cannot be used in ground containing boulders or other massive obstructions, nor can they be used in soils subject to ground heave.

Driven and cast-in-place piles which employ a withdrawable tube cannot be used for very deep penetrations because of the limitations of jointing and pulling out the driving tube. For such conditions, a driven pile would be suitable. For hard driving conditions, for example in glacial till (boulder clays) or gravelly soils, a thick-walled steel tubular pile or a steel H-section can withstand heavier driving than a precast concrete pile of solid or tubular section.

Some form of drilled pile, such as a drilled-in steel tube, would be used for piles taken down into a rock for the purpose of mobilising resistance to uplift or lateral loads.

When piling in *contaminated land* using boring techniques, the disposal of arisings to licensed tips and measures to avoid the release of damaging aerosols are factors limiting the type of pile which can be considered and can add significantly to the costs. Precautions may also be needed to avoid creating preferential flow paths while piling which could allow contaminated groundwater and leachates to be transported downwards into a lower aquifer. Tubular steel piles can be expensive for piling in contaminated ground when compared with other displacement piles, but they are useful in overcoming obstructions which could cause problems when driving precast concrete or boring displacement piles. Large-displacement piles are unlikely to form transfer conduits for contaminants, although untreated wooden piles may allow 'wicking' of volatile organics. Driving precast concrete piles will densify the surrounding soil to a degree and in permeable soil the soil-pile contact will be improved, reducing the potential for flow paths. End-bearing H-piles can form long-term flow conduits into aquifers (particularly when a driving shoe is needed), and it may be necessary for the piles to be hydraulically isolated from the contaminated zone.

The factor of *durability* affects the choice of material for a pile. Although timber piles are cheap in some countries, they are liable to decay above groundwater level, and in marine structures, they suffer damage by destructive mollusc-type organisms. Precast concrete piles do not suffer corrosion in saline water below the *splash zone*, and rich well-compacted concrete can withstand attack from quite high concentrations of sulphates in soils and groundwaters. Cast-in-place concrete piles are not so resistant to aggressive substances because of difficulties in ensuring complete compaction of the concrete, but protection can be provided against attack by placing the concrete in permanent linings of coated light-gauge metal or plastics. Checklists for durability of man-made materials in the ground are provided in EC2-1 and complementary concrete standards BS 8500 and BS EN 206; durability of steel is covered in EC3-1 and EC3-5.

Steel piles can have a long life in ordinary soil conditions if they are completely embedded in undisturbed soil, but the portions of a pile exposed to seawater or to disturbed soil must be protected against corrosion by cathodic means if a long life is required. Corrosion rates are provided in Clause 4.4 of EC3-5, and work by Corus Construction and Industrial[2.3,2.4] has refined guidelines for corrosion allowances for steel embedded in contaminated soil. The increased incidence of *accelerated low water corrosion* (ALWC) in steel piles in UK tidal waters is considered in Section 10.4. *Mariner grade* steel H-piles to ASTM standard can give performance improvement of two to three times that of conventional steels in marine splash zones.

Other factors influence the choice of one or another type of pile in each main classification, and these are discussed in the following pages, in which the various types of pile are described in detail. In UK practice, specifications for pile materials, manufacturing requirements (including dimensional tolerances), workmanship and contract documentation are given in the Specification for Piling and Embedded Retaining Walls published by Institution of Civil Engineers[2.5] (referred to as SPERW). This document is generally consistent with the requirements in EC7-1 and the associated standards for the 'Execution of special geotechnical works', namely,

- BS EN 1536:2010 Bored piles
- BS EN 12063:1999 Sheet piling
- BS EN 12699:2001 Displacement piles
- BS EN 14199:2005 Micropiles

Having selected a certain type or types of pile as being suitable for the location and type of structure, for the ground conditions at the site and for the requirements of durability, the final choice is then made on the basis of *cost*. However, the total cost of a piled foundation is not simply the quoted price per metre run of piling or even the more accurate comparison of cost per pile per kN of load carried. Consideration must also be given to the overall cost of the foundation work which will include the main contractor's on-site costs and overheads.

Depending on the contract terms, extra payment may be sought if the piles are required to depths greater than those predicted at the tendering stage. Thus, a contractor's previous experience of the ground conditions in a particular locality is important in assessing the likely pile length and diameter on which to base a tender. Experience is also an important factor in determining whether the cost of preliminary test piling can be omitted and testing limited to that of proof loading selected working piles. In well-defined ground conditions and relatively light structural loads, the client may rely on the contractor's warranty that the working piles meet the specified load-carrying capacity and settlement criteria. However, the potential to save costs by omitting preliminary pile tests will be limited by EC7-1 Clause 7.6.2, which requires that pile designs based on calculation using *ground test results* (i.e. the measurement of soil properties) or on dynamic impact tests must have been validated by previous evidence of acceptable performance in static load tests, in similar ground conditions.

A thorough ground investigation and preliminary pile tests are essential in difficult ground. If these are omitted and the chosen pile design and installation procedures are shown to be impractical at the start of construction, then considerable time and money can be expended in changing to another piling system or adopting larger-diameter or longer piles. The allocation of costs resulting from such disruption is likely to be contentious.

A piling contractor's resources for supplying additional rigs and skilled operatives to make up time lost due to unforeseen difficulties and his technical ability in overcoming these difficulties are factors which will influence the choice of a particular piling system.

As a result of the introduction of new and revised codes and standards, considerable cross-referencing is now necessary to produce compliant designs. While it is not possible to deal with all the implications, this chapter provides a summary of some of the main points from the standards concerned with piling.

## 2.2 DRIVEN DISPLACEMENT PILES

### 2.2.1 Timber piles

In many ways, timber is an ideal material for piling. It has a high strength-to-weight ratio, it is easy to handle, it is readily cut to length and trimmed after driving and in favourable conditions of exposure, durable species have an almost indefinite life. Timber piling is also a low-cost, sustainable resource and may become more widely used as an alternative 'environmentally friendly' material when compared with steel and concrete[2.6]. To demonstrate that timber products come from managed and sustainable forests, recognised forest management certification should be provided to the user together with *chain of custody* statement. Timber piles used in their most economical form consist of round untrimmed logs which are driven butt uppermost. The traditional British practice of squaring the timber can be detrimental to its durability since it removes the outer sapwood which is absorptive to liquid preservative as BS 8417 (see Section 10.2). The less absorptive heartwood is thus exposed, and instead of a pile being encased by a thick layer of well-impregnated sapwood, there is only a thin layer of treated timber which can be penetrated by the hooks or slings used in handling the piles or stripped off by obstructions in the ground.

Timber piles, when situated wholly below groundwater level, are resistant to fungal decay and have an almost indefinite life. However, the portion above groundwater level in a structure on land is liable to decay, and BS EN 12699 prohibits the use of timber piles above free-water level, unless adequate protection is used. The solution is to cut off timber piles just below the lowest predicted groundwater level and to extend them above this level in concrete (Figure 2.1a). If the groundwater level is shallow, the pile cap can be taken down below the water level (Figure 2.1b).

Timber piles in marine structures are liable to be severely damaged by the mollusc-type borers which infest seawater in many parts of the world, particularly in tropical seas. The severity of this form of attack can be reduced to some extent by using softwood impregnated

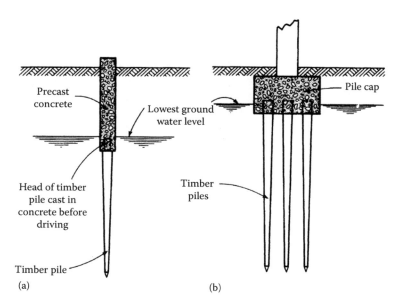

*Figure 2.1* Protecting timber piles from decay by (a) precast concrete upper section above water level and (b) by extending pile cap below water level.

with preservative or greatly minimised by the use of a hardwood of a species known to be resistant to borer attack. The various forms of these organisms, the form of their attack and the means of overcoming it are discussed in greater detail in Chapter 10.

Bark should be removed from round timbers where these are to be treated with preservative. If this is not done, the bark reduces the depth of impregnation. Also the bark should be removed from piles carrying uplift loads by shaft friction in case it should become detached from the trunk, thus causing the latter to slip. Bark need not be removed from piles carrying compression loads or from fender piles of untreated timber (hardwoods are not treated because they will not absorb liquid preservatives).

BS 5268-2, which provided the allowable design stresses for compression parallel to the grain for the species and grade of green timber being used, has been withdrawn. The replacement Eurocode EC5-1 provides common rules for calculating stresses which apply to the design of timber piling. Reference must also be made to BS EN 338 for characteristic values for all timber classes as described under common and botanical names in BS EN 1912. The design load and design compressive stress parallel to the grain are then calculated using the EC5 National Annex partial factors for timber for verification against failure. (See McKenzie and Zhang[2.7].)

Examples of commercially available timbers which are suitable for piling are shown in Table 2.1. The values given for hardwoods, such as greenheart, are considerably higher than those of softwoods, and generally, timber suitable for piles is obtained from SS grades or better. The timber should be straight-grained and free from defects which could impair its strength and durability. To this end, the sectional dimensions of hewn timber piles must not change by more than 15 mm/m, and straightness shall not deviate more than 1% of the length.

The stresses quoted are for timber at a moisture content consistent with a temperature of 20°C and relative humidity of 65%. Timber piles are usually in a wet environment requiring the application of reduction factors ($k_{mod}$, see Section 7.10) to convert the code stress properties to the wet conditions. When calculating the stresses on a pile, allowance must be made for

*Table 2.1* Summary of characteristic values of some softwoods and tropical hardwoods suitable for bearing piles (selected from BS EN 1912 Table 1 and BS EN 338 Table 1)

| Standard name | Strength class | Grade | Bending parallel to grain ($f_{m,k}$) (N/mm²) | Compression parallel to grain ($f_{c,0,k}$) (N/mm²) | Shear parallel to grain ($f_{v,k}$) (N/mm²) | 5% modulus of elasticity ($E_{0.5}$) (kN/m²) |
|---|---|---|---|---|---|---|
| British spruce | GS | C14 | 14 | 16 | 3 | 4.7 |
| European redwood | GS | C16 | 16 | 17 | 3.2 | 5.4 |
| Canadian western red cedar | SS | C18 | 18 | 18 | 3.4 | 6.0 |
| British pine | SS | C22 | 22 | 20 | 3.8 | 6.7 |
| Douglas fir–larch, United States | SS | C24 | 24 | 21 | 4 | 7.4 |
| Jarrah | HS | D40 | 40 | 26 | 4 | 10.9 |
| Teak | HS | D40 | 40 | 26 | 4 | 10.9 |
| Ekki | HS | D70 | 70 | 34 | 5 | 16.8 |
| Greenheart | HS | D70 | 70 | 34 | 5 | 16.8 |

GS is visually graded *general structural* softwood to BS 4978:2007; HS is visually graded hardwood to BS 5756:2007; SS is visually graded *special structural* softwood to BS 4978:2007.

The UK gradings apply for timber used in the United Kingdom and abroad.

bending stresses due to eccentric and lateral loading and to eccentricity caused by deviations in the straightness and inclination of a pile. Allowance must also be made for reductions in the cross-sectional area due to drilling or notching and the taper on a round log.

Typical pile lengths are from 5 to 18 m carrying applied loads from 5 to 350 kN. The maximum capacity of the pile will be limited by the set achievable without causing damage. Large numbers of timber piles, mainly Norwegian spruce, are driven below the water table in the Netherlands every year for light structures, housing, roads and embankments.

As a result of improved ability to predict and control driving stresses, BS EN 12699 allows the maximum compressive stress generated during driving to be increased to 0.8 times the characteristic compressive strength measured parallel to the grain. While some increase in stress (up to 10%) may be permitted during driving if stress monitoring is carried out, it is advisable to limit the maximum load which can be carried by a pile of any diameter to reduce the need for excessively hard driving. This limitation is applied in order to avoid the risk of damage to a pile by driving it to some arbitrary *set* as required by a dynamic pile-driving formula and to avoid a high concentration of stress at the toe of a pile end bearing on a hard stratum. Damage to a pile during driving is most likely to occur at its head and toe. It is now common practice to use a pile driving analyser (PDA) which can measure the stress in the pile during driving to warn if damage is likely to occur.

The problems of splitting of the heads and unseen 'brooming' and splitting of the toes of timber piles occur when it is necessary to penetrate layers of compact or cemented soils to reach the desired founding level. This damage can also occur when attempts are made to drive deeply into dense sands and gravels or into soils containing boulders, in order to mobilise the required frictional resistance for a given uplift or compressive load. Judgement is required to assess the soil conditions at a site so as to decide whether or not it is feasible to drive a timber pile to the depth required for a given load without damage or whether it is preferable to reduce the applied load to a value which permits a shorter pile to be used. As an alternative, jetting or pre-boring may be adopted to reduce the amount of driving required. Cases have occurred where the measured set achieved per blow has been due to the crushing and brooming of the pile toe and not to the deeper penetration required to reach the bearing stratum.

Damage to a pile can be minimised by reducing as far as possible the number of hammer blows necessary to achieve the desired penetration and also by limiting the height of drop of the hammer to 1.5 m. This necessitates the use of a heavy hammer (but preferably less than 4 tonnes), which should at least be equal in weight to the weight of the pile for hard driving conditions and to one-half of the pile weight for easy driving. The lightness of a timber pile can be an embarrassment when driving groups of piles through soft clays or silts to a point bearing on rock. Frictional resistance in the soft materials can be very low for a few days after driving, and the effect of pore pressures caused by driving adjacent piles in the group may cause the piles already driven to rise out of the ground due to their own buoyancy relative to that of the soil. The only remedy is to apply loads to the pile heads until all the piles in the area have been driven.

Heads of timber piles should be protected against splitting during driving by means of a mild steel hoop slipped over the pile head or screwed to it (Figure 2.2a and b). A squared pile toe can be provided where piles are terminated in soft to moderately stiff clays (Figure 2.2a). Where it is necessary to drive them into dense or hard materials, a cast-steel point should be provided (Figure 2.2b). As an alternative to a hoop, a cast-steel helmet can be fitted to the pile head during driving. The helmet must be deeply recessed and tapered to permit it to fit well down over the pile head, allowing space for the insertion of hardwood packing.

Commercially available timbers are imported in lengths of up to 18 m. If longer piles are required, they may be spliced as shown in Figure 2.3. A splice near the centre of the length

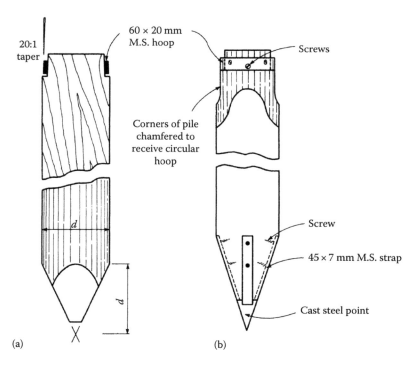

Figure 2.2 Protecting timber piles from splitting during driving. (a) Protecting head by mild steel hoop. (b) Protecting toe by cast-steel point.

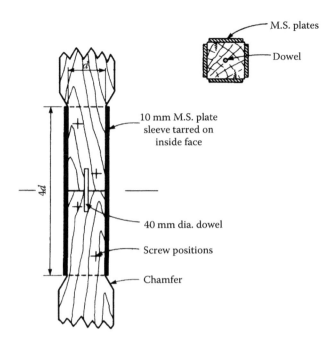

*Figure 2.3* Splice in squared timber pile.

of a pile should be avoided since this is the point of maximum bending moment when the pile is lifted from a horizontal position by attachments to one end or at the centre. Timber piles can be driven in very long lengths in soft to firm clays by splicing them in the leaders of the piling frame as shown in Figure 2.4. The abutting surfaces of the timber should be cut truly square at the splice positions in order to distribute the stresses caused by driving and loading evenly over the full cross section.

## 2.2.2 Precast concrete piles

Precast concrete piles have their principal use in marine and river structures, that is in situations where the use of driven and cast-in-place piles is impracticable or uneconomical. For land structures, unjointed precast concrete piles can be more costly than driven and cast-in-place types for two main reasons:

1. Reinforcement must be provided in the precast concrete pile to withstand the bending and tensile stresses which occur during handling and driving. Once the pile is in the ground, and if mainly compressive loads are carried, the majority of this steel is redundant.
2. The precast concrete pile is not readily cut down or extended to suit variations in the level of the bearing stratum to which the piles are driven.

However, there are many situations for land structures where the precast concrete pile can be the more economical, especially where high-quality concrete is required. Where large numbers of piles are to be installed in easy driving conditions, the savings in cost due to the rapidity of driving achieved may outweigh the cost of the heavier reinforcing steel necessary. Reinforcement may be needed in any case to resist bending stresses due to lateral loads or tensile stresses from uplift loads. Where high-capacity piles are to be driven to a hard

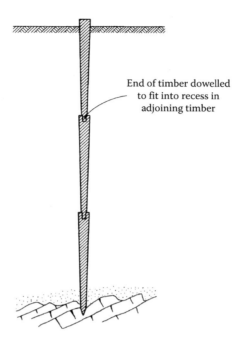

End of timber dowelled
to fit into recess in
adjoining timber

*Figure 2.4* Splicing timber piles in multiple lengths.

stratum, savings in the overall quantity of concrete compared with cast-in-place piles can be achieved since higher stresses can be used. Where piles are to be driven in sulphate-bearing ground or into aggressive industrial waste materials, the provision of sound, high-quality dense concrete is essential. The problem of varying the length of the pile can be overcome by adopting a jointed type as Section 2.2.3.

Piles can be designed and manufactured in ordinary reinforced concrete or in the form of pretensioned or post-tensioned prestressed concrete members. The ordinary reinforced concrete pile is likely to be preferred for a project requiring a fairly small number of piles, but prestressed piles may be required for hard driving conditions. Precast concrete piles in ordinary reinforced concrete are usually square or hexagonal and of solid cross section for units of short or moderate length, but for saving weight, long piles can be manufactured with a hollow core in hexagonal, octagonal or circular sections. The interiors of these piles can be filled with concrete after driving to avoid bursting where piles are exposed to severe frost action. Alternatively, drainage holes can be provided to prevent water accumulating in the hollow interior. Hollow-core piles can be readily inspected for breakages in difficult driving and can be strengthened by infilling with structural reinforced concrete when considered for reuse. Where piles are designed to carry the applied loads mainly in end bearing, for example piles driven through soft clays into medium-dense or dense sands, economies in concrete and reductions in weight for handling can be achieved by providing the piles with an enlarged toe, up to 1.6 times the shaft width with a minimum length of 500 mm or equal to the width of the enlargement.

Precast and prestressed piles have to be designed not only to withstand the loads from the structure but also to meet the stresses and other serviceability requirements during handling, pitching and driving and in service as stated in the relevant material Eurocodes and the associated National Annexes. To avoid excessive flexibility while handling and driving, the usual maximum unjointed lengths of square section piles and the range of load-bearing

Table 2.2 Typical capacity and maximum lengths for ordinary precast concrete piles of square section (subject to reinforcement)

| Pile size (mm²) | Applied load (kN) | Maximum length (m) |
|---|---|---|
| 250 | 200–300 | 12 |
| 300 | 300–450 | 15 |
| 350 | 350–600 | 18 |
| 400 | 450–750 | 21 |
| 450 | 500–900 | 25 |

capacities applicable to each size are shown in Table 2.2. (See also Figure 7.2 for maximum lengths at various lifting points.)

EC2-1 provides common rules for concrete for building and civil engineering which are not very different from the withdrawn BS 8110 in terms of general design approach, but the replacement codes contain significant cross-references which now have to be considered for concrete design. Concrete performance, quality and production are subject to BS EN 206-1, which must be read in conjunction with the United Kingdom's complementary rules for strength and exposure classes, cover, etc. in BS 8500-1 and BS 8500-2 as designated in Table 2.3. The minimum concrete class for precast and prestressed piles specified in BS EN 12794 clause 4.2.2.1 is C35/45 and can be deemed suitable for hard driving conditions. (Note the strength classification in EC2 is based on denoting the minimum characteristic strength of a *cylinder* at 28 days/minimum characteristic *cube* strength at 28 days in N/mm², i.e. $f_{ck\ cyl}$ and $f_{ck\ cube}$ represented, e.g. as C35/45.) BS 8500 recommends strength classes of concrete C45/55 in tidal splash zones as in Table 2.4. The strengths in BS EN 13369 dealing in general with precast concrete products are not appropriate for most piling applications, but the reinforcement requirements have to be adhered to (as below).

Table 2.3 Summary of exposure classes as BS 8500-1

| Exposure class | Class description | Examples applicable to piling |
|---|---|---|
| XO | No risk of corrosion or attack | Reinforced concrete exposed to very dry conditions |
| XC | Carbonation-induced corrosion | Reinforced concrete buried in soil Class AC-1 |
| XD | Chloride-induced corrosion (not from seawater) | Reinforced concrete immersed in chloride conditions |
| XS | Chloride-induced corrosion (from seawater) | Reinforced concrete below mid tide level |
| XF | Freeze–thaw attack | Concrete subjected to frequent splashing with water and exposed to freezing |

Note: Each class is subdivided depending on the severity of attack as shown in Table 2.4.

Table 2.4 Typical concrete grades and cover suitable for exposures

| Strength class | Exposure class | Water/cement ratio | Cement content (kg/m³) | Nominal cover (mm) |
|---|---|---|---|---|
| 25/30 | XC2 (non-aggressive) | 0.65 | 260 | 25–50 + $\Delta_c$ |
| 35/45 | XS1 (airborne salt) | 0.45 | 360 | 35 + $\Delta_c$ |
| 45/55 | XS3 (intertidal wet/dry) | 0.35 | 380 | 45 + $\Delta_c$ |

BS EN 12794 Table 3 gives detailed production tolerances and defines two classes of precast piles – *Class 1* with distributed reinforcement or prestressed piles and *Class 2* with a single central reinforcing bar. Foundations in naturally aggressive ground conditions/brownfield sites/contaminated land are not covered in EC7-1, and the recommendations in BRE Special Digest 1[(2.8)] (SD1) and BS 8500-1 should be followed for both in situ foundation concrete and precast units.

High stresses, which may exceed the handling stresses, can occur during driving, and it is necessary to consider the serviceability limit of cracking. EC2-1 Clause 7.3 allows for maximum crack widths of 0.3 mm in reinforced concrete elements taking account of the proposed function of the structure and exposure of precast and prestressed elements. It has been UK practice to require cracks to be controlled to maximum widths close to the main reinforcement ranging from 0.3 mm down to 0.15 mm in an aggressive environment, important when considering laterally loaded and tension piles. Annex ZA to BS EN 12794 deals with the CE marking of foundation piles and the presumption of fitness for the intended use. (All timber, precast and steel piles will have to be so marked for use on European construction sites from 2013.)

In EC2-1 Clause 4.4, nominal cover to reinforcement is defined as $c_{nom} = c_{min} + \Delta c_{dev}$ where $c_{min}$ is dependent on bond requirements or environmental conditions as detailed in Tables 4.1 through 4.5 of EC2. $\Delta c_{dev}$ allows for deviations, set at 10 mm in EC2 NA, but may be reduced where strict QA/QC procedures are in force. Cover required in BS EN 12794 is $c_{min}$ but the value of $\Delta c$ to satisfy the environmental conditions defined in BS 8500-1 and BS EN 206-1 is shown in Table 2.4 for two classes of concrete specified for precast piles with an intended life of 50 years and 20 mm maximum aggregate. UK practice would indicate that for well-controlled production, $\Delta c$ should be 5 mm generally and 10 mm in marine exposures.

Although the XC2 classification in BS 8500 for reinforced concrete in non-aggressive ground allows a minimum strength of C25/30, this is not appropriate for piles as noted earlier. The durability of concrete in aggressive ground is considered in Section 10.3.1.

Concrete made with ordinary Portland cement (CEM 1) is generally suitable for precast piles at the above-mentioned strengths in normal exposures. Table 1 of BS EN 197-1 gives the composition of the main types of cement which address all the exposure classes, and the groups in Table A1 of BS 8500-2 show the comparisons with the SD1 *ACEC* exposure grades. For example, cement to address Class XS3 given earlier is limited to types CEM 1, IIA (with fly ash), IIBS (with ground granulated blast furnace slag), and SRPC. Note the codes no longer refer to pfa (*pulverised fuel ash*) and 'flyash' may be other ash from power stations, not necessarily pfa.

BS EN 12794 (Annex B9) states that for Class 1 piles, longitudinal reinforcement shall be a minimum diameter of 8 mm with at least one bar placed in the corner of square piles; circular section piles shall have at least 6 bars 8 mm diameter placed evenly around the periphery. Transverse reinforcement must be at least 4 mm diameter depending on the pile diameter, and the pile head must have a minimum of 9 links in 500 mm. Percentages of transverse steel are specified for hollow-core piles. BS EN 12794 refers to BS EN 13369 for the quality of reinforcement and prestressing steel to be used, which in turn refers to other standards, such as BS EN 10080 steel for reinforcement of concrete and BS 5896 for prestressing wire and strand. The specification and grades of steel given in BS 4449 steel for the reinforcement of concrete, as revised in 2009, complement BS EN 10080. EC2-1-1 in Annex C states that the code applies only to reinforcement with characteristic yield strength ($f_{yk}$) in the range 400–600 N/mm². Other steels, including plain bars, may be used provided they conform to Annex C requirements. Ribbed bars in 500 N/mm² steel, classified as A, B or C depending on the steel ductility and the ratio of $f_{tk}/f_{yk}$, are the most common grade used in the United Kingdom. Users of reinforcement are referred to data sheets provided by UK CARES, the third-party certifying body for reinforcing steels, for additional clarification.

The diameter of main reinforcing steel in the form of longitudinal bars may have to be increased depending on the bending moments induced when the pile is lifted from its casting bed to the stacking area. The magnitude of the bending moments depends on the number and positioning of the lifting points (see Table 7.2). Design data for various lifting conditions are dealt with in Section 7.2. In some cases, the size of the externally applied lateral or uplift loads may necessitate the provision of more main steel than is required by lifting considerations. In hard driving conditions, it is advantageous to place additional transverse steel in the form of a helix at the head of the pile to prevent shattering or splitting. The helix should be about two pile widths in length with a pitch equal to the spacing of the link steel at the head. A design for a precast concrete pile for use in easy driving conditions is shown in Figure 2.5a. A design for a longer octagonal pile suitable for driving to end bearing on rock is shown in Figure 2.5b. The design of a typical prestressed concrete pile in accordance with UK practice is shown in Figure 2.6. Square and octagonal piles are usually fabricated up to 600 mm wide.

Prestressed concrete piles have certain advantages over those of ordinary reinforced concrete. Their principal advantage is in their higher strength-to-weight ratio, enabling long

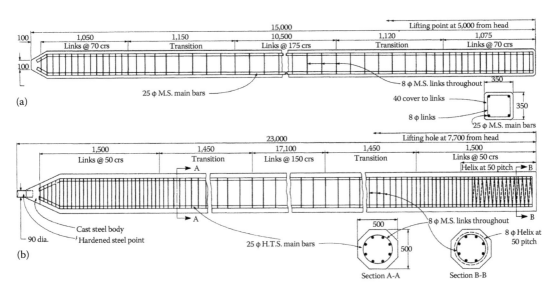

**Figure 2.5** Design for precast concrete piles (a) 350 mm square pile, 15 m long (b) 500 mm octagonal pile, 23 m long.

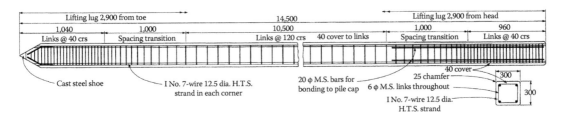

**Figure 2.6** Design for prestressed concrete pile.

slender units to be lifted and driven. However, slenderness is not always advantageous since a large cross-sectional area may be needed to mobilise sufficient resistance in shaft friction and end bearing and additional lifting points required for pitching. The second main advantage is the effect of the prestressing in closing up cracks caused during handling and driving. This effect, combined with the high-quality concrete necessary for economic employment of prestressing, gives the prestressed pile increased durability which is advantageous in marine structures and corrosive soils. Prestressed concrete piles of hollow cylindrical section are manufactured by centrifugal spinning in diameters ranging from 900 to 2100 mm and lengths up to 40 m. For optimum driving performance, the prestressing force, after losses, is usually between 7 and 10 N/mm$^2$.

Prestressed concrete piles should be made with designed concrete mixes of at least Class C35/45, but as noted earlier, account should be taken of the special exposure conditions quoted in BS 8500 and BS EN 206-1. Minimum percentages of prestressing steel stipulated in BS EN 12794 are 0.1% of cross-sectional area in mm$^2$ for piles not exceeding 10 m in length, 0.01% cross-sectional area × pile length for piles between 10 and 20 m long, and 0.2% for piles greater than 20 m long. The high concrete strength required for prestressed piles means that they can withstand hard driving and achieve high bearing capacity. However, it may be desirable to specify a maximum load which can be applied to a precast concrete pile of any dimensions. As in the case of timber piles, this limitation is to prevent unseen damage to piles which may be overdriven to achieve an arbitrary set given by a dynamic pile-driving formula. BS EN 12699 limits the calculated stress (including any prestress) during driving of precast piles to 0.8 times the characteristic concrete strength in compression at time of driving; a 10% increase is permitted if the stresses are monitored during driving (e.g. with a PDA).

Metal shoes are not required at the toes of precast concrete piles where they are driven through soft or loose soils into dense sands and gravels or firm to stiff clays. A blunt pointed end (Figure 2.7a) appears to be just as effective in achieving the desired penetration in these soils as a more sharply pointed end (Figure 2.7b), and the blunt point is better for maintaining alignment during driving. A cast-iron or cast-steel shoe fitted to a pointed toe may be

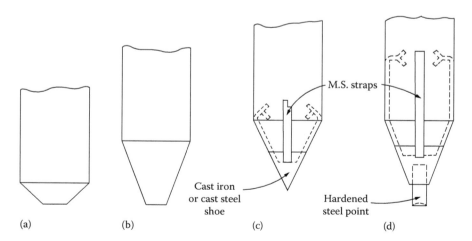

Figure 2.7 Shoes for precast (including prestressed) concrete piles. (a) For driving through soft or loose soils to shallow penetration into dense granular or firm to stiff clays. (b) Pointed end suitable for moderately deep penetration into medium-dense to dense sands firm to stiff clays. (c) Cast-iron or cast-steel shoe for seating pile into weak rock or breaking through cemented soil layer. (d) *Oslo* point for seating pile into weak rock.

used for penetrating rocks or for splitting cemented soil layers. The shoe (Figure 2.7c) serves to protect the pointed end of the pile.

Where piles are to be driven to refusal on a sloping hard rock surface, the *Oslo point* (Figure 2.7d) is desirable. This is a hollow-ground hardened steel point. When the pile is judged to be nearing the rock surface, the hammer drop is reduced and the pile point is seated on to the rock by a number of blows with a small drop. As soon as there is an indication that a seating has been obtained, the drop can be increased and the pile driven to refusal or some other predetermined set. The Oslo point was used on the piles illustrated in Figure 2.5b, which were driven on to hard rock at the site of the Whitegate Refinery, Cork. A hardened steel to BS 970 with a Brinell hardness of 400–600 was employed. The 89 mm point was machined concave to 12.7 mm depth and embedded in a chilled cast-iron shoe. Flame treatment of the point was needed after casting into the shoe to restore the hardness lost during this operation.

The strict requirements imposed by BS EN 12699 and BS EN 12794 mean that precast and prestressed piles are now usually made in factory conditions using precision steel moulds on firm reinforced concrete beds. Distortion in timber forms and when tier casting (Figures 2.8 and 2.9) and the difficulty in squaring the drive end can then be eliminated. Moulds can be stripped as soon as crushing tests on cylinders/cubes (cured using the same methods as for the pile) indicate that the piles have reached 60% of the required 28-day strength. For example, Aarsleff Piling produced 600 mm square precast piles up to 14.3 m long for the Channel Tunnel Rail Link (CTRL) using purpose-built steel moulds in their factory in Newark. The sides of the moulds were locked together using a combination of cams and hydraulic rams which, after the concrete had reached an initial set of 24–28 N/mm² in 21 h, were operated to release the 12.5 tonne pile. A typical steel mould is shown in Figure 2.10.

There are situations when it is appropriate to set up pile production on a construction site, for example where established factories are remote from the site, where the number of piles justifies the costs of setting up a casting yard, or where there are transportation restrictions. In Bangkok, 17,000 × 500 mm diameter prestressed, precast hollow cylindrical piles, 10–14 m long with 100 mm thick wall, were required for the depot of the new Mass Rail Transit system[2.9]. A casting yard was established adjacent to the site to fabricate the pile elements, using centrifugal spinning and 24 h autoclave curing followed by a period of ambient wet curing to give minimum strength of 50 N/mm². At peak production, 19 rigs were on-site driving 95 piles per day. Another type of prestressed pile was used for the Oosterschelde

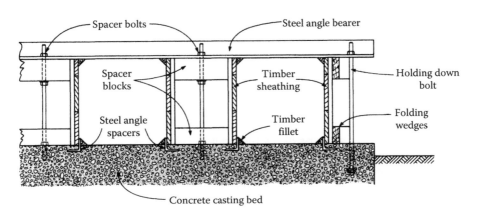

*Figure 2.8* Timber formwork for precast concrete piles.

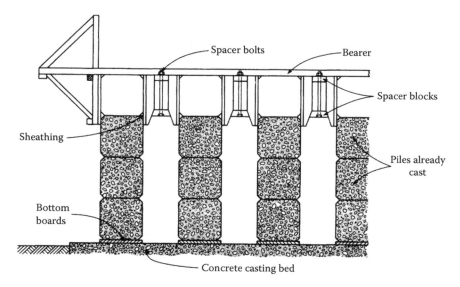

*Figure 2.9* Casting precast concrete piles in tiers.

*Figure 2.10* Steel moulds in pile casting yard.

Bridge in the Netherlands. Here, 4 m diameter prestressed concrete cylinder piles were made as vertically cast segments and then joined longitudinally to form 60 m long piles for installation by crane barge and caisson-sinking methods.

All precast piles should be clearly marked with a reference number, length and date of casting at or before the time of lifting, to ensure that they are driven in the correct sequence. Timber bearers should be placed between the piles in the stacks to allow air to circulate around them. They should be protected against too-rapid drying in hot weather by covering the stack with a tarpaulin or polyethylene sheeting. Care must be taken to place the bearers

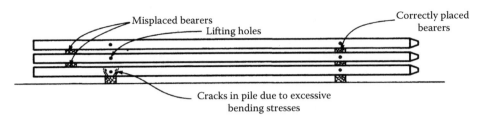

*Figure 2.11* Misplaced packing in stacks of precast concrete piles.

only at the lifting positions, as, if they are misplaced, there could be a risk of excessive bending stresses developing and cracking occurring (Figure 2.11).

One of the principal problems associated with precast concrete piles is unseen breakage
due to hard driving conditions. Jointed precast concrete piles when driven through soft or
loose soils on to hard rock are particularly susceptible to damage. On some sites, the rock
surface may slope steeply, causing the piles to deviate from a true line and break into short
sections near the toe. Accumulations of boulders over bedrock can also cause the piles to
be deflected with consequent breakage. Where such conditions are expected, it is advisable
to provide a central inspection hole in test piles and sometimes in a proportion of the working piles. A check for deviation of the pile from line can be made by lowering a steel tube
down the hole. If the tube can be lowered to the bottom of the hole under its own weight,
the pile should not be bent to a radius which would impair its structural integrity. If the
tube jams in the hole, an inclinometer is used to record the actual deviation and hence to
decide whether or not the pile should be rejected and replaced. The testing tube also detects
deviations in the position or alignment of a jointed pile with a central hole. Deviation from
the production straightness of the axis of the pile should be limited to a maximum of 0.2%
of the pile length.

Breakages are due either to tensile forces caused by easy driving with too light a hammer
in soft or loose soils or to compressive forces caused by driving with too great a hammer drop on to a pile seated on a hard stratum; in both situations, the damage occurs in
the buried portion of the pile. In the case of compression failure, it occurs by crushing or
splitting near the pile toe. Such damage is not indicated by any form of cracking in the
undriven portion of the pile above ground level. The use of the PDA will assist in determining actual stresses along the pile (Figure 7.3b) for comparison with the calculated stresses;
remedial actions then include changing the hammer, reducing the stroke and changing the
cushioning.

The precautions for driving precast concrete piles are described in Section 3.4.2, and the
procedures for bonding piles to caps and ground beams and lengthening piles are described
in Sections 7.6 and 7.7.

## 2.2.3 Jointed precast concrete piles

The disadvantages of having to adjust the lengths of precast concrete piles either by cutting
off the surplus or casting on additional lengths to accommodate variations in the depth to
a hard bearing stratum will be evident. These drawbacks can be overcome by employing
jointed piles in which the adjustments in length can be made by adding or taking away short
lengths of pile which are jointed to each other by devices capable of developing the same
bending and tensile resistance as the main body of the pile. BS EN 12794 defines pile joints
in four classes, Class A to Class D, depending on whether the pile is used in compression,

*Table 2.5* Dimensions and properties of square section piles
as manufactured by Balfour Beatty Ground Engineering
in the United Kingdom

| Square section (mm) | Maximum section length (m) | Typical applied load (kN) |
|---|---|---|
| 190 | 8 | 350 |
| 235 | 14 | 500 |
| 270 | 15 | 800 |
| 350 | 13.5 | 1200 |

Note: Resistance to applied load is dependent on dimensions of pile and soil
properties.

tension or bending and the impact load test to be applied to verify the static design calculations. If the pile joint satisfies the impact and bending tests, then the ultimate capacity of the
joint is 'identical' to the calculated static bearing capacity. A segment length is chosen for
the initial driving which is judged to be suitable for the shallowest predicted penetration in
a given area. Additional lengths are locked on if deeper penetrations are necessary or if very
deep penetrations requiring multiples of the standard lengths are necessary. It is possible to
drive the jointed piles to 40 m in soft ground.

Balfour Beatty Ground Engineering produces and installs typical Class 1 precast piles in
a range of segment lengths and square sections as shown in Table 2.5 normally in C45/55
concrete. The precast concrete units are locked together by a steel bayonet-type joint to
obtain the required bending and tensile resistance, and a rock shoe incorporating an Oslo
point may be used (Figure 2.7d).

Other types of jointed precast concrete piles include the *Centrum* pile manufactured
and installed by Aarsleff Piling in the United Kingdom using C40/50 concrete and rigid
welded reinforcement cages in varying lengths from 4 to 13 m in square sections from 200
to 400 mm. Lengths greater than 4 m for the 200 and 250 mm sections can be jointed using
a single locking pin driven horizontally into locking rings in the joint box. The *multi-lock*
ABB joint with four bayonet locking pins is used for the larger sections and provides a degree
of pretensioning to the joint (Figure 2.12). Depending on the length, section and joint used

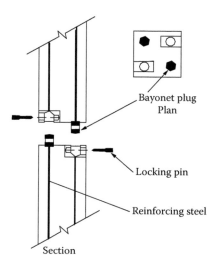

Bayonet plug
Plan

Locking pin

Reinforcing steel

Section

*Figure 2.12* Typical locking pin joint for precast concrete pile.

and the ground conditions, capacities up to 1200 kN in compression and 180 kN in tension are possible. In addition to the above-mentioned 14.3 m long 600 mm square piles, Aarsleff produced 600 mm square jointed segmental piles up to 3.5 m long for low-headroom work on CTRL.

*RB* precast square concrete piles with a single central bar (as Class 2 given earlier) are made and installed by Roger Bullivant Ltd. They are available in a range of capacities (depending on ground conditions) from 200 kN for the nominal 150 mm square section to 1200 kN for the 355 mm square pile, in lengths of 1.5, 3 and 4 m. The standard joint for the limited tensile and bending capability is a simple spigot and socket type bonded with epoxy resin with each pile length bedded on a sand/cement mortar. Special joints (such as the Emeca joint) and pile reinforcement can be provided as needed to resist bending moments and tension forces.

Precast concrete piles which consist of units joined together by simple steel end plates with welded butt joints are not always suitable for hard driving conditions or for driving on to a sloping hard rock surface. Welds made in exposed site conditions with the units held in the leaders of a piling frame may not always be sound. If the welds break due to tension waves set up during driving or due to bending caused by any deviation from alignment, the pile may break up into separate units with a complete loss of bearing resistance (Figure 2.13). This type of damage can occur with keyed or locked joints when the piles are driven heavily, for example in order to break through thin layers of dense gravel. The design of the joint is, in fact, a critical factor in the successful employment of these piles, and tests to check bending, tension and compression capabilities should be carried out for particular applications. However, even joints made from steel castings require accurate contact surfaces to ensure that stress concentrations are not transferred to the concrete.

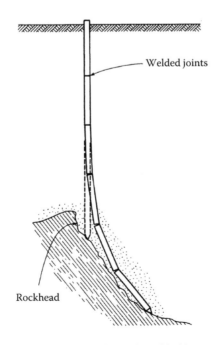

Figure 2.13 Unseen breakage of precast concrete piles with welded butt joints.

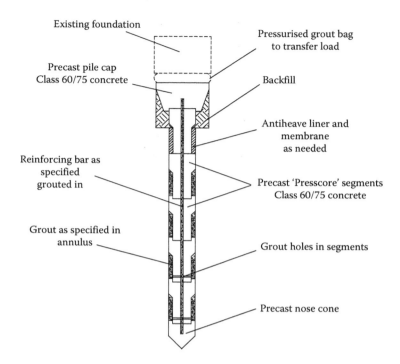

Existing foundation

Pressurised grout bag
to transfer load

Precast pile cap
Class 60/75 concrete

Backfill

Antiheave liner and
membrane
as needed

Reinforcing bar as
specified
grouted in

Precast 'Presscore' segments
Class 60/75 concrete

Grout as specified in
annulus

Grout holes in segments

Precast nose cone

*Figure 2.14* Presscore pile. (Courtesy of Abbey Pynford Foundation Systems Ltd., Watford, England.)

The *Presscore* pile developed and installed by Abbey Pynford PLC is a jointed precast concrete pile consisting of short units which are jacked into the soil. The concrete in the pile units and precast pile cap is 60 N/mm², and a reinforcing bar can be placed through the centre of the units (Figure 2.14). On reaching the required bearing depth, the annulus around the pile is grouted through ports in the units. The use of jacked-in piles for underpinning work is described in Chapter 9.

A high-strength cylindrical precast pile, 155 mm diameter and 1 m long, was developed in Canada for underpinning a 90-year-old building in Regina[2.10]. The segments were cast using steel fibre-reinforced concrete with a 28-day compressive strength of 90 N/mm² and steel fibre content of 40 kg/m³. Each segment was reinforced with four steel wires (9 mm) welded to a steel wire circumferential coil. Recesses were provided at each end of the segment and stainless steel rods connected each segment to form the joint. Hydraulic jacks with a capacity of 680 kN reacted against a new pile cap, and as each segment was jacked down, the next segment was screwed and tensioned on to the connecting rod. The required 600 kN pile capacity was achieved at depths ranging from 11 to 13 m.

## 2.2.4 Steel piles

Steel piles have the advantages of being robust, easy to handle, capable of carrying high compressive loads when driven on to a hard stratum, and capable of being driven hard to a deep penetration to reach a bearing stratum or to develop a high frictional resistance, although their cost per metre run is high compared with precast concrete piles. They can be designed as small-displacement piles, which is advantageous in situations where ground heave and lateral displacement must be avoided. They can be readily cut down and extended

*Figure 2.15* Box piles using Z-sheet pile sections in fabrication yard. (Courtesy of Maxx Piling Ltd., Shenfield, UK.)

where the level of the bearing stratum varies; also the head of a pile which buckles during driving can be cut down and re-trimmed for further driving. They have a good resilience and high resistance to buckling and bending forces.

Types of steel piles include plain tubes, box sections, box piles built up from sheet piles, H-sections and tapered and fluted tubes. Hollow-section piles can be driven with open ends as Figure 2.15. If the base resistance must be eliminated when driving hollow-section piles to a deep penetration, the soil within the pile can be cleaned out by grabbing, by augers, by reverse water-circulation drilling or by airlift (see Section 3.4.3). It is not always necessary to fill hollow-section piles with concrete. In normal undisturbed soil conditions, they should have an adequate resistance to corrosion during the working life of a structure, and the portion of the pile above the seabed in marine structures or in disturbed ground can be protected by cathodic means, supplemented by bituminous or resin coatings (Section 10.4). Concrete filling may be undesirable in marine structures where resilience, rather than rigidity, is required to deal with bending and impact forces.

Where hollow-section piles are required to carry high compressive loads, they may be driven with a closed end to develop the necessary end-bearing resistance over the pile base area. Where deep penetrations are required, they may be driven with open ends and with the interior of the pile closed by a stiffened steel plate bulkhead located at a predetermined height above the toe. An aperture should be provided in the bulkhead for the release of water, silt or soft clay trapped in the interior during driving. In some circumstances, the soil plug within the pile may itself develop the required base resistance (Section 4.3.3).

The facility of extending steel piles for driving to depths greater than predicted from soil investigation data has already been mentioned. The practice of welding on additional lengths of pile in the leaders of the piling frame is satisfactory for land structures where the quality of welding may not be critical, but testing should be carried out as required in

BS EN 12699. A steel pile supported by the soil can continue to carry high compressive loads even though the weld is partly fractured by driving stresses. However, this practice is not desirable for marine structures where the weld joining the extended pile may be above seabed level in a zone subjected to high lateral forces and corrosive influences. Conditions are not conducive to first-class welding when the extension pile is held in leaders or guides on a floating vessel or on staging supported by piles swaying under the influence of waves and currents. It is preferable to do all welding on a prepared fabrication bed with the pile in a horizontal position where it can be rotated in a covered welding station. The piles should be fabricated to cover the maximum predicted length and any surplus length cut off rather than be initially of only medium length and then be extended. Cut-off portions of steel piles usually have some value as scrap, or they can be used in other fabrications. However, there are many situations where in situ welding of extensions cannot be avoided. The use of a stable jack-up platform (Figure 3.7) from which to install the piles is then advantageous.

Long lengths of steel tubular piles for offshore petroleum production platforms can be handled in a single length on large crane barges. Where this is not practical, they can be driven by underwater hammers, but for top-driven sectional piles, a pile connector is a useful device for joining lengths of pile without the delays which occur when making welded joints. The Frank's Double Drive Shoulder Connector (Figure 2.16) was developed in the United States for joining and driving lengths of oil well conductor pipe and can be adapted for making connections in piles up to 914 mm diameter. It is a pin and box joint which is flush with the outside diameter (OD) and inside diameter (ID) of the pile, with interlocking threads which pull the pin and box surfaces together. The joint is usually welded on to the steel pipe, not formed on the pipe ends. Long steel tubular piles driven within the tubular members of a jacket-type structure are redundant above their point of connection by annular grouting to the lower part of the tubular sleeve. This redundant part of the pile, which acts as a follower for the final stages of driving, can be cut off for reuse.

Where large steel tubular piles need to be spliced to drive below ground level and are required to carry compressive loads only, splicing devices such as those manufactured

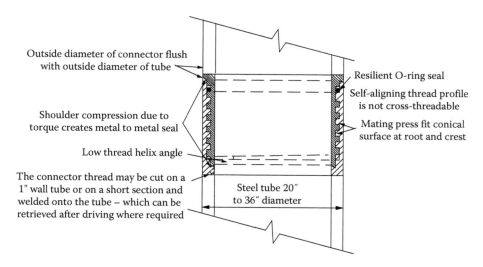

*Figure 2.16* Schematic arrangement of Frank's *Double Drive Shoulder Connector.*

by the Associated Pile and Fitting Corporation of the United States (APF) or Dawson Construction Plant in the United Kingdom can be used. The splicer consists of an external collar which is slipped on to the upper end of the pile section already driven and is held in position by an internal lug. The next length of pile is then entered into the collar and driven down. The APF splicer can also be used for cylindrical precast piles. Splicers are also available for H-piles in compression and consist of a pair of channel sections set on the head of the pile length already driven to act as a guide for placing and then welding on the next length.

Steel tubular piles are the preferred shape when soil has to be cleaned out for subsequent placement of concrete, since there are no corners from which the soil may be difficult to dislodge by the cleaning out. They are also preferred for marine structures where they can be fabricated and driven in large diameters to resist the lateral forces in deep-water structures. The circular shape is also advantageous in minimising drag and oscillation from waves and currents (Sections 8.1.3 and 8.1.4). The hollow section of a tubular pile is also an advantage when inspecting a closed-end pile for buckling. A light can be lowered down the pile and if it remains visible when lowered to the bottom, no deviation has occurred. If a large deviation is shown by complete or partial disappearance of the light, then measures can be taken to strengthen the buckled section by inserting a reinforcing cage and placing concrete.

Steel tubes are manufactured to order in Britain by Deepdale Engineering in a range of ODs up to 4000 mm in standard carbon steel and high-tensile steels to BS EN 10025-2 with wall thickness from 10 to 50 mm. ArcelorMittal produces a standard range of piles up to 3 m diameter and 25 mm wall thickness and up to 53 m long (without splices). The tubes are manufactured as either seamless, spirally welded or longitudinally welded units. There is nothing to choose between the latter two types from the aspect of strength to resist driving stresses. In the spiral welding process, the coiled steel strip is continuously unwound and spirally bent cold into the tubular. The joints are then welded from both sides. In the longitudinally welding process, a steel plate is cut and bevelled to the required dimensions and then pressed or rolled into tubular form and welded along the linear joints. The spiral method has the advantage that a number of different sizes can be formed on the same machine, but there is a limitation on the plate thickness that can be handled by particular machines. There is also some risk of weld *unzipping* from the pile toe under hard driving conditions. This can be prevented by a circumferential shoe of a type described below. Piles driven in exposed deep-water locations are fabricated from steel plate in thicknesses up to 62 mm by the longitudinal welding process. Special large-diameter piles can be manufactured by the process.

Economies in steel can be achieved by varying the wall thickness and quality of the steel. Thus, in marine structures, the upper part of the pile can be in mild steel which is desirable for welding on bracing and other attachments; the middle section can be in high-tensile steel with a thicker wall where bending moments are greatest, and the lower part, below seabed, can be in a thinner mild steel or high-tensile steel depending on the severity of the driving conditions. The 1.3 m OD steel tubular piles used for breasting dolphins for the Abu Dhabi Marine Areas Ltd. tanker berth at Das Island were designed by BP to have an upper section 24 mm in thickness, a middle section 30 mm in thickness, and a lower section of 20 mm in thickness. The overall length was 36.6 m. As an economic alternative to tubular steel piles for turbine bases at a wind farm on a reinstated open-cast coal site in County Durham, Aarsleff installed 36 340 mm OD recycled, high-grade oil well casings through unpredictable backfill to toe into sandstone bedrock at each base. The additional stiffness of the casings allowed the use of a 4 tonne accelerated impact hammer to overcome obstructions to

*Table 2.6* Dimensions and nominal applied loads for typical concrete-filled cased piles using light-gauge tubes

| Internal diameter (mm²) | Area of concrete (mm²) | Typical capacity (kN) for ordinary soil[a] | Typical capacity (kN) for rock[b] |
|---|---|---|---|
| 254 | 50,670 | 150 | 200 |
| 305 | 72,960 | 300 | 350–460 |
| 356 | 99,300 | 400 | 500–650 |
| 406 | 129,700 | 500 | 600–850 |
| 457 | 164,100 | 650 | 800–1,000 |
| 508 | 202,700 | 800 | 1,000–1,300 |
| 559 | 245,200 | 1,000 | 1,250 |
| 610 | 291,800 | 1,200 | 1,500 |

[a] Ordinary soil – sand, gravel or very stiff clay.
[b] Rock, very dense sand or gravel, very hard marl or hard shale.

driving and achieve a set of 25 mm in 10 blows. Sections of the threaded and collared casing could be joined to produce the maximum depth of 21 m.

Light spirally welded mild steel tubular piles in the range of sizes and typical capacity listed in Table 2.6 are widely used for lightly loaded structures, usually driven by a drop hammer acting on a plug of concrete in the bottom of the pile (see Section 3.2). These piles, known as *cased piles*, are designed to be filled with concrete after driving. Extension tubes can be welded to the driven length to increase penetration depth. Roger Bullivant Ltd. provides thicker wall tubes for cased piles from 125 to 346 mm diameter with up to 10 mm wall section for top driving of the pile. If piles have to be spliced, a special compression joint is needed for driving. Pile capacities claimed range from 350 to 1250 kN depending on ground conditions. In countries where heavy timbers are scarce, cased piles have replaced timber piling for temporary stagings in marine or river work. Here, the end of each pile is closed by a flat mild steel plate welded circumferentially to the pile wall.

Concrete-filled steel tubular piles need not be reinforced unless required to carry uplift or bending stresses which would overstress a plain concrete section cast in the lighter gauges of steel. Continuity steel is usually inserted at the top of the pile to connect with the ground beam or pile cap.

*Steel box piles* are fabricated by welding together trough-section sheet piles such as the CAZ and CAU sections made by ArcelorMittal in double, triple or quadruple combinations or using specially rolled trough plating. Larssen U-section piles and Hoesch Z-sections, both rolled by Hoesch, are also suitable for box piles. The types fabricated from sheet piles are useful for connection with sheet piling forming retaining walls, for example to form a wharf wall capable of carrying heavy compressive loads in addition to the normal earth pressure. However, if the piles rotate during driving, there can be difficulty in making welded connections to the flats. Plain flat steel plates can also be welded together to form box piles of square or rectangular section.

The *MV* pile consists of either a steel box section (100 mm) or H-section fitted with an enlarged steel shoe to which a grout tube is attached. The H-pile is driven with a hammer or vibrator, while grout is injected at the driving shoe. This forms a fluidised zone along the pile shaft and enables the pile to be driven to the deep penetration required for their principal use as anchors to retaining walls. The hardened grouted zone around the steel provides the necessary frictional resistance to enable them to perform as anchors.

*H-section piles*, hot rolled in the United Kingdom to BS 4-1 as universal bearing piles (Figure 2.20a), have a small volume displacement and are suitable for driving in groups at close centres in situations where it is desired to avoid substantial ground heave or lateral displacement. The Steel Construction Institute's *H-Pile Design Guide*, 2005,[2.11] is based on limit state design as provided in the Eurocodes and, in addition to describing H-piles in detail, makes reference to the offshore industry's recommended practice for steel tubular piles based on North Sea experience as described in the *ICP Design Methods for Driven Piles in Sands and Clays* (see Section 4.3.7).

Corus (part of the Tata Group) produces a range of broad flange H-piles in sizes from 203 mm × 203 mm × 45 kg/m to 356 mm × 358 mm × 174 kg/m; the ArcelorMittal HP range is similar. They can withstand hard driving and are useful for penetrating soils containing cemented layers and for punching into rock. Their small displacement makes them suitable for driving deeply into loose or medium-dense sands without the *tightening* of the ground that occurs with large-displacement piles. They were used for this purpose for the Tay Road Bridge pier foundations, where it was desired to take the piles below a zone of deep scour on the bed of the Firth of Tay. Test piles 305 × 305 mm in section were driven to depths of up to 49 m entirely in loose becoming medium-dense to dense sands, gravels, cobbles and boulders, which is indicative of the penetrating ability of the H-pile.

The ability of these piles to be driven deeply into stiff to very stiff clays and dense sands and gravels on the site of the Hartlepool Nuclear Power Station is illustrated in Figure 2.17. On this site, driving resistances of 355 × 368 mm H-piles were compared with those of precast concrete piles of similar overall dimensions. Both types of pile were driven by a Delmag D-25 diesel hammer (see Table 3.4). Although the driving resistances of both types were roughly the same to a depth of about 14 m (indicating that the ends of the H-piles were plugged solidly with clay) at this level, the heads of the concrete piles commenced to spall and they could not be driven below 14.9 m, whereas the H-piles were driven on to 29 m without serious damage, even though driving resistance had increased to 0.5 mm/blow at the end of driving. Three of the H-piles were loaded to 3000 MN without failure, but three of the precast concrete piles failed at test loads of between 1100 and 1500 MN.

Because of their relatively small cross-sectional area, H-piles cannot develop a high end-bearing resistance when terminated in soils or in weak or broken rocks. In Germany and Russia, it is frequently the practice to weld short H-sections on to the flanges of the piles near their toes to form *winged piles* (Figure 2.18a). These provide an increased cross-sectional area in end bearing without appreciably reducing their penetrating ability. The bearing capacity of tubular piles can be increased by welding T-sections onto their outer periphery when the increased capacity is provided by a combination of friction and end bearing on the T-sections (Figure 2.18b). This method was used to reduce the penetration depth of 1067 mm OD tubular steel piles used in the breasting dolphins of the Marine Terminal in Cromarty Firth. A trial pile was driven with an open end through 6.5 m of loose silty sand for a further 16 m into a dense silty sand with gravel and cobbles. The pile was driven by a MENCK MRB 1000 single-acting hammer with a 1.25 m drop of the 10 tonne ram. It will be seen from Figure 2.19 that there was only a gradual increase in driving resistance finishing with the low value of 39 blows/200 mm at 22.6 m penetration. The pile was then cleaned out and plugged with concrete but failed under a test load of 6300 kN.

It was evident from the driving records that the plain piles showed little evidence of developing base resistance by plugging and would have had to be driven much deeper to obtain

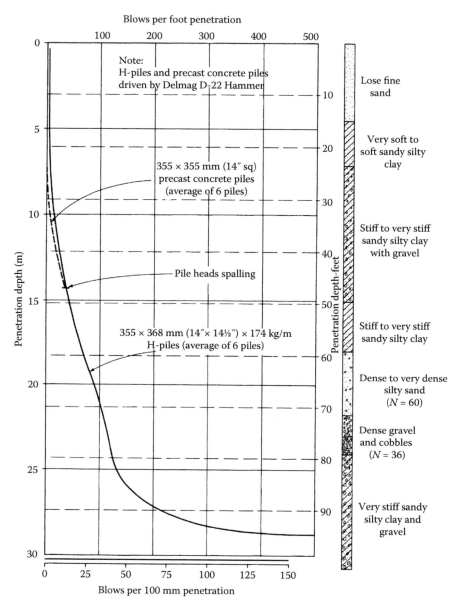

*Figure 2.17* Comparison of driving resistances of 355 × 355 mm precast concrete piles and 355 × 368 mm H-section piles driven into glacial clays, sands and gravels in Hartlepool Nuclear Power Station.

the required bearing capacity. In order to save the cost and time of welding on additional lengths of pile, it was decided to provide end enlargements in the form of six 0.451 × 0.303 × 7.0 m long T-sections welded to the outer periphery in the pattern shown in Figure 2.18b. The marked increase in driving resistance of the trial pile is shown in Figure 2.19. The final resistance was approaching refusal at 194 blows/200 mm at 19 m below seabed. The winged pile did not fail under the test load of 6300 kN.

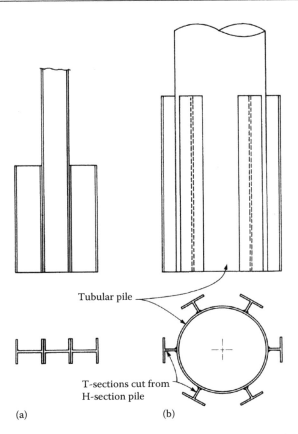

Figure 2.18 Increasing the bearing capacity of steel piles with welded-on wings (a) H-section wings welded to H-section pile and (b) T-section wings welded to tubular pile.

A disadvantage of the H-pile is a tendency to bend about its weak axis during driving. The curvature may be sharp enough to cause failure of the pile in bending. From his research, Bjerrum[2.12] recommended that any H-pile having a radius of curvature of less than 366 m after driving should be regarded as incapable of carrying load. A further complication arises when H-piles are driven in groups to an end bearing on a dense coarse-grained soil (sand and gravel) or weak rock. If the piles bend during driving so that they converge, there may be an excessive concentration of load at the toe and a failure in end bearing when the group is loaded. A deviation of about 500 mm was observed of the toes of H-piles after they had been driven only 13 m through sands and gravels to an end bearing on sandstone at Nigg Bay in Scotland. Such damage can be limited by careful monitoring during driving using a PDA. EC3-5 defines the slenderness criteria for assessing buckling where the soil does not provide sufficient lateral restraint.

The curvature of H-piles can be measured by welding a steel angle or channel to the web of the pile. After driving, an inclinometer is lowered down the square-shaped duct to measure the deviation from the axis of the pile. This method was used by Hanna[2.13] at Lambton Power Station, Ontario, where 305 and 355 mm H-piles that were driven through 46 m of clay into shale had deviated 1.8–2.1 m from the vertical with a minimum radius of

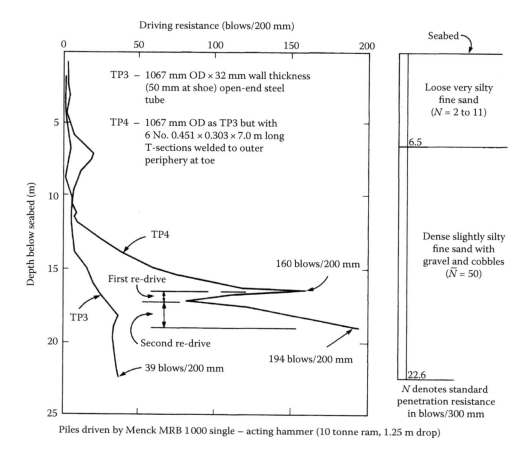

Figure 2.19 Comparison of driving resistance of open-ended plain and winged tubular steel piles at Britoil Tanker Terminal, Cromarty Firth.

curvature of 52 m. The piles failed under a test load, and the failure was attributed to plastic deformation of the pile shaft in the region of maximum curvature.

H-piles can be spliced on-site, either horizontally prior to installation to produce the desired length or to extend a driven section, using 100% butt weld to ensure full development of the strength of the section. End preparation using oxy-cutting to form either V or X bevels depending on alignment is usually acceptable[2.14]. The reuse of extracted H-piles is allowed under BS EN 12699, provided that the material complies with the design requirements, particularly in respect of durability and being undamaged.

*Peine piles* are broad-flanged H-sections rolled by Hoesch with bulbs at the tips of the flanges (Figure 2.20b). Loose clutches ('locking bars') are used to interlock the piles into groups suitable for dolphins or fenders in marine structures. They can also be interlocked with the Hoesch–Larssen sections to strengthen sheet pile walls. The ArcelorMittal HZ piles have tapered flange tips for interlocking.

*The Monotube pile* fabricated by the Monotube Pile Corporation of the United States is a uniformly tapering hollow steel tube. It is formed from steel which is cold-worked to a fluted section having a tensile yield strength of 345 N/mm² or more. The strength of the fluted section is adequate for the piles to be driven from the top by hammer without an internal mandrel or concrete filling. The tubes have a standard tip diameter of 203 mm,

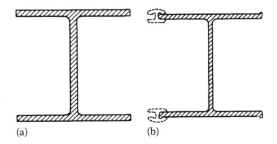

*Figure 2.20* Types of H-section steel piles. (a) Universal bearing pile (UK, European and US manufacture). (b) Peine pile (Hoesch).

and the shaft diameter increases to 305, 356, 406 or 457 mm at rates of taper which can be varied to suit the required pile length. An upper section of uniform diameter can be fitted (Figure 2.21), which is advantageous for marine work where the fluted section has satisfactory strength and resilience for resisting wave forces and impact forces from small- to medium-size ships. The tubes are fabricated in 3, 5, 7 and 9 gauge steel, and taper lengths can be up to 23 m. The heavier gauges enable piles to be driven into soils containing obstructions without the tearing or buckling which can occur with thin steel shell piles.

The *Soilex* system, developed in Sweden, uses the patented expander body to form an enlarged bulb to displace and compact the soil. The expander body consists of a thin folded sheet metal tube which, after insertion into the soil, is inflated by injecting concrete or grout under controlled pressure to form a bulb 5–10 times the original diameter. Installation may be by conventional drilling, driving, jacking or vibration methods or placement in a preformed hole, the pile shaft geometry above the bulb being determined by the method

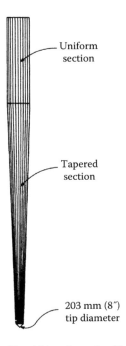

Uniform section

Tapered section

203 mm (8″) tip diameter

*Figure 2.21* Union Monotube pile. (Union Metal Manufacturing Co., Canton, OH.)

of installation. The tube dimensions before expansion range from 70 to 110 mm square up to 3 m long which following inflation provides end-bearing areas of 0.12–0.5 m². In Borgasund, Sweden, fifty-seven 11 m long Soilex piles using a 110 mm expander body welded to 168 mm diameter thick-walled tube were installed by a vibrator in a predrilled hole in medium-dense sand below new railway bridge abutments. Approximately 0.5 m³ of concrete was used to inflate the expander body to form an 800 mm diameter bulb producing a pile which had an estimated ultimate capacity of 1100 kN, limited by the strength of the concrete-infilled steel shaft. The system is also useful for underpinning where short piles are appropriate and as tension ground anchors; in all cases, the spacing of piles is critical to avoid interference.

## 2.2.5 Shoes for steel piles

No shoes or other strengthening devices at the toe are needed for tubular piles driven with open ends in easy to moderately easy driving conditions. Where open-ended piles have to be driven through moderately resistant layers to obtain deeper penetrations or where they have to be driven into weak rock, the toes should be strengthened by welding on a steel ring. The internal ring (Figure 2.22a) may be used where it is necessary to develop the full external frictional resistance of the pile shaft. An external ring (Figure 2.22b) is useful for reducing the friction to enable end-bearing piles to be driven to a deep penetration, but the uplift resistance will be permanently reduced. Hard driving through strongly resistant layers or to seat a pile onto a rock may split or tear the ring shoe of the type shown in Figure 2.22a and b. For hard driving, it is preferable to adopt a welded-on thick plate shoe designed so that the driving stresses are transferred to the parent pile over its full cross-sectional area (Figure 2.22c).

A shoe of this type can be stiffened further by cruciform steel plates (Figure 2.23a). Buckling and tearing of an external stiffening ring occurred when 610 mm OD steel tube piles were driven into the sloping surface of strong limestone bedrock (Figure 2.23b).

Steel box piles can be similarly stiffened by plating unless they have a heavy wall thickness such that no additional strengthening at the toe is necessary. Steel tubular or box piles designed to be driven with closed ends can have a flat mild steel plate welded to the toe (Figure 2.24a) when they are terminated in soils or weak rocks. The flat plate can be stiffened by vertical plates set in a cruciform pattern. Where they are driven on to a sloping hard rock surface, they can be provided with Oslo points as shown in Figure 2.24b.

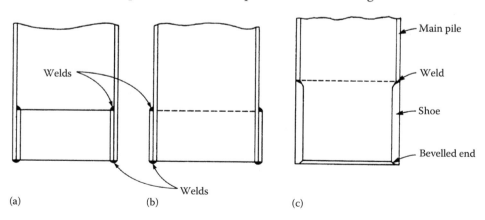

*Figure 2.22* Strengthening toe of steel tubular piles. (a) Internal stiffening ring. (b) External stiffening ring. (c) Thick plate shoe.

*Figure 2.23* (a) Strengthening shoe of tubular steel pile by cruciform plates. (b) Buckling and tearing of welded-on external stiffening ring to tubular steel pile driven onto sloping rock surface.

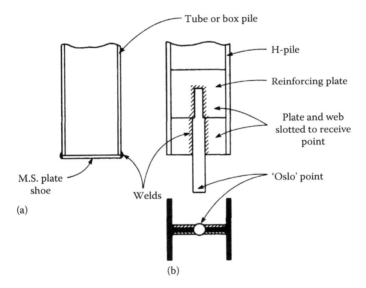

*Figure 2.24* Shoes for steel piles. (a) Flat plate for tubular or box pile. (b) *Oslo* point for H-section pile.

Steel H-piles may have to be strengthened at the toe for situations where they are to be driven into strongly cemented soil layers or soil containing cobbles and boulders. The strengthening may take the form of welding on steel angles (Figure 2.25a) or purpose-made devices such as the *Pruyn Point* manufactured in the United States by APF (Figure 2.25b). Dawson Construction Plant Ltd. manufactures a range of shoes for steel and timber piles.

## 2.2.6 Yield stresses for steel piles

As with other Eurocodes, EC3 makes no reference to allowable working stresses. Nominal and ultimate yield strengths applicable to steel bearing piles are those for steel structures generally given in Table 3.1 of EC3-1-1 and the BS ENs noted in Tables 2.7 through 2.9. EC3-5 (for steel piling) refers to EC3-1-1 for the strengths of bearing piles, but the GP grades provided for steel sheet pile sections quoted in EC3-5 are different from EC3-1-1, which must be noted when designing box piles. These nominal values should be used as the characteristic

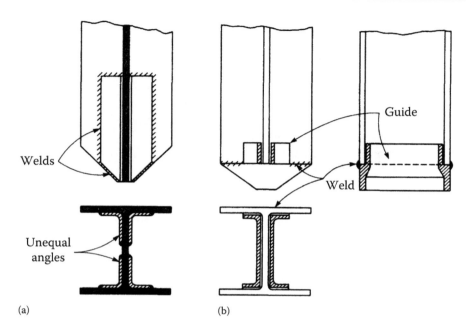

Figure 2.25 Strengthening toe of H-section pile. (a) Welded-on steel angles. (b) Pruyn Point. (Associated Pile and Fitting Corporation, Parsippany, NJ.)

Table 2.7 Summary of BS EN codes for the production and composition of steel and manufacture of steel sections by hot rolling and cold forming which apply to bearing and sheet pile design

| BS EN | Type of non-alloy steel | Use in piles |
|---|---|---|
| 10024:1995 | Hot-rolled structural steel | Taper flange I-sections |
| 10025-2:2004 | Hot-rolled structural steel | Tubular and H-piles |
| 10210-1:2006 | Hot-finished structural hollow sections | Tubular piles |
| 10219-1:2006 | Cold-formed welded hollow sections | Tubular piles |
| 10248-1:1996 | Hot-rolled sheet piling | Sheet piles/box piles |
| 10249-1:1996 | Cold-formed sheet piling | Sheet piles/box piles |
| 10025-6:2004 | Hot-rolled structural steel flats | Pile bracing |

Table 2.8 Summary of BS EN 10027 rules for designating the type and yield strength of steels in the above standards and the abbreviated identification code

| Type | Description |
|---|---|
| S | Structural steel |
| E | Engineering steel |
| 275 and 355 | Minimum yield strength in N/mm$^2$ |
| W | Improved atmospheric corrosion resistance |
| N | Normalised |
| Q | Quenched and tempered |
| H | Hollow section |
| G | General purpose |
| P | Sheet piles |

*Table 2.9* Further series of designations required to describe the fracture toughness of the steel in tension to resist impacts at normal and low temperature using the Charpy V impact test values in Table 2.1 of EC3-1-10 as summarised

| Subgrade | JR | J0 | J2 | K2 | N | NL |
|---|---|---|---|---|---|---|
| Charpy impact value | 27 J at 20°C | 27 J at 0°C | 27 J at −20°C | 40 J at −20°C | 40 J at >−20°C | 27 J at >−50°C |

Note:  Different test temperatures are applied to sheet piles as EC3-5, Table 3.3.

The Charpy test is defined in BS EN ISO 148-1.

values in design calculations to determine stresses in steel piles using the partial factors in EC3 National Annex, subject to the nominal thickness of the structural element (see Section 7.10). For example, S275 grade steel has a characteristic value of $f_y = 275$ N/mm$^2$ for a nominal thickness ≤40 mm, but this reduces to 255 N/mm$^2$ for nominal thickness between 40 and 80 mm. For the *tougher* steels (see below), the reduction required in yield stress is greater for thickness >40 mm. EC3-1-1 covers steel in the range S235 to S460.

The limitations on stress in BS EN 12699, Clause 7.7.3, apply to the calculated stress in piles during driving. For steel piles, the calculated driving stresses are permitted to be 0.9 times the characteristic yield strength of the steel, and 'these may be 20% higher than the above values' if the stresses are monitored during driving as noted for other driven piles (this would imply a stress of 108% of the yield strength). The American Petroleum Institute[4.15] (API) in specification RP2A states that the dynamic stresses during driving should not exceed 80%–90% of yield strength depending on specific circumstances such as previous experience and confidence in the method of analysis.

The selection of a grade of steel for a particular task depends on the environmental conditions as well as on the calculated stresses. For piles wholly embedded in the ground, or for piles in river and marine structures which are not subjected to severe impact forces, particularly in tropical or temperate waters, a mild steel grade S275G (minimum yield strength 275 N/mm$^2$) or a high-tensile steel S355G (minimum yield strength 355 N/mm$^2$) is satisfactory. Corus (Tata) produces hot-finished tubular sections suitable for general piling in grades S355JOH and S355J2H (for more exposed conditions). The ArcelorMittal cold-formed tubular pile range is S235JRH to S460MH (M indicating 'thermo-mechanically' rolled), with special grades to order for additional corrosion resistance. Steel grades for hot-rolled sheet piles used to form box piles range from S240GP to S430GP. Tubular steel piles are also produced to API 5L[2.15] grades X52 to X80.

Piles for deep-water platforms or berthing structures for large vessels are subjected to high dynamic stresses from berthing impact and wave forces. In water at zero or sub-zero temperatures, there is a risk of brittle fracture under dynamic loading, and the effects of fatigue damage under large numbers of load repetitions and also of saltwater corrosion need to be considered. The lowest service temperature to be taken into account for fracture toughness in steel piles is −15°C as given in the NA to EC3-5, and steels must be selected to have a high impact value when tested at low temperatures as given previously. Steel grade S235 is only produced in Charpy subgrades JR, J0 and J2, whereas the higher grades can be provided in the all the subgrades noted in Table 2.9. Piles or bracing members for deep-water structures may be required to be fabricated from plates 30 mm or more in thickness. The steel for such plates should have a brittle fracture resistance at low temperatures, and note must be taken of the maximum thicknesses allowed in EC3-1-10 for each grade of steel at normal and lower temperatures. High-tensile steel conforming to grades above S460Q with mechanical and chemical properties superior to BS EN 10210 and a Charpy impact value of 60 J at −50°C can be produced in order to meet these special requirements.

## 2.3 DRIVEN AND CAST-IN-PLACE DISPLACEMENT PILES

### 2.3.1 General

Driven and cast-in-place piles are installed by driving to the desired penetration a heavy-section steel tube with its end temporarily closed by a sacrificial end cap. A reinforcing cage is placed in the tube which is then filled with concrete, either as the tube is withdrawn or following withdrawal. Thin steel shell piles (similar to the preceding cased pile) are driven by means of an internal mandrel, and concrete is placed in the permanent shells after withdrawing the mandrel; reinforcement can be installed before or after concreting. The driven concrete shell pile is no longer viable economically as a result of improved driving/withdrawal plant for drive tube and steel shell methods.

Driven and cast-in-place piles have the principal advantage of being readily adjustable in length to suit the desired depth of penetration. Thus, in the withdrawable-tube types, the tube is driven only to the depth required by the ground conditions. Another advantage, not enjoyed by all types of shell pile, is that an enlarged base can be formed at the toe. BS EN 12699 gives specific procedures for concreting in dry tubes and allows the use of a *tremie* pipe in clean wet conditions and, by implication, in stable bores where the drive tube has been withdrawn. Some specifications forbid the use of a wholly uncased shaft for all forms of driven and cast-in-place pile in conditions such as soft to firm clays or in loose to medium-dense sands and materials such as uncompacted fill. These restrictions are designed to prevent lifting of the concrete while pulling out the driving tube and squeezing 'waisting' the unset concrete in the pile shaft where this is formed in soft clays or peat. One of the techniques to avoid these problems is to insert permanent light-gauge steel shells before placing the concrete and withdrawing the tube. Such expedients increase the cost of the withdrawable-tube piles to the extent that their advantage in price over shell piles may be wholly or partially lost. The soundness of the uncased type of pile depends on the skill and integrity of the operatives manning the piling rig.

The withdrawable-tube or thin-shell pile types are unsuitable for marine structures, but they can be employed in marine situations if they are extended above the seabed as columns or piers in steel or precast concrete. As with all forms of driven pile, noise abatement procedures must be followed (Section 3.1.7). When driving heavy-duty thick-walled tubes in urban environments, the cost advantages of the method can be negated.

When installing driven and cast-in-place piles in groups, it is advisable to limit the distance centre to centre of adjacent uncased piles to not less than 6 pile diameters until the concrete has reached adequate strength. This distance should be increased if the undrained shear strength of the soil is less than 50 kN/m$^2$. Ground heave problems are considered in Section 5.7.

### 2.3.2 Withdrawable-tube types

In conditions favourable for their employment, where the required penetration depth is within the capability of the piling rig to pull out the tube and there are no restrictions on ground heave or vibrations, withdrawable-tube piles can be installed more cheaply than any other type of driven or bored pile for comparable capacities.

The installation methods for the various types of driven and cast-in-place piles described in Construction Industry Research and Information Association (CIRIA) report PG1[2.16] have changed in many respects as a result of the improved pulling capacity of mobile self-erecting rigs (see Table 3.6) and cranage. The original pile of this type, the *Franki* pile, employs an internal drop hammer (2–8 tonnes) acting on a plug of gravel or dry concrete at

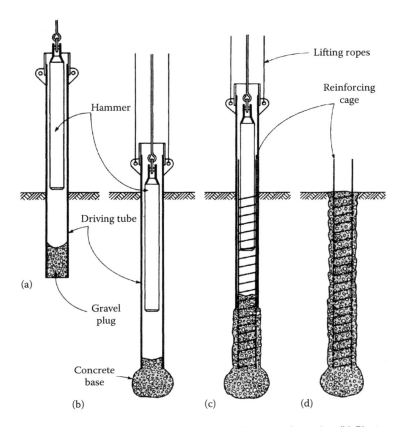

*Figure 2.26* Stages in installing an open-ended Franki pile. (a) Driving piling tube. (b) Placing concrete in piling tube. (c) Compacting concrete in shaft. (d) Completed pile.

the bottom of a thick-walled drive tube, 248–610 mm diameter. The drive tube is carried down with the plug until the required toe level is reached when the tube is restrained from further penetration by rope tackle. The gravel plug and batches of dry concrete are then hammered out to form a bulb or enlarged base to the pile. The full-length reinforcing cage is inserted, followed by placing a semi-dry concrete in batches as the drive tube is pulled out in stages. After each stage of withdrawal, the concrete is compacted by the internal hammer (Figure 2.26). Depths up to 30 m have been achieved, capable of carrying loads up to 2000 kN, subject to ground conditions. Franki piles may be raked up to 1:3 in special cases, but insertion of the reinforcement and concrete needs careful control. Driving by internal hammer and concreting in stages are slower than the top driving method on heavy-duty tube. Hence, these techniques are used only when there are economic advantages, for example when the enlarged base adds appreciably to the bearing capacity of the pile.

In a variation of the Franki technique, the gravel plug (or dry concrete plug) can be hammered out at several intermediate stages of driving to form a shell of compact material around the pile shaft. This technique is used in very soft clays which are liable to squeeze inwards when withdrawing the tube. Composite Franki piles are formed by inserting a pre-cast concrete pile or steel tube into the driving tube and anchoring it to the base concrete plug by light hammer blows. The drive tube is then withdrawn.

In the now conventional withdrawable-tube pile, the thick-walled section tube has its lower end closed by an expendable steel plate or shoe (capable of keeping out groundwater) and is driven from the top by a 5-tonne hydraulic hammer. On reaching the required toe level, as predetermined by calculation or as determined by measurements of driving resistance, the hammer is lifted off and a reinforcing cage is lowered down the full length of the tube. A highly workable self-compacting concrete is then placed in the tube through a hopper, and the tube raised by a hoist rope operated from the pile mast or frame and, where needed, vibrating the tube. The tube may be filled completely with concrete before it is lifted or it may be lifted in stages depending on the risks of the concrete jamming in the tube. The length of the pile is limited by the ability of the rig to pull out the drive tube. This restricts the length to about 20–30 m. Pile diameters range from 285 to 525 mm with load capacity up to 1500 kN.

Although BS EN 12699 allows for unreinforced piles in certain ground conditions, a full-length reinforcing cage is advisable in the driven and cast-in-place pile. It acts as a useful tell-tale against possible breaks in the integrity of the pile shaft caused by arching and lifting of the concrete as the tube is withdrawn. A shorter cage (4 m) may be used in vertical piles, subject to the bending and tensile stresses in the pile, and inserted into the wet concrete. Where reinforcement is designed, BS EN 12699 for driven displacement piles generally follows BS EN 1536 requirements for bored piles (see Table 2.11). Minimum cover should be 50 mm where the casing is withdrawn, 75 mm where reinforcement is installed after concreting (or where subject to ground contaminants) and 40 mm where there is permanent lining. The spacing of bars in the reinforcing cage should give ample space for the flow of concrete through them. Transverse reinforcement should also be as stated in BS EN 1536.

The problem of inward squeezing of soft clays and peats or of bulging of the shafts of piles from the pressure of fluid concrete in these soils is common to cast-in-place piles both of the driven and bored types. As noted earlier, a method of overcoming this problem is to use a permanent light-gauge steel lining tube to the pile shaft. However, great care is needed in withdrawing the drive tube to prevent the permanent liner being lifted with the tube. Even a small amount of lifting can cause transverse cracks in the pile shaft of sufficient width to result in excessive settlement of the pile head under the applied load. The problem is particularly difficult in long piles when the flexible lining tube tends to snake and jam in the drive tube. Also where piles are driven in large groups, ground heave can lift the lining tubes off their seating on the unlined portion of the shaft. Snaking and jamming of the permanent liner can be avoided by using spacers such as rings of sponge rubber.

In most cases, the annulus left outside the permanent liner after pulling the drive tube will not close up. Hence, there will be no frictional resistance available on the lined portion. This can be advantageous because downdrag forces in the zone of highly compressible soils and fill materials will be greatly reduced. However, the ability of the pile shaft to carry the applied load as a column without lateral support below the pile cap should be checked. Problems concerned with the installation of driven and cast-in-place piles are discussed further in Section 3.4.5.

Apart from the dry mix for the Franki pile as noted, the stresses on the shafts of these piles are determined by the need to use easily workable self-compacting mixes. BS EN 12699 requires the rules on the concreting of bored piles using self-compacting concrete as stated in BS EN 1536 to apply to all cast-in-place displacement piles unless otherwise specified. BS 8500 designates a self-compacting mix as S4, with a slump in the range of 180 ± 30 mm and cement content ≥325 kg/m$^3$ to make allowances for possible imperfections in the concrete placed in unseen conditions. Henderson et al.[2.17] in CIRIA Report C569 make recommendations for the coarse and fine aggregates content. When semi-dry concrete is tamped during installation, the concrete class should be at least C25/30 with a minimum cement content of 350 kg/m$^3$.

Table 2.10 Some typical load capacities for driven and
cast-in-place piles of various shaft diameters

| Nominal shaft diameter (mm) | Typical load capacity (kN) |
|---|---|
| 300 | 350–500 |
| 350 | 450–700 |
| 400 | 600–900 |
| 450 | 800–1000 |
| 500 | 1000–1400 |
| 600 | 1400–2000 |

The higher ranges in the Table 2.10 should be adopted with caution, particularly in difficult ground conditions.

The *Vibrex pile* installed by Fundex, Verstraeten BV, employs a diesel or hydraulic hammer to drive the tube which is closed at the end by a loose sacrificial plate. An external ring vibrator is then employed to extract the tube after the reinforcement cage, and concrete has been placed. A variation of the technique allows an enlarged base to be formed by using the hammer to drive out a charge of concrete at the lower end of the pile. The Vibrex pile is formed in shaft diameters from 350 to 600 mm.

The speciality of the *Vibro pile* (not to be confused with the vibro concrete column in Section 2.3.7) is the method used to compact the concrete in the shaft utilising the alternate upward and downward blows of a hydraulic hammer on the driving tube. Once the drive tube reaches the required level, the upward blow of the hammer operates on links attached to lugs on top of the tube. This raises the tube and allows concrete to flow out. On the downward blow, the concrete is compacted against the soil. The blows are made in rapid succession which keeps the concrete *alive* and prevents jamming of the tube as it is withdrawn. Diameters up to 600 mm with 740 mm shoes are available.

## 2.3.3 Shell types

Types employing a *metal shell* generally consist of a permanent light-gauge steel tube in diameters from 150 to 500 mm with wall thickness up to 6 mm and are internally bottom driven by a drop hammer acting on a plug of dry concrete (care being taken not to burst the tube). The larger-diameter tubes are usually fabricated to the estimated length and handled into a piling frame with a crane. Smaller-diameter, spirally welded tube can be manually placed on the rig leader and welded in sections to produce the required depth during installation. On reaching the bearing layer, the hammer is removed, any reinforcement inserted and a high slump (S4) concrete placed to produce the pile. Capacities up to 1200 kN are possible.

In France, cased piles varying in diameter from 150 to 500 mm are installed by welding a steel plate to the base of the tubular section to project at least 40 mm beyond the outer face of the steel. As the pile is driven down, a cement/sand mortar with a minimum cement content of 500 kg/m³ is injected into the annulus formed around the pile by the projecting plate through one or more pipes having their outlet a short distance above the end plate. The rate of injection of the mortar is adjusted by observing the flow of mortar from the annulus at the ground surface. The steel section is designed to carry the applied load. The calculated stress permitted of 160 N/mm² is higher than the value normally accepted for steel piles using EN24-1 steel, because of the protection given to the steel by the surrounding mortar. Steel H- or box sections can be given mortar protection in a similar manner.

*Figure 2.27* The TaperTube pile.

The *TaperTube* pile (Figure 2.27), a steel shell similar to the Monotube but without the flutes, has been developed by DFP Foundation Products and Underpinning & Foundation Constructors of the United States. It uses a heavier wall thickness of 9.5 mm in 247 N/mm² grade hot-rolled steel to form a 12-sided polygon tapering from 609 to 203 mm at the cast-steel point over lengths of 3–10 m. Where tube extensions are needed, the top of the polygon can be formed into a circle for butt welding; this provides improved axial uplift resistance. After top driving is completed, the tapered shell pile is filled with concrete. Ultimate bearing capacities up to 4000 kN and lateral resistance to 200 kN have been determined in pile tests.

*Fibre-reinforced polymer* and fibreglass composite tubes can be considered as shell piles and, as they have high resistance to corrosion, rot and marine borers, are used for light marine structures. They can be drilled, and in suitable soft soil, thick-walled 400 mm tubes can be driven to depths of 20 m; they can have a steel tube core and be infilled with concrete to improve bearing and compression resistance. Pearson Pilings of Massachusetts produces fibreglass piles up to 400 mm diameter with claimed axial capacities up to 600 kN.

### 2.3.4 Stresses on driven and cast-in-place piles

A common feature of nearly all the driven and cast-in-place pile types is an interior filling of concrete placed in situ, which forms the main load-carrying component of the pile (Table 2.10). Whether or not any load is allowed to be carried by the steel shell depends on its thickness and on the possibilities of corrosion or tearing of the shell.

Structural design stresses in EC7 are required to conform to EC1, EC2, EC3 and EC4 for the relevant material. The specified concrete strength grades in BS EN 12699 are C20/25 to C45/55, as for bored piles. Depending on the installation method used, the reduction factors noted for bored piles in Section 2.4: may need to be applied to allow for possible deficiencies in workmanship during placing the concrete or reductions in section of the pile shaft due to 'waisting' or buckling of the shells. Where steel tubes or sections are used as part of the load-carrying capability or reinforcement of the pile, BS EN 12699 requires EC4-1 rules to be applied.

### 2.3.5 Rotary displacement auger piles

*Auger displacement piles and screw piles* are drilled piles, but the soil is displaced and compacted as the auger head is rotated into the ground to form the stable pile shaft, with little soil being removed from the hole. The methods were mainly developed in the 1960s in Belgium from continuous flight auger techniques (see Section 2.4.3) and are now widely available. The original proprietary system is the cast-in-place *Atlas* pile in which the special dual flight auger head is screwed into the ground on a thick-walled steel tube. The helical shape of the pile shaft produced by screwing in the auger flange is maintained as the auger is back-screwed to form a stable hole into which the reinforcement cage is placed prior to concreting. Other proprietary rotary displacement piles such as the *ScrewSol* pile by Bachy Soletanche (Figure 2.28a and b), which produces a helical flanged pile shaft in weak soils, also use specially shaped augers on the end of the drill tube to compact the soil and inject concrete. Reinforcement is generally inserted into wet concrete. The benefits of the technique are reduced spoil at the surface, improved pile shaft capacity and in certain conditions reduced length and diameter for an equivalent bored pile.

The rotation of the auger flights on the end of the *Omega* cylindrical pile (Figure 2.29) breaks up the soil which is then displaced laterally and compacted by the cylindrical body above the auger. Concrete is injected at the auger base during extraction, and the reverse flights above the compacting cylinder ensure the hole remains stable until the concrete supports the bore to form the pile.

(a)                                          (b)

*Figure 2.28* (a) The ScrewSol tapered auger and tight-fit follower tube. (b) Cleaned-off section of an excavated ScrewSol pile.

*Figure 2.29* Omega displacement pile auger. (Courtesy of Malcolm Drilling Company, San Francisco, CA.)

The *Fundex pile* and the *Tubex pile* are forms of displacement pile. A helically screwed drill point is held by a bayonet jointed to the lower end of the piling tube. The tube is then rotated and forced down by hydraulic rams on the drill rig. On reaching founding level, a reinforcing cage and concrete are placed in the tube which is then withdrawn leaving the sacrificial drill point in the soil. The piling tube is left in place in the Tubex pile when used in very soft clays to avoid *waisting* of the shaft. The tube can be drilled down in short lengths, each length being welded to the one already in place. Thus, the pile is suitable for installation in conditions of low headroom, for example for underpinning work. This pile can also be installed with simultaneous grout injection which leaves a skin of grout around the tube and increases bearing capacity.

Rigs for the displacement auger piles are similar to the high-torque, instrumented CFA pile units (Table 3.6), but the power required to install screw piles can be 20% greater than that required for equivalent CFA piles; additional pull-down is usually necessary. As only a small amount of material is removed as the auger is initially inserted, the screw pile is particularly useful for foundations in contaminated ground.

Design of displacement screw piles should be based on a detailed knowledge of the ground using pressuremeter tests, cone penetration tests (CPT) and standard penetration tests (SPT) and pile test data in the particular soil. Care is required in selecting the effective diameter of the helical shaft for determination of shaft friction and end-bearing capacity. Bustamante and Gianeselli[2.18] have provided a useful simplified method of predetermining the carrying capacity of helical shaft piles based on a series of tests and recommend that a design diameter of 0.9 times the OD of the auger flange should be used for calculating both base and shaft resistance for thin flanges. For thick flanges (say 40 mm deep, 75 mm wide), the OD of the helix is appropriate. Depending on the ground conditions and the size of the helical flanges formed, savings of 30% in concrete volume compared with the equivalent bored pile are claimed. Typical pile dimensions are 500 mm outside auger diameter and 350 mm shaft diameter, and lengths of 30 m are possible. The technique is best suited to silty sands and sandy gravels with SPT *N-values* between 10 and 30; for $N > 50$, there is likely to be refusal with currently available rigs, and unacceptable heave and shearing can occur in clays.

Guidance on installation of displacement screw piles in BS EN 12699 is limited, but comprehensive trials of different types of pile at Limelette[2.19] in Belgium during 2000 and 2002 in stiff dense sand, together with earlier trials in stiff clay, have produced significant data on design, installation and performance of screw piles (including references to EC7 design procedures and CPT testing). Two main conclusions were that the bearing capacity is of similar magnitude as that for full displacement piles and the prediction of bearing capacity was in good agreement with load tests, irrespective of the method used.

### 2.3.6 Helical plate screw piles

These piles, although not strictly speaking displacement piles, have been used for many years to support light structures, gantry masts, as underpinning (Section 9.2.2) and in a variety of soils. They comprise either solid steel shafts or tubular shafts up to 320 mm OD with two or three helical steel plates between 200 and 1000 mm diameter attached at intervals in excess of three times the plate diameter along the shaft (limited to avoid heave). The number of helices will depend on the bearing capacity of each plate determined from the soil parameters; the depth can be increased by plain follower sections to ensure the bearing layer is achieved. The pile is screwed into the soil by hydraulic top-drive rig, usually attached to an excavator, or by handheld units with around 4 kNm torque – resisted by a torque bar for the small-diameter helix. Axial bearing capacity of up to 3000 kN is claimed in appropriate conditions; shaft resistance is usually ignored on the smaller-diameter shafts and lateral resistance is limited. Torque correlations should only be used as confirmation of resistance at the target stratum depth. Care is needed during design and installation to consider the effects of groundwater around the shaft, buckling and corrosion, particularly where high organic soil and landfill may be expected. Black and Pack[2.20] describe screw pile foundations in collapsible and expansive soils where load capacities of up to 890 kN in compression or tension were achieved and downdrag reduced.

### 2.3.7 Vibrated concrete columns

Vibrated concrete columns (VCCs) are a development of the bottom-feed vibro stone column technique used for ground improvement. They act as cast-in-place displacement piles in that little spoil is brought to the surface and is therefore useful for deep load-bearing foundations on brownfield sites, where the removal of contaminated arisings would be a problem, and in peat and organic soils. The poker is similar to that used for stone columns, but for a VCC, the poker is charged with concrete before commencing penetration of the soil. The poker is then vibrated to the required depth and the concrete is pumped out to form a bulb, with the poker raised and lowered into the bulb, while pumping additional concrete until the set resistance is achieved (Figure 3.15). The poker is then withdrawn at a controlled rate, while concrete pumping continues to form the shaft, monitored by data logging. Enlarged heads can be formed by reinserting the poker and injecting additional concrete, and reinforcement can be inserted on completion. By forming the end bulb, it is possible to achieve the required resistance at shallower depths in weak ground compared with conventional piling, typically 3–10 m. VCC shaft diameters range from 400 to 600 mm with a base bulb and enlarged heads of 1000 mm in soils with shear strengths of 15–60 kN/m² are usual. Depending on soil conditions, applied axial loads up to 900 kN are possible but lateral loading is limited. In variable strata, there is a risk of waisting of the shaft.

## 2.4 REPLACEMENT PILES

### 2.4.1 General

Replacement piles are installed by first removing the soil by a drilling process and then constructing the pile by placing concrete or some other structural element in the drilled hole. The simplest form of construction consists of drilling an unlined hole and filling it with concrete. However, complications may arise such as difficult ground conditions, the presence of groundwater or restricted access. Such complications have led to the development of specialist piling plant for drilling holes and handling lining tubes, but unlike the driven and cast-in-place piles, very few proprietary piling systems have been promoted. This is because the specialist drilling machines are available on sale or hire to any organisation which may have occasion to use them. The resulting pile as formed in the ground is more or less the same no matter which machine, or method of using the machine, is employed.

### 2.4.2 Bored and cast-in-place piles

In stable ground, an unlined hole can be drilled by mechanical auger or, in rare cases nowadays, by hand auger. If reinforcement is required, a light cage is then placed in the hole, followed by the concrete. In loose or water-bearing soils and in broken rocks, casing is needed to support the sides of the borehole, this casing being withdrawn during or after placing the concrete. In stiff to hard clays and in weak rocks, an enlarged base can be formed to increase the end-bearing resistance of the piles. The enlargement is formed by a rotating expanding tool. Hand excavation is rarely carried out because of stringent statutory health and safety regulations. A sufficient cover of stable fine-grained soil must be left over the top of the enlargement in order to avoid a 'run' of loose or weak soil into the unlined cavity, as shown in Figure 2.30.

Bored piles drilled by light cable percussive tripod rigs (up to 600 mm) are rarely used now for even lightly loaded buildings. As noted in the *Design Guide* for piles to low-rise housing prepared by the National House Building Council (NHBC)[2.21], small-diameter piles produced by modern hydraulic equipment can be an effective means of producing efficient foundations and reducing $CO_2$ emissions, when compared with deep trench-fill foundations. The guide also points out that the amount of material required in piles is likely to be less and spoil disposal reduced.

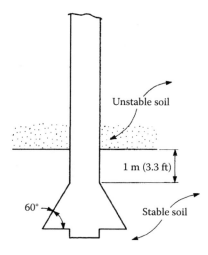

*Figure 2.30* Under-reamed base enlargement to a bored and cast-in-place pile.

Bored piles drilled by mechanical spiral-plate or bucket augers or by grabbing rigs can drill piles with a shaft diameter up to 7.3 m. Rotary drilling equipment consisting of drill heads with multiple rock roller bits have been manufactured for drilling shafts up to 7 m in diameter (e.g. the LDD7000 rig, a larger version of that shown in Figure 3.31). Under-reaming tools can further enlarge these shafts in stable soils to allow casings to be inserted. Standard plate auger boring tools for use with kelly bar rigs (see Section 3.3.4) range from 600 to 3650 mm. Rigs with telescopic kelly bars can reach 70 m depth and 102 m exceptionally.

When using bentonite or other drilling fluids[2.22] to support the sides of boreholes or diaphragm walls, the bond of the reinforcement to the concrete may be affected. Research by Jones and Holt[2.23] comparing the bond stresses in reinforcement placed under bentonite and polymer fluids indicated that it is acceptable to use the BS 8110 values of ultimate bond stress provided that the cover to the bar is at least twice its diameter when using deformed bars under bentonite. The results for the polymers investigated showed that the code bond stresses could be reduced by a divisor of 1.4. EC2-1-1 Clause 4 includes for a minimum cover factor dependent on bond requirements, and Clause 8 gives a reduction factor of 0.7 to apply to the ultimate bond stress where 'good' bond conditions do not exist – compatible with the Jones and Holt data for polymers. It also covers laps between bars using the reduced bond stress as appropriate, although good bond conditions may be available where the cover to the main bar is twice the main bar diameter. BS EN 1536 for bored piles states that only ribbed bars shall be used for main reinforcement where a stabilising fluid, bentonite or polymer, is used. Section 3.3.8 describes the use of stabilising fluids generally.

Unreinforced bored and cast-in-place piles can be considered as conforming to Clause 12 of EC2, subject to serviceability and durability requirements. Tension piles and piles in swelling/shrinking clays should always be fully reinforced, and for piles in axial compression, the reinforcement should extend over the length in compression. Where bending moments due to load transfer from ground beams, pile caps and rafts may occur, the upper part of the pile shaft should be reinforced to withstand such bending. A full-length cage is a useful guide to upward movement of the concrete when temporary casing has to be moved from the bore, as noted in Section 2.3.2.

Ample space between the bars to allow the flow of concrete is essential, and PD 6687-1 considers the problem of restrictions where bars have to be lapped. Concrete cover to the bars is detailed in BS EN 1536 which requires 60 mm cover for piles greater than 600 mm diameter and 50 mm for piles less than 600 mm, all increased to 75 mm in uncased bores in 'soft soil', for environmental exposures, and where the cage is inserted following concrete placement. Where reinforcement is designed, BS EN 1536 follows EC2 Clause 9.8.5 rules for longitudinal reinforcement areas for bored piles depending on the pile cross-sectional area (Table 2.11). Reinforcement grades are as for precast piles and follow BS 4449 and BS EN 10080 general requirements. Large welded cages are usually manufactured off-site to

*Table 2.11* Minimum amount of longitudinal reinforcement (as Table 5, BS EN 1536)

| Pile cross section, $A_c$ | Minimum area of longitudinal reinforcement, $A_s$ |
|---|---|
| $A_c \leq 0.5$ m² | $A_s \geq 0.005\, A_c$ |
| $0.5$ m² $< A_c \leq 1.0$ m² | $A_s \geq 25$ cm² |
| $A_c > 1.0$ m² | $A_s \geq 0.0025\, A_c$ |

Note: EC2-1 National Annex limits the area of steel at bar laps to $0.084A_c$.

BS EN ISO 17660-1 standards. Non-symmetrical cages should be avoided, unless they can be accurately positioned and restrained.

BS EN 1536 stipulates a minimum of four 12 mm diameter longitudinal bars and spaced at centres greater than 100 mm (80 mm when using <20 mm aggregate). The diameter of transverse reinforcement depends on the form: that is links should be ≥6 mm, welded wire mesh ≥5 mm and flat steel strips ≥3 mm thick, all spaced as for the main bars. However, EC2 Clause 9.8.5 requires a minimum of six 16 mm diameter longitudinal bars at less than 200 mm spacing to provide the area in Table 2.11. It is recommended that the EC2 provision is applied where there is design shear or bending in the pile.

Concrete grade may be between C20/25 and C45/55 and must be self-compacting and free flowing, with a minimum cement content of 325 kg/m$^3$ in dry conditions and 375 kg/m$^3$ in submerged conditions. Water/cement ratio is limited to 0.6 and the slump should be 200 ± 20 mm when placed under supporting fluid. In a stable dry bore, concreting is carried out from a hopper over the pile with a short length of pipe to direct flow into the centre of the reinforcement, ensuring that segregation does not occur. When concreting boreholes under flooded conditions or under stabilising fluid, a full-length tremie pipe as described in Section 3.4.8 is essential. For reasons of economy and the need to develop shaft friction, it is the normal practice to withdraw the casing during or after placing the concrete. As in the case of driven and cast-in-place piles, this procedure requires care and conscientious workmanship by the operatives in order to prevent the concrete being lifted by the casing, resulting in voids in the shaft or inclusions of collapsed soil.

Structural design stresses in the concrete are calculated using the characteristic strength of a concrete cylinder and the material partial factor ($\gamma_C$) in EC2 as given in Table 7.3 of Section 7.10. To allow for potential necking or waisting of bored piles which are not permanently cased, Clause 2.3.4.2 of EC2-1-1 requires a reduction in the nominal diameter ($d_{nom}$) as Table 4.6, Section 4.1.4. For the same reason, Clause 2.4.2.5 stipulates an increase in $\gamma_C$ when checking material ULS for uncased piles. Bored piles should preferably be concreted on the same day as they are bored; if not, it is advisable to extend the hole before placing concrete to avoid compromising end bearing.

Over 1100 large-diameter bored piles were installed at Canary Wharf by Bachy Soletanche in London Docklands ranging from 900 to 1500 mm and to depths of 30 m through terrace gravels, Lambeth Group clays, sands and gravels, and Thanet sands. It was possible to bore the piles without the aid of drilling fluids due to the low water table in the Thanet beds. Once the piles had reached the required depth using temporary casing, the shaft was filled with bentonite slurry to minimise the risk of pile collapse during concreting operations. The reinforcement cage was inserted to which were attached *tubes à manchette* (TaM) for pile base grouting 2 days after concreting.

A casing oscillator and crane-supported grab can be an economical method of boring large-diameter piles (up to 3000 mm) in gravels and cobbles, where a heavy chisel is needed to break up boulders and rockhead (Figure 3.32). The method is essentially the same as the basic bored and cast-in-place as described earlier, with the reinforcement cage installed to full depth and concrete placed by hopper or tremie as appropriate. Depths up to 50 m are feasible.

*Barrettes* can be an alternative to large-diameter bored and cast-in-place piles where in addition to vertical loads, high lateral loads or bending moments have to be resisted. They are constructed using diaphragm wall techniques to form short discrete lengths of rectangular wall and interconnected Ell- and Tee-shapes and cruciforms to suit the loading conditions in a wide variety of soils and rock to considerable depths. The *hydrofraise* or hydromill reverse-circulation rig (see Section 3.3.5) is particularly well adapted to form barrettes, as verticality is accurately controlled and the time for construction is reduced compared with grab rigs, thereby reducing the potential for the excavation to collapse. Barrettes are usually only economical when the rig is mobilised for the construction of other basement walls.

### 2.4.3 Continuous flight auger piles

*Continuous flight auger or auger-injected piles*, generally known as CFA piles, are installed by drilling with a rotary CFA to the required depth. They are now the most popular type of pile in the United Kingdom, used in a variety of ground conditions for bearing piles and as contiguous/secant pile walls. They are, however, best suited for ground conditions where the majority of the applied load is resisted by shaft friction and the ground is free from large cobbles and boulders. The CFA pile has considerable advantage over the conventional bored pile in water-bearing and unstable soils in that temporary casing is not usually needed, although, as noted below, the range of soil conditions which can now be augered has increased with the application of simultaneous casing methods.

The established practice is to bore the shaft using a CFA with a hollow stem temporarily closed at the bottom by a plug. After reaching the final level, a high slump concrete is pumped down the hollow stem displacing the plug, and once sufficient pressure has built up, the auger is withdrawn at a controlled rate, removing the soil and forming a shaft of fluid concrete extending to ground level (Figure 2.31) or lower cut-off level. Thus, the walls of the borehole are continually supported either by the spiral flights and the soil within them or by the concrete. Self-compacting concrete with grades as described for the above-mentioned bored piles is used with a plasticiser added to improve its 'pumpability', in accordance with the rules in BS EN 206-9. If concrete flow is not achieved, it is necessary to remove the auger

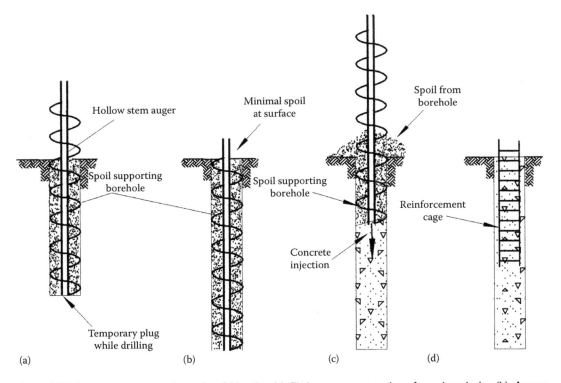

*Figure 2.31* Stages in construction of a CFA pile. (a) Flight auger rotated to form borehole. (b) Auger reaches required depth. (c) Concrete injected as auger rotated from hole. (d) Reinforcement cage inserted into wet concrete.

and re-drill (possibly after backfilling) to 0.5 m below the initial depth before recommencing concreting. The reinforcing steel cage, complying with the requirements for bored piles, can be pushed into the fluid concrete to a depth of about 15 m. Vibrators may be used to assist penetration. The shaft diameters range from the 100 mm micropile sections (in which sand–cement grout may be injected in place of concrete) up to 1200 and 1500 mm exceptionally. Load capacities up to 7500 kN and depths up to 34 m are now feasible (see Table 3.6), depending on ground conditions and pile dimensions.

In stable ground above the water table, it may be advantageous to remove the auger and place high slump concrete as in an unlined cast-in-place bored pile. The auger should never be withdrawn before concreting in unstable or water-bearing soils. BS EN 1536 requires that where unstable soil conditions are expected, a trial bore should be drilled, unless experience of the same conditions shows the CFA method is feasible.

The drilling operations are reasonably quiet and vibrations are low making the method suitable for urban locations (although the larger rigs can exceed 100 dBA when installing casing). As with any other in situ type of pile, the CFA pile depends for its integrity and load-bearing capacity, on strict control of workmanship. This is particularly necessary where a high proportion of the load is to be carried in end bearing. Because it is not possible to check the stratification and quality of the soil during installation as with conventional bored piles, considerable research and development has been undertaken by piling companies into the use of computerised instrumentation to monitor the process and ensure the quality and integrity of CFA piles. A computer screen is positioned in the drilling rig cab in front of the operator which continuously displays the boring and concreting parameters. During the boring operation, the depth of auger, torque applied, speed of rotation and penetration rate are displayed. During concreting, a continuous record of concrete pumping pressure and flow rate is shown, and on completion, the results are provided on a printout of the pile log which records the construction parameters and under- or oversupply of concrete (Figure 2.32). Most specifications for CFA piles[2.5] require the rig to be provided with such automated instrumentation to control the process, relieving the operator of some of the decision-making. Regular checks to ensure the reliability of the controls are essential. Even with this monitoring, doubts may exist in certain ground conditions as to whether or not the injected material has flowed out to a sufficient extent to cover the whole drilled area at the pile toe. For this reason, it may be advisable either to assume a base diameter smaller than that of the shaft or to adopt a conservative value for the end-bearing resistance. Farrell and Lawler[2.24] describe the need to reduce the bearing capacity factor in some stiff glacial tills. In addition, 'polishing' of the shaft can occur in stiff clays due to over-rotation and 'over-flighting' (i.e. vertical movement of the soil on the auger relative to the soil on the wall of the borehole resulting in local shaft distortion) which, in loose silty sands where over-rotation disturbs the surrounding soil, can reduce shaft resistance by 30%.

To address the problem of overflighting and loss of ground in soft soil leading to settlement and concreting difficulties, techniques have been developed to install temporary casing while simultaneously advancing the auger to the foundation depth, the *cased auger pile* as described in Section 3.3.3. The casing is normally withdrawn during concreting, but permanent steel liners can be installed to reduce downdrag and protect concrete in aggressive ground. Where the concrete has to be cut off below ground level, concreting through the auger stem is stopped at a level above the cut-off. The auger is then removed from the temporary casing at the required level using a flap valve to retain the soil on the auger to leave an open cased hole. The reinforcement is pushed into the concrete and the casing withdrawn while backfilling the hole above the concrete. As the cased CFA pile can be installed more accurately than the standard CFA method, it is increasingly used for constructing secant pile walls. Bustamante et al.[2.25] have also shown that the cased CFA system can effectively

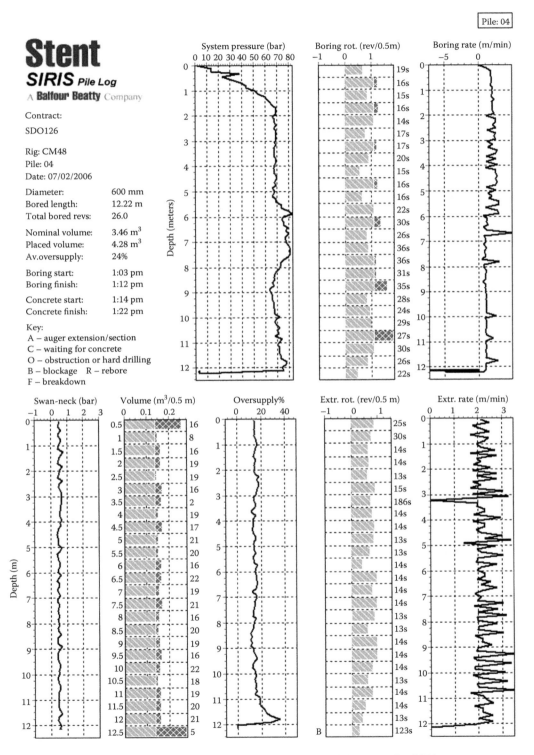

Figure 2.32 Pile log for CFA pile. (Courtesy of Stent Foundations Ltd., Basingstoke, UK.)

and accurately penetrate stiff marl which the standard CFA system may have difficulty in penetrating, resulting in 'refusal' before reaching the design depth.

The shaft friction resistance of CFA piles in chalk has been assessed by Lord et al.[2.26] It is considered that there should be little difficulty in forming satisfactory CFA piles in better-quality structured chalk, but in chalks with low penetration resistance, there may be problems of softening and hole instability, particularly below water table. Further information on installation and monitoring of CFA piles is given in a paper by Fleming[2.27], and potential risks in CFA piling are outlined by Windle and Suckling[2.28].

### 2.4.4 Drilled-in tubular piles

The essential feature of the drilled-in tubular pile is the use of a tube with a medium-to-thick wall, which is capable of being rotated into the ground to the desired level and is left permanently in the ground with or without an infilling of concrete. Soil is removed from within the tube as it is rotated down, by various methods including grabbing, augering and reverse circulation, as described in Section 3.3.5. The tube can be continuously rotated by a hydraulically powered rotary table or by high-torque rotary drill head or be given a semi-rotary motion by means of a casing oscillator.

The drilled-in tubular pile is a useful method for penetrating ground containing boulders or other obstructions, heavy chisels being used to aid drilling. It is also used for founding in hard formations, where a rock socket capable of resisting uplift and lateral forces can be obtained by drilling and grouting the tubes into the rock, under-reaming as necessary. In this respect, the drilled-in tubular pile is a good type for forming berthing structures for large ships. These structures have to withstand high lateral and uplift loads for which a thick-walled tube is advantageous. In rock formations, the resistance to these loads is provided by injecting a cement grout to fill the annulus between the outside of the tube and the rock forming the socket.

Where a rock socket is predrilled into which a tubular steel pile is driven and sealed, care must be taken not to over-drive the pile. 'Curtain folds' and ovality can occur (even in dense chalk), potentially compromising the load-bearing capacity, and are difficult to rectify to produce an acceptable pile. It is preferable to use an under-reamer or hole opener to match the OD of the pile before finally driving to seal the tube. Annex D2 of EC3-5 provides a method of verifying a pile which has buckled or become oval.

In the United States, *caisson piles*, comprising steel H-sections, lowered inside drilled-in tubes which are then infilled with concrete to ensure full interaction between the elements, provide high end-bearing capacity on strong rock.

## 2.5 COMPOSITE PILES

Various combinations of materials in driven piles or combinations of bored piles with driven piles can be used to overcome problems resulting from particular site or ground conditions. The problem of the decay of timber piles above groundwater level has been mentioned in Section 2.2.1. This can be overcome by driving a composite pile consisting of a precast concrete upper section in the zone above the lowest predicted groundwater level, which is joined to a lower timber section by a sleeved joint of the type shown in Figure 2.1. The same method can be used to form piles of greater length than can be obtained using locally available timbers.

Alternatively, a cased borehole may be drilled to below water level, a timber pile pitched in the casing and driven to the required depth, and the borehole then filled with concrete.

Another variation of the precast concrete–timber composite pile consists of driving a hollow cylindrical precast pile to below water level, followed by cleaning out the soil and driving a timber pile down the interior.

In marine structures, a composite pile can be driven that consists of a precast concrete upper section in the zone subject to the corrosive influence of seawater and a steel H-pile below the soil line. The H-section can be driven deeply to develop the required uplift resistance from shaft friction.

Generally, composite piles are not economical compared with those of uniform section, except as a means of increasing the use of timber piles in countries where this material is readily available. The joints between the different elements must be rigidly constructed to withstand bending and tensile stresses, and these joints add substantially to the cost of the pile. Where timber or steel piles are pitched and driven at the bottom of drilled-in tubes, the operation of removing the soil and obtaining a clean interior in which to place concrete is tedious and is liable to provoke argument as to the standard of cleanliness required.

The uniform section of a prefabricated steel–concrete composite pile can be economical in conditions requiring improved durability. This type of pile comprises a thin-walled steel casing with a hollow spun concrete core with ODs up to 1400 mm and can be either driven to depth or driven in the base of an augered hole. The National Composites Network has reviewed composite piles using a fibre-reinforced polymer tube either with concrete infill or with an infilled internal steel tube core. Lengths up to 20 m and 600 mm diameter are available and have been used in aggressive marine conditions, mainly in the United States, as noted for shell piles in Section 2.3.3; data on geotechnical performance is limited.

When the *top-down* construction method is required to start the superstructure before the basement is excavated, the *plunge column* provides the future permanent stanchion support. The schematic in Figure 2.33 shows a typical installation. In London, in a 52 m

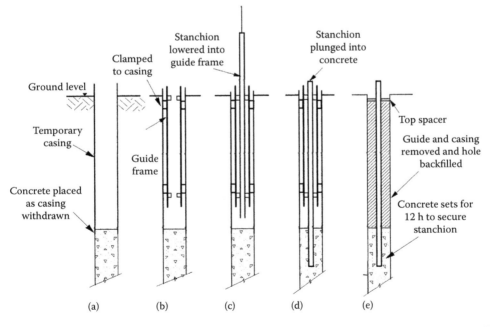

*Figure 2.33* Schematic of plunge column installation. (a) Hole bored and cased; reinforcement placed and concrete poured to above trim level as casing withdrawn. (b) Guide frame inserted and clamped. (c) Steel stanchion aligned in guide frame. (d) Stanchion plunged into wet concrete. (e) Top spacer fitted, concrete sets, guide frame removed and casing removed as bore backfilled.

deep bore, Bauer/Keller installed a 27 m reinforced concrete pile with a 33 m long steel section embedded 5 m into the concrete. The depth of embedment is determined by the load transfer required and the verticality controlled to conform to the superstructure codes: 1 in 400 for tall buildings. EC4 requirements apply to the design of composite steel and concrete structures.

## 2.6 MINIPILES AND MICROPILES

The definition of these piles has become somewhat blurred as size and capacity have increased; *micropiles* are generally defined as bored piles having a diameter between 90 and 300 mm and driven piles are <150 mm in width; axial capacities are in the range of 50–500 kN and up to 1000 kN when installed using pressure-grouting techniques. They are capable of being installed through existing structures to interact with the ground to provide axial resistance without a separate load transfer structure. The larger *minipiles*, with bored diameters from 200 to 600 mm and driven displacement piles of around 300 mm, are used where higher capacities are needed, but access is restricted allowing only smaller drill rigs to be deployed. They can be further defined as piles which require a load transfer structure. Both types have applications in supporting new structures, arresting settlement, excavation support and underpinning as described in Section 9.2.

### 2.6.1 Minipiles

Limit state principles in accordance with EC7, based on conventional total and effective stress methods, are applicable to the design of minipiles installed with regular plant, using the procedures described in Chapters 4, 5 and 6. However, the first step when faced with a restricted site (following a detailed ground investigation) is to decide what plant is capable of efficiently installing deep foundations. This will dictate the diameter and depth of pile available to support the superstructure, bearing in mind that if the bearing stratum is deeper than estimated, possibly by only a metre or so, the restricted access plant may not be able to cope, leading to considerable extra cost for alternatives.

The choice of the load transfer structure required to support the applied load is then considered. Where a single pile and cap is not feasible, it will be necessary to use a group of small-diameter minipiles with the load transferred to axially loaded piles through a rigid or flexible slab, the thickness possibly being governed by the available construction methods. Group action effects are therefore more commonly seen with minipiles than with larger-diameter piles, and checks must be made as noted in Section 4.9.2 and Chapter 5. Group effects can be an advantage in some instances. For example, axially loaded friction minipiles in over-consolidated clays may be more efficient at a spacing of 2.5 times diameter than the usual $3 \times D$ (see also Section 4.9.5 in respect of downdrag). Similarly, a group of driven minipiles designed as end bearing in dense to very dense coarse-grained soil or weathered rock may combine to provide enhanced performance with greater group end-bearing resistance than the sum of the individual piles, subject to adequate penetration into the bearing stratum to create the confinement.

Minipiles can also be used as retaining structures where the plant can install piles with adequate depth and diameter capability to produce the required wall stiffness. In addition, it is essential that a reinforcement cage can be installed to the depth needed to provide the design bending resistance.

The methods of installing minipiles and the precautions needed are similar to those for replacement and displacement piling as described earlier and in Chapter 3 on unrestricted sites.

Where the rigs are required to drill in confined spaces with low headroom either vertically or inclined (Table 3.7), some special requirements and precautions arise:

- Compact, short-masted rigs weighing 5–11 tonne and using conventional rotary boring techniques are suitable for drilling open holes up to 600 mm diameter down to 15 m in stable soils. Casing depths are more limited. Full-depth reinforcement cages can be inserted, with short sections spliced, and in situ concrete poured.
- The sectional flight auger (SFA) is the main drilling method for restricted access working in headroom as low as 2.5 m using compact rigs. Rigs weighing as little as 2 tonne with a separate power pack are available for confined spaces (Figure 2.34). Auger lengths are 1–2 m and depths around twice that for similar sized rotary bores are possible in a wide range of soil and weathered rock conditions where a stable hole can be formed. Caution is necessary when considering small-diameter SFA drilling in water-bearing uniform silts and sands, as even when cased, the removal of spoil as liquefied slurry will be difficult and the auger will be jammed; in such cases, the base resistance may be negligible. Loss of soil is another potential difficulty and can have serious consequences on the building being treated and adjacent structures.
- The CFA technique using short lengths of sectional hollow stem augers is suitable for water-bearing sands and gravels and soft fine-grained soils where loss of ground and vibration must be controlled. The compact rotary drills are not usually equipped with the instrumentation deployed on the large CFA rigs, requiring special attention to installation technique. The depth of reinforcement cage may also be limited as splicing bars before inserting into wet concrete will be difficult.
- Thin-walled steel shells (up to 323 mm diameter) driven by an internal 500–1000 kg drop hammer acting on an internal plug are filled with concrete and left in place. Short shell sections may be welded. They are predominantly end bearing and are useful in difficult conditions such as brownfield sites, peaty soils and soft clay, subject to founding on a competent stratum. Empirical dynamic formulae may be used for design,

*Figure 2.34* Klemm 702 restricted access drill with separate power pack installing bored piles. (Courtesy of Malcolm Drilling Company, San Francisco, CA.)

but note the precautions stated in Section 1.4. Such piles will have limited tension capacity and should not be used in heaving ground or where vibration may cause problems.

- Top-driven temporary casing, into which the reinforcement is placed, is withdrawn as grout is injected under pressure to produce compacted zones in the soil; reinforcement bars can be inserted after concreting. These piles and top-driven tubular steel piles are not favoured in the United Kingdom for constricted sites due to noise and vibration.

### 2.6.2 Micropiles

BS EN 14199 requires micropiles, as defined earlier, to be designed in accordance with the principles of the Eurocodes and materials standards. However, the design will be influenced by the installation method, particularly where piles are pressure grouted. Again the limitations of the drilling plant have to be considered as part of the design.

Micropiles generally have small base areas and rely mainly on shaft resistance for load bearing unless a base enlargement can be constructed. Otherwise, additional penetration may be needed in yielding strata to pick up sufficient shaft friction to avoid excessive settlement, and this may make the system uneconomical. They have little resistance to lateral loading and shear forces if unreinforced. Micropiles can be installed in most ground conditions and at any inclination, but the potential for obstructions to cause deviation of the small-diameter drill must be investigated. In weak ground, they should usually be lined, and in soils with $c_u$ < 10 kN/m², a check on buckling should be made. Ground anchor design methodology as given in BS 8081 and BS EN 1537 procedures can be a useful alternative approach to the BS EN 14199 guidance for grouted piles, especially when piles are subject to tension and compression loads. The construction of a closely spaced network of multiple micropiles, referred to as *reticulated piles*, will form a reinforced soil mass to underpin existing buildings or support new structures. Depending on the application, the structural load is applied either to the whole reinforced mass or to the individual piles.

Micropiles can be installed to depths up to 20 m by a variety of methods using drilling and driving plant similar to the small rigs as mentioned above. In addition, top-drive rotary-percussive drills or down-the-hole drills using the under-reaming Odex system or the more recent Symmetrix system are versatile tools to install casings in difficult soil conditions. The pile stiffness at the head of a compression pile may need to be improved with a steel over-tube. Also, where micropiles are used for underpinning in clays susceptible to heave and shrinkage, it is advisable to insert a sleeve into a pre-bored hole over the top 2–3 m of the shaft. In this case, the pile must be considered as a column over the sleeved length and designed accordingly. It is usually not economic to carry out static load tests or apply integrity testing on working micropiles; hence, quality control of the installation processes is most important. Pre-contract load testing is advisable and is required when a new technique is to be employed.

Some examples of installation techniques for micropiles are as follows:

- The *Grundomat* system, in which 150 mm diameter steel tubes are driven by a pneumatic hammer (*mole*) acting on a plug of dry concrete. Extension tubes 2–3 m long with watertight joints, where needed, give maximum depths of around 8 m. These piles are considered end bearing, based on empirical dynamic design formula, and have been used for many years in underpinning.
- Jacked-down steel tubes, steel box sections or precast concrete sections are useful as end-bearing piles where vibration has to be minimised. The sections may be joined by sleeving or dowelling.

- *Self-drilling* pressure-grouted hollow steel bars, such as the *Dywidag* ribbed and threaded bar and the Ischebeck *Titan* anchor bars, are installed by rotary-percussive drills in coarse-grained soils. Design methods are given in BS EN 14199 and address the small-scale soil–pile interface connection which is a critical aspect of micropile design. The pile diameter depends on the minimum cover specified and on the grouting pressure which can be applied. Loss of steel in aggressive soil can be 3 mm over 50–60 years. The expendable bit is liable to become blocked with drill cuttings when adding bars.
- The *Pali Radice* (*root pile*) system is one of the earliest forms of reticulated micropiles, extensively used for underpinning through existing structures and to minimise the size of a load transfer structure. Also it is useful where congested services have to be avoided for new foundations.
- Helical plate screw piles are used to support light structures, particularly where rapid installation is required, and as underpinning.
- *Soilex* piles as described earlier.
- Jet grouting, where the soil is mixed with injected grout to produce an in situ column into which, on withdrawal of the jetting lance, a reinforcing bar can be inserted.
- *Tubes à manchette* allow for the repeated injections of the micropile where the tube forms part of the pile; grout pressure based on the Ménard pressuremeter limit.

## 2.7 PRE-PACKED PILES

Although mentioned in BS EN 1536, piles formed of gravel placed in the borehole and then grouted are rarely used as reliable pile-grade concrete is difficult to achieve. The main requirements are clean coarse aggregate >25 mm, adequate grout pipes to the bottom of the hole and a flowable grout which will permeate the aggregate to produce concrete on setting. Grout pipes may be removed during injection.

## 2.8 FACTORS GOVERNING CHOICE OF TYPE OF PILE

The selection of an appropriate type of pile is one of the most important design decisions and is best made on the basis of experience in similar ground conditions. Piling contractors maintain a database of previous works detailing load testing and pile capacity of their systems; hence, early involvement of the contractor in the foundation design is of benefit to the project. The advantages and disadvantages of the various forms of pile described in Sections 2.2 through 2.7 affect the choice of pile for any particular foundation project, and these are summarised as follows:

### 2.8.1 Driven displacement piles

*Advantages*

1. Material forming pile can be inspected for quality and soundness before driving.
2. Not liable to 'squeezing' or 'necking'.
3. Construction operations not affected by groundwater.
4. Projection above ground level advantageous to marine structures.
5. Can be driven in long lengths; H-piles up to 50 m; tubular piles up to 40 m; jointed precast piles may be up to 40 m.

6. Can be designed to withstand high bending and tensile stresses.
7. Can be re-driven if affected by ground heave.
8. Pile loads over 10,000 kN are feasible for large-diameter steel piles and up to 15,000 kN with a solid concrete core.
9. Jointed types can be adapted for use in low headroom.

*Disadvantages/problems*

1. Unjointed types cannot readily be varied in length to suit varying level of bearing stratum.
2. May break during driving, necessitating replacement piles; splitting of timber piles.
3. May suffer unseen damage which reduces carrying capacity.
4. Uneconomical if cross section is governed by stresses due to handling and driving rather than by compressive, tensile or bending stresses caused by working condition.
5. Noise and vibration due to driving may be unacceptable.
6. Displacement of soil during driving may lift adjacent piles or damage adjacent structures.
7. Difficult to correct deviations once driving started.
8. End enlargements, if provided, destroy or reduce shaft friction over shaft length.
9. Driving H-piles in chalk may cause breakdown of layer of rock around pile.
10. Jointed precast piles may not be suitable for tension and lateral loads.

## 2.8.2  Driven and cast-in-place displacement piles

*Advantages*

1. Length can easily be adjusted to suit varying level of bearing stratum.
2. Driving tube driven with closed end to exclude groundwater.
3. Driving records give check of stiffness of bearing stratum.
4. Enlarged base possible.
5. No spoil to remove; important on contaminated sites.
6. Formation of enlarged base does not destroy or reduce shaft friction.
7. Material in pile not governed by handling or driving stresses.
8. Noise and vibration can be reduced in some types by driving with internal drop hammer.
9. Reinforcement determined by compressive, tensile or bending stresses caused by working conditions.
10. Concreting can be carried out independently of the pile driving.
11. Pile lengths up to 30 m and pile loads to around 2000 kN are common.

*Disadvantages/problems*

1. Concrete in shaft liable to be defective in soft squeezing soils or in conditions of artesian water flow where withdrawable-tube types are used.
2. Concrete cannot be inspected after installation.
3. Concrete may be weakened if artesian groundwater causes piping up shaft of pile as tube is withdrawn.
4. Length of some types limited by capacity of piling rig to pull out driving tube.
5. Displacement may damage fresh concrete in adjacent piles, or lift these piles or damage adjacent structures.

6. Once cast cannot be re-driven to deal with heave.
7. Noise and vibration due to driving may be unacceptable.
8. Cannot be used in river or marine structures without special adaptation.
9. Cannot be driven with very large diameters.
10. End enlargements are of limited size in dense or very stiff soils.
11. When light steel sleeves are used in conjunction with withdrawable driving tube, shaft friction on shaft will be destroyed or reduced.

### 2.8.3 Bored and cast-in-place replacement piles

*Advantages*

1. Length can readily be varied to suit variation in level of bearing stratum.
2. Soil or rock removed during boring can be inspected for comparison with site investigation data.
3. In situ loading tests can be made in large-diameter pile boreholes or penetration tests made in small boreholes.
4. Very large (up to 7.3 m diameter) bases can be formed in favourable ground.
5. Drilling tools can break up boulders or other obstructions which cannot be penetrated by any form of displacement pile.
6. Material forming pile is not governed by handling or driving stresses.
7. Can be installed without appreciable noise or vibration.
8. No ground heave.
9. Can be installed in conditions of low headroom.
10. Pile lengths (drilled shafts) up to 50 m over 3 m in diameter with capacities over 30,000 kN are feasible.

*Disadvantages/problems*

1. Concrete in shaft liable to squeezing or necking in soft soils; poor concrete mix design.
2. Lateral pressure on soft soil from fresh concrete causing bulges in shaft.
3. Special techniques needed for concreting in water-bearing soils, for example, tremie pipe.
4. Local slumping of open bore due to groundwater seepage.
5. Concrete cannot be inspected after installation.
6. Poorly designed reinforcement cages; preventing flow of concrete; displacement of main steel.
7. Enlarged bases cannot be formed in coarse-grained soils.
8. Cannot be extended above ground level without special adaptation.
9. Low end-bearing resistance in coarse-grained soils due to loosening by conventional drilling operations.
10. Drilling a number of piles in a group can cause loss of ground and settlement of adjacent structures.
11. Possible overflighting of CFA piles reducing shaft resistance; softening of chalk.
12. Necking of a CFA or screw pile due to poor control of extraction rate and concrete injection.

### 2.8.4 Choice of pile materials

*Timber* is cheap relative to concrete or steel. It is light, easy to handle and readily trimmed to the required length. It is very durable below groundwater level but is liable to decay above this level. In marine conditions, softwoods and some hardwoods are attacked by wood-boring organisms, although some protection can be provided by pressure impregnation. Timber piles are unsuitable for heavy applied loads, typical maximum being 500 kN. Due to depletion of supplies of the well-known hardwoods quoted previously, TRADA[2.29] is recommending the use of lesser-known species such as *angelim vermelho* and *tali* which compare favourably with greenheart for durability in marine conditions although somewhat lower in strength class.

*Concrete* is adaptable for a wide range of pile types. It can be used in precast form in driven piles or as insertion units in bored piles. Dense, well-compacted good-quality concrete can withstand fairly hard driving, and it is resistant to attack by aggressive substances in the soil or in seawater or groundwater. However, concrete in precast piles is liable to damage (possibly unseen) in hard driving conditions. Concrete with good workability, using plasticisers as appropriate, should be placed as soon as possible after boring cast-in-place piles. Weak, honeycombed concrete in cast-in-place piles is liable to disintegration when aggressive substances are present in soils or in groundwater.

BS 8500-1 provides three basic methods of specifying concrete – *designated concrete*, *designed concrete* and *prescribed concrete*. *Designated* concretes are identified by the application for which they will be used to satisfy requirements for strength and durability. The concrete should be specified by the designer in accordance with the exposure conditions in BS 8500-1 (see Tables 2.3 and 2.4) and materials as required by BS 8500-2. This will essentially mean giving the contractor the concrete designation (e.g. C25/30) and the maximum size of aggregate, with the contractor providing the concrete producer with the consistency and other information, such as the method of placing and testing regime. *Designed* concretes require the designer to be more specific: in addition to the basic designation, he should state the chemical resistance needed, cement content and types, water/cement ratio and chloride class. It is usually the producer's responsibility to prepare a mix design to meet this specification. For *prescribed* concrete, the specifier gives the producer full details of the constituents, their properties and quantities to provide a concrete with the specified performance. The specifier alone is responsible for conformance. Comprehensive guidance on specifying concrete is given in the UK National Structural Concrete Specification for Building Construction[2.30].

*Steel* is more expensive than timber or concrete, but this disadvantage may be outweighed by the ease of handling steel piles, by their ability to withstand hard driving, by their resilience and strength in bending and by their capability to carry heavy loads. Limit state design and recent research into pile behaviour indicate that steel is becoming more economic. Steel piles can be driven in very long lengths and cause little ground displacement. They are liable to corrosion above the soil line and in disturbed ground, and they require cathodic protection if a long life is desired in marine structures. Long steel piles of slender section may suffer damage by buckling if they deviate from their true alignment during driving.

## REFERENCES

2.1   Love, J.P. The use of settlement reducing piles to support a flexible raft structure in West London, *Proceedings of the Institution of Civil Engineers, Geotechnical Engineering*, 156, 2003, 177–181.

2.2   Raison, C.A. North Morecambe Terminal, Barrow: Pile design for seismic conditions, *Proceedings of the Institution of Civil Engineers, Geotechnical Engineering*, 137, 1999, 149–163.

2.3   Marsh, E. and Chao, W.T. The durability of steel in fill soils and contaminated land, Corus Research, Development & Technology, Technology Centre, Rotherham UK. Report No. STC/CPR OCP/CKP/0964/2004/R, 2004.

2.4    A corrosion protection guide for steel bearing piles in temperate climates, Corus Construction and Industrial, Scunthorpe, England, 2005.

2.5    The Institution of Civil Engineers. *Specification for Piling and Embedded Retaining Walls* (SPERW), 2nd ed. Thomas Telford Ltd, London, UK, 2007.

2.6    Reynolds, T. and Bates, P. The potential for timber piling in the UK, *Ground Engineering*, 42 (1), 2009, 31–34.

2.7    McKenzie, W.M.C. and Zhang, B. *Design of Structural Timber to Eurocode 5*, 2nd ed. Palgrave MacMillan, Basingstoke, England, 2009.

2.8    Building Research Establishment (BRE). *Concrete in Aggressive Ground*, Special Digest 1:2005, BRE, Watford, England, 2005.

2.9    Roohnavaz, C. Driven pile construction control procedures and design, *Proceedings of the Institution of Civil Engineers, Geotechnical Engineering*, 163 (5), 2010, 241–255.

2.10   Hassam, T., Rizkalla, S., Rital, S. and Parmentier, D. Segmental precast concrete piles – A solution for underpinning, *Concrete International*, August, 2000, 41.

2.11   Biddle, A.R. *H-Pile Design Guide*. SCI Publication P355. Steel Construction Institute, Ascot, England, 2005.

2.12   Bjerrum, L. Norwegian experiences with steel piles to rock, *Geotechnique*, 7 (2), 1957, 73–96.

2.13   Hanna, T.H. Behaviour of long H-section piles during driving and under load, *Ontario Hydro Research Quarterly*, 18 (1), 1966, 17–25.

2.14   HP Bearing Piles – Execution details. Profilarbed S.A. ArcelorMittal Group.

2.15   *API 5L: Specifications for Steel Line Pipe*, 44th ed. American Petroleum Institute, Washington, DC, 2007.

2.16   Wynne, C.P. A review of bearing pile types, Construction Industry Research and Information Association (CIRIA). Report PG1, 2nd ed., London, UK, 1988.

2.17   Henderson, N.A., Baldwin, N.J.R., McKibbins, L.D., Winsor, D.S. and Shanghavi, H.B. Concrete technology for cast in-site foundations, Construction Industry Research Information Association (CIRIA). Report C569, London, UK, 2002.

2.18   Bustamante, M. and Gianeselli, L. Installation parameters and capacity of screwed piles, in *Proceedings of Third International Seminar on Deep Foundations on Bored and Augered Piles*, W.E. Van Impe and A.A. Haegeman (ed.), AA Balkema, Rotterdam, the Netherlands, 1998, 95–108.

2.19   Holeyman, A. and Charue, N. International pile capacity prediction event at Limelette, in *Belgian screw pile technology design and recent developments. Proceedings of Second Symposium on Screw Piles, Brussels*, J. Maertens and N. Huybrechts (ed.), Swets and Zeitlinger, Lisse, the Netherlands, 2003, 215–234.

2.20   Black, D.R. and Pack, J.S. Design and performance of helical screw piles in collapsible and expansive soils in and regions of the United States, *Proceedings of Ninth International Conference on Piling and Deep Foundations*, Presses du l'école nationale des Ponts et Chaussées, Champs-sur-Marne, France, 2002, 469–476.

2.21   NHBC Foundation. *Efficient Design of Piled Foundations for Low-Rise Housing*. Design Guide NF21. IHS BRE Press, Watford, England, 2010.

2.22   *Bentonite Support Fluids in Civil Engineering*, 2nd ed. Federation of Piling Specialists, Bromley, England, 2006.

2.23   Jones, A.E.K. and Holt, D.A. Design of laps for deformed bars in concrete under bentonite and polymer drilling fluids, *The Structural Engineer*, 80 (18), 2004, 32–38.

2.24   Farrell, E.R. and Lawler, M.L. CFA behaviour in very stiff lodgement till, *Proceedings of the Institution of Civil Engineers, Geotechnical Engineering*, 161 (1), 2008, 49–57.

2.25   Bustamante, M., Gianeselli, L. and Salvador, H. Double rotary CFA piles: Performance in cohesive soils, *Proceedings of Ninth International Conference on Piling and Deep Foundations*, Presses de l'école nationale des Ponts et Chaussées, Champs-sur-Marne, France, 2002, pp. 375–381.

2.26   Lord, J.A., Hayward, T. and Clayton, C.R.I. Shaft friction of CFA piles in chalk, Construction Industry Research Information Association (CIRIA). Project Report 86, London, UK, 2003.

2.27 Fleming, W.G.K. The understanding of continuous flight auger piling, its monitoring and control, *Proceedings of the Institution of Civil Engineers, Geotechnical Engineering*, 113, 1995, 157–165.

2.28 Windle, J. and Suckling, T. CFA piling: A cheap solution or a problem waiting to happen, in *Forensic Engineering from Failure to Understanding*, B.S. Neale (ed.), Thomas Telford, London, UK, 2008, pp. 335–342.

2.29 TRADA. Wood Information Sheet WIS 2/3-66. Specifying timber species in marine and freshwater construction. Timber Research and Development Association, High Wycombe, England, 2011.

2.30 *UK National Structural Concrete Specification for Building Construction*, 4th ed. The Concrete Centre, London, 2010.

# Chapter 3

# Piling equipment and methods

The development and availability of larger drilling rigs and more efficient impact and vibratory hammers continue to promote the use of new methods to install larger piles to greater depths accompanied by reduced environmental impact. Improved mobility and speed of operation together with in-cab instrumentation and precision setting out with the global positioning system (GPS) have all added to the expansion and reliability of piling operations. Satellite links from the rig to the company office allow the foundation designer to monitor installation in difficult environmental conditions where ground investigations have been limited. The amount of data now being produced could eventually lead to the practical application of pile–soil interaction theory to the determination of bearing capacity as the pile is installed.

The development of piling equipment has proceeded on different lines in various parts of the world, depending mainly on the influence of the local ground conditions: high groundwater levels in Holland, stiff clays in the United States and karstic conditions in Europe. In the United Kingdom, with a wide variety of soil types and as demand for new heavier infrastructure has grown, the full range of piling equipment and techniques has been adopted, from continuous flight augers (CFAs), vibrated concrete columns and press-in piles to large piling hammers to install large-diameter monopiles for offshore wind farms.

The manufacturers of piling equipment and the range of types they produce are too numerous for all makes and sizes to be described in this chapter. Health and safety requirements and environmental legislation as well as commercial pressures all mean that piling plant and methods are constantly changing. Noise abatement in particular influenced the trend away from diesel hammers towards forms of pile that are installed by drilling, vibration and pressing methods. Landfill taxes have been a major influence on limiting spoil from boreholes leading to the use of auger screw displacement piles, particularly on contaminated brownfield sites.

The principal types of current equipment in each category are described below, but the reader should refer to manufacturers' handbooks and their comprehensive websites for the full details of their dimensions and performance. The various items of equipment are usually capable of installing more than one of the many piling systems which are described in Chapter 2. Installation methods of general application are described in the latter part of this chapter.

All piling equipment should comply with the requirements in BS EN 996 Piling equipment, Safety requirements; BS EN 791 Drill rigs, Safety and the various parts of BS EN 16228 in preparation for other foundation equipment.

## 3.1 EQUIPMENT FOR DRIVEN PILES

### 3.1.1 Piling frames

The piling frame has the function of guiding the pile at its correct alignment from the stage of first pitching in position to its final penetration. It also carries the hammer and maintains it in position coaxially with the pile. The essential parts of a piling frame are the *leaders* or *leads*, which guide and support the hammer and pile. They are stiff members constructed of channel, box or tubular section held by a lattice or tubular mast that is in turn supported at the base by a moveable carriage and at the upper level by backstays. The latter can be adjusted in length by a telescopic screw device or by hydraulic rams, to permit the leaders to be adjusted to a truly vertical position or to be raked forwards, backwards or sideways. Where piling frames are mounted on elevated stagings, *extension leaders* can be bolted to the bottom of the main leaders in order to permit piles to be driven below the level of the base frame.

The *piling winch* is mounted on the base frame or carriage. This may be a double-drum winch with one rope for handling the hammer and one for lifting the pile. A three-drum winch with three sheaves at the head of the piling frame can lift the pile at two points using the outer sheaves and the hammer by the central sheave. Some piling frames have multiple-drum winches which, in addition to lifting the pile and hammer, also carry out the duties of operating the travelling, slewing and raking gear on the rig.

Except in special conditions, say for marine work, stand-alone piling frames have largely been replaced by the more mobile self-erecting hydraulic leaders on tracked carriages or by the crane-mounted fixed or hanging leaders offered by the major piling hammer manufacturers. In Europe, the pile hammer usually rides on the front of the leader (*spud* type), whereas in the United States, the practice is to guide the pile between the leaders (*U* type). The pile head is guided by a cap or helmet which has jaws on each side that engage with U-type leaders. The hammer is similarly provided with jaws. The leaders are capable of adjustment in their relative positions to accommodate piles and hammers of various widths.

*Self-erecting leaders* on powerful hydraulic crawler carriages can be configured for a variety of foundation work (Table 3.1). Initial erection and changing from drilling to driving tools can be rapidly accomplished, and with the electronic controls now available, the mast can be automatically aligned for accurate positioning. Some crawlers have expandable tracks to give added stability and can handle pile hammers with rams up to 12 tonne at 1:1 back rake.

Note that the information given in Tables 3.1 through 3.7 is only a selective summary of the range of equipment and the manufacturers should be contacted for full details and when making assessments of performance for particular applications. Technical information on the equipment is also readily available online. Because of market changes, some equipment will be obsolescent, but well-maintained used hammers not in current production may be available.

The ABI Mobilram TM series of telescopic leader masts (Figure 3.1) has been designed to handle pile driving with impact and vibratory hammers; the torsional rigidity also makes the rig suitable for pile drilling and pressing. The Banut 555 and 650 piling rigs (Figure 3.2) are primarily designed to drive precast concrete piles with diesel or hydraulic impact hammers but are also effective for installing most bearing and sheet piles. The hydraulic stays attached to the crawler enable forward rakes of up to 18° and 45° back rakes, together with lateral movement of up to 14° available on both units. The usable length given for the 650 unit relates to the Banut SuperRAM 6000 hydraulic hammer (see Table 3.2).

The Junttan PM hydraulic piling rigs with fixed leaders can drive piles ranging from 16 to 36 m long (with telescopic leader extensions and HHK hydraulic hammers), using hammer

Table 3.1 Characteristics of some crawler-mounted pile-driving rigs

| Maker | Type | Usable leader length (m) | Maximum capacity (pile plus hammer) (tonne) | Pile winch capacity (tonne) |
|---|---|---|---|---|
| ABI Mobilram[a] | TM13/16 SL | 15.7 | 9 | 5 |
| (Germany) | TM18/22 | 25.3 | 12 | 5 |
| | TM20/25 | 28.8 | 15 | 5 |
| | TM22 | 24.7 | 15 | 5 |
| Banut (Germany) | 555 | 15.0 | 12 | 6 |
| | 655 | 15.0 | 12 | 8.5 |
| Junttan (Finland) | PM16 | 16.0 | 8 | 5 |
| | PMx20 | 13.8 | 13 | 8 |
| | PMx22 | 20.0 | 16 | 10 |
| | PM25H | 25.0 | 20 | 10 |
| | PM30 | 32 | 35 | 12 |
| Liebherr (Austria) | LRB 125 | 12.5 | 12 | 6 |
| | LRB 155 | 24.0 | 15.0 | 8 |
| | LRB 255 | 30.0 | 30.0 | 20 |

[a] Telescopic mast.

Figure 3.1 ABI Mobilram with telescopic leader fully extended driving tubular pile. (Courtesy of ABI GmbH, Niedernberg, Germany.)

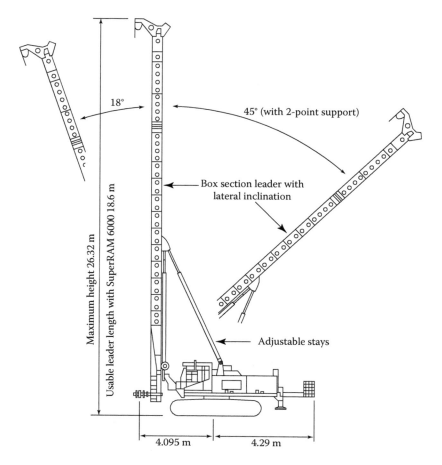

18°

45° (with 2-point support)

Box section leader with
lateral inclination

Maximum height 26.32 m

Usable leader length with SuperRAM 6000 18.6 m

Adjustable stays

4.095 m    4.29 m

*Figure 3.2* Banut 650 piling rig. (Courtesy of ABI GmbH, Niedernberg, Germany.)

rams from 3,000 to 12,000 kg. Fore and aft rakes are available subject to recommendations by the maker. Liebherr provides fixed leaders mounted on their own and others' crawler carriages. The LRB series can operate as pile-driving rigs and rotary drills for CFA and kelly bored piles with fore and aft inclinations.

### 3.1.2 Crane-supported leaders

Although the hydraulic piling rig with its base frame and leaders supported by a stayed mast provides a reliable means of ensuring stability and control of the alignment of the pile, there are many conditions which favour the use of leaders suspended from a standard crawler crane. Rigs of this type have largely supplanted the frame-mounted leaders for driving long piles on land in Europe and the United States.

*Fixed leaders* are rigidly attached to the top of the crane jib by a swivel and to the lower part of the crane carriage by a *spotter* or stay. Hydraulic spotters can extend and retract to control verticality and provide fore and aft raking; they can also move the leader from side to side. The International Construction Equipment (ICE) heavy-duty spotter provides 6 m of hydraulic movement fore and aft and an optional 35° leader rotation (Figure 3.3). In *fixed extended* arrangements, the leaders extend above the top of the jib with a connector which

Table 3.2 Characteristics of some hydraulic impact hammers

| Maker | Type | Mass of ram (kg) | Maximum energy per blow (kJ) | Striking rate at maximum stroke height (blows/min) |
|---|---|---|---|---|
| American Piledriving | X13 | 35,896 | 88 | 28 |
| Equipment (United States) | 7.5[a] | 5,443 | 32.5 | 40 |
| | 7.5[b] | 4,626 | 27.6 | 40 |
| | 7.5[c] | 3,446 | 20.6 | 40 |
| | 9.5[a] | 6,712 | 68.5 | 40 |
| | 400U | 36,287 | 488 | 30 |
| | 500U | 54,431 | 369 | 28 |
| | 750 | 54,431 | 847 | 20 |
| BSP International | CX50 | 4,000 | 51 | 46 |
| Foundations (United | CX60 | 5,000 | 60 | 45 |
| Kingdom) | CX85 | 7,000 | 83 | 42 |
| | CX110 | 9,000 | 106 | 36 |
| | CG180 | 12,000 | 176 | 34 |
| | CG210 | 14,000 | 206 | 36 |
| | CG240 | 16,000 | 235 | 34 |
| | CG300 | 20,000 | 294 | 32 |
| | CGL370 | 22,500 | 370 | 32 |
| | CGL440 | 27,000 | 440 | 32 |
| | CGL520 | 31,400 | 520 | 32 |
| Banut SuperRAM | 5000 | 5,060 | 59 | 100 |
| (Germany) | 6000 | 6,075 | 71 | 100 |
| | 6000XL | 6,110 | 71 | 100 |
| | 8000XL | 8,020 | 94 | 100 |
| | 10000XL | 10,000 | 118 | 80 |
| IHC Hydrohammer | S40 | 2,235 | 40 | 45 |
| (Netherlands) | S90 | 4,572 | 90 | 46 |
| | S150 | 7,620 | 150 | 44 |
| | S200 | 10,160 | 200 | 45 |
| | S500 | 25,400 | 500 | 45 |
| | S600 | 30,480 | 600 | 42 |
| | S900 | 43,690 | 900 | 38 |
| | S1200 | 60,960 | 1,200 | 38 |
| | S1800 | 91,440 | 1,800 | 35 |
| | S2300 | 116,840 | 2,300 | 30 |
| | SC75[b] | 5,791 | 75 | 50 |
| | SC110[b] | 8,026 | 110 | 45 |
| | SC150[b] | 11,176 | 150 | 45 |
| | SC200[b] | 13,818 | 200 | 45 |
| Junttan[c] (Finland) | HHK 4A | 4,000 | 47 | 40–100 |
| | HHK 5A | 5,000 | 59 | 40–100 |
| | HHK 7A | 7,000 | 82 | 40–100 |
| | HHK 12A | 12,000 | 141 | 40–100 |
| | HHK 14A | 14,000 | 165 | 40–100 |

(continued)

*Table 3.2 (continued)* Characteristics of some hydraulic impact hammers

| Maker | Type | Mass of ram (kg) | Maximum energy per blow (kJ) | Striking rate at maximum stroke height (blows/min) |
|---|---|---|---|---|
| Junttan[c] (Finland) | HHK 5S | 5,000 | 74 | 30–100 |
| | HHK 7S | 7,000 | 103 | 30–100 |
| | HHK 14S | 14,000 | 206 | 30–100 |
| | HHK 16S | 16,000 | 235 | 30–100 |
| | HHK 25S | 25,000 | 368 | 30–100 |

[a] Free-fall hammer (many hammers now have assisted acceleration).
[b] SC series more suited to driving concrete piles.
[c] Extensions can be provided to increase ram weight and energy.

*Figure 3.3* ICE 225 spotter with optional front lead rotation. (Courtesy of International Construction Equipment, Charlotte, NC.)

allows freedom of movement. Leaders are usually provided in top and intermediate sections about 5 and 2.5 m long jointed together to provide the required leader height. As an alternative to spotters, hydraulic telescopic rams are used to enable raking piles to be driven, with a bottom stabbing point on the leader to fix the pile location. BSP International Foundations Ltd. produces fixed extended leaders in lattice sections 610 and 850 mm², with lengths of 7.5 and 10 m, respectively. The respective maximum lengths under the cathead are 22.5 and 38 m, subject to crane jib length. The maximum load for pile and hammer at a back rake of 1:12 with the 610 mm section is 12 tonne and 18 tonne for the 835 mm section at a back rake of 1:10 using standard stays.

*Swinging leaders* are suspended from the crane rope and usually are 5 m shorter than the jib. They are mainly used for driving vertical piles, but because of the freedom of movement, they have to be used with a pile guide or template. *Hanging leaders* are similar to swinging leaders but with a connection to the top of the crane jib and a head block which allows movement fore and aft from the crane. The bottom of the leader is attached to the crane chassis with a fixed strut or spotter. As an example, the Liebherr LRH 600, 50 m long hanging leader has a maximum capacity of 65 tonne when used with the Liebherr HS 895 HD carriage, but as with all the leaders, account has to be taken of bending moments induced by the weight of the piling hammer when driving raked piles (Figure 3.4). The Delmag MS, MU and MH swinging and hanging spud-type leaders are designed for use with Delmag diesel hammers up to 12 tonnes on a 30 m long leader. The EU-type

*Figure 3.4* Liebherr LRH 400 48 m long swinging leader on HS 885 HD crane. (Courtesy of Liebherr Great Britain Ltd, Biggleswade, UK.)

offshore lead fits on to the top of tubular piles (2200 mm maximum diameter) pitched in a frame or the platform jacket pile sleeve.

Swinging leads can be attached to most models of crane with suitable capacity. Where the lead is not connected to the carriage base by a spotter, it can be rotated 360° around its vertical axis allowing piles to be driven at rakes up to 1:1 (Figure 3.5). There is a practical limit to the length of pile which can be driven by a given type of rig, and this can sometimes cause problems when operating the rig in the conventional manner without the assistance of a separate crane to lift and pitch the pile. The conventional method consists of first dragging the pile in a horizontal position close to the piling rig. The hammer is already attached to the leader and drawn up to the cathead. The pile is then lifted into the leaders using a line from the cathead and secured by toggle bolts. The helmet, dolly and packing (see Section 3.1.8) are then placed on the pile head, and the assembly is drawn up to the underside of the hammer. The carriage of the piling rig is then slewed round to bring the pile over to the intended position, and the stay and angle of the crane jib are adjusted to correct for verticality or to bring the pile to the intended rake.

In determining the size of the leader whether rig-mounted, fixed or hanging, it is always necessary to check the available height beneath the hammer when it is initially drawn up to the cathead. Taking the example of leaders with a usable height of 20.5 m in conjunction with a hammer with an overall length of 6.4 m, after allowing a clearance of 1 m between the lifting lug on the hammer and the cathead and about 0.4 m for the pile helmet, the maximum length of pile which can be lifted into the leaders is about 12.7 m.

*Figure 3.5* US style 26 in. swinging leader supporting a Dawson HPH2400 hammer driving a 305 mm H-pile on 2:3 rake. (Courtesy of Dawson Construction Plant Ltd, Milton Keynes, UK.)

Guyed leaders independent of any base machine are rarely used, even for two or three preliminary test piles, as they are cumbersome to erect and move and need a separate winch to operate the hammer.

### 3.1.3 Trestle guides

Another method of supporting a pile during driving is to use guides in the form of a move-able trestle. The pile is held at two points, known as *gates*, and the trestle is designed to be moved from one pile or pile-group position to the next by crane (Figure 3.6). The hammer is supported only by the pile and is held in alignment with it by leg guides on the hammer (similar to the EU lead noted above) extending over the upper part of the pile shaft. Because of flexure of the pile during driving, there is a greater risk, especially with raking piles, of the hammer losing its alignment with the pile during driving than in the case of piling frames which support and guide the hammer independently of the pile. For this reason, the method of supporting the hammer on the pile in conjunction with trestle guides is usually confined to steel piles where there is less risk of damage to the pile head by eccentric blows. When driving long steel raking piles in guides, it is necessary to check that the driving stresses combined with the bending stress caused by the weight of the hammer on the pile are within allowable limits.

Pile guides which are adjustable in position and direction to within very close limits are used on jack-up barges for marine piling operations. A travelling carriage or gantry is cantilevered from the side of the barge or spans between rail tracks on either side of the barge *moon pool*. The travelling gear is powered by electric motor and final positioning by hydraulic rams. Hydraulically operated pile clamps or gates are mounted on the travelling carriage at two levels and are moved transversely by electric motor, again with final adjustment by hydraulic rams allowing the piles to be guided either vertically or to raking positions. Guides provided by hydraulic clamps on a guide frame fixed to the side of a piling barge are shown in Figure 3.7.

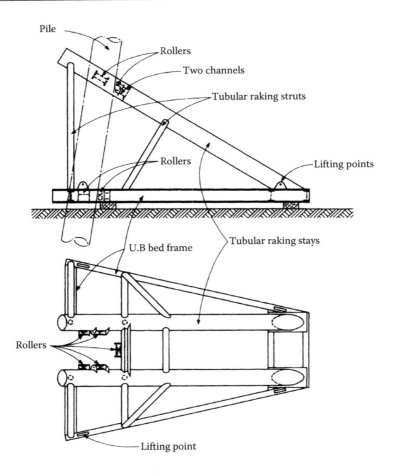

*Figure 3.6* Trestle guides for tubular raking pile.

Trestle guides can be usefully employed for rows of piles that are driven at close centres simultaneously. The trestle shown in Figure 3.8 was designed for the retaining wall foundations of Harland and Wolff's shipbuilding dock at Belfast[3.1]. Three rows of five 356 × 368 mm H-piles were pitched into the guides and were driven by a Delmag D22 hammer.

Guides can be used in conjunction with piling frames for a two-stage driving operation, which may be required if the piles are too long to be accommodated by the available height of frame. Guides are used for the first stage of driving, the piles carrying the hammer which is placed and held by a crane. At this stage, the pile is driven to a penetration that brings the head to the level from which it can be driven by the hammer suspended in the piling frame. Figure 3.9 shows a crane-mounted hammer driving piles through a guide from initial pitching to final level in stages.

### 3.1.4 Piling hammers

The simplest form of piling hammer is the winch-raised *drop hammer*, which is guided by lugs or jaws sliding in the leaders. The basic winch-operated drop hammer consists of a solid mass or assemblies of forged steel, the total mass ranging from 1 to 5 tonne. The drop ranges from 0.2 to 2 m and the weight needed is between half and twice the pile weight. The striking speed is slower than in the case of single- or double-acting hammers, and when

*Figure 3.7* Installing a 4 m diameter monopile foundation for North Hoyle offshore wind farm with pile-top rig and specially designed leader leg pile frame. (Courtesy of Fugro Seacore Ltd, Falmouth, UK.)

drop hammers are used to drive concrete and timber piles, there is a risk of damage to the pile if an excessively high drop of the hammer is adopted when the driving becomes difficult. To avoid such damage, the drop of each blow of the winched hammer has to be carefully coltrolled by the operator. However, for driving all types of pile in stiff to hard clays, a heavy blow with a small drop is more efficient and less damaging to the pile than a large number of lighter blows.

Drop hammers also include those raised by steam, air and hydraulic pressure, generally permitting a higher hammer energy, and may be free falling or assisted by pressure on the downstroke to give a bigger and more controllable impact blow to the pile head as described below. The original Vulcan winched hammer was developed into a series of large steam-activated free-fall hammers up to the Vulcan 6300 with a weight of 140 tonnes and 2440 kJ energy (production ceased in the 1980s). Drop hammers can be adapted to operate within a sound-proofed box to comply with noise abatement regulations (see Section 3.1.7).

The wide range of modern hydraulic *single-acting hammers* is indicated in Table 3.2. The ram is raised by hydraulic fluid under high pressure to a predetermined height and then allowed to fall under gravity, or as in the BSP CX series (Figure 3.10), some have the option of additional acceleration by pressurising the *equalising housing* above the piston, thereby increasing the energy by up to 20%. The hammer stroke and blow rate are controlled by instrumentation so that at the required stroke height, the flow of the hydraulic fluid is cut off. Pressures within the actuator then equalise allowing the ram to decelerate

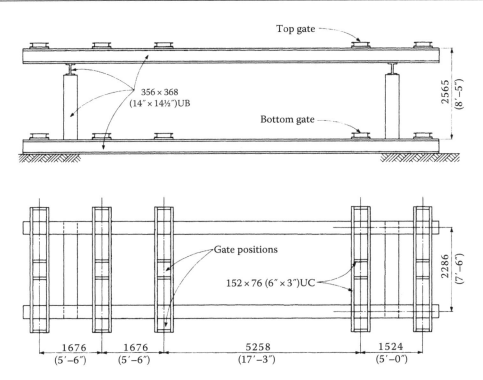

356 × 368
(14″ × 14½″)UB

Top gate

Bottom gate

2565
(8′–5″)

Gate positions

152 × 76 (6″ × 3″)UC

2286
(7′–6″)

| 1676 | 1676 | 5258 | 1524 |
| (5′–6″) | (5′–6″) | (17′–3″) | (5′–0″) |

*Figure 3.8* Trestle guides for multiple vertical piles.

*Figure 3.9* BSP CG300 hydraulic hammer suspended from a Kobelco CKEE2500 crane driving tubular piles in stages through trestle guides. (Courtesy of Steel Pile Installations Ltd, Bolton, UK.)

as it approaches the top of its stroke. The falling hammer repositions the piston rod for the next stroke. These hammers can deliver an infinitely variable stroke and blow rate within the limits stated so that the energy matches the driving conditions. The latest models from the main manufacturers can be fitted with instrumentation giving a continuous display of depth, driving resistance and set and are relatively quiet to operate.

*Figure 3.10* BSP CX110 hydraulic piling hammer on Hitachi crane-mounted leader. (Courtesy of BSP International Foundations Ltd, Ipswich, England.)

For driving precast concrete piles, a hammer mass of 4000 kg is appropriate for a 800 kN applied load. High bearing-capacity driven steel piles will require hammer mass of 10,000 kg for a load of 3,000 kN and much greater for the large-diameter offshore piles now being installed. For example, the MENCK MRBS offshore pneumatic hammers (Figure 3.11) have masses ranging from 8.6 to 125 tonne with a maximum stroke of 1.75 m. They are fully automatic with infinitely variable stroke. By adding a belled-out section beneath the hammer, Seacore has developed a rig capable of driving piles up to 4 m diameter into predrilled holes for the *monopile* foundations for offshore wind turbine towers as in Figure 3.12.

Hydraulic hammers, driven by a separate power pack, produce no exhaust at the hammer and therefore have the advantage of being able to operate underwater. Large underwater hydraulic hammers have been designed especially for driving piles in deep-water locations. The MENCK MHU double-acting hammer range in Table 3.3 is designed specifically for underwater work: the S hammer series is for water depth up to 400 m and the T series for 3000 m. The MHU 3000 S with a ram weight of 180 tonne and 3000 kJ energy is one of the largest piling hammers ever constructed. A nitrogen shock absorber ring protects the hammer from rebound forces and shock loads and will largely eliminate a tension wave in the pile (see Section 7.3). The MHU hammers are designed either to operate as free-riding units mounted on the pile with a slack lifting line or to reduce weight on the guides so that they can be suspended from the crane with a heave compensator to counteract wave action and so

*Figure 3.11* MRBS air/steam single-acting hammer with stabilising cage driving 54 in. diameter piles in legs of offshore jacket platform. (Courtesy of MENCK GmbH, Kaltenkirchen, Germany.)

*Figure 3.12* Driving a 4 m diameter monopile foundation for North Hoyle offshore wind farm using a MENCK MHU 500T hammer with large-diameter pile sleeve and anvil adapter. (Courtesy of Fugro Seacore Ltd, Falmouth, UK.)

*Table 3.3* Characteristics of some double-acting and differential-acting piling hammers

| Maker | Type | Mass of ram (kg) | Maximum energy per blow (kJ) | Maximum striking rate (blows/min) |
|---|---|---|---|---|
| BSP International Foundations (United Kingdom) | LX30[a] | 2,500 | 30 | 65 |
| | LX50[a] | 4,000 | 50 | 60 |
| | CX110 | 9,000 | 106 | 36 |
| | SL20da | 1,500 | 20 | 90 |
| | SL30da | 2,500 | 30 | 84 |
| Dawson Construction Plant (United Kingdom) | HPH1200 | 1,040 | 12 | 80–120 |
| | HPH2400 | 1,900 | 24 | 80–120 |
| | HPH4500 | 3,500 | 45 | 80–120 |
| | HPH6500 | 4,650 | 65 | 80–120 |
| | HPH9000 | 4,750 | 90 | 60–90 |
| MKT (United States) | 9B3 | 725 | 6 | 145 |
| | 10B3 | 1,360 | 10 | 105 |
| | 11B3 | 2,270 | 14 | 95 |
| MENCK[b] (Germany) | MHU100C | 5,000 | 100 | 50 |
| | MHU300S[c] | 16,200 | 300 | 40 |
| | MHU440S | 24,300 | 440 | 38 |
| | MHU550S | 30,200 | 550 | 38 |
| | MHU800S | 45,400 | 820 | 38 |
| | MHU1200S | 66,000 | 1,200 | 38 |
| | MHU1900S | 95,000 | 1,900 | 32 |
| | MHU3000S | 180,000 | 3,000 | 32 |

[a]  Single-acting versions with lower energy available.
[b]  Differential-acting.
[c]  S denotes version for use in shallow water or onshore use; the T version with similar energies is designed for deep water.

maintain constant tension in the lifting line. Other pile-top hammers operate with a follower attached above the structural pile, and slender hydraulic hammers can operate inside the pile.

MENCK has developed a deep-water system hydraulic power pack which sits directly on the MHU pile hammer within the pile, needing only a single umbilical to provide energy, air and communications from the surface; it can operate in water depths of 2000 m.

*Double-acting* (or *differential-acting*) hammers are either hydraulically or air operated with control valves to apply pressure on both the upstroke and downstroke, designed to impart a rapid succession of small-stroke blows to deliver higher energy to the pile. The double-acting hammer exhausts air on both the up- and downstrokes. In the case of the differential-acting hammer, however, the cylinder is under equal pressure above and below the piston and is exhausted only on the upward stroke. The downward force is a combination of the weight of the ram and the difference in total force above and below the piston, the force being less below the piston because of the area occupied by the piston rod. These hammers are most effective in granular soils where they keep the ground *live* and shake the pile into the ground, but they are not so effective in clays. The characteristics of a selection of hammers are shown in Table 3.3. The BSP hydraulic double-acting LX series are used mainly for driving steel sheet piles and small bearing piles with blow rates of 90 blows per minute and can be provided as single acting.

*Diesel hammers* are suitable for all types of ground except soft clays. They have the advantage of being self-contained without the need for separate power packs, air compressors or steam generators. They work most efficiently when driving into stiff to hard clays, and with their high striking rate and high energy per blow, they are favoured for driving all types of bearing piles up to about 2.5 m in diameter. The principle of the diesel hammer is that as the falling ram compresses air in the cylinder, diesel fuel is injected into the cylinder and this is atomised by the impact of the ram on the concave base. The impact ignites the fuel and the resulting explosion imparts an additional *kick* to the pile, which is already moving downwards under the blow of the ram. Thus, the blow is sustained and imparts energy over a longer period than the simple blow of a drop or single-acting hammer. The ram rebounds after the explosion and scavenges the burnt gases from the cylinder. The well-known Delmag series of hammers (Figure 3.13) ranges from the D6 with a ram mass of 600 kg suitable for driving piles up to 2000 kg to the 20 tonnes ram of the D200 with a drop height of 3.4 m suitable for piles weighing up to 250 tonnes. The characteristics of various makes of diesel hammer are shown in Table 3.4.

A difficulty arises in using the diesel hammer in soft clays or weak fills, since the pile yields to the blow of the ram and the impact is not always sufficient to atomise the fuel. Bermingham of Ontario has developed a high-injection-pressure, 'smokeless' diesel hammer which virtually eliminates the problem. The more resistant the ground, the higher the rebound of the ram, and hence the higher the energy of the blow. This can cause damage to precast concrete piles when driving through weak rocks containing strong bands. Although the height of drop can be controlled by adjusting the amount of fuel injected, this control cannot cope with random hard layers met at varying depths, particularly when these

*Figure 3.13* Delmag D30-20 diesel hammer on American-style leaders with helmet for driving steel H-piles.

*Table 3.4* Characteristics of some diesel piling hammers

| Maker | Type | Mass of ram (kg) | Maximum energy per blow (kJ) | Maximum striking rate (blows/min) |
|---|---|---|---|---|
| Berminghammer (Canada) | B9 | 910 | 18.5 | 37–54 |
| | B32 | 3,200 | 110 | 34 |
| | B64 | 6,400 | 220 | 35–56 |
| | B5505 | 4,180 | 146 | 35–56 |
| | B6505 | 8,000 | 275 | 35–56 |
| | B6505HD | 10,000 | 300 | 35–56 |
| Delmag (Germany) | D6-32 | 600 | 19 | 38–52 |
| | D8-22 | 800 | 27 | 36–52 |
| | D12-42 | 1,280 | 46 | 35–52 |
| | D16-32 | 1,600 | 54 | 36–52 |
| | D19-42 | 1,820 | 66 | 35–42 |
| | D25-32 | 2,500 | 90 | 35–52 |
| | D30-32 | 3,000 | 103 | 36–52 |
| | D36-32 | 3,600 | 123 | 36–53 |
| | D46-32 | 4,600 | 166 | 35–53 |
| | D62-22 | 6,200 | 224 | 35–50 |
| | D100-13 | 10,000 | 360 | 35–45 |
| | D150-42 | 15,000 | 512 | 36–45 |
| | D200-42 | 20,000 | 683 | 36–52 |
| MKT (United States) | DE-33/20/20C | 1,495 | 23 | 40–50 |
| | DE-42/35 | 1,905 | 30 | 40–50 |
| | DE-70/50C | 3,175 | 50 | 40–50 |
| | DE-150/110C | 6,804 | 107 | 40–50 |
| ICE International Construction Equipment (United States) | I-8V2 | 800 | 25.3 | 36–52 |
| | I-30V2 | 3,000 | 94.8 | 35–52 |
| | I-80V2 | 8,000 | 282 | 35–45 |
| | I-100V2 | 10,000 | 353 | 35–45 |
| | I-160V2 | 16,000 | 580 | 35–45 |
| | 32S | 1,364 | 43.0 | 41–60 |
| | 60S | 3,175 | 98.9 | 41–59 |
| | 100S | 4,535 | 162.7 | 38–55 |
| | 120S | 5,440 | 202.0 | 38–55 |
| | 205S | 9,072 | 284.7 | 40–55 |

are unexpected. The diesel hammer operates automatically and continuously at a given height of drop unless the injection is adjusted, whereas with the hydraulic hammer every blow is controlled in height.

Because of difficulties in achieving a consistent energy of blow, due to temperature and ground resistance effects, the diesel hammer is being supplanted to a large extent by the hydraulic hammer, particularly when being used in conjunction with the pile driving analyser (see Section 7.3) to determine driving stresses. In addition, their use in the United Kingdom and elsewhere has declined as a result of environmental restrictions on the exhaust and noise.

Manufacturers and suppliers of impact hammers in the United States provide tables of bearing capacity based on the efficiency, hammer energy and final set per blow, usually based on a modification of the Hiley 1925 formula.

### 3.1.5 Piling vibrators

Vibrators, consisting of one or two pairs of exciters rotating at the same speed in opposite directions, can be mounted on piles where their combined weight and vibrating energy cause the pile to sink down into the soil (Figure 3.14). The two types of vibratory hammers, either mounted on leaders or as free hanging units, operate most effectively when driving small displacement piles (H-sections or open-ended steel tubes) into loose to medium-dense granular soils. Ideally, a pile should be vibrated at or near to its natural frequency, which requires 100 Hz for a 25 m steel pile. Thus, only the high-frequency vibrators are really effective for long piles as summarised by Holeyman et al.[3.2], and while resonant pile-driving equipment is costly, high penetration rates are possible. The modern resonant drivers are compact units and operate at frequencies from 80 to 150 Hz, automatically tuning to the natural frequency of the pile, with little ground vibration and no start-up/shutdown problems. The resonant driver uses a cylinder–piston mechanism to deliver the force to the pile through a specialised clamp. It operates at high accelerations (180 g at 150 Hz) and low amplitudes (8 mm at 80 Hz), controlled by proprietary algorithms. As an example, resonance drivers can drive HP360 piles to 36 m and 600 mm open-ended tubular piles to 16 m. However, most types of vibrators operate in the low- to medium-frequency range (i.e. 10–39 Hz). Vibrators mounted on the dipper arm of hydraulic excavators have high power-to-weight ratios and are useful for driving short lengths of small tubular section and H-piles, limited by the headroom under the bucket arm, say 6 m at best.

Rodger and Littlejohn[3.3] proposed vibration parameters ranging from 10 to 40 Hz at amplitudes of 1–10 mm for granular soil when using vibrators to drive piles with low point resistance, to 4–16 Hz at 9–20 mm amplitude for high-point-resistance piles. Vibrators are not very effective in firm and stiff clay where frequencies in excess of 40 Hz and high amplitude will be needed; when used in other fine-grained soils, care must be exercised because of the potential changes in soil properties such as remoulding, liquefaction and thixotropic transformation. Predicting the performance of vibratory pile driving is still not very reliable.

*Figure 3.14* Driving a pile casing with a PVE 200 m free hanging vibrator.

Where specific test data are not available for the vibrator installing bearing piles or the pile is not bearing on a consistent rockhead, it is advisable to use the vibrator to install the pile to within 3 m of expected penetration and then, subject to environmental considerations, use an impact hammer to drive to required set in the bearing layer. Vibrators can be used in bored pile construction for sealing the borehole casing into clay after predrilling through the granular overburden soils. After concreting the pile, the vibrators are used to extract the casings and are quite efficient for this purpose in all soil types (see Section 3.4).

Vibrators have an advantage over impact hammers in that the impact noise and shock wave of the hammer striking the anvil is eliminated. They also cause less damage to the pile and have a very fast rate of penetration in favourable ground. Provided that the electric generator for the exciter motor is enclosed in a well-designed acoustic chamber, the vibrators can be used in urban areas with far lower risk of complaints arising due to noise and shock-wave disturbance than when impact hammers are used. However, standard vibrators with constant eccentric moment have a critical frequency during starting and stopping as they change to and from the operating frequency, which may resonate with the natural frequency of nearby buildings. This can cause a short period of vibrations which are quite alarming to the occupants. The development of high-frequency (greater than 30 Hz), *resonance-free* (RF) vibrators with automatic adjustment has virtually eliminated this start-up and shut-down 'shaking zone', reducing peak particle velocity (ppv) to levels as low as 3 mm/s at 2 m from the pile (see Section 3.1.7). These vibrators are more powerful than the lower-frequency variable moment (VM) vibrators, generating greater driving force and displacement amplitude to overcome the toe resistance when driving longer and larger displacement piles[3.4]. Types of vibrators suitable for driving bearing piles are shown in Table 3.5.

Vibrating pokers or *vibroflots*, which are used extensively for improving the bearing capacity and settlement characteristics of weak soils by vibro-compaction or vibro-replacement techniques, have been adapted to construct vibro concrete columns (VCCs). As described in Section 2.3.7, concrete is injected into the hole at the tip of the poker and vibrated as it is withdrawn to provide a pile capable of carrying light vertical loading in weak soils. Figure 3.15 shows the simple set-up for concreting.

### 3.1.6 Selection of type of piling hammer

The selection of the most suitable type of hammer for a given task involves a consideration of the type and weight of the pile and the characteristics of the ground into which the pile is to be driven. Single- and double-acting hammers and hydraulic and diesel hammers are effective in all soil types, and the selection of a particular hammer for the given duty is based on a consideration of the value of energy per blow, the striking rate and the fuel consumption. The noise of the pile-driving operation will also be an important consideration in the selection of a hammer. This aspect is discussed in Section 3.1.7.

Knowledge of the value of energy per blow is required to assess whether or not a hammer of a given weight can drive the pile to the required penetration or ultimate resistance without the need for sustained hard driving or risk of damage to the pile or hammer and the possible injury to the operator. The use of a dynamic pile-driving formula to provide a rough assessment of the ability of a hammer to achieve a specific ultimate pile capacity has largely been replaced by the application of data from a large number of instrumented pile-driving tests undertaken to assess hammer capabilities and related pile performance. As a result, the manufacturer's rated energy per blow is now more reliable, and the efficiency of hammers has been improved significantly. Vibratory hammers will operate at 90%–100% efficiency on sheet piles, and well-maintained, modern hydraulic hammers with internal ram velocity measurements can operate at efficiencies approaching 100%. Diesel hammers can operate

Table 3.5 Characteristics of some pile-driving and extracting vibrators

| Maker | Type | Frequency range (Hz) | Mass (kg) | Minimum power supply (KVA) |
|---|---|---|---|---|
| ABI Gruppe (Germany) | HVR45 | 41 | 800 | 65 |
| | HVR75 | 41 | 1,400 | 130 |
| | MRZV 17VV | 30–43 | 2,105 | 257 |
| | MRZV 36VV | 32–43 | 4,043 | 470 |
| | MRZV 10V | 36 | 2,170 | 155 |
| | MRZV 36V | 33 | 4,000 | 465 |
| | MRZV12V[a] | 0–35 | 2,560 | 205 |
| | MRZV30V[a] | 0–25 | 4,280 | 490 |
| | MRZV36V[a] | 0–23 | 4,280 | 465 |
| American Piledriving Equipment (United States) | 3 | 0–50 | 204 | 7 |
| | 50 | 0–38 | 2,064 | 202 |
| | 200 | 0–30 | 1,183 | 438 |
| | 600 | 0–23 | 22,000 | 883 |
| | 120 VM | 0–38 | 3,402 | 276 |
| | 170 VM | 0–38 | 4,037 | 276 |
| | 250 VM | 0–38 | 6,985 | 515 |
| Dawson Construction Plant (United Kingdom) | EMV70[b] | 50 | 410 | 12 |
| | EMV300A[b] | 40 | 625 | 60 |
| | EMV450[b] | 41 | 1,008 | 88 |
| | EMV550[b] | 42 | 1,150 | 120 |
| Dieseko PVE (Holland and United States) | 25M | 28 | 2,900 | 272 |
| | 38M | 28 | 3,000 | 295 |
| | 52M | 28 | 4,000 | 434 |
| | 110M[c] | 28 | 7,000 | 558 |
| | 200M | 23 | 21,000 | 980 |
| | 300M | 23 | 27,250 | 1,633 |
| | 2312VM | 38 | 2,050 | 152 |
| | 2319VM | 38 | 2,675 | 291 |
| | 2335VM | 38 | 4,400 | 590 |
| | 2070VM | 33 | 6,800 | 913 |
| ICE – International Construction Equipment (Holland and United States) | 14C | 32 | 1,716 | 168 |
| | 55B | 25 | 5,740 | 444 |
| | 84C | 25 | 7,240 | 597 |
| | 84C/1200 | 27 | 7,240 | 887 |
| | 14RF | 38 | 2,420 | 213 |
| | 28RF | 38 | 3,800 | 431 |
| | 64RF | 32 | 5,000 | 683 |
| | 416L | 27 | 2,350 | 209 |
| | 55NF | 28 | 3,580 | 360 |
| | 1412C | 23 | 6,400 | 525 |
| | 625 | 42 | 685 | 117 |
| | 3220 | 33 | 3,850 | 285 |

(continued)

*Table 3.5* (continued)  Characteristics of some pile-driving and extracting vibrators

| Maker | Type | Frequency range (Hz) | Mass (kg) | Minimum power supply (KVA) |
|---|---|---|---|---|
| PTC (France) | 8HFV | 39 | 1,402 | 113 |
| | 18H2 | 27 | 2,450 | 112 |
| | 14HFV | 35 | 3,590 | 148 |
| | 30HV | 28 | 4,400 | 220 |
| | 45HV | 28 | 7,000 | 298 |
| | 52HV | 28 | 7,070 | 323 |
| | 75HD | 25 | 11,800 | 360 |
| | 120 HD | 23 | 1,330 | 481 |
| | 200HD | 23 | 19,540 | 709 |
| | 265HD | 24 | 27,450 | 1131 |
| Soilmec (Italy) | VS-2 | 30 | 1,138 | 106 |
| | VS-4 | 30 | 1,901 | 200 |
| | VS-8 | 30 | 3,500 | 450 |

V, generally denotes VM vibrator.
[a]  Leader mounted.
[b]  Mounted on excavator dipper arm.
[c]  A modular hammer, several of which can be mounted around a tubular pile (10 m maximum diameter).

*Figure 3.15* Installation of VCC showing concreting hose connected to vibrator. (Courtesy of Vibro Ménard, part of the Bachy Soletanche Group, Ormskirk, UK.)

at 80% but are likely to exhibit the widest efficiency variations, particularly in difficult driving conditions. In all cases, the energy delivered by the hammer to the pile depends on the accuracy of alignment of the hammer, the type of packing inserted between the pile and the hammer, and the condition of the packing material after a period of driving.

The GRLWEAP® software from Pile Dynamics Inc (see Appendix C) contains a large database of hammer performance which enables the piling engineer to predict driveability, optimise the selection of hammer, select an energy level which will not damage the pile and

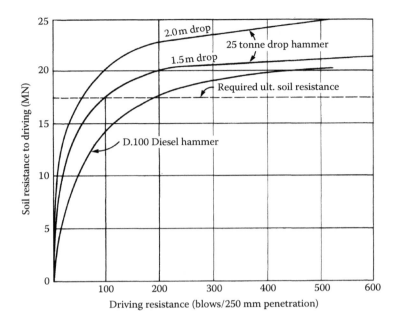

*Figure 3.16* Pile driveability curves.

ensure that the correct dolly and adapters are used. When used in conjunction with a pile-driving analysis program based on the Smith wave equation (see Section 7.3), the designer can receive outputs showing driving stresses and hammer performance in real time.

The curves of the type in Figure 3.16 show the results of an investigation into the feasibility of using a D100 diesel hammer to drive 2.0 m outside diameter (OD) by 20 mm wall thickness steel tube piles through soft clay into a dense sandy gravel. The piles were to be driven with closed ends to overcome a calculated soil resistance of 17.5 MN at the final penetration depth. Figure 3.16 shows that a driving resistance (blow count) of 200 blows/250 mm penetration would be required at this stage. This represents a rather severe condition. A blow count of 120–150 blows/250 mm is regarded as a practical limit for sustained driving of diesel or hydraulic hammers. However, 200 blows/250 mm would be acceptable for fairly short periods of driving.

The American Petroleum Institute (API)[4.15] suggests that if no other provisions are included in the construction contract, pile-driving *refusal* is defined as the point where the driving resistance exceeds either 300 blows per foot (248 blows/250 mm) for 1.5 consecutive metres or 800 blows per foot (662 blows/250 mm) for 0.3 m penetration. Figure 3.16 also shows the driving resistance curves for a 25 tonne drop hammer with drops of 1.5 or 2.0 m to be used as a standby to achieve the required soil resistance if this could not be obtained by the diesel hammer.

Vibratory hammers are very effective in loose to medium-dense granular soils, and the high rate of penetration of low-displacement steel piles driven by vibratory hammers may favour their selection for these conditions. The drawback is that there is no reliable correlation between pile refusal under vibration and the dynamic resistance of the soil.

### 3.1.7 Noise and vibration control in pile driving

The control of noise on construction sites is a matter of increasing importance in the present drive to improve environmental conditions, and the 'Control of Noise at Work Regulations

2005' implements the European directive for the protection of workers from the risks related to the exposure to noise. The requirements for employers to make an assessment of noise levels and take action to eliminate and control noise are triggered by three action levels: daily or weekly (5 days of 8 h) personal noise exposures of 80 dBA as the lower level, 85 dBA as the upper level and a peak (single loud noise) of between 135 and 137 dBC weighted. The exposure limit values are 87 dBA and 140 dBC at peak; the method of calculating the various exposure levels is defined in the regulations. If these levels are exceeded, then employers are required to reduce noise at source by using appropriate working methods and equipment, but if noise levels cannot be controlled below the upper action level by taking reasonably practicable measures, suitable personal hearing protection which eliminates the risk must be provided. It should be noted that the noise is measured on a logarithmic scale – a reduction in noise of 3 dB is equivalent of reducing the intensity of the noise by half. As a guide, if it is necessary to shout to be heard 2 m away, then the noise level is likely to be above 85 dBA. As the regulations do not apply to the control of noise to prevent annoyance or hazards to the health of the general public outside the place of work, the Environment Protection Act (EPA) and Control of Pollution Act provide the general statutory requirements to control noise and vibrations which are considered to be a legal nuisance.

Code of practice BS 5228-1 (Noise) gives best practice recommendations for noise control onsites and guidance on predicting and measuring noise. It also covers the procedures for obtaining consent from the local authority under sections 61 of the Control of Pollution Act for proposed noise control measures. It is recognised that the noise from many pile-driving methods will exceed 85 dBA, but as the operations are not continuous through the working day, the observed noise level (or 'basic sound power level' as given in the Code) can be converted to an 'equivalent continuous sound pressure level' that takes into account the duration of the noise emission, distance from the source, screening and reflection[3.5]. For example, in Table C12 of the Code, a Junttan PM25 4 tonne hydraulic hammer driving cast-in-place piles has a sound power level of 103 dB, which, if operated for 65% of the site day, reduces to an equivalent continuous sound pressure level of 84 dB at 10 m.

Local authorities are empowered under the EPA and Control of Pollution Act to set their own standards of judging noise nuisance, and maximum daytime and night-time noise levels of 70 and 60 dBA respectively, are frequently stipulated for urban areas (and as low as 40 dBA in sensitive areas – the typical sound level of rainfall). The higher of these values can be compared with field observations of pile-driving noise obtained from a number of sources and shown in Figure 3.17. Other information has shown that the attenuation of pile-driving impact noise to the 70 dBA level from the noisiest of the hammers requires a distance of more than 1000 m from the sound. Thus, if a maximum sound level of 70 dBA is stipulated by a local authority, it is necessary to adopt some means of controlling noise emission in order to protect the general public whose dwellings or place of work is closer to the construction operations[3.6]. Methods include enclosing the pile and hammer within an acoustic shroud, hanging flexible acoustic screens, using the appropriate dolly (cap), and changing the piling system to push-in or vibration. As an example of an acoustic shroud, *Hoesch* steel sandwich panels (from ThyssenKrupp) were used to form a tower comprising an outer 2 mm steel plate, a plastics layer 0.4 mm thick and an inner 1.5 mm steel plate jointed by a rubber insertion material, with a lid incorporating a sound-proofed air exhaust. This box reduced the noise from a Delmag D12 diesel hammer driving a sheet pile from 119 dBA at 7 m to 87–90 dBA at the same distance. Figure 3.18 shows a typical stand-alone shroud. The MENCK noise reduction shroud which is mounted directly onto the MHU hydraulic hammer can reduce the noise level by 10–12 dBA. In sensitive areas in the United States, noise-absorbing blankets have to be placed around the piling works.

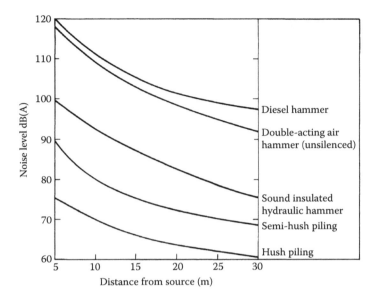

Figure 3.17 Typical noise levels for various pile-driving techniques.

Figure 3.18 Noise suppression shroud around tubular steel pile driven by a hydraulic impact hammer.

Crane-mounted augers using kelly bars for bored piles (see Section 3.3.4) and large CFA rigs can produce sound power levels as high as 108 dBA and are usually operated between 85% and 100% of the shift. This results in equivalent continuous sound pressure levels in excess of 80 dBA at 10 m. Acoustic enclosures are essential for ancillary plant. The use of vibratory hammers for driving steel bearing piles has increased, and although noise is generally less than that produced by impact hammers, basic sound levels can still be around 115 dBA in difficult driving conditions. Even with conversions to equivalent sound, noise abatement measures are usually necessary.

There is little evidence to show that ground-borne vibrations from well-controlled construction operations cause structural damage to buildings in good repair[3.7]. However, if there is a concern, then steps must be taken to survey buildings and measure vibrations induced by construction activity. BS 7385 describes methods of assessing vibrations in buildings and gives guidance on potential damage levels. The limits for transient vibration above which non-structural ('cosmetic') damage could occur are given in Code of practice BS 5228-2 (Vibration). For example, the limits for residential property are a ppv of 15 mm/s at 4 Hz to 50 mm/s at 50 Hz and for heavy and stiff buildings 50 mm/s at 4 Hz and above. Protected buildings and buildings with existing defects and statutory services undertakings will be subject to specific lower limits, and under the Control of Pollution Act, local authorities need to give prior consent for piling work which may cause vibrations. EC3-5 requires that vibration limits to suit connected or adjacent structures are taken into account for serviceability.

The human response, which can be sensitive to vibration well below that needed to cause damage, should also be considered; BS 6472-1 advises on the assessment of the 'vibration dose value' for night-time and daytime working. Transmission of vibrations during piling depends on the strata, size and depth of pile and hammer type, and predictions of the resulting ground frequency and ppv at distance from the source are difficult, as can be seen from the historical data given in Table D of BS 5228. Monitoring of noise and vibration is now regularly applied on urban piling sites, with the data recorded electronically and reported in real time to interested parties.

The acoustic measurements given in the COWRIE reports[3.8] on environmental effects of impact piling for offshore structures have revealed that the noise generated can affect marine life for several kilometres from the site. The mitigation measures studied include bubble curtains (limited effect), with preference for smaller piles and vibratory piling. New wind farm developments in the North Sea are likely to consider shallow gravity foundations to avoid disturbance from piling.

*Press-in* drivers (or vibration-less hydraulic jacking) such as the Dawson *push–pull* unit with 2078 kN pressing force are becoming more common particularly for sheet piling, but many of the units can be adapted for installing box-type bearing piles, tubular piles and H-pile groups. The advantages of these powerful, high-pressure hydraulic drivers using two to four cylinders are the low noise levels (around 60 dBA) and the speed and vibration-less installation and extraction of piles. Figure 3.19 shows the push–pull unit, mounted on a leader with a supporting piling frame, installing a box pile comprising 4 sheet piles clutched together to a depth of 14 m through stiff boulder clay; the leader is capable of providing additional pull-down where needed. The applied load was 2760 kN. The upper part of the piles were exposed and filled with concrete to form permanent bridge piers. Figure 3.20 shows the four-cylinder unit suspended from a crane pressing a box pile through temporary aligning casing. In both cases, the reaction for the push-in ram is provided by clamping the adjacent rams to the driven sheet piles of the box. Cleaning out the soil plug to allow bonding of concrete to a sheet pile box pile is not feasible.

Figure 3.19 Push–pull piler installing box piles for bridge piers. (Courtesy of Dawson Construction Plant Ltd, Milton Keynes, UK.)

Figure 3.20 Crane-suspended push–pull piler. (Courtesy of Dawson Construction Plant Ltd, Milton Keynes, UK.)

## 3.1.8 Pile helmets and driving caps

When driving precast concrete piles, a helmet is placed over the pile head for the purpose of retaining in position a resilient *dolly* or cap block that cushions the blow of the hammer and thus minimises damage to the pile head. The dolly is placed in a recess in the top of the helmet (Figure 3.21). For easy driving conditions, it can consist of an elm block, but for rather harder driving, a block of hardwood such as oak, greenheart, pynkado or hickory is

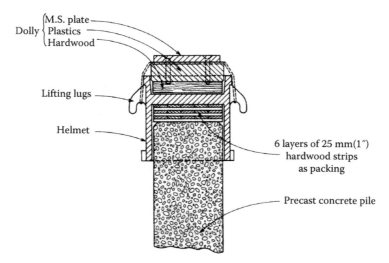

*Figure 3.21* Dolly and helmet for precast concrete pile.

set in the helmet end onto the grain. Plastic dollies are the most serviceable for hard driving concrete or steel piles. The Micarta dolly consists of a phenolic resin reinforced with laminations of cross-grain cotton canvas; its modulus of elasticity of 3200 MN/m$^2$ is 10 times that of a hardwood (oak) dolly. Layers of these laminates can be bonded to aluminium plates or placed between a top steel plate and a bottom hardwood pad. The helmet should not fit tightly onto the pile head but should allow for some rotation of the pile, which may occur as it strikes obstructions in the ground.

Packing is placed between the helmet and the pile head to cushion further the blow on the concrete. This packing can consist of coiled rope, hessian packing, thin timber sheets, coconut matting or wallboards. Asbestos fibre packing, while resistant to the heat generated, is no longer acceptable. The packing must be inspected at intervals and renewed if it becomes heavily compressed and loses its resilience. Softwood packing should be renewed for every pile driven.

Driving caps are used for the heads of steel piles, but their function is more to protect the hammer from damage than to protect the pile. The undersides of the caps for driving box or H-section piles have projecting lugs to receive the head of the pile. Those for driving steel tubular piles (Figure 3.22) have multiple projections that are designed to fit piles over a range of diameters. They include jaws to engage the mating hammers.

Plastic dollies of the Micarta type have a long life when driving steel piles to a deep penetration into weak rocks or soils containing cemented layers. These can last 40 times longer than elm blocks, for example when driving precast piles, and hence are more economic. Thick cushion blocks of softwood, further softened by soaking, have been used for each pile to avoid damage when driving prestressed concrete piles. However, for economy, contractors often cushion the pile heads with scrap wire rope in the form of coils or in short pieces laid crosswise in two layers. These are replaced frequently as resilience is lost after a period of sustained driving and noise levels increase significantly. If dollies have to be changed while driving a pile, the blow count could change significantly.

### 3.1.9 Jetting piles

Water jets can be used to displace granular soils from beneath the toe of a pile. The pile then sinks down into the hole formed by the jetting, so achieving penetration without the use of

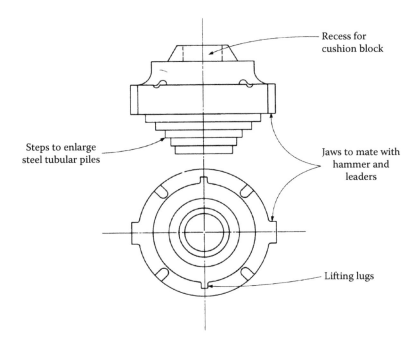

*Figure 3.22* Vulcan driving cap for steel tubular pile.

a hammer. Jetting is a useful means of achieving deep penetration into a sandy soil in conditions where driving a pile over the full penetration depth could severely damage it. Jetting is ineffective in firm to stiff clays however, and when used in granular soils containing large gravel and cobbles, the large particles cannot be lifted by the wash water. Nevertheless, the sand and smaller gravel are washed out, and penetration over a limited depth can be achieved by a combination of jetting and hammering. Air can be used for jetting instead of water, and bentonite slurry can be also used if the resulting reduced shaft friction is acceptable.

For jetting piles in clean granular soils, a central jetting pipe is the most effective method, as this helps to prevent the pile from deviating off line. A 25–50 mm nozzle should be used with a 50–75 mm pipe. The quantity of water required for jetting a pile of 250–350 mm in size ranges from 15 to 60 litres/s for fine sands through to sandy gravels. A pressure at the pump of at least 5 bar is required. The central jetting pipe is connected to the pump by carrying it through the side of the pile near its head. This allows the pile to be driven down to a *set* on to rock or some other bearing stratum immediately after shutting down the jetting pump. When using jets to assist driving of prestressed piles, it is essential that water from the internal jetting pipe does not make contact with the body of the pile, as this may enter any rebound tension cracks resulting in the compression blow damaging the pile.

A central jetting pipe is liable to blockage when driving through sandy soils layered with clays, and the blockage cannot be cleared without pulling out the pile. A blockage can result in pipe bursting if high jetting pressures are used. Open-ended steel tubular piles and box piles can be jetted by an independent pipe worked down the centre or the outside of the pile, and H-piles can be similarly jetted by a pipe operated between the flanges but rigging the system can cause delays to pile driving. Large-diameter tubular piles can have a ring of peripheral jetting pipes to assist in breaking up a soil plug. For example, Gerwick[3.9] has described the system for jetting 4 m diameter tubular steel piles with 50 mm wall thickness for a marine terminal. Sixteen 100 mm pipes were permanently installed around the inner periphery of the pile with the nozzles cut away at each side to direct the flow to the pile tip. He gives the following typical requirements for jetting large-diameter piles:

| | |
|---|---|
| Jet pipe diameter | 40–50 mm |
| Pressure | 20–25 bar (at pump) |
| Volume | 12 litres/s per jet pipe |

The large volume of water used in jetting can cause problems by undermining the piling rig or adjacent foundations as it escapes towards the surface. It can also cause a loss of shaft friction in adjacent piles in a group, and external jetting for marine piles will reduce lateral resistance. Where shaft friction must be developed in a granular soil, the jetting should be stopped when the pile has reached a level of about 1 m above the final penetration depth, the remaining penetration then being achieved by hammering the pile down. The jetting method is best suited to piles taken down through a granular overburden to end bearing on rock or some other material resistant to erosion by wash water.

Water jetting is also used in conjunction with press-in and vibratory piling techniques to assist penetration of sheet piles in dense granular soil. A lance is fitted inside the pile pan and both are driven simultaneously into the ground. On reaching the required depth, the lance is removed for reuse. Low injection rates are used at high pressure (5 litres/s at 150 bar).

## 3.2 EQUIPMENT FOR INSTALLING DRIVEN AND CAST-IN-PLACE PILES

The rigs used to install driven and cast-in-place piles are similar in most respects to the types described in Sections 3.1.1 through 3.1.3, but the firms who install proprietary types of pile usually make modifications to the rigs to suit their particular systems. The piling tubes are of heavy section, designed to be driven from the top by drop, single-acting or diesel hammers, but the original Franki piles (Figure 3.23 and Section 2.3.2) are driven by an internal drop hammer. The internal hammer mass will be between 2 and 8 tonne for pile tubes of 248–610 mm diameter. The leaders of the piling frames are often adapted to accommodate guides for a concreting skip (Figure 3.24).

Thick-walled steel cased piles designed to be filled with concrete are driven more effectively by a hammer operating on the top than by an internal drop hammer acting on a plug of concrete at the base. This is because a hammer blow acting on top of the pile causes the tube to expand and push out the soil at the instant of striking, followed by a contraction of the tube. This frees the tube from some of the shaft friction as it moves downwards under the momentum of the hammer. The flexure of the pile acting as a long strut also releases the friction at the moment of impact. However, when using an internal drop hammer, tension is induced in the upper part of the pile and the diameter contracts, followed by an expansion of the soil and an increase in friction as the pile moves downwards. Flexure along the piling tube does not occur when the hammer blow is at the base, and thus there is no reduction in friction from this cause. Tension caused by driving from the bottom can cause the circumferential cracking of hollow-core reinforced concrete and thin-walled steel tubular piles.

Top driving has another advantage in allowing the pile to be driven with an open end, thus greatly reducing the end-bearing resistance during driving, but the soil plug will have to be drilled out if the concrete pile is to be cast in place as the tube is withdrawn.

*Figure 3.23* Franki pile-driving rig.

Also top-driven thick-walled drive tubes with expendable end plate/shoes produce a dry hole for concreting as the tube is withdrawn. In easy driving conditions, bottom driving on a plug will give economy in the required thickness of the steel and considerable reduction in noise compared with top driving. For example, Cementation Foundations Skanska installed 508 mm diameter bottom-driven thin-walled (6 mm) steel piles up to 15 m long in Cardiff Bay in preference to thicker-walled, top-driven, cased piles to reduce disturbance to residents. A 4 tonne drop hammer was used to drive the bottom plug to found in Mercia Mudstone; concreting was direct from the mixer truck or by skip.

Great care is necessary to avoid bursting of the tube by impact on the concrete when bottom driving through dense granular soil layers or into weak rocks containing bands of stronger rock. The concrete forming the plug should have a compacted height of not less than 2.5 times the pile diameter. In calculating the quantity of concrete required, allowance should be made for a volume reduction of 20%–25% of the uncompacted height. The concrete should be very dry with a water/cement ratio not exceeding 0.25 by weight. A hard aggregate with a maximum size of 25 mm should be used.

At least 10 initial blows should be given with hammer drops not exceeding 300 mm and then increasing gradually. The maximum height of drop should never exceed the maximum specified for the final set which is usually between 1.2 and 1.8 m. Driving on a plug should not exceed a period of 1½ h. After this time, fresh concrete should be added to a height of not less than the pile diameter, and driving continued for a period of not more than 1½ h before a further renewal. For prolonged hard driving, it may be necessary to renew the plug every three-quarters of an hour.

*Figure 3.24* Discharging concrete into the driving tube of a withdrawable-tube pile. Concreting skip travelling on pile frame leaders.

## 3.3 EQUIPMENT FOR INSTALLING BORED AND CAST-IN-PLACE PILES

### 3.3.1 Power augers

Power-driven rotary auger drills are suitable for installing bored piles in clay soils. A wide range of machines is available using drilling buckets, plate and spiral augers, and CFAs, mounted on trucks, cranes and crawlers to bore open holes. This allows for the installation of a full-length reinforcement cage where needed – say in tension piles. The range of diameters and depths possible is considerable, from 300 mm to over 5000 mm and to depths of 100 m. Hydraulic power is generally used to drive either a rotary table, a rotating kelly drive on a mast or a top-drive rotary head; some tables are mechanically operated through gearing. Most units have additional pull-down or crowd capability to apply pressure to the bit. The soil is removed from spiral-plate augers by spinning them after withdrawal from the hole and from buckets either by spinning or through a single or double bottom opening. It is an EU mandatory safety requirement that spoil from an auger should be removed at the lowest possible level during extraction to ensure that debris from the flights cannot fall onto personnel or damage machinery and to avoid rig instability. Hydraulically operated cleaners which can be rapidly adjusted to suit CFA diameters from 400 to 2000 mm are available.

As well as being used for producing under-reamed or belled pile bases, large-diameter bored piles have facilitated the construction of high-capacity piles incorporating the *plunge column* technique, allowing top-down construction of basements (Section 2.5).

*Figure 3.25* Watson 5000 crane attachment power auger on elevated platform on a 40 tonnes crane with 200 mm telescopic kelly for installing 2440 mm casings.

The use of crane-mounted attachments for boring piles with a kelly, rotated by either a top-drive hydraulic unit or a mechanical/hydraulic rotary table, has declined considerably in recent years with the introduction of the more mobile and powerful top-drive units. Watson continues to produce the 5000 model (Figure 3.25) which has a rotary torque of 153.7 kNm capable of running 3000 mm boring tools using quadruple kellys. The truck-mounted unit (Figure 3.26) is a self-contained drill for 1800 mm diameter bores up to 18 m deep using a telescopic kelly. The largest Watson crawler drill with rotary torque of 244 kNm is specifically designed to bore shafts up to 3660 mm diameter to depths of 41 m. The range of Calweld drilling machines has also been eclipsed by the modern mobile rig, but there are many lorry-mounted bucket drills, crane attachments and rotary drive table units on the resale and hire market, particularly in the United States.

Soilmec produces a limited range of crane-mounted rigs; the RT3-ST, which has a mechanically driven rotary table with a maximum torque of 210 kNm, can bore 3000 mm diameter holes to a depth of 42 m with a standard kelly and to 120 m with a special quadruple kelly. The largest unit is the SA40 which has a hydraulically powered rotary table producing up to 452 kNm torque capable of drilling 5000 m diameter holes to 90 m, mounted on a 90 tonne crane.

Bauer has developed a powerful bucket auger unit (the *Flydrill System* in Figure 3.27) which integrates the hydraulic power packs and the rotary drive on one platform for mounting on top of a partially driven tubular pile. The rotary drive produces a torque of 462 kNm at 320 bar, and two hydraulic crowd cylinders provide a pull-down of 40 tonne. The clamping device can exert a total force of 90 tonnes to resist the torque and apply the pull-down. The system operates a triple telescope kelly with 3 and 4.4 m diameter buckets and was used for cleaning out and reaming below 4.75 m diameter tubular monopile foundations to allow driving to be completed to a depth of 61 m at the offshore wind farm in the Irish Sea off

*Figure 3.26* Watson 2100 truck-mounted auger drill.

*Figure 3.27* *Flydrill 5500* with bucket auger removing spoil from 4.75 m diameter monopiles at the Barrow offshore wind farm site. (Courtesy of Bauer Maschinen GmbH, Schrobenhausen, Germany.)

Barrow-in-Furness. Leffer has produced a crane-suspended, down-the-hole hydraulic power swivel which clamps to the cased pile bore and sits directly above the auger bucket. The largest unit will operate in 3000 mm casing at a torque of 30 kNm.

The range and capabilities of crawler-mounted hydraulic rotary piling rigs have increased significantly in recent years. The rigs in Table 3.6 are usually capable of installing CFA and rotary displacement piles as well as standard bored piles, but the height of the mast

Table 3.6 Some hydraulic self-erecting crawler rigs

| Maker | Type | Standard stroke (m) | Main winch capacity (kN)[a] | Maximum diameter (mm) | Typical maximum depth (m) | Maximum torque (kNm) |
|---|---|---|---|---|---|---|
| American | SA 12 | 4.0 | 135 | 1500 | 41 | 160 |
| Piledriving | SA 20 | 5.0 | 180 | 2000 | 50 | 225 |
| Equipment | SA 25 | 6.0 | 250 | 2500 | 86 | 280 |
| (United States) | BG 15H | 12 | 110 | 1500 | 40 | 151 |
| Bauer | BG 20H | 15 | 170 | 1500 | 51 | 200 |
| (Germany) | BG 24H | 15.4 | 200 | 1700 | 54 | 222 |
| | BG 28H | 18.4 | 250 | 1900 | 71 | 270 |
| | BG 40 | 19.7 | 300 | 3000 | 80 | 390 |
| | BG 50 | 19.5 | 500 | 3000 | 82 | 468 |
| | RG 18S | 18.0 | 170 | Driven piles | 18 | 200 |
| | RG 22S | 22.0 | 55 | " | 22 | 200 |
| | RG 25S | 25.0 | 200 | | 25 | 275 |
| Casagrande | B125 XP | 12.6 | 160 | 1500 | 50 | 125 |
| (Italy) | B200 XP | 13.7 | 214 | 2200 | 67 | 210 |
| | B300 XP | 15.0 | 270 | 2500 | 90 | 300 |
| | B400 HT | 13.5 | 320 | 3000 | 87 | 358 |
| | B450XP | 21.5 | 420 | 3000 | 110 | 420 |
| | C850 H50 | 14 | 320 | 3000 | 87 | 545 |
| | C850 DH | 19.1 | 250 | 1000 | 18.6/24.5 | 358/421 |
| | C850[b] | 34 | 320 | 1000 | 35 | 545 |
| Delmag | RH12 | 12 | 200 | 1450 | 18 | 120 |
| (Germany) | RH14 | 12.5 | 200 | 1580 | 23 | 144 |
| | RH20 | 14.2 | 300 | 1830 | 30 | 206 |
| | RH26 | 15 | 420 | 1960 | 36 | 265 |
| Liebherr | LB 16 | | 200 | 1500 | 34 | 161 |
| (Germany and | LB 20 | | 200 | 1500 | 46 | 200 |
| United States) | LB 28 | | 250 | 2500 | 70 | 280 |
| | LB 36 | | 300 | 3000 | 88 | 366 |
| Soilmec (Italy) | SR 30 | 3.5 | 135 | 1500 | 48 | 130 |
| | SR 50 | 11.0 | 185 | 2000 | 61 | 180 |
| | SR 70 | 6.5 | 240 | 2500 | 77 | 271 |
| | SR 100 | 21.7 | 270 | 3500 | 28 | 480 |
| | SF50[b] | 19.5 | NA | 900 | 25 | 100 |
| | SF70[b] | 22.5 | NA | 1000 | 28 | 165 |
| | SF140[b] | 27.9 | NA | 1400 | 34 | 305 |

[a] Pulling force.
[b] Rigged for CFA drilling.

and stroke available may limit the depth achievable; hence, the major manufacturers produce special long stroke rigs for CFA piles up to 34 m deep. For bored piles, many rigs
can accommodate casing oscillators and most have rams or winches to provide additional
crowd and extraction forces, requiring robust masts and extendable tracks for stability.
The major manufacturers also produce double rotary heads (usually capable of rotating in
opposite directions) as attachments for the more powerful piling rigs which enable casing
up to 1000 mm diameter to be installed with the lower drive while augering with the top
drive. The dual-rotary system from Foremost Industries of Canada operating on their DR
40 crawler rig provides 30 kNm torque through the top drive for boring and 339 kNm
torque on the lower rotary table for simultaneous casing up to 1000 mm diameter. The
Liebherr pile-driving rigs (see Section 3.1.1) have the option of running double rotary top
drive or kelly tools for bored and CFA piles. In-cab electronic instrumentation and read-
out to control positioning and drilling parameters are standard on most modern rigs.

A major benefit of the modern self-erecting boring rig is the ability to change tools quickly
to suit changing ground conditions. These units can be rigged in a variety of ways for CFA,
kelly and rotary-percussive boring and pile driving. In addition, the larger rigs are enhanced
with electronic systems and on-board telemetry which improve accuracy of pile installation
and reduce noise emissions. The depths, diameters and strokes quoted in Table 3.6 depend
on the drilling method used and whether extended leaders are added.

Various types of equipment are available for use with rotary augers. The standard and
rock augers (Figure 3.28a and b) have scoop-bladed openings fitted with projecting teeth.
The coring bucket is used to raise a solid core of rock (Figure 3.28c), and the bentonite
bucket (Figure 3.28d) is designed to retain the stabilising *filter cake* which forms on the
borehole wall (see Section 3.3.8). Both types of bucket augers are available in diameters up

(a)                          (b)                          (c)

(d)                          (e)

*Figure 3.28* Types of drilling tools. (a) Standard auger. (b) Rock auger. (c) Coring bucket. (d) Bentonite
bucket. (e) Chisel.

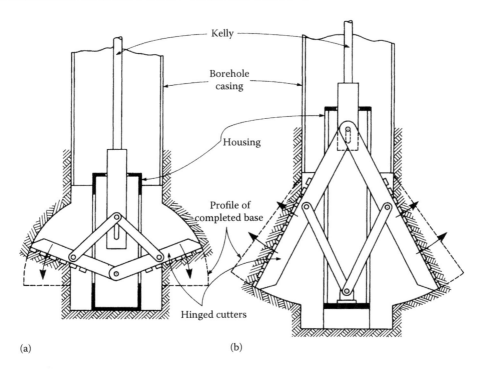

Kelly

Borehole
casing

Housing

Profile of
completed base

Hinged cutters

(a)                                    (b)

*Figure 3.29* Under-reaming tools. (a) Bottom hinge. (b) Top hinge.

to 2500 mm and can be configured to rip hard soil and medium rock; they are effective in fine- and coarse-grained soil, with borehole support where necessary. They are also effective below water table.

Enlarged or under-reamed bases can be cut by rotating a belling bucket within the previously drilled straight-sided shaft. The bottom-hinged bucket (Figure 3.29a) cuts to a hemispherical shape, and because it is always cutting at the base, it produces a clean and stable bottom. However, the shape is not as stable as the conical form produced by the top-hinged bucket (Figures 3.29b and 3.30), and the bottom-hinged arms have a tendency to jam when raising the bucket. The arms of the top-hinged type are forced back when raising the bucket, but this type requires a separate cleaning-up operation of the base of the hole after completing the under-reaming. Belling buckets normally form enlargements up to 3.7 m in diameter in shafts of at least 760 mm but can excavate to a diameter of 7.3 m with special attachments in large-diameter bores.

The optimum condition for the successful operation of a rotary auger rig is a fine-grained soil which will stand without support until a temporary steel tubular liner is lowered down the completed hole or a granular soil supported by bentonite slurry or other stabilising fluid. In these conditions, fast drilling rates of up to 7 m per hour are possible for the smaller shaft sizes. The use of sectional flight augers (SFAs) to install temporary casing in water-bearing uniform sands is not advisable, because as water drains, a solid plug can form in the casing jamming the auger. Methods of installing piles with these rigs are described in Section 3.4.6.

Figure 3.31 shows the LD5000 reverse-circulation pile-top drill and 4.3 m under-reaming bit, both designed, built and operated by Large Diameter Drilling Ltd., mobilising for installing monopiles at the Gwynt y Mor offshore wind farm. The monopiles will generally be driven to target depth (up to 64 m) with the LD5000 deployed to replace the hammer when needed in hard ground to under-ream the pile for further driving. Golightly[3.10] comments

*Figure 3.30* Top-hinged under-reaming bucket.

*Figure 3.31* LDD reverse-circulation drill bit 4.3 m diameter with expandable under-reamers to maximum 6 m for drilling inside and below tapered piles. (Courtesy of Large Diameter Drilling Ltd, Penryn, UK.)

on the problems of constructing ever-larger monopiles – for example 6.5 m diameter with D/t ratios up to 100, 70 m deep in water depths of 40 m – such as severe tip buckling and adverse tilt and settlement where piles are not end bearing on hard dense soils or bedrock.

### 3.3.2 Boring with casing oscillators

In difficult drilling conditions through loose sands, gravels and broken rock formations, the pile borehole is likely to require continuous support by means of casing. In such conditions, it is advantageous to use an oscillator mechanism which imparts a semi-rotating motion (or fully rotating in special applications) to the casing through clamps. Vertical rams attached to the clamps enable the temporary, double-walled casing with carbide shoes, to be forced down to follow the drilling tool. The semi-rotating motion is continuous (usually through 25°), which prevents the casing from becoming 'frozen' to the soil, and it is continued while extracting the casing after placing the concrete. Typical jointed casings (e.g. the Bauer and Casagrande types) have male/female joints which are locked by inserting and tightening bolts manually (which can have safety implications) or by an automatic adapter lock to resist the high rotating or oscillating forces.

Hydraulic casing oscillators are available from most of the large rig manufacturers to attach to crane-mounted rigs or to rotary drills with diameters from 1000 to 3800 mm and torque capability up to, for example, 8350 kNm from the Soilmec 3000, which has a clamping force of 478 kN and lifting force of 725 kN. The material has to be broken up and excavated from within the pile casing with ancillary equipment, and various methods are used; these include a hammer grab hanging from a crane, removal by augers and down-the-hole hammers on crawler rigs. The Malcolm Drilling Company used the large Leffer VRM 3800 oscillator, capable of applying torque up to 12,620 kNm, to install permanent 3.7 m diameter, 38 mm thick welded casings to rockhead 52 m deep, for the foundation shafts of the Doyle Drive Viaduct in San Francisco. Excavation of the highly variable overburden in the casing was by a 40 tonnes spherical grab (Figure 3.32), and a 3 m diameter, 14 m deep rock socket into the complex subducted Franciscan beds was rotary drilled using a Bauer BG40 rig. Dense reinforcement and a self-compacting concrete were required for the length of the shaft to meet the extreme seismic conditions.

Drilling and installing casing simultaneously ('duplex' drilling) through cobbles, boulders and rubble using special casing shoes and casing under-reamers attached to top-drive, down-the-hole compressed air hammers has advanced significantly. For example, Numa hammers of the United States manufacture a range of drills capable of installing casing up to 1219 mm diameter to 15 m deep using a rotary-percussive under-reamer which can be retracted to allow concreting of the pile as the casing is withdrawn (Figure 3.33).

### 3.3.3 Continuous flight auger drilling rigs

A typical CFA rig is shown in Figure 3.34. Drilling output with the rigs in Table 3.6 is greater than that achievable with standard bored piles as the pile is installed in one continuous pass; hence, the mast must have an adequate stroke for the auger under the rotary head. A kelly may be inserted through the rotary head to increase depth on some rigs. Most CFA rigs have crowd capability to assist in penetrating harder formations, and augers should be designed to suit the high torques available. Possible diameters range from 500 to 1400 mm to a maximum depth of 34 m.

Cased CFA piles have become more popular with the development of cleaners/collectors operating at the top of the casing which discharge spoil into telescopic chutes for removal at ground level (Figure 3.35). With suitable auger extensions and a robust drilling mast, it is

*Figure 3.32* Leffer VRM 3800 casing oscillator and spherical grab installing 3.7 m diameter casings for the Doyle Drive Viaduct. (Courtesy of Malcolm Drilling Company, San Francisco, CA.)

possible to simultaneously drill and case CFA 1200 mm diameter piles to about 20 m. It is essential that the spoil cleaning and collecting system at the top of the casing does not hinder the drilling stroke. The more powerful rigs referenced in Table 3.6 have separate drive heads for the casing and auger to rotate the casing and auger in counter directions and can move on the mast independently of the casing. For example, the SR-100 rig can be configured with a 330 kNm torque for the upper rotary auger head and 448 kNm on the casing driver. The auger drive has to accommodate the concreting swivel.

Displacement auger piling is carried out with rigs similar to the high-torque CFA equipment, but the diameter is limited to less than 600 mm by the shape of the displacement tool; maximum depth is around 30 m (Section 2.3.5).

### 3.3.4 Drilling with a kelly

The kelly is a square or circular drill rod made of high-tensile steel which is driven by keying into a rotary table fixed either to the rig near the ground surface or to a crane attachment. As a result of the improvements in rig stability and mast rigidity, the most usual rotation method now is by a moveable hydraulic drive head on the mast. The full range of drilling tools, plate and bucket augers, drag bits, compound rotary drill plate bits and tricone bits can be rotated by the kelly in most drillable ground conditions, subject to the available power. The kelly may be in sections or, more usefully, telescopic to make up the required length of drill string. Typical torque range is 100–400 kNm and lengths up to

*Figure 3.33* Numa hammer with extending under-reaming drill bit for simultaneously drilling and inserting casing. (Courtesy of Numa Hammers, Thompson, CT.)

70 m are available. Boreholes can be drilled as open holes or supported either by excess hydrostatic head using support fluids (see Section 3.3.8) or by casing. The casing can be installed by oscillators or by the rotary drive with some of the larger rigs. Under-reamers and belling tools are expanded by an upward or downward force from the rotating kelly. Grabs can also be operated from the kelly bar.

## 3.3.5 Reverse-circulation drilling rigs

Reverse-circulation drilling rigs operate on the principle of the airlift pump. Compressed air is injected near the base of the centrally placed discharge or riser pipe, above the drill cutting head. As the air rises and expands in the discharge pipe, the density of the fluid in the riser decreases creating a pressure differential between the internal fluid and the fluid in the hole. This causes the higher-density outer column to be sucked into the riser through the cutting head opening. A reverse-circulation system is shown in Figure 3.36. The casing tubes and airlift riser pipe may be rotated together or separately by means of a hydraulic rotary table as shown or, more usually, by a top-drive power swivel. The airlift riser comprises dual drill pipes, maximum bore 330 mm, either flange jointed or flush; air is delivered through an air/discharge swivel at the drive head, down the annulus between the inner and outer tubes. The riser is maintained centrally in the casing by one or more stabilisers, and the soil boring is effected by rock roller bits mounted on a cutter head, ranging from 0.76 to 8.0 m diameter. The injected airflow and pressure and the point of injection all affect the

*Figure 3.34* Installation of CFA piles in chalk with crane handling reinforcement cage. (Courtesy of Cementation Skanska, Rickmansworth, UK.)

*Figure 3.35* Self-erecting drill rigged for installing cased CFA piles with telescopic spoil chute. (Courtesy of Bachy Soletanche Ltd, Ormskirk, UK.)

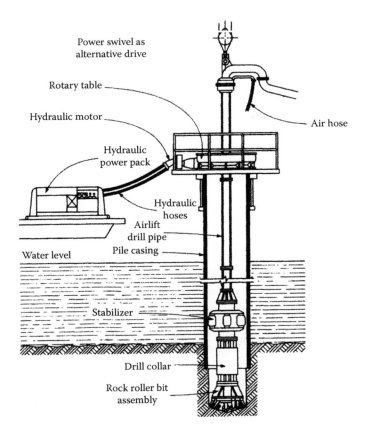

*Figure 3.36* Rotary table drill rigged for reverse circulation.

efficiency of cuttings removal; air injection rate is up to 130 m³/min at a pressure of 12 bar, requiring large air compressors. At maximum airflow and injection pressure, mud/spoil can be discharged at rates up to 2500 m³/h, depending on the delivery head.

For offshore work, the hole will be kept full of seawater, but on land, drilling mud is used to remove the cuttings necessitating the use of mud tanks and cleaners (see Section 3.3.8). Also on land, the reverse-circulation system with mud may maintain a stable hole without the use of casing for cast-in-place piles. The more powerful self-erecting crawler rigs with dual-rotary drive heads in Table 3.6 can be rigged for reverse circulation for holes up to 3300 mm in diameter to 100 m in depth.

Pile-top rigs such as the LD5000 (Figure 3.31) and the Seacore Ltd. *Teredo* units (Figure 3.37) using powerful top-drive swivels are more versatile than large rotary tables for over-water work. The Teredo rig, equipped with a 460 kNm power swivel, is capable of rock drilling up to 7 m diameter. The Bauer power auger in Figure 3.27 can be classified as a pile-top rig but has to be handled off the pile to discharge the bucket, requiring continuous service by a suitable crane.

Reverse-circulation rigs can drill at a fast rate in a wide range of ground conditions including weak rocks. They are most effective in granular soils and the large diameter of the airlift pipes enables them to lift large gravel, cobbles, and small boulders when drilling in glacial soils or in jointed rocks which are broken up by the rock roller bits. Under-reamed bases can be provided in stiff clays or weak rocks by means of a hydraulically operated rotary enlarging tool mounted above the cutter head.

*Figure 3.37* Pile-top rig drilling 3.8 m diameter piles for foundation strengthening to the Richmond-San Raphael Bridge, California. (Courtesy of Fugro Seacore Ltd, Falmouth, UK.)

### 3.3.6 Large grab rigs

The use of diaphragm wall grabs to form barrettes in preference to large-diameter bored pile groups is well established. The grabs may be suspended from cranes or mounted on purpose-built crawlers and excavate a square hole, Ell-, Tee- or rectangular slots under bentonite or other support fluid. The *hydromill* or *hydrofraise* rig as developed by Bachy Soletanche is a reverse-circulation down-the-hole milling machine with two contra-rotating cutter drums powered by hydraulic motors mounted on a heavy steel frame as in Figure 3.38. The cuttings are removed from the slot in a bentonite or polymer slurry by a pump fitted above the drums to the de-sanding and cyclone plant at the surface where the slurry is reconditioned for reuse. Overbreak is minimal and the absence of vibration makes the system suitable for urban sites and operating close to existing buildings. Standard width is 600 mm but greater widths are possible for depths to 60 m. Walls have been constructed to 150 m deep, and low headroom versions are available.

### 3.3.7 Tripod rigs

Small-diameter piles with diameters from 300 to 600 mm, installed in soils which require continuous support by lining tubes, can be drilled by tripod rigs. The drilling is performed in clays by a clay cutter, which is a simple tube with a sharpened cutting edge, the tube

*Figure 3.38* Hydromill for forming barrettes. (Courtesy of Sound Transit, Seattle, WA.)

being driven down under the impact of a heavy drill stem. The soil which jams inside the tube is prised out by spade when the cutter is raised to the surface. Drilling is effected in coarse-grained soils by means of a baler or 'shell', which is again a simple tube with a cutting edge and flap valve to retain the soil, the soil being drawn into the baler by a suction action when the tool is raised and lowered. If no groundwater is present in the pile borehole, water must be poured in, or a bentonite slurry may be used. This suction action inevitably causes loosening of the soil at the base of the pile borehole, thus reducing the base resistance. The loosening may be accompanied by settlement of the ground surface around the pile borehole. Rocks are drilled by chiselling and using a baler to raise the debris. These rigs are mainly used in situations where low headroom or difficult access would prevent the deployment of lorry-mounted or track-mounted augers.

*Table 3.7* Some compact low-headroom rigs for limited access situations

| Maker | Type | Feed stroke (m) | Weight (tonne) | Maximum diameter (mm) | Maximum torque (kNm) |
|---|---|---|---|---|---|
| GP Services (United Kingdom) | D1000 (Drop hammer) | 2.48–3.48 | 2.4 | | |
| | T3000[a] | 1.35 | 1.3 | 300 | 3.15 |
| Hutte (Germany) | 203[a] | 1.2 | 2.3 | 250 | 26.4 |
| Klemm[b] (Germany) | 702[a] | 1.2–2.2 | 3.6 | 356 | 27 |
| | 704Electro | 2.15–3.25 | 4.5 | 356 | 15 |
| Mait[b] (United States) | Baby drill | 1.1 | 5.3 | 600 | 17.7 |
| Toa-Tone[b] (Japan) | EP-26 (sonic drilling) | 1.4 | 2.6 | 150 | 3.4 |

[a] Separate power pack.
[b] Radio remote controls available.

### 3.3.8 Drilling for piles with bentonite slurry and support fluids

Lining tubes or casings to support the sides of pile boreholes are a requirement for most of the bored pile installation methods in coarse-grained soils using equipment described in Sections 3.3.1 through 3.3.7. However, even in stiff fine-grained soils, it may be necessary to use casings for support since these soils are frequently fissured or may contain pockets of sand which can collapse into the boreholes, resulting in accumulations of loose soil at the pile toe or discontinuities in the shaft.

The use of casings may be avoided by providing support to the pile borehole in the form of a slurry of bentonite clay or polymer drilling fluid; but note that BS EN 1536 requires that the borehole under support fluid shall be protected by a lead-in tube or guide wall (for a barrette). Bentonite, or other montmorillonite clay with similar characteristics, has the property of remaining in suspension in water to form a stiff *gel* when allowed to become static. When agitated by stirring or pumping, however, it has a mobile fluid consistency – that is it is *thixotropic*. In a granular soil, the slurry penetrates the walls of the borehole and gels there to form a strong and stable filter cake. In a clay soil, there is little penetration of the slurry but the hydrostatic pressure of the fluid, which has an initial density of around 1040 kg/m$^3$, prevents collapse where the soil is weakened by fissures. The slurry also acts as the flushing medium and carries the drill cuttings to the surface where they are removed in separation plants. The rheological properties which govern performance of the fluids for use in pile bores are given in BS EN 1536, and the Federation of Piling Specialists[3.11] provide detailed information on the preparation, use and testing of suitable slurries.

When used in conjunction with auger or grab-type rigs, the slurry is maintained in a state of agitation by the rotating or vertical motion of the drilling tools. When it becomes heavily contaminated with drill cuttings or diluted by groundwater, the filter cake is weakened and the slurry must be replaced by pumping in fresh or reconditioned slurry to maintain hole support. Toothed or bladed augers with double-helix configurations and a flap in the carriage area help to retain spoil as the auger is withdrawn through the slurry. A support fluid is used most efficiently in conjunction with reverse-circulation rigs (see Section 3.3.5). Here, the slurry is pumped into the outer annulus and the slurry–soil mixture that is discharged from the airlift riser pipe is allowed to settle in lagoons or tanks to settle soil particles before skimming-off cleaned slurry for return to the hole. On large projects, further cleaning to remove ultra-fine particles will be economical using separation plants comprising vibrating screens, hydro-cyclones and centrifuges which deliver the output fluid to storage tanks where gelling aids may be added before the reconditioned slurry is returned to the pile borehole.

If a bentonite slurry becomes overloaded with solids from the excavation, the resulting thick filter cake is not as effective in supporting the soil and may not be removed by scouring during concreting. In such cases, it will be necessary to use a mechanical scraper to remove the excess filter cake prior to concreting. Reese et al.[3.12] recommend a minimum diameter of 600 mm for piles installed using bentonite slurry techniques, to avoid some of the problems associated with the method. Another potential cost is that waste bentonite slurry has to be treated as *hazardous* under pollution control regulations and disposed of accordingly, whereas polymers can be neutralised and, subject to de-sanding and approval from the water company, may be disposed of to existing drains.

Where a relatively small layer of coarse-grained soil lies over a stiff end-bearing soil and support from casing is needed, it is not cost-effective to bring in high-speed mixers, slurry tanks, pumps and reconditioning plant for the normal employment of bentonite slurry techniques for short-term support. Instead, a few bags of dry bentonite are dumped into the pile borehole and mixed with the groundwater, or added water, to form a crude slurry which is adequate to smear the wall of the borehole. After drilling through the granular overburden under the thick slurry, the casing is lowered and pushed or vibrated to seal it into the stiff fine-grained soil below. This technique is known as 'mudding-in' the casing.

Some problems caused when placing concrete in bentonite slurry supported bores, with or without casing, and the means of overcoming them are described in Section 3.4.8; the effects of a bentonite slurry on shaft friction and end-bearing resistance of piles are discussed in Section 4.2.3.

Polymer support fluids, which are available in a wide range of commercial products from the basic natural gums (e.g. xanthan) to complex copolymers, have several advantages over bentonite as borehole support fluids but need care in application. Pure biopolymers have been used in place of the civil engineering grade of sodium bentonite, giving better solids carrying capacity in sands and gravels, but can degrade unless treated with biocides leading to potential environmental concerns on disposal. Polymers are added to sodium bentonite formulations by manufacturers to improve rheology, but adding polymer to bentonite slurry on site can give unreliable results. Research into the more complex synthetic polymers has led to increased use over the past 10 years, and although more expensive than bentonite as an initial cost, economies result as less polymer powder is required, mixing is easier and time required for de-sanding of slurry is minimised. They are better suited to drilling large-diameter piles and shafts where the hole has to be supported for up to 36 h of drilling time. The filter-cake formation on the sides of the hole is much thinner and therefore more easily scoured when placing concrete. Also, the sides do not soften to the same extent as with bentonite slurry support, and clay swelling is controllable.

Longer-chain synthetic polymers (e.g. *partially hydrolysed polyacrylamides*) now being developed can give improved foundation performance and are easier to mix and handle on site; cleaning is done in a settling tank and de-sanded diluted fluid can be disposed of to foul sewers (subject to approval). The drawbacks are that properties are lost with repeated circulation by centrifugal pumps which break up the polymer chain and the polymer is sorbed onto soils. This requires fresh polymer to be added regularly in order to maintain viscosity to ensure the hole remains stable. As the fluid density (~1020 kg/m$^3$) is much lower than for bentonite, stability of the bore relies on an excess head, and coarser drill cuttings will settle out. Lam et al.[3.13] report on a field trial in London which tested three instrumented piles drilled in the Lambeth Group/Thanet sand under bentonite or polymer fluid. They found that the two polymer piles outperformed the bentonite pile under the maximum proof load for load/settlement behaviour and no adverse effect was caused by the deliberately extended soil–fluid exposure time. The auger was designed to avoid suction developing as it was withdrawn and vigorous base cleaning was carried out. They also comment that although no detrimental effects were observed in the concrete exposed to the polymer, more research is desirable into the effect of intermixing of fluid and concrete.

### 3.3.9 Base and shaft grouting of bored and cast-in-place piles

When bored and cast-in-place piles are installed in granular soils, the drilling operation may loosen the soil surrounding the shaft and beneath the base of the pile borehole. Such loosening below the base can cause excessive working load settlements when the majority of the load is carried by end bearing. Base grouting is a means of restoring the original in situ density and reducing settlements. Bolognesi and Moretto[3.14] described a method of grouting the disturbed soil below 1 and 2 m diameter piles bored under bentonite, using a metal basket filled with uniform gravel which was attached to the base of the pile reinforcement (Figure 3.39). The pile was concreted and, after a period of hardening, the basket injected with cement grout, the potential uplift being resisted by the pile shaft friction and pile cap. High-pressure grout will flow up the sides of the shaft increasing resistance. The *flat-jack* method

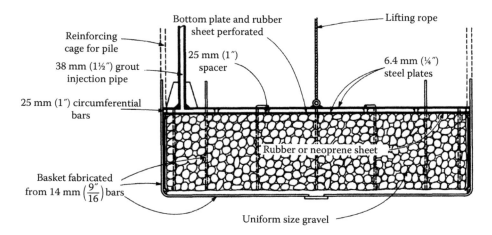

Reinforcing cage for pile

38 mm (1½") grout injection pipe

Bottom plate and rubber sheet perforated

25 mm (1") spacer

Lifting rope

6.4 mm (¼") steel plates

25 mm (1") circumferential bars

Rubber or neoprene sheet

Basket fabricated from 14 mm $\left(\frac{9''}{16}\right)$ bars

Uniform size gravel

*Figure 3.39* Preloading cell for compressing loosened soil beneath base of bored piles by grouting. (After Bolognesi, A.J.L. and Moretto, O., Stage grouting preloading of large piles in sand, *Proceedings of the Eighth International Conference*, ISSMFE, Moscow, Russia, Vol. 2.1, pp. 19–25, 1973.)

of pressure grouting at the base of the shaft is similar. Here, a circular steel plate is attached to the base of the reinforcement cage, and a flexible metal sheet covers the underside of this plate. Grout pipes are connected to the gap between the plate and the sheet and also around the periphery of the cage to a given height above the base. After concreting and allowing a hardening period, the peripheral grout pipes are injected with cement grout, and after this peripheral grout has hardened, the gap between the base plates is injected.

Both these methods are difficult to control and have been largely replaced by the sleeve tube or *tube-à-manchette* (TaM) technique as described by Yeats and O'Riordan[3.15] for the 1.2 m diameter piles for an office block in London. The 38.2 m deep test pile shaft was drilled by rotary auger under a bentonite slurry through the alluvium and stiff to hard clays of the London Clay and Woolwich and Reading Formation (Lambeth Group) into very dense Thanet sands. The upper 31 m of the shaft was supported by casing. After completing the drilling, four separate grout tube assemblies as shown in Figure 3.40 were lowered to the base of the borehole. The injection holes in the tubes were sleeved with rubber to form the TaM. The pile shafts were then concreted under bentonite, and 24 h after this, water was injected to crack the concrete surrounding the grout tubes. Base grouting commenced 15 days after concreting. The injections were undertaken in stages with pressures up to 60 bar, and frequent checks to ensure the pile head did not lift by more than 1 mm. Similar base-grouting techniques were used at six sites in the Docklands area of London beneath piles with diameters in the range of 0.75–1.5 m[3.16]. The general procedure for base grouting with TaM is to limit the volume of grout injected in the first phase and apply the limiting pressure for the second phase; a total injection is usually specified at 25–35 litres/m² of pile surface. Uplift of the pile and the residual pressure in the grout tubes is recorded. Exceptionally, remedial base grouting may be carried out through grout pipes drilled through the set concrete.

Part of the internal plugs to the 2.50 and 3.13 m OD driven tubular steel piles for the Jamuna River Bridge[4.42] were cleaned out by airlifting which loosened the soil at the base. In order to reconsolidate the remaining plug of sand, a grid of TaMs was placed in the hole above the plug and a layer of gravel placed by tremie to cover the grout tubes. A 7 m plug of concrete was placed over the gravel, and 12 h later, water was injected at a pressure of 20 bar to crack open the sleeves. Cement grout (40 litres of water, 50 kg cement, 0.35 kg bentonite

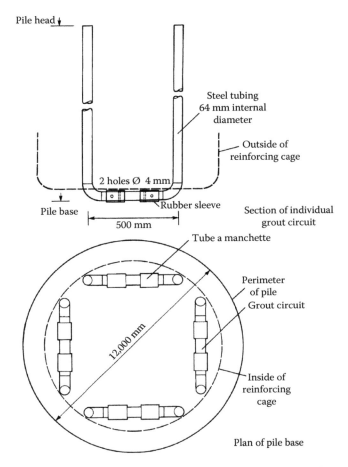

Figure 3.40 Arrangement of circuits for base-grouting of piles. (After Yeats, J.A. and O'Riordan, N.J., The design and construction of large-diameter base-grouted piles in Thanet Sand, London, *Proceedings of the International Conference on Piling and Deep Foundations*, Vol. 1, Balkema, Rotterdam, the Netherlands, pp. 455–461, 1989.)

and 0.5 kg plasticiser) was then injected into the gravel plug. Grouting was terminated when the pressure reached 50 bar, in order to ensure that uplift of the pile would not occur, or when 1000 litres of grout had been injected to limit hydrofracture of the soil below the gravel.

Shaft grouting of cast-in-place piles and barrettes entails rupturing the outer skin of the pile and pushing it against the surrounding soil. This increase in lateral pressure is intended to cause local increases in the soil density which had become loosened or softened by the pile construction and thereby enhance the shaft resistance of the pile. When shaft grouting in granular soils, cementation of the soil particles may occur and voids and fissures become filled giving improved contact between pile and soil. The usual technique is to install 50 mm diameter steel TaMs around the outside perimeter of the reinforcement cage for the depth to be treated, with return connections to the surface. The sleeves on the tubes at 1 m centres are staggered around the cage to form a spiral injection track. After allowing the concrete to cure for 24 h, the sleeves are cracked at pressures up to 80 bar and flushed with water; each sleeve is pressure grouted 10–15 days thereafter using double packers. Two-phase injections at each sleeve may be needed depending on the injection pressure relative to the overburden pressure, requiring water flushing of the tubes between phases. Littlechild et al.[3.17] report on a series of tests on 20 shaft grouted, cast-in-place piles in soft marine clay underlain by alluvial deposits of stiff clay and dense to very dense sand in Bangkok. The measured shaft resistances for the shaft grouted piles, ranging from 150 to 320 $kN/m^2$, were approximately double those without shaft grouting. The test piles were reloaded more than 1 year after grouting and showed no loss of resistance in either the clay or sands. Core samples along the pile–grout interface showed grout infilling cracks and fissures in the concrete and a grouted zone 20–30 mm around the pile with some cementation of the sands.

Suckling and Eager[3.18] compare the results published for base-grouted and non-base-grouted bored piles bearing in Thanet sand, including the Yeats and O'Riordan data. They show that the ultimate end-bearing capacity ranged from 12,000 to 17,000 $kN/m^2$ for non-grouted pile bases and from 17,000 to 21,000 $kN/m^2$ for base-grouted piles. They conclude that, given sound construction, base grouting in this formation is unnecessary except when considering exceptional loading. The Shard tower piles in London were such a case as described by Beadman et al.[3.19] The ultimate end-bearing capacity of the base-grouted 1.8 m diameter piles was limited to 20,000 $kN/m^2$ as proposed by Suckling and Eager. The bearing capacity factor of $N_q^* = 47$ used for the pile calculations for the 46 m depth to the Thanet sands was confirmed by load tests on a preliminary 1.2 m diameter pile which indicated that the base capacity was about 22,500 $kN/m^2$. ADSC (The International Association of Foundation Drilling) is due to publish a major report on the increasing use of base grouted piles.

## 3.4 PROCEDURE IN PILE INSTALLATION

Each class of pile employs its own basic type of equipment, and hence the installation methods for the various types of pile in each class are the same. Typical methods are described below to illustrate the use of the equipment described in the preceding sections of this chapter. Particular emphasis is given to the precautions necessary if piles are to be installed without unseen breakage, discontinuities or other defects. The installation methods described in this section are applicable mainly to vertical piles. The installation of raking piles whether driven or bored is a more difficult operation and is described in Section 3.4.11.

BS EN 1536 and BS EN 12699 deal with the execution of bored and displacement piles respectively. However, in many respects, the guidance on installation in these new codes is not as comprehensive as that contained in withdrawn BS 8004. For example, BS EN 12699 does not comment on appropriate installation procedures, simply requiring that a suitable hammer or vibrator be used to achieve the required depth or resistance without damage to the pile. As noted in Section 3.1.6 to avoid overstressing of a pile during driving, assessment of driveability is necessary followed by stress wave measurements on preliminary test piles.

One of the major factors in producing a stable bored pile or accurately aligned driven pile is the setting up of the rig on a firm level base and the attention paid to maintaining verticality of the drill mast. Tilting of the rig or violent operation of an auger leads to misalignment and the need for corrective action by reaming the pile sides; hammer blows which are not hitting the pile centrally will cause damage and compromise bearing capacity. The report 'Working Platforms for Tracked Plant'[3.20] from the Building Research Establishment (BRE) provides guidance for the design and construction of ground-supported platforms for piling plant. In a 2011 review into the use of alternative approaches to the design of platforms, BRE found that the use of structural geosynthetic reinforcement is acceptable provided that safety is preserved and the approach is based on credible and representative research. But as pointed out by Fountain and Suckling[3.21], the assessment of the platform subgrade and the selection of design parameters to provide realistic mat thicknesses is still a problem. It is suggested that ground-probing radar and plate loading tests are performed on site to assist in the design. The requirements for the safety of operatives should be rigorously followed as detailed in the British Drilling Association Health and Safety Manual[3.22]. Casings protecting open pile boreholes should extend above ground level and should be provided with a strong cover.

### 3.4.1 Driving timber piles

Timber piles are driven by drop hammer or single-acting hammer after pitching them in a piling frame, in crane-suspended leaders or in trestle guides. A hammer with a minimum mass of 1 tonne is advisable with a maximum of 1.5 times the mass of pile and helmet. Diesel hammers, unless they are of the light type used for driving trench sheeting, are too powerful and are liable to cause splitting at the toe of the pile. The heads of squared piles are protected by a helmet of the type shown in Figure 3.21. Round piles are driven with their heads protected by a steel hoop. A cap is used over the pile head and hoop, or packing can be placed directly on the head. Care should be taken in the use of slings and hooks to prevent damage to protective treatments.

### 3.4.2 Driving precast (including prestressed) concrete piles

The methods of handling the piles after casting and transporting them to the stacking area are described in Section 2.2.2. They must be lifted from the stacking positions only at the prescribed points as designed. If designed to be lifted at the quarter or third points (Section 7.2), they must not at any stage be allowed to rest on the ground on their end or head until in the leader. Particular care should be taken to avoid overstressing by impact if the piles are transported by road vehicles. Additional support points should be introduced if necessary, particularly important for long prestressed piles.

A helmet of the type shown in Figure 3.21 and its packing are carefully centred on the pile, and the hammer position should be checked to ensure that it delivers a concentric blow. The hammer should preferably weigh not less than the pile. The guidance in BS 8004 is relevant for driving precast piles, that is the mass and power of the hammer should be such to ensure a final penetration of about 5 mm/blow unless the rock has been reached. The stroke of a single-acting hammer should ideally not be greater than 1 m. The hammer mass will be between 2 and 4 tonnes, with the 2 tonnes unit suitable for 10 m maximum length of pile and applied load of 450 kN; the 4-tonne hammer is used for long piles in compact soils with applied loads up to 1200 kN. Further specific recommendations are given in the GRLWEAP database. It is preferable to use the heaviest recommended hammer and to limit the stroke to avoid damage and limit tensile stress in the pile.

The driving of the piles should be monitored, and where piles rotate or move off line, any bindings should be eased. After the completion of driving, the pile heads should be prepared for bonding into the pile caps as described in Section 7.7. Hollow piles with a solid end may burst under the impact of the hammer if they become full of water, and holes should therefore be provided to drain off accumulated water. Where a soil plug is formed at the toe of an open-ended pile, water accumulation or arching of the soil within the pile may also result in bursting during driving. Further general guidance is given in CIRIA Report PG8[3.23].

### 3.4.3 Driving steel piles

Because of their robustness, steel piles can stand up to the high impact forces from a diesel hammer without damage other than the local distortion of the pile head and toe under hard driving. Open-ended tubular or box piles or H-piles can be driven to a limited penetration by a vibrator.

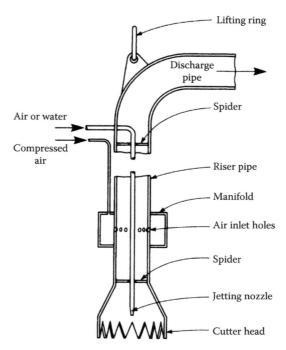

Lifting ring

Discharge pipe

Spider

Air or water

Compressed air

Riser pipe

Manifold

Air inlet holes

Spider

Jetting nozzle

Cutter head

*Figure 3.41* Airlift for cleaning-out soil from steel tubular piles.

By using rolled steel corner sections, plugged tube bearing piles can be formed by driving a number of interlocking U-section sheet piles sequentially. As the resistance to driving is less than for welded box piles, vibrators or press-in pilers (Figures 3.19 and 3.20) can be used to install high-capacity piles to greater depths at sensitive sites where impact driving cannot be tolerated.

To achieve the required depth of penetration, it is sometimes necessary to reduce the base resistance by removing the soil plug which forms at the bottom of an open-ended tubular or box pile. A sandy-soil plug can be removed by simple water jetting. A plug of clay or weak broken rock can be removed by lowering the airlift device shown in Figure 3.41 down the tube, the soil or broken rock in the plug being loosened by dropping or rotating the riser pipe. A reverse-circulation rig with a rotating cutter (Figure 3.36) is an efficient means of removing soil if justified by the number and size of the piles. Crane-mounted power augers of the type shown in Figure 3.25 can only be used for cleaning after the pile has been driven down to its final level where there is space for the crane carrying the auger to be manoeuvred over the pile head. The self-erecting crawler rigs are more manoeuvrable and with the other methods described earlier can be used to under-ream the pile toe and so ease the driving resistance. However, drilling below the toe also reduces the shaft friction, and the method may have to be restricted to end-bearing piles. This aspect is discussed further in Section 8.3 on piling for marine structures. Because of the delays involved in alternate drilling and driving operations, it is desirable that any drilling to ease the driving resistance should be restricted to only one operation on each pile.

Difficulties arise when it is necessary to place a plug of concrete at the toe of the cleaned-out pile to develop high end-bearing resistance or to transfer uplift loads from the super-structure to the interior wall of the hollow pile through a reinforcing cage. In such cases, a good bond must be developed between the concrete filling and the interior of the steel pile. Any remaining adherent soil must be cleaned off the pile wall. A sandy soil can be effectively removed by water jetting or by airlifting as mentioned above. However, readily available equipment has not been developed which will quickly and effectively remove adherent clay from the wall to a standard which will allow good bonding between the concrete and steel. Various one-off devices have been used with varying degrees of success, for example high-pressure water jets around a central airlift pipe together with wire strand brushes attached to a base plate rotated to scour the pile wall. The procedure for placing the concrete plug in the cleaned-out pile or for completely filling a steel tubular or box pile is similar to that described below for shell piles.

### 3.4.4  Driving and concreting steel shell piles

Steel shell piles are usually driven by internal drop hammers acting on a concrete base plug. Problems can arise with heave when driving shell piles in groups and distortion or collapse of the shells when driving past obstructions. Shell piles have the advantage that the interior of the shell can be inspected before concrete is placed. Distortion of the shells can be detected by light reflected down the pile, by lowering a lamp to the toe or by CCTV. To correct distortion, it may be necessary to pull up the shells and re-drive them or, in the case of tapered shells, insert and re-drive a new tapered shell assembly. The problem of heave is discussed in Sections 5.7 through 5.9.

Sometimes leakage of groundwater occurs through shells in quantities which do not justify replacing the damaged units. The water can be removed from the shells before placing the concrete by pumping (if the depth to the pile toe is within the suction lift of the available pump), by an airlift or by baling. If, after removing the water, the depth of inflow is seen to be less than a few centimetres in 5 min, the collected water can again be removed and concrete placed quickly to seal off the inflow. For higher rates of seepage, the water should be allowed to fill the pile up to its rest level, and the concrete should then be placed by tremie pipe as described in Section 3.4.8.

Concrete placed in dry shell piles is merely dumped in by barrow or chute. It should be reasonably workable with a slump of 100–150 mm to avoid arching as it drops down a tapered shell or onto the reinforcing cage. The cement content should be such as to comply with the requirements in BS EN 1536 or with any special requirements for durability. The American Concrete Institute[6.12] states that vibration due to driving adjacent piles has no detrimental effect on fresh concrete in shell piles. Therefore, concreting can proceed immediately after driving the shell even though adjacent shells are being driven, provided there are no detrimental effects due to ground heave or relaxation.

### 3.4.5  Installation of withdrawable-tube types of driven and cast-in-place piles

There are no standard procedures for installing driven and cast-in-place piles of the types which involve the driving and subsequent withdrawal of a casing tube. However, BS EN 12699 requires that cast-in-place displacement piles shall be concreted in the dry using high-workability concrete or semi-dry concrete as appropriate to the methods for each type of pile as described in Section 2.3.2. Where the concrete is compacted by internal drop hammer, a mix is required that is drier than that which is suitable for compaction by vibrating the piling tube. Depending on the durability designation of the concrete as given in BS 8500, the workability and mix proportions of the concrete may be decided by the designer or the contractor in accordance with the UK Concrete Specification[2.30] and BS EN 1536.

The procedures to be adopted for avoiding waisting or necking of the shaft, or the inclusion of silt pockets and laitance (a surface skin of weak cement), are similar to those adopted for bored and cast-in-place piles and are described in the following section of this chapter.

Precautions against the effects of ground heave are described in Section 5.8. Because the casing/drive tube is, in all cases, driven down for the full length of the pile, it is essential to ensure that the interior of the tube is free of any encrustations of hardened concrete. Even small encrustations can cause the concrete to arch and jam as the tube is withdrawn. If the reinforcing steel is lifted with the tube, the pile shaft is probably defective and should be rejected. Further guidance is given in CIRIA Report PG8[3.23].

### 3.4.6 Installation of bored and cast-in-place piles by power auger equipment

The employment of a power auger for the drilling work in bored and cast-in-place piles presupposes that the soil is sufficiently cohesive to stand unsupported, at least for a short time. Any upper soft or loose soil strata or water-bearing layers are 'cased off' by drilling down a casing or pushing the tubes down into the predrilled hole by vibrator or the crowd mechanism on the kelly bar. If necessary, 'mudding-in' techniques are used at this stage (see Section 3.3.8). After the auger has reached the deeper and stiffer fine-grained soils, the borehole is taken down to its final depth without further support, until the stage is reached when a loosely fitting tube is lowered down the completed hole. This loose liner may be required for safety purposes when inspecting the pile base before placing the concrete; or if an enlarged base is required, the lining prevents the clay collapsing around the shaft over the period of several hours or more required to drill the under-ream. The loose liner may not be needed for straight-sided piles in weak rocks or in stable unfissured clays, where there is no risk of collapse before or during the placing of the concrete. However, if the clays are in any degree fissured, there is a risk of the walls collapsing during concreting, thus leading to defects of the type shown in Figure 3.42. Lining tubes must be inserted in potentially unstable soils if a remote *visual* inspection is to be made of the pile base. *Manned* inspections of bores and under-reams are not permitted in current UK specifications, notwithstanding the updating of the relevant BS 8008 in 2008 for

*Figure 3.42* Defective shaft of bored pile caused by collapse of clay after lifting casing.

'descent into machine-bored shafts for piling' (see Figure 11.6). High-resolution colour CCTV inspection is appropriate provided good lighting is available, and the absence of remoulding on the shelf of the under-ream should be checked by a sampling device or penetrometer. Concreting should commence within 2 h of the inspection of the under-ream.

The final cleaning-up operation before placing concrete in a bored pile consists of removing large crumbs of soil or trampled puddled clay from the pile base. Any lumps of clay adhering to the walls of the borehole or to the lining tubes should be cleaned off. The reinforcing cage can then be placed and concreting commenced. The time interval between the final cleaning-up and placing concrete should not exceed 6 h. If there is any appreciable delay, the depth of the pile bottom should be checked against the measured drilled depth before placing the concrete to ensure that no soil has fallen into the hole. If the reinforcing cage is to extend only part way down the hole, it should be suspended from the top of the pile shaft before commencing to place the concrete.

The concrete used in the pile base and shaft should be easily workable with a slump of 180 mm as recommended in BS EN 1536. As the concrete is placed in 'free fall' from a chute or hopper over the bore and vibration in the bore is not feasible, the mix must be self-compacting, designated S4 in BS 8500-1, preferably using rounded aggregate. In addition, the mix proportions should comply with the requirements for strength and minimum cement content in BS EN 1536, and care is needed when considering mix design for durability (see Section 10.3.1). A dry mix should be used for the first few charges of concrete if the pile base is wet. After completing concreting, the lining tubes are withdrawn. If a loose liner is used inside an upper casing, the former is lifted out as soon as the concrete extends above the base of the outer tube. A vibrator of the type described in Section 3.1.5 is a useful expedient for extracting the upper casings used to support soft clays or loose sand. The quantity of concrete placed in the shaft should allow for the outward slumping which takes place to fill the space occupied by the tube and any overbreak of the soil outside it. Concreting should be continuous so that laitance does not form at the top of a batch, causing weakness within the shaft. Laitance on top of the shaft on completion is inevitable. This laitance may be contaminated with water and silt expelled from around the casing as the concrete slumps outwards to fill the gap. Thus, the level of the concrete should be set high so that this weak laitance layer can be broken away before bonding the pile head onto its cap. The terms of the contract should make it clear whether or not this removal should be performed by the piling contractor.

The concrete in a pile shaft may be required to be terminated at some depth below ground level, for example, when constructing from ground surface level, piles designed to support a basement floor. It is a matter of some experience to judge the level at which the concrete should be terminated, and it is difficult to distinguish between fluid concrete and thick laitance when plumbing the level with a float. Where the piles are to support plunge columns, the casting level will be considerably lower than the piling platform; the concrete mix must be designed for an extended period of workability and maximum cohesion to reduce the need for removing a thick layer of laitance at basement level.

There is little guidance in either current standard specifications or BS EN 1536 on casting tolerances, but in general, it is better to leave finished pile heads high. The following suggestions by Fleming and Lane[3.24], while somewhat conservative for all conditions, are indicative:

Concrete cast under water +1.5 to +3 m
Concrete cast in dry uncased holes +75 to +300 mm
Concrete cast in cased holes, the greater of
    a.  +75 to +300 mm + (cased length)/15
    b.  +75 to + 300 mm + [(depth to casting level – 900 mm)/10]

The important criterion is that when the pile is trimmed to the required cut-off level, a sound connection can be made with the pile cap. The tolerance for this construction joint is from +40 to –70 mm. The reinforcement in the section to be removed may be debonded from the concrete. Trimming of the pile must be carried out using a suitable breaking method which avoids causing damaging vibration. There are several alternatives to the handheld pneumatic hammers such as hydraulic croppers and breakers and 'hydrodemolition' as described in the guidance issued by the Federation of Piling Specialists[3.25].

The use of a permanent casing in the form of a light-gauge metal sleeve surrounding a pile shaft in soft clays or peats was described in Section 2.4.2. This sleeving cannot be used within a temporary lining tube where the latter has to be withdrawn in a long length by means of a vibrator or by jacking. This involves the risk of distortion or jamming of the sleeve, which is then lifted while raising the temporary tube with disastrous effects on the concrete in the pile shaft. The sleeve can be used within an outer temporary liner where the depth of soft clay is shallow, and it can be used in conjunction with a casing oscillator which keeps the outer tube free of any jamming by the sleeve. There are no problems of using the light-gauge sleeve where power auger drilling can be performed to produce a stable hole without employing a temporary outer lining tube.

Unfortunately, defects in a pile shaft of the type shown in Figure 3.43 are by no means uncommon, even when placing a workable concrete in the dry open hole of a large-diameter bored pile. Defects can take the form of large unfilled voids or pockets of clay and silt in the concrete. Some causes of these defects are listed as follows:

1. Encrustations of hardened concrete or soil on the inside of the lining tubes can cause the concrete to be lifted as the tubes are withdrawn, thus forming gaps in the concrete. *Remedy*: The tubes must be cleaned before they are lowered down the borehole.
2. The falling concrete may arch and jam across the lining tube or between the tubes and the reinforcement. *Remedy*: Use a concrete of sufficient workability to slump easily down the hole and fill all voids. Ensure the concrete chute or hopper is centrally placed. Consider tamping or vibration.

*Figure 3.43* Defective shaft of bored pile caused by cement being washed out of unset concrete.

3. The falling concrete may jam between the reinforcing bars and not flow outwards to the walls of the borehole. *Remedy*: Ensure a generous space between the reinforcing bars (between 80 and 100 mm depending on concrete aggregate size and workability). The cage should be stiff enough to prevent it twisting or buckling during handling and subsequent placing of concrete. Widely spaced stiff hoops are preferable to helical binding, particularly in tension piles with a large amount of main reinforcement. Check that the bars have not moved together before the cage is lowered down the hole.

4. Lumps of clay may fall from the walls of the borehole or lining tubes into the concrete as it is being placed. *Remedy*: Always use lining tubes if the soil around the borehole is potentially unstable, and do not withdraw them prematurely. Ensure that adhering lumps of clay are cleaned off the tubes before they are inserted and after completing drilling.

5. Soft or loose soils may squeeze into the pile shaft from beneath the base of the lining tubes as they are withdrawn, forming a *waisted* or *necked* shaft. *Remedy*: Ensure sufficient head of concrete in casing (but not so high that when removing the casing it will lift the concrete, important in small diameter piles). Check the volume of concrete placed against the theoretical volume and take remedial action (removal and replacement of the concrete) if there is a significant discrepancy.

6. If bentonite has been used for support, the hydrostatic pressure of the bentonite in the annulus, which is disturbed on lifting the casing, may be higher than that of the fluid concrete, thus causing the bentonite to flow into the concrete. This is a serious defect and is difficult to detect. It is particularly liable to happen if the concrete is terminated at some depth below the top of the casing. *Remedy*: Keep a careful watch on the level and density of the bentonite gel when the casing is lifted. Watch for any changes in level of the concrete surface and for the appearance of bentonite within the concrete. If inflow of the bentonite has occurred, the defective concrete must be removed and replaced, and the slurry support technique must be abandoned.

7. Infiltration of groundwater may cause gaps or honeycombing of the concrete. *Remedy*: Adopt the techniques for dealing with groundwater in pile boreholes described in Section 3.4.8.

## 3.4.7 Installing continuous flight auger piles

CFA piles can be installed in a variety of soils, dry or waterlogged, loose or cohesive, and through weak rock. The soil is loosened on insertion of the auger, and the borehole walls are supported by the auger flights filled with drill cuttings; bentonite support slurry is not used. The pile is concreted through a bottom or side exit at the tip of the hollow stem auger (100 or 127 mm bore) using a concrete pump connected by hose to a swivel on the rotary head as the auger is slowly rotated and withdrawn. Soil is brought to the surface on the auger blades. The concrete flow rate and feed pressure are continuously measured at the tip; reinforcement is pushed or vibrated into the fresh concrete.

The main problems with CFA pile construction are overflighting and polishing (see Section 2.4.2), too rapid withdrawal of the auger initially causing reduction in end-bearing capacity, and too rapid withdrawal when nearing the top of pile causing contamination with soil. In order to avoid these problems, reliable instrumentation in the operator's cab showing auger rotation, injection pressure and volume injected in real time and experienced operators are essential.

For rotary displacement auger piles, the displacement tool, which is mounted at the bottom of a drill tube, is rotated by the high-torque top drive and forced into the ground by the rig crowd, thereby compacting the wall of the hole. To form the various types of screw piles, discussed in Section 2.3.5, the thick-flanged continuous auger is screwed into the ground with limited crowd applied, although for less cohesive soil, more thrust will be necessary to reach the required depth. The auger is rotated out of the hole as concrete is pumped through the tip to fill the helical profile of the pile, with only minimal soil being brought to the surface. As with CFA piles, the rig must be instrumented to ensure auger extraction and concreting are compatible with the formation of the required pile profile.

### 3.4.8 Concreting pile shafts under water

Groundwater in pile boreholes can cause serious difficulties when placing concrete in the shaft. As noted in Section 3.4.4, an inflow of only a few centimetres deep in, say, 5 min which has trickled down behind the lining tubes or has seeped into the pile base can be readily dealt with by baling or pumping it out and then placing dry concrete to seal the base against any further inflow. However, larger flows can cause progressive increases in the water content of the concrete, weakening it and forming excess laitance.

A strong flow can even wash away the concrete completely. The defective piles shown in Figure 3.43 were caused by the flow of water under an artesian head from a fissured rock on which the bored piles were bearing after the boreholes had been drilled through a soft clay overburden. The lined boreholes were pumped dry of water before the concrete was placed, but the subsequent 'make' of water was sufficiently strong to wash away some of the cement before the concrete has set. The remedial action in this case was to place dry concrete in bags at the base of the pile borehole and then to drive precast concrete sections into the bags.

In all cases of strong inflow, the water must be allowed to rise to its normal rest level and topped up to at least 1.0 m above this level to stabilise the pile base. BS EN 1536 requires that a tremie pipe be used for concreting in submerged conditions (water or support fluid). The maximum OD of the tremie pipe should be less than 0.35 times the pile diameter or the inner diameter of the casing or 0.6 times the inner width of the reinforcement cage. Consideration must also be given to matching the tremie internal diameter with the size of aggregate – six times the maximum size of aggregate or 150 mm whichever is the greater. The tremie pipe must be cleaned and lowered to the bottom of the pile and lifted slightly to start concrete flow. A flap valve should be used on the end of the tremie pipe rather than a plug or polyethylene 'go-devil'. During concreting, the tremie tip must always be immersed in the concrete: 1.50 m below concrete surface for piles less than 1200 mm diameter and 2.50 m for piles greater than 1200 mm. If immersion is lost during concreting, special precautions are required before placement can continue; for example, steps must be taken to re-immerse the tremie so that any contamination will be above the final cut-off level. The tremie should be fed by a concrete pump as a surface hopper is unlikely to provide sufficient differential head.

A bottom-opening bucket should not be used instead of a tremie pipe for placing concrete in pile boreholes, even large-diameter shafts. This is because the crane operator handling the bucket cannot tell, by the behaviour of the crane rope, whether or not he has lowered the bucket to the correct level into the fluid concrete before he releases the hinged flap. There may be a case for using the bucket method in special conditions in marine piling, but generally the tremie must be preferred.

The procedure for drilling pile boreholes with support by a bentonite slurry or polymer fluid is described in Section 3.3.8. In both cases, concrete must be placed using a tremie as described earlier, with sufficient hydrostatic pressure of concrete in the pipe above bentonite level to overcome the external head of the slurry and the friction in the tremie pipe. Where the slurry becomes flocculated and heavily charged with sand (i.e. has a density greater than 1300–1400 kg/m³), it should be replaced by a lighter mud before placing the concrete.

Sometimes a dispersing agent is added to the bentonite to break down the gel before placing the concrete. These measures will not deal with a thick filter cake on the sides and base of the pile, and it should be removed mechanically as the upward flow of concrete is unlikely to scour the sides completely to ensure optimum concrete–soil contact and maintain concrete cover to reinforcement. To minimise restriction to upward flow of concrete, circumferential steel should be kept to a minimum. Concrete mixes are designed with plasticisers and retarders to ensure appropriate flow characteristics ($200 \pm 20$ mm slump) and avoid segregation. Caution is required when designing slurry and concrete mixes for use in high ground temperatures to avoid jamming in the tremie pipe. The use of synthetic polymer support fluids produces only limited (or no) cake on the bore sides, and the tremie concreting is effective in displacing the polymer. Cleaning of coarse particles from the base prior to concreting is essential.

### 3.4.9 Installation of bored and cast-in-place piles by grabbing, vibratory and reverse-circulation rigs

The use of either grabbing, vibratory, and reverse-circulation machines for drilling pile boreholes can involve continuous support by lining tubes which may or may not be withdrawn after placing the concrete. In all three methods, the tubes may have to follow closely behind the drilling in order to prevent the collapse of the sides and the consequent weakening of shaft friction. For reverse circulation, the boreholes must be kept topped up with fluid to provide the flushing medium. In other cases, this is necessary to avoid *blowing* of the pile bottom due to upward flow of the groundwater and when drilling through water-bearing sand layers interbedded with impervious clays.

Grabbing in weak rocks can cause large accumulations of slurry in the boreholes which make it difficult to assess whether the required termination level of the pile in sound rock has been achieved. The slurry should be removed from time to time by baling or by airlift pump with a final cleaning-up before placing the concrete.

The techniques of placing concrete are the same as described in Sections 3.4.6 and 3.4.8.

### 3.4.10 Installation of bored and cast-in-place piles by tripod rigs

When boring in stiff clay, water should not be poured down the hole to assist in advancing the bore or used to aid removal of the clay from the cutter as this causes a reduction in shaft friction. When drilling in granular soils, the lining tubes should follow closely behind the drilling to avoid overbreak, and the addition of water may be needed to prevent *blowing* and to facilitate the operation of the baler or shell. Piles drilled by tripod rigs are relatively small in diameter, requiring extra care when placing the concrete as this is more likely to jam in the casing tubes when they are lifted.

### 3.4.11 Installation of raking piles

BS EN 1536 (Clause 8.2.3) states that pile bores, whether drilled or driven, should be cased throughout their length if the rake is flatter than 1 horizontal to 15 vertical unless it can be shown that an uncased pile bore will be stable. Similarly, stabilising fluids should not be used if the rake is flatter than 1 in 15 unless precautions are taken when inserting casing and concreting.

The advantages of raking piles in resisting lateral loads are noted in Chapters 6 and 8. However, the installation of such piles may result in considerable practical difficulties, and they should not be employed without first considering the method of installation and the ground conditions. If the soil strata are such that the piles can be driven to the full penetration depth without the need to drill out a soil plug or to use jetting to aid driving, then it should be feasible to adopt raking piles up to a maximum rake of 1 to 2. However, the efficiency of the hammer is reduced due to the friction of the ram in the guides. It may therefore be necessary to use a more powerful hammer than that required for driving vertical piles to the same penetration depth with implications for stresses in the pile head. Casing oscillators are available from major manufacturers which can operate on a modest rake to assist casing insertion.

The vertical load caused by the inclined pile and hammer on the leaders of the piling frame must be taken into consideration. It is not usual to drive raking piles in guides without the use of leaders, as the bending stresses caused by the weight of the hammer on the upper end of the pile must be added to the driving stresses and a check should be made to ensure that the combined stresses are within allowable limits.

The principal difficulties arise when it is necessary to drill ahead of a driven, open-ended raked pile to clear boulders or other obstructions, using the methods described in Section 3.3.5. When the drill penetrates below the shoe of the pile tube, it tends to drop by gravity and it is then likely to foul the shoe as it is pulled out to resume further driving. Similarly, under-reaming tools are liable to be jammed as they are withdrawn. The risks of fouling the drilling tool are less if the angle of rake is small (say 1 in 10, 84° or more) and the drill string is adequately centralised within the piling tube. However, the drill must not be allowed to penetrate deeply below the toe of the pile. This results in frequent alternations of drilling and driving with consequent delays as the hammer is taken off to enter the drill, followed by delays in entering and coupling up the drill string and then removing it before replacing the hammer.

Difficulties also arise when installing driven and cast-in-place piles by means of an internal drop hammer, due to the friction of the hammer on the inside face of the driving tube. Installers of these piles state that a rake not flatter than 1 in 3.7 (75°) is possible.

Power augers operating on self-erecting leader rigs as shown in Figure 3.2 are capable of drilling open bores at rakes up to 1 in 1 exceptionally. Rakes of 1 in 2 are feasible in good soil conditions, but to satisfy BS EN 1536 tolerance limits, casing is necessary to support the pile borehole. A drill mast rigged with a dual-drive head which can bore and case simultaneously should avoid the difficulties of jamming of the drill tool under the toe of the casing. Rotary-percussive drills which also drill and case simultaneously are useful in these conditions.

Problems can occur when placing concrete in raking piles. Internal ramming is not reliable as the rammer catches on the reinforcing cage. High-slump concrete should be pumped through a tremie pipe, with special precautions being taken to prevent the reinforcement being lifted with the lining tubes.

The American Concrete Institute[6.12] recommends using an over-sanded mix for placing concrete in raking pile shells or tubes. A concrete mix containing 475 kg/m$^3$ of coarse aggregate with a corresponding increase in cement and sand to give a slump of 100 mm is recommended. This mix can be pumped down the raking tube.

### 3.4.12 Withdrawal of temporary casings

The withdrawal or extraction of temporary casings is a feature of many of the piling methods covered earlier and in Chapter 2 and must always be undertaken with care. The Federation of Piling Specialists has produced *Notes for Guidance*[3.26] on this matter detailing the potential factors which have to be considered, assessment of the extraction load and the method of extraction, whether by the rig pull-out system, vibrator or crane.

### 3.4.13 Positional tolerances

It is impossible to install a pile, whether by driving, drilling or jacking, so that the head of the completed pile is always exactly in the intended position or that the axis of the pile is truly vertical or at the specified rake. Driven piles tend to move out of alignment during installation due to obstructions in the ground or the tilting of the piling frame leaders. Driving piles in groups can cause horizontal ground movements which deflect the piles. (Note the marker pins for piles in a group may also be displaced by driving adjacent piles). In the case of bored piles, the auger can wander from the true position, or the drilling rig may tilt due to the wheels or tracks sinking into a poorly prepared platform. However, controlling the positions of piles is necessary since misalignment affects the design of pile caps and ground beams (see Sections 7.8 and 7.9), and deviations from alignment may cause interference between adjacent piles in a group or dangerous concentrations of load at the toe. Accordingly, execution codes specify tolerances in the position of pile heads or deviations from the vertical or intended rake. If these are exceeded, action is necessary either to redesign the pile caps or to install additional piles to maintain the design loads. The higher tolerances for raked piles reflect the potential problems of maintaining alignment, particularly in soft soils at the pile head and when the use of long leaders is necessary. The significance of positional tolerance to piling beneath deep basements is noted in Section 5.9.

Some codes of practice requirements are as follows:

*BS EN 1536*: Plan location tolerances are given in Clause 8.1, 100 mm for pile diameters of vertical and raking bored piles less than 1000 mm, 0.1× diameter (or width) for piles between 1000 and 1500 mm and 150 mm for piles greater than 1500 mm. Deviation in inclination of vertical bored piles and bored piles designed for a rake less than 1 in 15 (86°) is limited to 20 mm/m run of pile. For piles designed with a rake of between 1 in 4 and 1 in 15, the deviation is limited to 40 mm/m.

*BS EN 12699*: The plan location tolerance (at working level) given in Clause 7.3 for vertical and raking displacement piles is 100 mm. Deviation for vertical and raking piles is 40 mm/m. The deviations in this code must be taken into account in the design.

*BS EN 14199*: The plan location tolerance (at working level) given in Annex B for micropiles is 50 mm. Deviation from the axis varies from 2% of the length for vertical piles to 6% for inclined piles. Radius of curvature should be 200 m depending on buckling conditions. These BS EN codes allow other tolerances to be specified.

*BS 6349-2 Clause 8.13*: A deviation of up to 1 in 100 is permitted for vertical piles driven in sheltered waters or up to 1 in 75 for exposed sites. The deviation for raking piles should not exceed 1 in 30 from the specified rake for sheltered waters or 1 in 25 for exposed sites. The centre of piles at the junction with the superstructure should be within 75 mm for piles driven on land or in sheltered waters. Where piles are driven through rubble slopes, the code permits a positional tolerance of up to 100 mm, and for access trestles and jetty heads, a tolerance of 75–150 mm is allowed depending on the exposure conditions.

*Institution of Civil Engineers*[2.5]: Plan position – maximum deviation of centre point of pile to be not more than 75 mm in any direction, but additional tolerance allowed for raking piles with cut-off below ground level. Verticality – maximum deviation of finished pile from the vertical is 1 in 75 at any level. Maximum deviation of finished pile from the specified rake is 1 in 25 for piles raking up to 1:6 and 1 in 15 for piles raking more than 1:6. The preceding limits apply to bearing piles and may be varied in the project specification, subject to design implications of this action. Other more stringent tolerances are specified for secant and contiguous piles in retaining walls. Note these tolerances are different from those given in the BS ENs stated earlier, which also allow for variations in the project specification.

*American Concrete Institute Recommendations*[6.12]: The position of the pile head is to be within 75–150 mm for the normal usage of piles beneath a structural slab. The axis may deviate by up to 10% of the pile length for completely embedded vertical piles or for all raking piles, provided the pile axis is driven straight. For vertical piles extending above the ground surface, the maximum deviation is 2% of the pile length, except that 4% can be permitted if the resulting horizontal load can be taken by the pile-cap structure. For bent piles, the allowable deviation is 2%–4% of the pile length depending on the soil conditions and the type of bend (e.g. sharp or gentle). Severely bent piles must be evaluated by soil mechanics calculations or checked by loading tests.

## 3.5 CONSTRUCTING PILES IN GROUPS

So far, only the installation of single piles has been discussed. The construction of groups of piles can have cumulative effects on the ground within and surrounding the pile group. These effects are occasionally beneficial (as in reticulated minipile groups) but more frequently have deleterious effects on the load/settlement characteristics of the piles and can damage surrounding property. Precautions can be taken against these effects by the installation methods and sequence of construction adopted. BS EN 1536 Clause 8.2.1.12 stipulates that the centre-to-centre distance of bored piles should be greater than four times the pile width with a minimum of 2 m, where adjacent piles are less than 4 h old. The distance for driven cast-in-place piles with withdrawable tubes is increased to six times the diameter in BS EN 12699. Because the problems are more directly concerned with the bearing capacity and settlement of the group as a whole, rather than with the installation of the piles, they are discussed in Sections 5.7 through 5.9.

## REFERENCES

3.1   Geddes, W.G.N., Sturrock, K.R. and Kinder, G. New shipbuilding dock at Belfast for Harland and Wolff Ltd, *Proceedings of the Institution of Civil Engineers*, 51, 1972, 17–47.
3.2   Holeyman, A., Vanden Berghe, J.-F. and Charue, N. *Vibratory Pile Driving and Deep Soil Compaction*. Swets and Zeitlinger, Lisse, the Netherlands, 2002, pp. 233–234.
3.3   Rodger, A.A. and Littlejohn, G.A. Study of vibratory driving in granular soils, *Geotechnique*, 30 (3), 1980, 269–293.
3.4   Viking, K. The vibratory pile installation technique, Pile driver, Pile Driving Contractors Association, Orange Park, FL, Spring 2005, pp. 27–35.
3.5   Willis, A.J. and Churcher, D.W. How much noise do you make? A guide to assessing and managing noise on construction sites, Construction Industry Research and Information Association (CIRIA). Report PR 70, London, UK, 1999.
3.6   Weltman, A.J. (ed.). Noise and vibration from piling operations, Construction Industry Research and Information Association (CIRIA). Report PSA/CIRIA PG7, London, UK, 1980.
3.7   Building Research Establishment. Damage to structures from ground-borne vibration, *BRE Digest* 403, 1995.
3.8   Nedwell, J., Langworthy, J. and Howell, D. Assessment of sub-sea acoustic noise and vibration from offshore wind turbines and its impact on marine wildlife: Initial measurements of underwater noise during construction of offshore wind farms and comparison with background noise. COWRIE Report 544 R 0424, The Crown Estate, London, UK, 2006.
3.9   Gerwick, B.C. *Construction of Offshore Structures*, 3rd ed. Taylor & Francis Group, Boca Raton, FL, 2007.
3.10  Golightly, C. Tilting of monopiles. Long, heavy and stiff; pushed beyond their limits, *Ground Engineering*, 47 (1), 2014, 20–23.

3.11 *Bentonite Support Fluids in Civil Engineering*, 2nd ed. Federation of Piling Specialists, London, UK, 2006.

3.12 Reese, L.C., O'Neill, M.W. and Touma, F.T. Bored piles installed by slurry displacement, *Proceedings of the Eighth International Conference, ISSMFE*, Moscow, Russia, Vol. 2.1, 1973, pp. 203–209.

3.13 Lam, C., Troughton, V., Jefferis, S. and Suckling, T. Effect of support fluids on pile performance – A field trial in east London, *Ground Engineering*, 43 (10), 2010, 28–31.

3.14 Bolognesi, A.J.L. and Moretto, O. Stage grouting preloading of large piles in sand, *Proceedings of the Eighth International Conference, ISSMFE*, Moscow, Russia, Vol. 2.1, 1973, pp. 19–25.

3.15 Yeats, J.A. and O'Riordan, N.J. The design and construction of large diameter base-grouted piles in Thanet Sand, London, *Proceedings of the International Conference on Piling and Deep Foundations*, Vol. 1, Eds Burland J.B. and Mitchell, J.M., Balkema, Rotterdam, the Netherlands, 1989, pp. 455–461.

3.16 Sherwood, D.E. and Mitchell, J.M. Base grouted piles in Thanet Sands, London, *Proceedings of the International Conference on Piling and Deep Foundations*, Vol. 1, Eds Burland J.B. and Mitchell, J.M., Balkema, Rotterdam, the Netherlands, 1989, pp. 463–472.

3.17 Littlechild, B.D., Plumbridge, G.D. and Free, M.W. Shaft grouted piles in sand and clay in Bangkok, *Deep Foundations Institute Seventh International Conference on Piling and Deep Foundations*, DFI, Englewood Cliffs, NJ, 1998, pp. 1.7.1–1.7.8.

3.18 Suckling, T.P. and Eager, D.C. Non base grouted piled foundations in Thanet sands for a project in East India Dock, London, *Underground Construction Symposium*, Brintex, London, UK, 2001, pp. 413–424.

3.19 Beadman, D., Pennington, M. and Sharratt, M. Pile test at the Shard London Bridge, *Ground Engineering*, 45 (1), 2012, 24–29.

3.20 Skinner, H. *Working Platforms for Tracked Plant*. BR470 Building Research Establishment, Watford, England, 2004.

3.21 Fountain, F. and Suckling, T. Reliability in the testing and assessing of piling work platforms, *Ground Engineering*, 45 (11), 2012, 29–31.

3.22 British Drilling Association. *Health and Safety Manual for Land Drilling – Code of Safe Drilling Practice*. BDA, Daventry, UK, 2002.

3.23 Healy, P.R. and Weltman, A.J. Survey of problems associated with the installation of displacement piles, Construction Industry Research and Information Association (CIRIA), Report PG8, London, UK, 1980.

3.24 Fleming, W.G.K. and Lane, P.F. Tolerance requirements and constructional problems in piling, *Proceedings of the Conference on the Behaviour of Piles*, Institution of Civil Engineers, London, UK, 1970, pp. 175–178.

3.25 *Breaking down of piles*. Federation of Piling Specialists, London, UK, 2008.

3.26 *Notes for Guidance on the Extraction of Temporary Casings and Temporary Piles within the Piling Industry*, 1st ed. Federation of Piling Specialists, London, UK, 2010.

# Chapter 4

# Calculating the resistance of piles to compressive loads

## 4.1 GENERAL CONSIDERATIONS

### 4.1.1 Basic approach to the calculation of pile resistance

The numerous types of pile and the diversity in their methods of installation have been described in Chapters 2 and 3. Each different type and installation method disturbs the ground surrounding the pile in a different way. The influence of this disturbance on the shaft friction and end-bearing resistance of piles has been briefly mentioned (see Section 1.3). This influence can improve or reduce the bearing capacity of the piles, and thus a thorough understanding of how the piles are constructed is essential to the formulation of a practical method of calculating load-carrying capacity.

The basic approach used in this chapter to calculate the resistance of piles to compressive loads is the 'static' or soil mechanics approach as opposed to the use of dynamic formulae. Over the years, much attention has been given by research workers to calculation methods based on pure soil mechanics theory. But it was realised that in order to determine the interface friction on a pile shaft under load, the postulated simple relationships between the coefficient of earth pressure 'at rest', the effective overburden pressure and the drained angle of shearing resistance of the soil had to be modified by factors to take account of the installation method. The application of the undisturbed shearing resistance of the soil surrounding the pile toe to calculate the end-bearing resistance of a pile was also considered by the researchers in classical soil mechanics terms. The importance of the settlement of a pile or pile groups at the applied load was recognised as an important factor in the design, and calculations were developed based on elastic theory, taking into account the transfer of load in shaft friction from the pile to the soil.

The research into the behaviour of the two main pile groups, namely, driven and bored piles subjected to full-scale instrumented load tests, demonstrated the fundamental departures from classical soil mechanics theory and the all-important effects of installation procedures on pile behaviour were recognised such as the highly complex conditions which develop at the soil–pile interface and which are often quite unrelated to the original undisturbed state of the soil or even to the fully remoulded state. The pore-water pressures surrounding the pile can vary widely over periods of hours, days, months or years after installation, such that the simple relationships of shaft friction to effective overburden pressure are unrealistic. Similarly, when considering deformations of a pile group under its applied load, any calculations of the transfer of load that are based on elastic theory which do not take account of soil disturbance for several diameters around the pile shaft and beneath the toe are also unrealistic.

Hence, while the calculation of pile carrying capacity is based on soil mechanics considerations, the approach is empirical, relating known pile behaviour to simple soil properties such as relative density and undrained shearing strength. These can be regarded as properties to which empirical coefficients can be applied to arrive at unit characteristic values for the shaft friction and end-bearing resistances.

The long-term observations of full-scale pile loading tests revealed the complexities of the problems and have shown that there is no simple fundamental design method. The empirical or semi-empirical methods set out in this chapter have been proved by experience and load testing to be reliable for practical design of light to moderately heavy loadings on land-based or near-shore marine structures. These methods are also the basis of many computer-aided design methods for routine pile design. Special considerations using more complex design methods are required for heavily loaded offshore structures in deep water as described by Randolph and Gourvenec[8.1].

The designer is often presented with inadequate information on the soil properties. Until the introduction of the Eurocode procedures for pile design, a decision had to be made whether to base designs on conservative material values with an appropriate global safety factor without any check by load testing or to use the design methods to give a preliminary guide to pile diameter and length and then base the final designs on an extensive field testing programme with loading tests to failure. The use of partial factors on loads and materials and the definitions of characteristic material strengths under Eurocode 7 rules have formalised the decision-making process to a degree. The design must still be verified either by comparing loading tests to failure in similar conditions or by project-specific load testing – always justified on a large-scale piling project to produce economic designs. Proof-load testing as a means of checking workmanship is a separate consideration (Section 11.4).

Where the effective overburden pressure is an important parameter for calculating the bearing capacity of piles (as is the case for coarse-grained soils), account must be taken of the unfavourable effects of a rise in groundwater levels. This may be local or may be a general rise, due, for example, to seasonal flooding of a major river or a long-term effect such as the ongoing general rise in groundwater levels in Greater London.

## 4.1.2 Behaviour of a pile under load

For practical design purposes, engineers must base their calculations of pile capacity on the application of the load at a relatively short time after installation. The reliability of these calculations is assessed by a loading test which is again made at a relatively short time after installation. However, the effects of time on pile capacity must be appreciated, and these are discussed in Sections 4.2.4 and 4.3.8.

When a pile is subjected to a progressively increasing compressive load at a rapid or moderately rapid rate of application, the resulting load/settlement curve is as shown in Figure 4.1. Initially, the soil–pile system behaves elastically. There is a straight-line relationship up to some point A on the curve, and if the load is released at any stage up to this point, the pile head will rebound to its original level. When the load is increased beyond point A, there is yielding at, or close to, the soil–pile interface, and slippage occurs until point B is reached, when the maximum shaft friction on the pile shaft will have been mobilised. If the load is released at this stage, the pile head will rebound to point C, the amount of 'permanent set' being the distance OC. The movement required to mobilise the maximum shaft friction is quite small and is only of the order of 0.3%–1% of the pile diameter. The base resistance of the pile requires a greater downward movement for its full mobilisation, and the amount of movement depends on the diameter of the pile. It may be

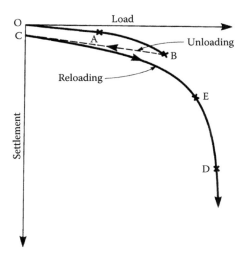

*Figure 4.1* Load/settlement curve for compressive load to failure on pile.

in the range of 10%–20% of the base diameter. When the stage of full mobilisation of the base resistance is reached (point D in Figure 4.1), the pile plunges downwards without any further increase of load, or small increases in load produce increasingly large settlements (a 'plunging failure').

If strain gauges are installed at various points along the pile shaft from which the compressive load in the pile can be deduced at each level, the diagrams illustrated in Figure 4.2 are obtained, which show the transfer of load from the pile to the soil at each stage of loading shown in Figure 4.1. Thus, when loaded to point A virtually, the whole of the load is carried by friction on the pile shaft, and there is little or no transfer of load to the toe of the pile (Figure 4.2). When the load reaches point B, the pile shaft is carrying its maximum frictional resistance and the pile toe will be carrying some load. At Point D, there is no further increase in the load transferred in friction, but the base load will have reached its ultimate value.

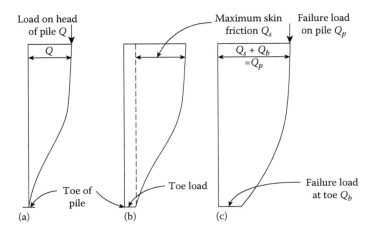

*Figure 4.2* Load transfer from head of pile to shaft at points A, B and D on load/settlement curve in Figure 4.1. (a) Load on pile shaft. (b) Maximum load on pile shaft. (c) Failure of pile base.

### 4.1.3 Determining allowable loads on piles using allowable stress methods

The loading corresponding to point D on the load/settlement curve in Figure 4.1 represents the ultimate resistance of the pile and is defined as the stage at which there is general shear failure of the soil or rock beneath the pile toe. However, this stage is of academic interest to the structural designer. A piled foundation has failed in its engineering function when the relative settlement between adjacent single piles or groups of piles causes intolerable distortion of the structural framework or damage to claddings and finishes. This stage may be represented by some point such as E on the load/settlement curve (Figure 4.1). Thus, structural failure will have occurred at a load lower than the ultimate resistance of the pile. Various criteria of assessing failure loads on piles from the results of loading tests are listed in Section 11.4.

The concept of the separate evaluation of shaft friction and base resistance forms the basis of all *static* calculations of pile bearing capacity. The basic equation is

$$Q_p = Q_b + Q_s - W_p \tag{4.1}$$

where
$Q_p$ is the ultimate resistance of the pile
$Q_b$ is the ultimate resistance of the base
$Q_s$ is the ultimate resistance of the shaft
$W_p$ is the net weight of the pile (i.e. the weight of the pile less the weight of soil displaced)

The components $Q_s$ and $Q_b$ of the failure load $Q_p$ are shown at the final loading stage in Figure 4.2. Usually, $W_p$ is small in relation to $Q_p$ and this term is generally ignored. However, it is necessary to provide for $W_p$ in such situations as piles in marine structures in deep water where a considerable length of shaft extends above seabed.

Allowable stress methods were applied in BS 8004, Foundations – now withdrawn. Here, the actual dead load of a structure and the most unfavourable combination of imposed loads were assumed to be applied to the ground. The foundation was assumed to be safe if the allowable stress on the soil or rock was not exceeded, taking into account the likely variable strength or stiffness properties of the ground and the effect of a varying groundwater level. In the case of piled foundations, uncertainty in the reliability of the calculation method was also taken into account. Because of the difficulty in predicting failure loads, the safety factors used to obtain the allowable load on a single pile from the calculated ultimate load were correspondingly high in order to cover a variety of uncertainties:

1. To provide for natural variations in the strength and compressibility of the soil
2. To provide for uncertainties in the calculation method used
3. To ensure that the design resistance of the material forming the pile shaft is within safe limits
4. To ensure that the total settlement(s) of the single isolated pile or the group of piles are within tolerable limits
5. To ensure that the differential settlements between adjacent piles or within groups of piles are within tolerable limits

As a result, for pile design to BS 8004, a global safety factor between 2 and 3 was generally adopted. Experience of a large number of loading tests on piles of diameter up to 600 mm taken to failure, both in sands and in clays, showed that if safety factor of

2.5 is taken on the ultimate resistance, then the settlement of the pile head at the applied load is unlikely to exceed 10 mm. For piles of diameters up to about 1000 mm, failure or ultimate loads as determined by loading tests were usually assumed to be the loads causing a pile head settlement of 10% of the base diameter. Eurocode EC7 retains this failure criterion at Clause 7.6.1.1(3). The Institution of Civil Engineers' Specification for Piling and Embedded Retaining Walls[2.5] (referred to as SPERW) further defines the ultimate capacity of a pile as 'the maximum load which can be applied achieving the specified settlement rate criteria' derived from a preliminary pile test. It also comments that the ultimate capacity is the 'maximum resistance offered by the pile when the strength of the soil is fully mobilised'.

When using allowable stress methods for piles in groups, it was accepted that a structure can suffer excessive distortion caused by group settlement long before an individual pile in the group has failed in bearing resistance. Hence, a separate calculation is made of group settlement based on a realistic assessment of dead load and the most favourable or unfavourable combinations of imposed loading, using unfactored values of the compressibility of the ground in the zone influenced by the group loading (see Chapter 5).

Where piles are end bearing on a strong intact rock, the concept of a global safety factor against ultimate failure is not appropriate, since it is likely that the pile itself will fail as a structural unit before shearing failure of the rock beneath the pile toe occurs. The applied loads are then governed by the safe working stress in compression and bending on the pile shaft and the settlement of the pile due to elastic deformation and creep in the rock beneath the base of the pile, together with the elastic compression of the pile shaft.

As described Section 4.1.4, Eurocode procedures abandon allowable stress design and present a unified set of limit state design principles for all structural design which avoid the problem of blurring allowable stresses and limit states, as occurred when designing foundations using BS 8004 (allowable stress for foundations) and BS 8110 and BS 5950 (both limit state codes for concrete and steel design, respectively, but now withdrawn). More precise identification of geotechnical material parameters is now required so that global factors of safety are not needed to cover the gathered-together uncertainties of loadings and strengths. The Eurocode limit state methodology makes foundation design compatible with the superstructure design.

## 4.1.4 Determining design loads and resistances in compression using the procedure in Eurocode BS EN 1997-1:2004 Geotechnical design

This account of design procedures adopted in this Eurocode (referred to as Eurocode 7 or EC7)[1.2] is only a brief review of a lengthy document containing many provisos, exceptions and cross-references to other Eurocodes referred to in Section 1.5. Several guides are available[1.3–1.5] to assist in the interpretation and application of EC7, and the text and worked examples given in this edition generally follow the procedures which were instituted with the initial adoption of the Eurocodes in the United Kingdom. The selection and application of partial factors for loads and resistances is the main issue to be addressed when using EC7 procedures, but in order to produce compliant designs, the designer must study the whole suite of documents.

The partial factors provided in EC7 have to cover the same uncertainties and variations which were used to decide the global safety factor approach in allowable stress design as noted earlier. When the factors are applied, the Eurocodes require a structure, including the foundations, not to fail to satisfy its design performance criteria as a result of exceeding various limit states:

The *ultimate limit state* (ULS) can occur under the following conditions:

1. Loss of equilibrium of the structure and the ground considered as a rigid body in which the strengths of the structural materials and the ground are insignificant in providing resistance (State EQU).
2. Internal failure or excessive deformation of a structure and its foundation (State STR).
3. Failure or excessive deformation of the ground in which the strengths of the soil or rock are significant in providing resistance (State GEO).
4. Loss of equilibrium of a structure due to uplift by water pressure or other vertical actions (State UPL).
5. Hydraulic heave, internal erosion and piping caused by hydraulic gradients (State HYD). State EQU could occur when a structure collapses due to a landslide or earthquake. This state is not considered further in this chapter. Design against occurrence of the other states listed earlier involves applying partial factors to the applied loads (actions) and to the ground resistance to ensure that reaching these states is highly improbable.

*Serviceability limit states* (SLSs) are concerned with ensuring that the deformations of a structure due to ground movements below the foundations do not reduce the useful life of the structure, do not cause discomfort to people or cause damage to finishes, non-structural elements, machinery or other installations in the structure.

Eurocodes require structures and their foundations to have sufficient *durability* to resist weakening from attack by substances in the ground or the environment.

As a preliminary, EC7 requires the structure to be considered in three categories of risk from the foundation aspect. Geotechnical Category 1 covers structures having negligible risk of failure or damage due to ground movements or where enough is known about the ground conditions to adopt a routine method of design, provided that there are no risk problems associated with excavation below groundwater level.

Category 2 includes conventional structures and their foundations with no exceptional risk or difficult ground or loading conditions. Structures requiring piling come into this category provided that there are adequate geotechnical data based on routine methods of ground investigation.

Category 3 applies to all categories not coming within the scope of 1 and 2. It includes very large or unusual structures and those involving abnormal risks or exceptionally difficult ground or loading conditions and also structures in highly seismic areas and areas of site instability. EC7 (Clause 2.2) lists 15 geological and environmental features which need to be considered generally in foundation design. All of these are relevant to piled foundations for which the code prescribes three basic approaches to design:

1. Empirical or analytical calculations
2. Static load tests
3. Dynamic load tests

*Geotechnical design by calculation* should be in accordance with BS EN 1990:2002, 'Basis of structural design', as for all structural design. It is emphasised that the quality of the information on the ground conditions is more significant than precision in calculation models and the partial factors employed. Accordingly, it is essential that the field operations and laboratory testing techniques should be undertaken in a thorough manner with the appropriate standard of quality (as EC7-2; see Section 11.1). Also the interaction between the structure and the ground should be considered to ensure that the strains in the structure are compatible with the ground movements resulting from the applied loading. Pile design by calculation, the preferred method in the United Kingdom, concentrates on avoiding ULS.

*Ground properties* are required to be obtained from field or laboratory tests, either directly or by correlation, theory or empiricism. The effects of time, stress level and deformation on the properties are to be taken into account.

*Characteristic values* of geotechnical parameters are selected as part of the design process from the available information, usually in the form of a site-specific ground investigation report. EC7 Clause 2.4.5.2(2)P requires that a 'cautious estimate' of the data be made within the zone influenced by stresses transmitted to the ground (including the zones beneath a pile group as shown in Figure 5.19) and must reflect the limit state being considered. The selected values may be lower ones which are less than the most probable ones (e.g. to estimate end-bearing resistance) or an upper range of values higher than the most probable ones. The latter selection would apply where high values have an unfavourable effect on foundation behaviour, for example when considering downdrag on piles or differential settlement. Statistical evaluation[1.3] of geotechnical data is permitted by EC7, but it is essential that different geologies are analysed separately. In practice, little difference is seen between the characteristic values for EC7 designs and those selected by engineering judgement when using the allowable stress calculations and global safety factors. In all cases, when selecting the characteristic parameter, it is essential to review case histories and local experience.

BS EN 1990 defines the *actions* on the foundations comprising *structural* actions, that is the loads transmitted from a structure directly to the pile head or through a raft, and *geotechnical* actions, which have to be assessed separately. Geotechnical actions (as listed in EC7 Clause 2.4.2(4)) include earth and groundwater pressures and ground movements such as soil swelling and shrinkage, frost action and downdrag. Duration of actions such as repetitive loading and time effects on soil drainage and compressibility have to be considered. Geotechnical actions can also occur from transversely applied loads such as those on piles supporting bridge abutments caused by surcharge from the adjacent approach embankments.

In Clause 7.3.2.1(3)P of EC7, the evaluation of geotechnical actions has to be determined in one of two ways:

a. By *soil–pile interaction* analyses when the degree of relative soil–pile movement is estimated and $t$–$z$ curves are produced by computer to give the corresponding strains and axial forces in the pile shaft (Section 4.6). In the case of transversely applied actions, a $p$–$y$ analysis is performed (Section 6.3.5). Alternatively, actions can be estimated from other forms of analysis, such as finite element analysis as summarised in Section 4.9.

b. By an *upper-bound force* exerted on the pile by the ground movement, calculated and treated as an action.

Method (b) when applied to actions resulting from downdrag can give over-conservative designs if due consideration is not given to variations in frictional forces over the depth of the pile shaft (Section 4.8).

*Design values of actions* are determined in accordance with BS EN 1990. The structural designer has to assess the permanent and variable actions (the dead and imposed loading) from the structure which have to be resisted by the foundations. These include *accompanying variable actions* and *transient actions* which can occur simultaneously such as wind load, snow load and earthquake. EC7 National Annex (NA) refers to Tables in the NA to BS EN 1990 for design values of such actions for buildings and bridges separately. In order to ensure that these action factors are not duplicated or factors omitted, it is essential that the structural engineer and pile designer liaise closely for the inputs to Equation 4.2. In the case of piled foundations, the design value ($F_d$) can be assessed directly or derived from representative values ($F_{rep}$) by the equation as EC7 Clause 2.4.6.1:

$$F_d = \gamma_F F_{rep} = \gamma_G G + \gamma_Q Q_1 \left( + \sum \gamma_{Qi} \psi_o Q_i \right)$$  (4.2)

where

    $G$ is the permanent action

    $Q_1$ is the leading variable action on the pile (the imposed load), with the relevant partial factor $\gamma_F$ for unfavourable or favourable action taken from the $A$ set factors in Table 4.1

    $\psi$ is a combination factor ($\leq 1.0$) from the NA to BS EN 1990 which is applied to the accompanying variable action $Q_i$ (*not* to a permanent action)

    $\Sigma$ indicates 'combined effect of'

In addition, it may be necessary to include an *accidental action* ($A_E$) for seismic and impact loading.

As noted for allowable stress in Equation 4.1, $F_d$ should in principle include the net pile weight; for piles in tension, the weight of the pile may be considered as an additional resistance. In this text, the term 'applied load' generally refers to the structural load prior to the application of the partial factors in Table 4.1.

*Design values of resistance* of the ground ($R_{cd}$) at the ULS have to be shown to be equal to or greater than the design value of the design action ($F_d$), that is,

$$R_{cd} \geq F_d$$  (4.3)

The design resistance to axial compression, $R_{cd}$, may be calculated using parameters obtained from ground tests or in situ tests and the results of pile loading tests. EC7 Clause 7.6.2.3(1)P requires that designs based on ground test results must have been established

Table 4.1  Partial factors on actions ($\gamma_F$) for STR and GEO limit states

| Action | Symbol | Set | |
|---|---|---|---|
| | | A1 | A2 |
| Permanent | | | |
|     Unfavourable | $\gamma_G$ | 1.35 | 1.0 |
|     Favourable | | 1.0 | 1.0 |
| Leading variable | | | |
|     Unfavourable | $\gamma_Q$ | 1.5 | 1.3 |
|     Favourable | | 0 | 0 |
| Accompanying variable | | | |
|     Unfavourable | $\gamma_{Qi}$ | $1.5\psi$ | $1.3\psi$ |
|     Favourable | | 0 | 0 |

The partial factors shown in Table 4.1 are the partial factors for buildings as Tables NA.A1.2(B) and (C) of BS EN 1990 for STR/GEO states. $\psi$ factors are given in Table NA.A1.1. Factors for bridge design are given in Tables NA.A2.4(B) and (C).

from pile load tests and comparable experience. Clause 7.6.2.3 provides for two methods of calculation from ground test results: the so-called *model pile* procedure and the *alternative* procedure.

The *model pile* method assumes that a pile of the same penetration depth and cross-sectional dimensions as proposed for the project is installed at the location of each borehole or in situ test. This is a cumbersome approach and it is assumed that it was intended for pile designs based on results from cone penetration tests (CPTs) and pressuremeter tests (PMTs) and not on results from laboratory tests on soil samples. As the method is rarely applied in the United Kingdom, it will not be considered in detail (but see Worked Example 4.6). Essentially the mean and minimum soil parameters for each test profile are used to calculate the shaft and base resistance ($R_{s\,cal}$ and $R_{b\,cal}$ respectively, using Equations 4.5a and b) from the in situ test data. The two components are then divided by a correlation factor ($\xi_3$ or $\xi_4$) given in Table A.NA.10 of the UK NA depending on the number of ground test profiles on the project site to give the characteristic design resistances $R_{bk}$ and $R_{sk}$. The lower of the characteristic resistances is then used to calculate the design resistance, $R_{cd}$, of the pile by applying the $R$ set partial factors from Tables 4.3 through 4.5, $\gamma_b$ and $\gamma_s$, to each component:

$$R_{cd} = R_{bd} + R_{sd} = \frac{R_{bk}}{\gamma_b} + \frac{R_{sk}}{\gamma_s} \tag{4.4}$$

If the superstructure or substructure supported by the piles is stiff enough to redistribute loads from the weaker to the stronger piles, Clause 7.6.2.3(7) allows the correlation factors $\xi_3$ and $\xi_4$ to be divided by 1.1 provided that $\xi_3$ is never less than 1.0.

The EC7 *alternative* to the model pile calculation is in line with the customary design method using the site-specific soil parameters and is generally the calculation method used in the United Kingdom. Characteristic values of the soil parameters over the penetration depth of the pile, as determined by field or laboratory testing, are used to obtain the components $R_{bk}$ and $R_{sk}$ characteristic of the whole site or homogeneous area of the site. The principle of the cautious estimate or statistical approach in achieving the *best-fit* curve for design at a particular limit state is important.

The ultimate base and shaft resistances are calculated using the standard equations (as used in allowable stress design):

$$R_b = q_b\,A_b \tag{4.5a}$$

$$R_s = q_s\,A_s \tag{4.5b}$$

where
  $q_b$ and $q_s$ are the unit base and shaft resistances (which can be determined from several sources and procedures as described later in this chapter)
  $A_b$ and $A_s$ are the base and shaft areas, respectively

These values are then divided by a *model factor*, $\gamma_{Rd}$, as described in Clause 7.6.2.3(8) of EC7, the purpose of which is to make the characteristic resistances $R_{bk}$ and $R_{sk}$ compatible with the model pile calculation:

$$R_{bk} = q_{bk}A_b = \frac{q_b}{\gamma_{Rd}}A_b \tag{4.6a}$$

$$R_{sk} = q_{sk}A_s = \frac{q_s}{\gamma_{Rd}}A_s \qquad\qquad (4.6b)$$

The characteristic resistances are inserted into Equation 4.4 to produce the total design resistance, $R_{cd}$, applying the $R$ set partial factors from Tables 4.3 through 4.5, $\gamma_b$ and $\gamma_s$, to each component as shown.

The NA (A.3.3.2) has set the model factor at 1.4, but this can be reduced to 1.2 'if the resistance is verified by a maintained load test taken to the calculated, unfactored ultimate resistance'. It can be implied that a reduction in $\gamma_{Rd}$ may also be made to calculations where there is a large database of test results. There is no recommendation for reducing the $\gamma_{Rd}$ factor for a 'stiff structure' as with the $\xi$ correlation factors, but if the structural engineer can confirm that loads are being distributed, this may be acceptable under EC7 Clause 7.6.2.1.(5)P. In such a case, a limit state will only occur if a significant number of piles fail together.

Partial factors of unity are used when checking a foundation design for compliance with SLS criteria.

The United Kingdom has adopted Design Approach 1 (DA1) in the NA for foundation design using the partial factors shown in Tables 4.1 through 4.5. Two different combinations of the $A$, $M$ and $R$ sets are stipulated to ensure that the inequality in Equation 4.3 is satisfied for an acceptable design and must be considered separately for each design combination:

DA1, combination 1 (DA1-1) uses sets $A1 + M1 + R1$.
DA1, combination 2 (DA1-2) uses sets $A2 + (M1 \text{ or } M2) + R4$.

The plus sign denotes 'combined with'. Design Approaches 2 and 3 (DA2 and DA3) are not considered in this text.

Taking the case of a pile loaded axially in compression and considering the limit states STR or GEO for DA1, Tables 4.2 through 4.5 show that the partial factors for ground properties and ground resistances are unity for approach DA1-1 and generally govern the STR limit state. DA1-2 provides for alternative material factors $M1$ or $M2$ and usually defines the critical geotechnical sizing (GEO state). $M1$ factors are used for structural actions, while $M2$ is applied to *unfavourable* geotechnical actions caused by ground movements, such as downdrag and transverse loading. $M2$ factors are not used to modify the adopted soil parameters for the design of axially loaded piles. The DA1-2 combination is frequently the governing situation and is worth checking first.

EC7 currently gives no guidance on the factors to be used to obtain the design value of $F_d$ where this is caused by geotechnical actions. The recommendations by Frank et al.[1.5] that the material and resistance factors as shown in Tables 4.2 through 4.5 should be applied as

Table 4.2 Partial factors for soil parameters ($\gamma_M$) for STR and GEO limit states (A.NA.4)

| Soil parameter | Symbol | Set M1 | Set M2 |
|---|---|---|---|
| Angle of shearing resistance[a] | $\gamma_{\phi'}$ | 1.0 | 1.25 |
| Effective cohesion | $\gamma_{c'}$ | 1.0 | 1.25 |
| Undrained shear strength | $\gamma_{cu}$ | 1.0 | 1.4 |
| Unconfined strength | $\gamma_{qu}$ | 1.0 | 1.4 |

[a] This factor is applied to $\tan \phi'$.

Also note that different partial factors are to be applied to soil parameters for design of piles for earthquake resistance as given in the NA to EC8-5.

*Table 4.3* Partial resistance factors ($\gamma_R$) for driven piles for STR and GEO limit states (A.NA.6)

| Resistance | Symbol | R1 | Set | |
|---|---|---|---|---|
| | | | R4 without explicit verification of SLS[A] | R4 with explicit verification of SLS[A] |
| Base | $\gamma_b$ | 1.0 | 1.7 | 1.5 |
| Shaft (compression) | $\gamma_s$ | 1.0 | 1.5 | 1.3 |
| Total/combined (compression) | $\gamma_t$ | 1.0 | 1.7 | 1.5 |
| Shaft in tension | $\gamma_{s;t}$ | 1.0 | 2.0 | 1.7 |

*Table 4.4* Partial resistance factors ($\gamma_R$) for bored piles for STR and GEO limit states (A.NA.7)

| Resistance | Symbol | R1 | Set | |
|---|---|---|---|---|
| | | | R4 without explicit verification of SLS[A] | R4 with explicit verification of SLS[A] |
| Base | $\gamma_b$ | 1.0 | 2.0 | 1.7 |
| Shaft (compression) | $\gamma_s$ | 1.0 | 1.6 | 1.4 |
| Total/combined (compression) | $\gamma_t$ | 1.0 | 2.0 | 1.7 |
| Shaft in tension | $\gamma_{s;t}$ | 1.0 | 2.0 | 1.7 |

*Table 4.5* Partial resistance factors ($\gamma_R$) for CFA piles for STR and GEO limit states (A.NA.8)

| Resistance | Symbol | R1 | Set | |
|---|---|---|---|---|
| | | | R4 without explicit verification of SLS[A] | R4 with explicit verification of SLS[A] |
| Base | $\gamma_b$ | 1.0 | 2.0 | 1.7 |
| Shaft (compression) | $\gamma_s$ | 1.0 | 1.6 | 1.4 |
| Total/combined (compression) | $\gamma_t$ | 1.0 | 2.0 | 1.7 |
| Shaft in tension | $\gamma_{s;t}$ | 1.0 | 2.0 | 1.7 |

Note A in Tables 4.3, 4.4 and 4.5; the lower $\gamma$ values in R4 set may be adopted:

(a) If serviceability is verified by load tests (preliminary and/or working) carried out on more than 1% of the constructed piles to loads not less than 1.5 times the representative load for which they were designed
(b) If settlement is explicitly predicted by means no less reliable than in (a)
(c) If settlement at the SLS is of no concern

(It is suggested that current empirical design methods would satisfy the requirement in (b) where data exist for comparable ground and pile type.)

*multipliers* to the characteristic values of the geotechnical actions to obtain the design values are considered to be overcautious. When considering downdrag due to soft clay as an action (see Section 4.8), care has to be taken in deciding how to apply the $\gamma_{cu}$ factor.

*Static pile loading tests* using procedures described in Section 11.4.2 can be used directly to obtain design resistance values as provided in EC7 Clause 7.6.2.2. In the United Kingdom, the pile test data are mainly used to verify the design resistances derived from ground test results or from empirical or analytical methods, rather than as the primary design tool. This clause also deals with trial piles 'tested in advance'. Again in the United Kingdom, it is rare to do more than one preliminary pile test on a site unless there are particular considerations arising from the geotechnical risk assessment and the linear nature of the site. Trial piles can also be used to check that the proposed installation method can achieve the design penetration depth without difficulty (particularly in the case of driven piles) and can produce a soundly constructed foundation. Loading tests are made on working piles at the project construction stage to confirm the experiences of pre-contract trials and as a routine check on the contractor's workmanship.

Whenever possible, maintained load (ML) static pile tests should be taken to failure or to the stage where a failure can be reliably extrapolated from the load/settlement diagram. In cases where the failure load or ULS resistance, $R_{cm}$, cannot be interpreted from a continuously curving load/settlement diagram, Clause 7.6.1.1(3) of EC7 permits $R_{cm}$ to be conservatively defined as the load applied to the pile head which causes a settlement of 10% of the pile diameter. Clause 7.5.2.1(4) recommends that tension tests should be taken to failure because of doubts about the validity of extrapolation in uplift loading. Note that the constant rate of penetration (CRP) test is excluded for use in design, as it tends to over-predict rate effects.

EC7 Clause 7.6.2.2 considers the situation where more than one pile test is carried out for a design. The characteristic resistances $R_{ck}$ have to be obtained using the model pile concept and applying the correlation factors shown in Table A.NA.9 of the UK NA to the resistances $R_{cm}$ obtained from each loading test to arrive at the design resistances $R_{cd}$. When instrumented piles are used to measure the separate components of base and shaft resistances ($R_{bk}$ and $R_{sk}$), the appropriate R set partial factors are used as shown in the tables.

*Dynamic impact loading tests* may also be used under EC7 Clause 7.6.2.4 to estimate design resistances to axial compression loads provided that there has been an adequate ground investigation. It is important that the method has been calibrated against static loading tests on the same type of pile and of similar length and cross section and in comparable soil conditions. The model pile correlation factors shown in Table A.NA.11 of the UK NA and the partial factors in Tables 4.3 through 4.5 are applied to obtain design resistances as for static load tests. The equipment used for dynamic testing and the method of interpretation are described in Sections 7.3 and 11.4.

*Geometrical data* are concerned with the cross-sectional dimensions of piles. In the case of precast concrete and manufactured steel sections, the dimensions are required to conform to manufacturing tolerances as set out in BS EN 1990 and summarised in Section 2.2.2. While these tolerances are insignificant in relation to the uncertainties involved with soil properties and design methods, they now comprise part of the mandatory 'fitness for purpose' regime. Bored piles in which the concrete is placed in unlined boreholes or driven and cast-in-place piles where the drive tube is extracted during or after placing the concrete may undergo reductions in shaft diameter caused by waisting or necking as described in Section 2.4.2. EC2-1-1 Clause 2.3.4.2(2) specifies that the diameters to be used in concrete design calculations for bored piles should be in accordance with the tolerances shown in Table 4.6. This is somewhat controversial in the United Kingdom as no supporting data are available and the clause allows for 'other provisions'. (See Section 2.3.5 for the design diameter of

Table 4.6 Structural design tolerances for diameters of uncased bored piles (as EC2-1-1)

| Nominal diameter ($d_{nom}$) | Design diameter (d) |
|---|---|
| <400 mm | $d = d_{nom} - 20$ mm |
| $400 \leq d_{nom} \leq 1000$ mm | $d = 0.95 d_{nom}$ |
| $d_{nom} > 1000$ mm | $d = d_{nom} - 50$ mm |

displacement auger piles). It is also necessary to consider the slope of the ground surface, groundwater levels and structural dimensions.

Designs by *prescription* and by the *observational method* are also referred to in the general part of the code. The prescriptive method applies to the tables of allowable bearing pressures for spread foundations in various classes of soils and rocks given in EC7-1 Annex G (previously quoted in BS 8004). Similar prescriptive tables are not generally available for piles except those giving allowable base pressures for pile bearing on rock. It is suggested that these tables should only be used for preliminary design purposes with a cautious approach to the values. Empirical prescriptive correlations, which refer to 'allowable stress' situations, are probably not compatible with EC7 rules.

The observational method is not usually relevant to piled foundation design. The method involves the observation during construction of the behaviour of the whole or part of the structure and its foundation. Typically, the total and differential settlements are measured as the loading increases, and any necessary modifications to the design are made if the movements are judged to be excessive. At this stage, the piling would have been long completed and too late to make any changes to the design without demolishing the superstructure or introducing underpinning piles. Clause 7.4.1 refers to design by 'observing the performance of a comparable foundation'.

*Experimental models* are not used in the day-to-day design of piled foundations. Scale models have their uses as a general research tool, provided that they reproduce the pile installation method, and the findings are verified by full-scale tests and by experience.

The following sections of Chapter 4 describe the use of partial factors in obtaining values for the separate components of base and shaft resistance of driven and bored piles in clays, sands and rocks. The procedure for pile groups is discussed in Chapter 5.

## 4.2 CALCULATIONS FOR PILES IN FINE-GRAINED SOILS

### 4.2.1 Driven displacement piles

When a pile is driven into a fine-grained soil (e.g. clays and clayey silts), the soil is displaced laterally and in an upward direction, initially to an extent equal to the volume of the pile entering the soil. The clay close to the pile surface is extensively remoulded and high pore-water pressures are developed. While it is not normal UK practice to drive piles to found in *soft clay*, it is worth noting that the high pore pressures developed may take weeks or months to dissipate. During this time, the shaft friction and end-bearing resistance, in so far as they are related to the effective overburden pressure (the total overburden pressure minus the pore-water pressure), are only slowly developed. The soft clay displaced by the pile shaft slumps back into full contact with the pile. The water expelled from the soil is driven back into the surrounding clay, resulting in a drier and somewhat stiffer material in contact with the shaft. As the pore-water pressures dissipate and the reconsolidation takes place, the heaved ground surface subsides to near its original level.

The effects in a *stiff clay* are somewhat different. Lateral and upward displacement again occurs, but extensive cracking of the soil takes place in a radial direction around the pile. The clay surrounding the upper part of the pile breaks away from the shaft and may never regain contact with it. If the clay has a fissured structure, the radial cracks around the pile propagate along the fissures to a considerable depth. Beneath the pile toe, the clay is extensively remoulded and the fissured structure destroyed. The high pore pressures developed in the zone close to the pile surface are rapidly dissipated into the surrounding crack system, and negative pore pressures are set up due to the expansion of the soil. The latter may result in an initially high ultimate resistance which may be reduced to some extent as the negative pore pressures are dissipated and relaxation occurs in the soil which has been compressed beneath and surrounding the lower part of the pile.

In allowable stress terminology, the unit *end-bearing resistance* of the displacement pile (for the term $Q_b$ in Equation 4.1) was calculated from the equation $q_b = N_c c_{ub}$. For EC7 designs, the *characteristic* base resistance obtained from ground parameters is the same but with the application of the model factor $\gamma_{Rd}$:

$$R_{bk} = A_b q_{bk} = A_b \frac{N_c c_{ub}}{\gamma_{Rd}} \tag{4.7}$$

where
  $N_c$ is the bearing capacity factor
  $c_{ub}$ is the 'cautious estimate' of undisturbed undrained shear strength representative of the strength at the pile toe (It may be advisable to use the fissured strength in stiff clays with distinct fissure planes.)
  $A_b$ is the cross-sectional area of pile toe

The bearing capacity factor $N_c$ is approximately equal to 9 provided that the pile has been driven at least to a depth of 5 diameters into the bearing stratum. It is not strictly correct to take the undisturbed strength for $c_{ub}$ since remoulding has taken place beneath the toe. However, the greater part of the failure surface in end bearing shown in Figure 4.3 is in soil which has been only partly disturbed by the penetration of the pile. In a stiff fissured clay, the gain in strength caused by remoulding is offset by the loss due to large-displacement strains along a fissure plane. In the case of a soft and sensitive clay, the full undisturbed cohesion should be taken only when the load is applied to the pile after the clay has had time to regain its original shearing strength (i.e. after full dissipation of pore pressures); the rate of gain in the carrying capacity of piles in soft clays is shown in Figure 4.4. It may be noted that a period of a year is required for the full development of carrying capacity in the Scandinavian *quick* clays. In any case, the end-bearing resistance of a small-diameter

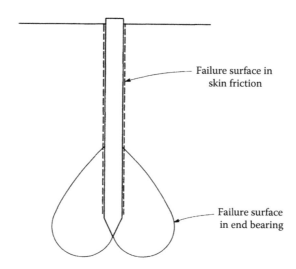

*Figure 4.3* Failure surfaces for compressive loading on piles.

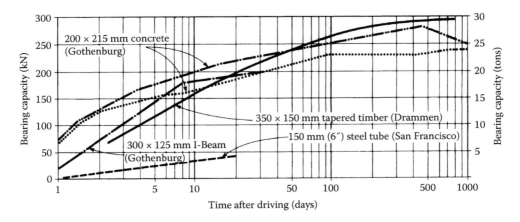

*Figure 4.4* Gain in bearing capacity with increasing time after driving of piles into soft clays.

pile in clay is only a small proportion of the total resistance, and errors due to the incorrect assumption of $c_{ub}$ on the failure surface are not of great significance.

In terms of pure soil mechanics theory, the *ultimate shaft friction* is related to the horizontal effective stress acting on the shaft and the effective interface angle of friction between the pile and the clay. Thus,

$$\tau_s = \sigma_h' \tan \delta_r \qquad (4.8)$$

where

$\tau_s$ is the unit shaft friction at any point
$\sigma_h'$ is the horizontal effective stress
$\delta_r$ is the effective remoulded angle of friction (taken as the interface friction)

A simplifying assumption is made that $\sigma_h'$ is proportional to the vertical effective overburden pressure $\sigma_{vo}'$. That is, $\sigma_h' = K\sigma_{vo}'$ so that

$$\tau_s = K\sigma_{vo}' \tan \delta_r \qquad (4.9)$$

The value of $K$, an *earth pressure coefficient*, is constantly changing throughout the period of installation of the pile and its subsequent loading history. In the case of a driven pile in a stiff clay, $K$ is initially very high, as a result of the energy transmitted by the hammer blows required to displace the clay around the pile. However, at this time, $\sigma_{vo}'$ is very low or even negative due to the high pore-water pressures induced by the pile driving. In the case of a bored pile, $K$ is low as the soil swells at the time of drilling the hole, but it increases as concrete is placed in the shaft. Because of these constantly changing values of $K$ and the varying pore pressures (and hence values of $\sigma_{vo}'$), pure soil mechanics methods cannot be applied to practical pile design for conventional structures without introducing empirical factors and simplified calculations to allow for these uncertainties.

A semi-empirical method based on cone-resistance values has been developed at Imperial College (IC), London, for determining the ultimate bearing capacity of piles driven into clays and sands. The method was developed primarily for piles carrying heavy compression and uplift loads on offshore platforms for petroleum exploration and production. The procedure for piles in clays is based on the use of rather complex and time-consuming laboratory tests, with the aim of eliminating many of the uncertainties inherent in the effective stress approach as noted earlier. It is particularly suitable for piles driven to a deep penetration in clays and sands and is briefly described in Section 4.3.7.

In the case of piles which penetrate a relatively short distance into the bearing stratum of firm to stiff clay, that is piles carrying light to moderate loading, a sufficiently reliable method of calculating the *unit shaft friction*, $q_s$, on the pile shaft in allowable stress terms was to use the equation $q_s = \alpha c_u$. For EC7 designs, the *characteristic* shaft resistance obtained from ground parameters is the same but with the application of the model factor $\gamma_{Rd}$:

$$R_{sk} = \sum A_s q_{sk} = \sum A_s \frac{\alpha c_u}{\gamma_{Rd}}$$
(4.10)

where
   $\alpha$ is an adhesion factor
   $c_u$ is the characteristic undisturbed undrained shear strength of each soil layer surrounding the pile shaft
   $A_s$ is the surface area of the pile shaft contributing to the support of the pile in shaft friction

(Note EC7 continues the traditional use of $c_u$ for undrained shear strength, but the alternative $S_u$ nomenclature is now used by some designers and academics and normally in the United States.)

The adhesion factor depends partly on the shear strength of the soil and partly on the nature of the soil above the bearing stratum of clay into which the piles are driven. Early studies[4.1] showed a general trend towards a reduction in the adhesion factor from unity or higher than unity for very soft clays to values as low as 0.2 for clays having a very stiff consistency. There was a wide scatter in the values over the full range of soil consistency, and these seemed to be unrelated to the material forming the pile.

Much further information on the behaviour of piles driven into stiff clays was obtained in the research project undertaken for the Construction Industry Research and Information Association (CIRIA)[4.2]. Steel tubular piles were driven into stiff to very stiff London Clay and were subjected to loading tests at 1 month, 3 months and 1 year after driving. Some of the piles were then disinterred for a close examination of the soil surrounding the interface. This examination showed that the gap, which had formed around the pile as the soil was displaced by its entry, extended to a depth of 8 diameters, and it had not closed up a year after driving. Between depths of 8 diameters and 14–16 diameters, the clay was partly adhering to the pile surface, and below 16 diameters, the clay was adhering tightly to the pile in the form of a dry skin 1–5 mm in thickness which had been carried down by the pile. Thus, in the lower part of the pile, the failure was not between the pile and the clay but between the skin and surrounding clay which had been heavily sheared and distorted. Strain gauges mounted on the pile to record how the load was transferred from the pile to the soil showed the distribution of load in Figure 4.5. It may be noted that there was no transfer of load in the upper part of the pile, due to the presence of the gap. Most of the load was transferred to the lower part where the adhesion was as much as 20% greater than the undrained strength of the clay. For structures on land, the gap in the upper part of the pile shaft is of no great significance for calculating pile capacity because the greater part of the shaft friction is provided at lower levels. In any case, much of the clay in the

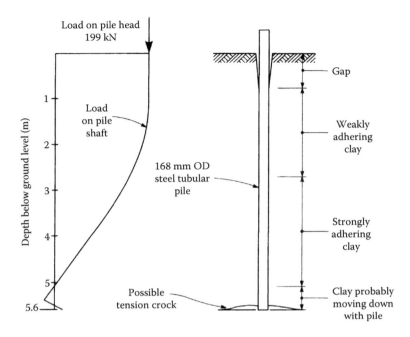

*Figure 4.5* Load transfer from pile to stiff clay at Stanmore. (From Tomlinson, M. J. The adhesion of piles in in stiff clay, Construction Industry Research and Information Association, Research Report No. 26, London, UK., 1970.)

region of the gap is removed when excavating for the pile cap. The gap may be significant for relatively short piles with shallow capping beams for house foundations where these are required as a precaution against the effects of soil swelling and shrinkage caused by vegetation (Section 7.9).

Research by Bond and Jardine[4.3] on extensively instrumented piles jacked into stiff London Clay confirmed the findings on the nature of the soil disturbance very close to the pile. Negative pore pressures were induced in the clay close to the pile wall and positive pressures further away from the pile. Equalisation of pore pressures after installation was very rapid occurring in a period of about 48 h. There was no change in shaft friction capacity after the equalisation period as observed by periodic first-time loading tests over a 3½-month period.

Earlier research, mainly in the field of pile design for offshore structures, has shown that the mobilisation of shaft friction is influenced principally by two factors. These are the over-consolidation ratio of the clay and the slenderness (or aspect) ratio of the pile. The over-consolidation ratio is defined as the ratio of the maximum previous vertical effective overburden pressure, $\sigma'_{vc}$, to the existing vertical effective overburden pressure, $\sigma'_{vo}$. For the purposes of pile design, Randolph and Wroth[4.4] have shown that it is convenient to represent the over-consolidation ratio by the simpler ratio of the undrained shear strength to the existing effective overburden pressure, $c_u/\sigma'_{vo}$. They also showed that the $c_u/\sigma'_{vo}$ ratio could be correlated with the adhesion factor, $\alpha$. A relationship between these two has been established by Semple and Rigden[4.5] from a review of a very large number of pile loading tests, the majority of them being on open-end piles either plugged with soil or concrete. This is shown in Figure 4.6a for the case of a rigid pile and where the shaft friction is calculated from the peak value of $c_u$. To allow for the flexibility and slenderness ratio of the pile, it is

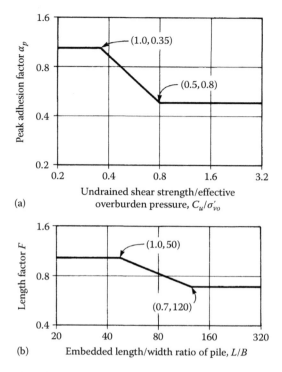

Figure 4.6 Adhesion factors for piles driven to deep penetration into clays. (a) Peak adhesion factor versus shear strength/effective overburden pressure. (b) Length factor. (After Semple, R.M. and Rigden, W.J., Capacity of driven pipe piles in clay, *Symposium on analysis and design of pile foundations*, American Society of Civil Engineers, San Francisco, CA, 59–79, 1984.)

necessary to reduce the values of $\alpha_p$ by a length factor, $F$, as shown in Figure 4.6b. Thus, the characteristic shaft resistance in EC7 terms is then

$$R_{sk} = F \sum A_s \frac{\alpha_p c_u}{\gamma_{Rd}} \tag{4.11}$$

The slenderness ratio, $L/B$, influences the mobilisation of shaft friction in two ways. First, a slender pile can 'whip' or flutter during driving causing a gap around the pile at a shallow depth. The second influence is the slip at the interface when the shear stress at transfer from the pile to the soil exceeds the peak value of shear strength and passes into the lower residual strength. This is illustrated by the shear/strain curve of the simple shear box test on a clay. The peak shear strength is reached at a relatively small strain followed by the much lower residual strength at long strain. It follows that when an axial load is applied to the head of a long flexible pile, the relative movement between the pile and the clay at a shallow depth can be large enough to reach the stage of low post-peak strength at the interface. Near the pile toe, the relative movement between the compressible pile and the compressible clay may not have reached the stage of mobilising the peak shear strength. At some intermediate level, the post-peak condition may have been reached but not the lowest residual condition. It is therefore evident that calculation of the shaft friction resistance from the results of the peak undrained shear strength, as obtained from unconfined or triaxial compression tests in the laboratory, may overestimate the available friction resistance of long piles. The length factors shown in Figure 4.6b are stated by Semple and Rigden to allow both for the flutter effects and the residual or part-residual shear strength conditions at the interface. The effect of these conditions on the settlement of single piles is discussed in Section 4.6.

The empirical $\alpha$ factor total stress approach to determine shaft friction is suitable where there is a good database of pile testing, as in London Clay, but several uncertainties are overlooked when considering $\alpha$ in soft clays. The effective stress principle in Equation 4.8 has been developed by means of a dimensionless shaft friction factor $\beta$, defined as $K_s \tan \delta'$, to calculate the unit shaft resistance as

$$q_s = \beta \sigma'_{vo} \tag{4.12}$$

While ideally $\beta$ can be precisely defined in soil mechanics terms, for practical application, it is necessary to use empirical values from direct measurement of $\tau_s$ in pile tests; judgement is therefore required when selecting a design value. In London Clay, $\beta$ values range from 0.8 to 1.2 (with $\delta_r$ between 17° and 24°) for piles less than 10 m deep, reducing with depth from 0.6 to 0.4 at 25 m deep. For driven piles, $\beta$ will be at the lower end of the range, and in soft clay, values of 0.15–0.25 are likely. In applying the $\beta$-method, it is assumed that the excess pore water is dissipated and loading takes place under fully drained conditions. The back analyses of pile tests in London Clay carried out by Bown and O'Brien[4.6] have shown that if the in situ horizontal effective stress, $\sigma'_h$, can be accurately measured, either by pile tests or by improved in situ soil testing, such as the self-boring pressuremeter, then Equation 4.8 can be applied directly to determine unit shaft friction in London Clay. It is recommended that an 'installation factor' of between 0.9 and 0.8 (decreasing with depth) is applied to $\tau_s$ in stiff clay.

In marine structures where piles may be subjected to uplift and lateral forces caused by wave action or the impact of berthing ships, it is frequently necessary to drive the piles to much greater depths than those necessary to obtain the required resistance to axial compression loading only. To avoid premature refusal at depths which are insufficient to obtain the required uplift or lateral resistance, tubular piles are frequently driven with open ends. At the early stages of driving, soil enters the pile when the pile is said to be 'coring'. As driving continues, shaft friction will build up between the interior soil and the pile wall. This soil is acted on by inertial forces resulting from the blows of the hammer. At some stage, the inertial forces on the core plus the internal shaft friction will exceed the bearing capacity of the soil at the pile toe calculated on the cross-sectional area of the open end. The plug is then carried down by the pile as shown in Figure 4.7a. However, on further driving and when subjected to the applied load, the pile with its soil plug does not behave in the same way as one driven to its full penetration with the tip closed by a steel plate or concrete plug. This is because the soil around and beneath the open end is not displaced and consolidated to the same extent as that beneath a solid-end pile.

Comparative tests on open-end and closed-end piles were made by Rigden et al.[4.7] The two piles were 457 mm steel tubes driven to a penetration of 9 m into stiff glacial till in Yorkshire. A clay plug was formed in the open-end pile and carried down to occupy 40% of the final penetration depth. However, the failure loads of the clay-plugged and steel plate closed piles were 1160 and 1400 kN respectively. Evaluation of the ultimate shaft friction and base resistances showed that the external shaft friction on the open-end piles was 20% less than that on the closed-end piles.

Accordingly, it is recommended that where field measurements show that a clay plug is carried down, the characteristic bearing resistance should be calculated as the sum of the base resistance, $R_{bk}$, (obtained from Equation 4.7) multiplied by a factor of 0.5 and the external shaft friction $R_{sk}$, (obtained from Equation 4.11 and Figure 4.6) multiplied by a factor of 0.8. Where an internal stiffening ring is provided at the toe of a steel pile, the base resistance should be calculated only on the net cross-sectional area of the steel. Attempts to clean out the core of soil from within the pile and replace it by a plug of concrete or cement–sand grout are often ineffective due to the difficulty of removing the strongly adherent clay

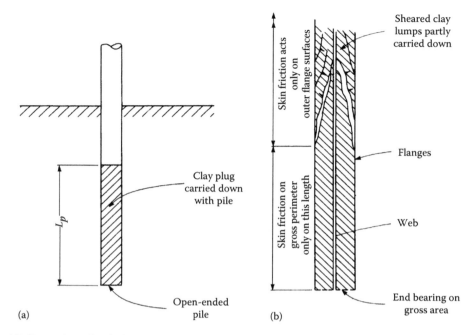

*Figure 4.7* Formation of soil plug at toe of small-displacement piles. (a) Open-ended tube. (b) H-section.

skin to provide an effective bond to the pile surface. Also on large-diameter piles, the radial shrinkage of the concrete or grout plug can weaken the bond with the pile. As already noted, the majority of the pile tests used to derive the relationships in Figure 4.6 were made on open-end piles plugged with soil or concrete. Hence, the shaft friction derived from them already incorporates the effect of the open end.

Plug formation between the flanges and web of an H-section pile is problematical. The possible plug formation at the toe of an H-pile is shown in Figure 4.7b. The mode of formation of a dragged-down soft clay or sand skin has not been studied. A gap has been observed around all flange and web surfaces of H-piles driven into stiff glacial till. An H-pile is not a good type to select if it is desired to develop shaft friction and end-bearing resistance in a stiff clay. It is recommended that the shaft friction is calculated on the outer flange surfaces only, but plugging can be allowed for by calculating the end-bearing resistance on the gross cross-sectional area of the pile. Because of the conservative assumptions of shaft friction and the relatively low proportion of the load carried in end bearing, the calculated resistance need not be reduced by the factor of 0.5 as recommended for tubular piles.

For design to EC7 rules in the United Kingdom, the characteristic base and shaft resistances in Equations 4.7, 4.10 and 4.11 or 4.12 are obtained from the cautiously assessed best-fit profile of all the ground test results[1.3], applying well-established practice, judgement and experience and $\gamma_{Rd}$ as shown. It is not usually necessary to use linear regression analysis to determine the best-fit line for the design profile. The design resistances are then calculated by applying the relevant partial resistance factors to each component as in Equation 4.4. The inequality in Equation 4.3 is checked for the two combinations of DA1. This preferred procedure using the combined profile of soil parameters is illustrated by Worked Example 4.1.

In view of the large amount of test data available to designers of piles in London Clay, there is a strong case for applying the reduced $\gamma_{Rd}$ of 1.2 to the calculated ultimate resistance. Also a revised model factor can be obtained from a statistical analysis of a large database of pile test results in other soils.

## 4.2.2  Driven and cast-in-place displacement piles

The end-bearing resistance of driven and cast-in-place piles terminated in clay can be calculated from Equation 4.7. Where the piles have an enlarged base formed by hammering out a plug of gravel or dry concrete, the area $A_b$ should be calculated from the estimated diameter of the base. It is difficult, if not impossible, for the designer to make this estimate in advance of the site operations since the contractor installing these proprietary piles makes his own decision on whether to adopt a fairly shallow penetration and hammer out a large base in a moderately stiff clay or whether to drive deeper to gain shaft friction, but at the expense of making a smaller base in the deeper and stiffer clay. In a hard clay, it may be impracticable to obtain any worthwhile enlargement over the nominal shaft diameter. In any case, the base may have to be taken to a certain minimum depth to ensure that settlements of the pile group are not exceeded (see Section 5.2.2). The decision as to this minimum length must be taken or approved by the designer.

The conditions for predicting shaft friction on the shaft are different from those with driven preformed piles in some important aspects. The effect on the soil of driving the piling tube with its end closed by a plug is exactly the same as with a steel tubular pile; the clay is remoulded, sheared and distorted, giving the same conditions at the pile–soil interface as with the driven preformed pile. The clay has no chance to swell before the concrete is placed and the residual radial horizontal stress in the soil closes up any incipient gap caused by shrinkage of the concrete. Also the gap which may form around the upper part of the driving tube (or down the full length of the driving tube if an enlarged detachable shoe is used to close its base) becomes filled with concrete. The tube, while being driven, drags down a skin of soft clay or sandy soil for a few diameters into the stiff clay, and it is quite likely that this skin will remain interposed between the concrete and the soil, that is the skin is not entirely pulled out by adhering to the tube. However, in one important aspect, there is a difference between the driven and the driven and cast-in-place pile in that water migrates from the unset concrete into the clay and softens it for a limited radial distance. This aspect is discussed in greater detail in Section 4.2.3. Thus, the adhesion factor for a driven and cast-in-place pile in a stiff clay may be slightly less than that for a driven pile in corresponding soil conditions. It will probably be greater over the length in a soft clay, however, the concrete slumps outwards as the tube is withdrawn, producing an increase in effective shaft diameter.

The results of a number of loading tests on driven and driven and cast-in-place piles in glacial till have been reviewed by Weltman and Healy[4.8]. There appeared to be little difference in the $\alpha$–$c_u$ relationship for either type of pile. They produced the design curves for the two types of driven pile shown in Figure 4.8, including a curve for piles driven a short penetration into stiff glacial till overlain by soft clay. Their review also included a study of the shaft friction on bored piles in glacial till. Trenter[4.9] recommended using the Weltman and Healy relationships and stated that it is essential to obtain 100 mm samples of the till suitable for strength tests.

The determination of the ULS resistance of driven and cast-in-place piles to EC7 rules should follow the procedure described in Section 4.2.1 using the model factor $\gamma_{Rd}$ to give the characteristic resistances.

## 4.2.3  Bored and cast-in-place non-displacement piles

The installation of bored piles using the equipment and methods described in Sections 3.3.1 through 3.3.6 and 3.4.6 causes changes in the properties of the soil on the walls of the pile borehole which have a significant effect on the frictional resistance of the piles. The effect of drilling is to cause a relief of lateral pressure on the walls of the hole. This results in swelling of the clay and there is a migration of pore water towards the exposed clay face. If the borehole intersects water-filled fissures or pockets of silt, the water will trickle down the

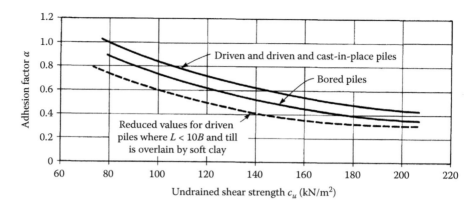

*Figure 4.8* Adhesion factors for piles in glacial till. (After Weltman, A.J. and Healy, P.R., Piling in 'boulder clay' and other glacial tills, Construction Industry Research and Information Association [CIRIA], Report PG5, London, UK, 1978.)

hole and form a slurry with the clay as the drilling tools are lowered down or raised from the hole. Water can also soften the clay if it trickles down from imperfectly sealed-off water-bearing strata above the clay or if hose pipes are carelessly used at ground level to remove clay adhering to the drilling tools.

The effect of drilling is always to cause softening of the clay. If bentonite drilling slurry is used to support the sides of the borehole, softening of the clay due to relief of lateral pressure on the walls of the hole will still take place, but flow of water from any fissures will not occur. There is a risk of entrapment of pockets of bentonite in places where overbreak has been caused by the rotary drilling operation. This would be particularly liable to occur in a stiff fissured clay.

After placing concrete in the pile borehole, water migrates from the unset concrete into the clay, causing further softening of the soil. The rise in moisture content due to the combined effects of drilling and placing concrete was observed by Meyerhof and Murdock[4.10], who measured an increase of 4% in the water content of London Clay close to the interface with the concrete. The increase extended for a distance of 76 mm from the interface.

This softening affects only the shaft. The soil within the zone of rupture beneath and surrounding the pile base (Figure 4.3) remains unaffected for all practical purposes, and the end-bearing resistance $R_{bk}$ can be calculated from Equation 4.7, the value of the bearing capacity factor $N_c$ again being 9. However, Whitaker and Cooke[4.11] showed that the fissured structure of London Clay had some significance on the end-bearing resistance of large bored piles, and they suggested that if a bearing capacity factor of 9 is adopted, the characteristic shearing strength should be taken along the lower range of the graph of shearing strength against depth. In other clays, if $c_{ub}$ is less than 96 kN/m² , then a pro rata reduction in $N_c$ to 8 at a $c_{ub}$ of 48 kN/m² could be considered. If bentonite drilling mud is used, slurry can be trapped beneath the pile base, and a reduction in end-bearing resistance will be needed as described by Reese et al.[3.12]

The effect of the softening on the shaft friction of bored piles in London Clay was studied by Skempton[4.12], who showed that the adhesion factor, $\alpha$, ranged from 0.3 to 0.6 for a number of loading test results. He recommended an average value of 0.45 for normal conditions where drilling and placing concrete followed a reasonably rapid sequence with a lower value of 0.3 in heavily fissured clay. The curve for bored piles in Figure 4.8 can be used to obtain the adhesion factor for very stiff to hard clays. Design charts for $\alpha$ have been based on mean $c_u$ values obtained from unconsolidated, undrained triaxial compression tests on 38 mm samples; if other sample sizes are used or different testing methods employed, then applying the traditional $\alpha$ value may not be appropriate. The London District Surveyors Association[4.13] makes this point and advises that the adhesion factor should be 0.5 in London Clay (using a mean value of $c_u$ not a characteristic value) with a limiting average $\alpha c_u$ value for unit shaft friction of 110 kN/m$^2$. A higher value can be used if verified by ML pile tests. A lower $\alpha$ value of 0.35 may be considered in *wet* shafts and at high length/diameter ratios. Viggiani et al.[4.14] summarise recent research into the estimation of the adhesion factor for displacement and replacement piles separately in respect of varying $c_u$ and $\sigma'_v$. The American Petroleum Institute (API)[4.15] has also adopted $\alpha$ values based on functions of $c_u$ and $\sigma'_v$ for displacement piles.

Where the planned construction programme will lead to long delays (say greater than 12 h) between drilling and placing the concrete, it is advisable to reduce the adhesion factor to account for the clay on the sides of the shaft swelling and softening. The use of a polymer drilling fluid to limit swelling can be considered as in Section 3.3.8.

Fleming and Sliwinski[4.16] observed little difference in the adhesion factor between bored piles drilled into clays in bentonite-supported holes and dry holes. This can be attributed to minimal time between drilling and concreting, the method of drilling – a plate auger causing scoring or gouging or a bucket auger smoothing the sides – or to the rising column of tremie-placed concrete sweeping a thin filter cake completely off the wall of the borehole. As noted in Section 3.3.8, where bentonite has been left in a bore for some time, the cuttings in suspension will lead to a thick filter cake (up to 15 mm) forming on the sides and base of the hole which should be cleared mechanically as it is unlikely to be scoured during tremie concreting. However, a reduction in the adhesion factor is not normally applied at design stage for bentonite-supported boreholes in London Clay based on the large number of load tests available. In other clays, it would be advisable to reduce the adhesion factor by 0.8 to allow for the effects of the filter cake, soil swelling and water from the concrete, unless a higher value can be demonstrated conclusively by preliminary loading tests. Cleaning of the base is also needed in these conditions.

The procedure for checking the ULS resistance of bored piles in clay when using the EC7 rules is the same as described in Section 4.2.1, applying the partial factors in Tables 4.4 and 4.5. When using the $\beta$ method to calculate $q_s$ in Equation 4.12 for bored piles in London Clay, the values given in Section 4.2.1 are used. In soft, normally consolidated clays, a value of 1.0 is suggested, subject to pile length and load testing.

The greater part of the resistance of bored piles in clay is provided by shaft friction. For the STR limit state, the partial factors in the $R1$ set for DA1 verification are unity in the preceding tables requiring the designer to give careful attention to the quality of field and laboratory testing and the selection of soil parameters. The higher values of the partial factors in set $R4$ for bored piles and continuous flight auger (CFA) piles for the GEO limit state compared with those for driven piles reflect the influence of the fissured structure of many stiff clays and also take into account possible inadequacies when cleaning out the base of the pile borehole before placing the concrete. There are also risks in soft clays of waisting or necking when placing concrete in uncased boreholes or when extracting temporary casing.

When enlarged bases are provided on bored piles in a fissured clay, there may be a loss of adhesion over part of the pile shaft in cases where appreciable settlements of the pile base are allowed to occur. The effect of such movements is to open a gap between the conical surface of the base and the overlying clay. The latter then slumps downwards to close the gap and this causes a downdrag on the pile shaft. Arching prevents slumping of the full thickness of clay from the ground surface to the pile base. It is regarded as overcautious to add the possible downdrag force to the applied load on the pile, but nevertheless it may be prudent to disregard the supporting action on the pile of shaft friction over a height of two shaft diameters above the top of base enlargement, as shown in Figure 4.9.

Disregarding shaft friction over a height of two shaft diameters and taking a reduced adhesion factor for the friction on the remaining length may make a pile with an enlarged base

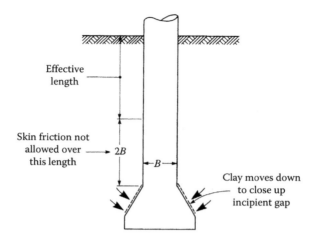

Figure 4.9 Effective shaft length for calculating friction on shaft of under-reamed pile.

an unattractive proposition in many cases when compared with one with a straight shaft. However, the enlarged base pile is economical if the presence of a very stiff or hard stratum permits the whole of the applied load to be carried in end bearing. These piles can also be advantageous where the concept of yielding or 'ductile' piles is adopted for the purpose of achieving load distribution between piles as discussed in Sections 5.2.1 and 5.10. Enlarged bases may also be a necessity to avoid drilling down to or through a water-bearing layer in an otherwise impervious clay.

Piles for marine structures are sometimes installed by driving a steel tube to a limited penetration below seabed, followed by drilling out the soil plug then continuing the drilled hole without further support by the pile tube, using bentonite where needed. On reaching the design penetration depth, a smaller-diameter steel tube insert pile is lowered to the bottom of the borehole, and a cement grout is pumped-in to fill the annulus around the insert pile and make a connection to the main pile (see Figure 8.18).

Kraft and Lyons[4.17] have shown that the adhesion factor used to calculate the shaft friction on the grout–clay interface is of the same order as that used for the design of conventional bored and cast-in-place concrete piles. Where bentonite is used as the drilling fluid, a reduction factor should be adopted as discussed earlier. A considerable increase in the adhesion factor can be obtained if grout is injected under pressure at the soil–pile interface after a waiting period of 24 h or more (see Section 3.3.9). Jones and Turner[4.18] report a two- to threefold increase in adhesion factor when post-grouting was undertaken around the shafts of 150 mm diameter micropiles in London Clay. However, the feasibility of achieving such increases should be checked by loading tests before using them for design purposes particularly if there are doubts about the ability of the grouting process to achieve full coverage of the shaft area. The post-grouting technique around the shafts of bored piles is used as a first step where base grouting is to be carried out as described in Section 3.3.9.

Bustamante and Gianeselli[4.19] presented a pile design method using CPT values, $q_c$, for application to fine-grained and coarse-grained soils, which can be expressed as $q_s = c_s \bar{q}_c$ for shaft resistance and $q_b = c_b q_{cb}$ for end-bearing resistance, where $c_s$ and $c_b$ are coefficients dependent on the soil type, pile roughness and installation method. $q_c$ is the average cone resistance for a layer and $q_{cb}$ is the average cone resistance within 1.5 pile diameters above and below the pile base. For soft clay, $c_s$ is quoted as 0.033 for bored and driven piles; the range for stiff clay is from 0.016 for bored piles to 0.008 for driven steel piles. For soft clay, $c_b$ is given as 0.4 and 0.5 and for stiff clay 0.45 and 0.55, both sets for bored and driven piles respectively.

### 4.2.4  Time effects on pile resistance in clays

Because the methods of installing piles of all types have such an important effect on the shaft friction, it must be expected that with time after installation, there will be further changes in the state of the clay around the pile, leading to an increase or reduction in the friction. The considerable increase in resistance of piles driven into soft sensitive clays due to the effects of reconsolidation has already been noted in Section 4.2.1.

Bjerrum[4.20] has reported on the effects of time on the shaft friction of piles driven into soft clays. He observed that if a pile is subjected to a sustained load over a long period, the shearing stress in the clay next to the pile is carried partly in effective friction and partly in effective cohesion. This results in a downward creep of the pile until such time as the frictional resistance of the clay is mobilised to a degree sufficient to carry the full shearing stress. If insufficient frictional resistance is available, the pile will continue to creep down-wards. However, the effect of long-term loading is to increase the effective shaft resistance as a result of the consolidation of the clay. It must therefore be expected that if a pile has an adequate resistance as shown by a conventional short-term loading test, the effect of the permanent (i.e. long-term) load will be to increase the resistance with time. However, Bjerrum further noted that if the load was applied at a very slow rate, there was a consider-able reduction in the resistance that could be mobilised. He reported a reduction of 50% in the adhesion provided by a soft clay in Mexico City when the loading rate was reduced from 10 to 0.001 mm/min and a similar reduction in soft clay in Gothenburg resulting from a reduction in loading rate from 1 to 0.001 mm/min. These effects must be taken into account when considering the application of partial factors and the model factor if a pile is required to mobilise a substantial proportion of the applied load in shaft friction in a soft clay.

Conclusive observations on the effects of sustained loading on piles driven in stiff clays have not appeared in the literature, but there may be a reduction in resistance with time. Surface water can enter the gap and radial cracks around the upper part of the pile caused by the entry of displacement piles, and this results in a general softening of the soil in the fissure system surrounding the pile. The migration of water from the setting and hardening concrete into the clay surrounding a bored pile is again a slow process, but there is some evidence of a reverse movement from the soil into the hardened concrete[4.21]. Some collected data on reductions in resistance with time for loading tests made at a rapid rate of applica-tion on piles in stiff clays are as follows:

| Type of pile | Type of clay | Change in resistance | Reference |
|---|---|---|---|
| Driven precast concrete | London | Decrease of 10%–20% at 9 months over the first test at 1 month | Meyerhof and Murdock[4.10] |
| Driven steel tube | London | Decrease of 4%–25% at 1 year over the first test at 1 month | Tomlinson[4.2] |

It is important to note that the same pile was tested twice to give the reductions shown above. Loading tests on stiff clays often yield load/settlement curves of the shape shown in Figure 11.16b (Section 11.4.2). Thus, the second test made after a time interval may merely reflect the lower *long-strain* shaft friction which has not recovered to the original peak value at the time of the second test. From the above-mentioned data, it is concluded that the fairly small changes in pile resistance for periods of up to 1 year after equalisation of pore pressure changes caused by installation are of little significance compared with other uncertain effects. An increase could be allowed in the case of soft clays sensitive to remoulding. For example, Doherty and Gavin[4.22] undertook a series of reload tests to examine 'aging effects' of driven piles in soft clay in Belfast. The tests on 10-year-old piles indicated an increase in capacity of

40%–50% compared with the previously established capacity. They conclude that where reuse of piles is an option for urban redevelopment, design loads in excess of the original capacity may be feasible but advise that further research into the underlying causes of aging is needed. Fleming et al.[4.23] also reviewed load test data for driven piles which show the changes in radial stresses (total and effective) with over-consolidation ratio, immediately after installation and after full equalisation of excess pore pressures. In this case, similar reductions to those quoted earlier may be inferred, but long-term *set-up* remains under-researched.

The Norwegian Geotechnical Institute is currently supervising long-term research into time effects on axial bearing capacity of piles with a view to incorporating the expected gain in capacity into the design of offshore and onshore structures. Test sites include soft and stiff clays and loose to medium-dense sands.

## 4.3 PILES IN COARSE-GRAINED SOILS

### 4.3.1 General

The allowable stress formulae for calculating the resistance of piles in coarse-grained soils follow the same form as Equation 4.1. The *characteristic* resistances are calculated using Equations 4.7 and 4.10 but applying the effective stress parameters of a coarse-grained soil ($c_u = 0$), as was the case for allowable stress design, namely, $q_b = N_q\sigma'_{vo}$ and $q_s = \sum K_s\sigma'_{vo}\tan\delta_f$ with the application of the model factor as before so that in EC7 terms the characteristic pile resistance is

$$R_{bk} = A_b q_{bk} = A_b \frac{N_q\sigma'_{vo}}{\gamma_{Rd}} \tag{4.13}$$

$$R_{sk} = A_s q_{sk} = \sum A_s \frac{K_s\sigma'_{vo}\tan\delta_f}{\gamma_{Rd}} \tag{4.14}$$

The design resistances $R_{bd}$ and $R_{sd}$ are then calculated as in Equation 4.4 with the relevant partial factors.

In these expressions, $\sigma'_{vo}$ is the effective overburden pressure at pile base level (or for the shaft the sum of the selected increments), $N_q$ is the bearing capacity factor and $A_b$ is the area of the base of the pile. $K_s$ is a coefficient of horizontal soil stress which depends on the relative density and state of consolidation of the soil, the volume displacement of the pile, the material of the pile and its shape. $\delta_f$ is the characteristic or average value of the angle of friction between pile and soil, and $A_s$ is the area of shaft in contact with the soil. The factors $N_q$ and $K_s$ are empirical and based on correlations with static loading tests: $\delta$ is obtained from empirical correlations with field tests and $N_q$ is derived from $\phi'$ using the relationship with cone penetration tests (CPT) or standard penetration tests (SPT).

The factor $N_q$ depends on the ratio of the depth of penetration of the pile to its diameter and on the angle of shearing resistance $\phi$ of the soil. The latter is normally obtained from the results on tests made in situ (see Section 11.1.4). The relationships between the standard penetration resistances N-*value* and $\phi$, as established by Peck et al.[4.24], and between the limiting static cone resistances $q_c$ and $\phi$, as established by Durgonoglu and Mitchell[4.25], are shown in Figures 4.10 and 4.11 respectively.

From tests made on instrumented full-scale piles, Vesic[4.26] showed that the increase of base resistance with increasing depth was not linear as might be inferred from Equation 4.13 but

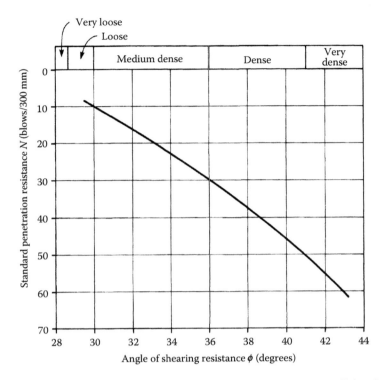

*Figure 4.10* Relationship between SPT *N*-values and angle of shearing resistance. (After Peck, R.B. et al., *Foundation Engineering*, 2nd ed., John Wiley, New York, 1974.)

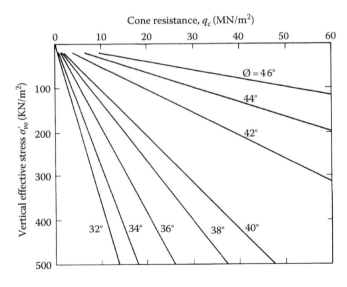

*Figure 4.11* Relationship between angle of shearing resistance and cone resistance for an uncemented, normally consolidated quartz sand. (After Durgonoglu, H.T. and Mitchell, J.K., Static penetration resistance of soils, *Proceedings of the Conference on In-Situ Measurement of Soil Properties*, American Society of Civil Engineers, Raleigh, NC, Vol. 1, pp. 151–188, 1975.)

that *rate* of increase actually decreased with increasing depth. For practical design purposes, it has been assumed that the increase is linear for pile penetrations of between 10 and 20 diameters and that below these depths, the unit base resistance has been assumed to be at a constant value. This simple design approach was adequate for ordinary foundation work where the penetration depths of closed-end piles were not usually much greater than 10–20 diameters. At these depths, practical refusal was usually met when driving piles into medium-dense to dense coarse soils. End-bearing displacement piles in dense coarse-grained soils which overlie weaker strata should be terminated at least 1 m above the weaker soils, and the stress on the lower strata should be checked when the pile toe is less than 2 m above this horizon.

However, the use of piled foundations for offshore petroleum production platforms and monopiles for offshore wind farms has necessitated driving hollow tubular piles with open ends to considerable depths below the seabed to obtain resistance in shaft friction to uplift loading. The assumption of a constant unit base resistance below a penetration depth of 10–20 diameters has been shown to be over-conservative (see Section 4.3.7).

The value of $N_q$ is obtained from the relationship between the drained angle of shearing resistance ($\phi'$) of the soil at the pile base and the penetration depth/breadth of the pile. The relationship developed by Berezantsev et al.[4.27] is shown in Figure 4.13. Vesic[4.26] stated that these $N_q$ values gave results which most nearly conform to the practical criteria of pile failure and are the most widely used for circular piles. The alternative is to use the Brinch Hansen[5.4] $N_q$ factors in Figure 5.6, multiplied by a shape factor (1.2) to convert them to a circular pile. These may be optimistic for $D/B$ ratios over 20 and $\phi'$ values greater than 35°. The Brinch Hansen factors have been adopted by API[4.15] with limiting values for shaft friction and end bearing. The values of $\phi'$ obtained from SPT N-values are not normally corrected for overburden pressure when relating them to the Brinch Hansen $N_q$ factors. However, Bolton[4.28] proposed that the Berezantsev $N_q$ value in Equation 4.13 should be limited to mean effective stress levels in excess of 150 kN/m², and below this, the value of $\phi'$ in sand should be corrected for mean stress level and a critical angle of friction, $\phi_{cr}$. Fleming et al.[4.23] give an iterative method of calculating the mean stress and provide useful design charts; the Oasys PILE program (Appendix C) also includes the Bolton method. Care is needed when dealing with multilayered soils.

The assumption of a constant unit base resistance below a penetration depth of 10–20 diameters has been shown to be over-conservative (see Section 4.3.7). The base resistance of open-end piles driven into sands is low compared with closed-end piles, except when a plug of sand formed at the toe is carried down during driving. The mechanics and effects of plug formation are discussed in Section 4.3.3.

Kulhawy[4.29] calculated the ultimate base resistance for very loose and very dense sands in dry and saturated conditions (i.e. in the absence of groundwater and piles wholly below groundwater level) for a range of depths down to a penetration of 30 m. Unit weights of 18.1 and 19.7 kN/m³ were used for the dry loose and dense sands respectively. These values shown in Figure 4.12 may be used for preliminary design purposes in uniform sand deposits. For densities between very loose and very dense, the base resistance values can be obtained by linear interpolation.

Reduction in the rate of increase in base resistance with increase in penetration depths is also shown by Berezantsev et al.[4.27] as shown in Figure 4.13. Cheng[4.30] has recalculated the Berezantsev depth factor and shown that $N_q$ can be increased by 4%–10% and is significant when $D/B$ is large. However, the revised Berezantsev values are still smaller than the corresponding Vesic values of $N_q$. Ultimate base resistance values using the original factors have been calculated for a closed-end pile of 1220 mm diameter driven into loose sand having a uniform unit submerged weight of 7.85 kN/m³ in Figure 4.14. The angle of shearing resistance of the sand has been assumed to decrease from 30° at the soil surface to 28° at 30 m depth. It will be seen that the Berezantsev $N_q$ values gave lower base resistance than those of Kulhawy.

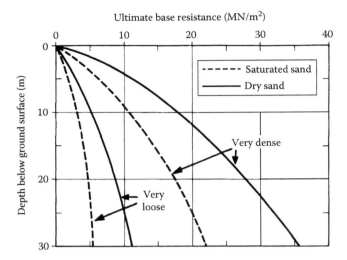

*Figure 4.12* Approximate ultimate base resistance for foundations in sand. (After Kulhawy, F.H., Limiting tip and side resistance, fact or fallacy, *Proceedings of the Symposium on Analysis and Design of Pile Foundations*, American Society of Civil Engineers, San Francisco, CA, pp. 80–98, 1984.)

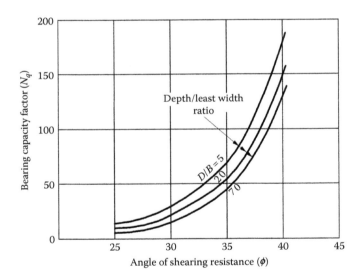

*Figure 4.13* Bearing capacity factors. (From Berezantsev, V.G. et al., Load bearing capacity and deformation of piled foundations, *Proceedings of the Fifth International Conference, ISSMFE*, Paris, France, Vol. 2, pp. 11–12, 1961.)

A similar comparison was made for the 1220 mm pile driven into a dense sand having a uniform unit submerged density of 10.8 kN/m³. The angle of shearing resistance was assumed to decrease from 40° at the soil surface to 37° at 30 m. Figure 4.14 shows that the Kulhawy base resistance values in this case were lower than those of Berezantsev. The penetration depths in Figure 4.14 have been limited to 20 m for dense sands. This is because the pile capacity as determined by the base resistance alone exceeds the value to which the pile can be driven without causing excessive compression stress in the pile shaft. For example, taking a heavy section tubular pile with a wall thickness of 25 mm in high-yield steel and,

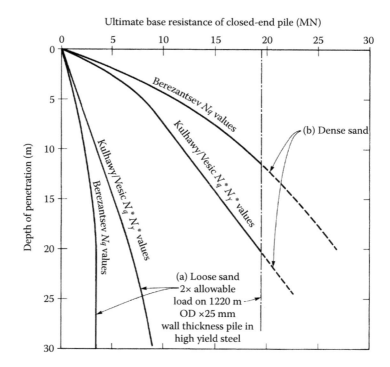

*Figure 4.14* Base resistance versus penetration depth for 1220 mm diameter closed-end pile driven into sand. (a) Loose sand. (b) Dense sand.

in allowable stress terms, limiting the compression stress to twice the value given by the allowable working stress of 0.3 times the yield stress, the ultimate pile load is 9.7 MN. This is exceeded at 12 and 20 m penetration using the Berezantsev and Kulhawy factors respectively. The high base resistances which can be obtained in dense sands often make it impossible to drive piles for marine structures to a sufficient depth to obtain the required resistances to uplift and lateral loading. This necessitates using open-end piles, possibly with a diaphragm across the pile at a calculated height above the toe as described in Section 2.2.4.

When piles are driven into coarse-grained soils (gravels, sands and sandy silts), significant changes take place around the pile shaft and beneath its base. Loose soils are readily displaced in a radial direction away from the shaft. If the loose soils are water bearing, vibrations from the pile hammer cause the soils to become *quick* and the pile slips down easily. The behaviour is similar with bored piles, when the loosened sand (which may initially be in a dense state) slumps into the borehole. When piles are driven into medium-dense to dense sands, radial displacement is restricted by the passive resistance of the surrounding soil resulting in the development of a high interface friction between the pile and the sand. Continued hard driving to overcome the build-up of frictional resistance may cause degradation of angular soil particles with consequent reduction in their angle of shearing resistance. In friable sands, such as the detritus of coral reefs, crushing of the particles results in almost zero resistance to the penetration of open-end piles.

Driving a closed-end pile into sand displaces the soil surrounding the base radially. The expansion of the soil mass reduces its in situ pore pressure, even to a negative state, again increasing the shaft friction and greatly increasing the resistance to penetration of the pile. Tests on instrumented driven piles have shown that the interface friction increases exponentially with increasing depth as shown in Figure 4.15.

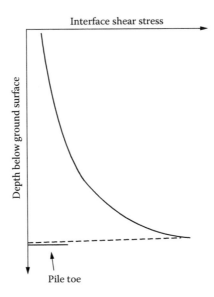

*Figure 4.15* Distribution of interface friction on shaft of pile driven into sand.

Friction on the pile shaft in coarse-grained soil is calculated using the simplified effective stress Equation 4.9 as given in the second term in Equation 4.14. The factor $K$ in Equation 4.9 is denoted by $K_s$, which is related to $K_0$, to the type of pile and to the installation method as shown in Table 4.7. The value of $K_s$ is critical to the evaluation of the shaft friction and is the most difficult to determine reliably because it is dependent on the stress history of the soil and the changes which take place during installation of the pile. In the case of driven piles, displacement of the soil increases the horizontal soil stress from the original $K_0$ value. Drilling for bored piles can loosen a dense sand and thereby reduce the horizontal stress.

The factor $K$ is governed by the following influences:

1. The stress history of the soil deposit, characterised by its coefficient of earth pressure at rest, $K_0$, in an undisturbed state
2. The ratio of the penetration depth to the diameter of the pile shaft
3. The rigidity and shape of the pile
4. The nature of the material forming the pile shaft

$K_0$ is measured by field tests such as the SPT or the CPT and by the pressuremeter (Section 11.1). In normally consolidated soils, $K_0$ is constant with depth and depends on

*Table 4.7* Values of the coefficient of horizontal soil stress, $K_s$

| Installation method | $K_s/K_0$ |
|---|---|
| Driven piles, large displacement | 1–2 |
| Driven piles, small displacement | 0.75–1.25 |
| Bored and cast-in-place piles | 0.70–1 |
| Rotary displacement piles | 0.7–1.2 |
| CFA piles | 0.5–0.9 |

Note: The values $K_s/K_0$ in Table 4.7 for CFA and rotary displacement piles in sands are dependent on the installation equipment and technique.

the relative density of the deposit. Some typical values for a normally consolidated sand are as follows:

| Relative density | $K_0$ |
|---|---|
| Loose | 0.5 |
| Medium-dense | 0.45 |
| Dense | 0.35 |

If the soil deposits are over-consolidated, that is if they have been subjected to an over-burden pressure at some time in their history, $K_0$ can be much higher than the values shown above, say of the order of 1–2 or more. It is possible to determine whether or not the soil deposit is over-consolidated by reference to its geological history or by testing in the field using SPTs or static cone tests. Normally consolidated soils show low penetration values at the surface increasing roughly linearly with depth. Over-consolidated soils show high values at shallow depths, sometimes decreasing at the lower levels.

The angle of interface friction $\delta_r$ in Equation 4.14 is obtained by applying a factor to the average effective angle of shearing resistance ($\phi'$) of the soil as determined from its relationship with SPT or CPT values as shown in Figures 4.10 and 4.11. The factor to obtain $\delta_r$ from the design $\phi'$ depends on the surface material of the pile. Factors established by Kulhawy[4.29] are shown in Table 4.8. They apply both to driven and bored piles. In the latter case, $\phi'$ depends on the extent to which the soil has been loosened by the drilling process (Section 4.3.6). The CFA type of bored pile (Section 2.4.2) is advantageous in this respect.

Use of the $K_s/K_0$ relationship in Table 4.7 to determine the characteristic shaft resistance of a pile driven into sand when using Equation 4.14 does not reflect the exponential distribution of intergranular friction shown in Figure 4.15. Fleming et al.[4.23] comment that $K_s$ may be estimated from $K_s = N_q/50$, which would not be linear. Poulos and Davis[4.31] also provide an empirical non-linear relationship, $\beta = K_s \tan \delta$ as a function of the initial $\phi'$, for application in Equations 4.9 and 4.14 to driven and bored piles in normally consolidated sands. Some suggested values for $\beta$ are as follows:

| Initial angle of internal friction $\phi'$ | Driven piles | Bored piles |
|---|---|---|
| 33° | 0.4 | 0.15 |
| 35° | 0.75 | 0.2 |
| 37° | 1.2 | 0.4 |

Table 4.8 Values of the angle of pile to soil friction for various interface conditions

| Pile–soil interface condition | Angle of pile–soil friction, $\delta$ |
|---|---|
| Smooth (coated) steel/sand | $0.5\bar{\phi} - 0.7\bar{\phi}$ |
| Rough (corrugated) steel/sand | $0.7\bar{\phi} - 0.9\bar{\phi}$ |
| Precast concrete/sand | $0.8\bar{\phi} - 1.0\bar{\phi}$ |
| Cast-in-place concrete/sand | $1.0\bar{\phi}$ |
| Timber/sand | $0.8\bar{\phi} - 0.9\bar{\phi}$ |

EC7 rules require that the base resistance of tubular piles driven with open ends having an internal diameter greater than 500 mm should be the lesser of the shearing resistance between the soil plug and the pile interior and the base resistance of the cross-sectional area of the pile at the toe.

### 4.3.2 Driven piles in coarse-grained soils

Driving piles into loose sands densifies the soil around the pile shaft and beneath the base. Increase in shaft friction can be allowed for by using the higher values of $K_s$ related to $K_0$ from Table 4.7. However, it is not usual to allow any increase in the $\phi$ values and hence the bearing capacity factor $N_q$ caused by soil compaction beneath the pile toe. The reduction in the rate of increase in end-bearing resistance with increasing depth has been noted earlier. A further reduction is given when piles are driven into soils consisting of weak friable particles such as calcareous soils consisting of carbonate particles derived from disintegrated corals and shells. This soil tends to degrade under the impact of hammer blows to a silt-sized material with a marked reduction in the angle of shearing resistance, shaft friction and end-bearing.

Because of these factors, published records for driven piles which have been observed from instrumented tests have not shown values of the ultimate base resistance much higher than 11 MN/m². This figure is proposed for closed-end piles as a practical peak value for ordinary design purposes, but it is recognised that higher resistances up to a peak of 22 MN/m² may be possible when driving a pile into a dense soil consisting of hard angular particles. While modern UK practice has generally moved away from limiting values of end-bearing pressure, such high values should not be adopted for design purposes unless proved by loading tests. Figure 4.14 shows that the base resistance of a closed-end pile driven into a dense sand can reach the maximum compressive stress to which the pile can be subjected during driving at a relatively short penetration. Whichever bearing capacity approach is used, with or without a depth factor, a maximum value of base resistance is reached at a penetration of 10–20 pile diameters and is unlikely to be exceeded no matter how much deeper the pile is driven into medium-dense to dense soils to gain a small increase in shaft friction. There is also the risk of pile breakage.

H-section piles are not economical for carrying high-compression loading when driven into sands. Plugging of the sand does not occur in the area between the web and flanges. The base resistance is low because of the small cross-sectional area. Accordingly, the pile must be driven deeply to obtain worthwhile shaft friction. The latter is calculated on the total surface of the web and flanges in contact with the soil. At Nigg in Scotland, soil displacements of only a few centimetres were observed on each side of the flanges of H-piles driven about 15 m into silty sand, indicating that no plugging had occurred over the full depth of the pile shaft. The base resistance of H-piles can be increased by welding short stubs or wings (see Figure 2.18a) at the toe. Some shaft friction is lost on the portion of the shaft above these base enlargements.

The exponential distribution of interface friction shown in Figure 4.15 has been shown by the Imperial College research to be a function of the length-to-diameter ratio or in the terms of the researchers the ratio of the height above the toe to the pile radius ($h/R$). It follows that it is more advantageous to use a large-diameter pile with a relatively short embedment depth rather than a small diameter with a deep penetration, but in some circumstances, however, it may be necessary to drive deeply to obtain the required resistance to uplift or lateral loading.

When applying EC2 material factors (see Section 7.10.1) to proprietary types of precast concrete piles, the design compressive strength is in the range of 14–20 MN/m². Therefore, if a peak base resistance of 11 MN/m$^2$ is adopted, the piles will have to develop substantial shaft friction to enable the maximum applied load to be utilised. This is feasible in loose to medium-dense sands but impracticable in dense sands or medium-dense to dense sandy gravels. In the latter case, peak base resistance values higher than 11 MN/m$^2$ may be feasible, particularly in flint gravels.

When using the EC7 rules, the design $\phi'$ values obtained from the best-fit test profiles could be divided by the 'calibration factor' of 1.05 to derive the $N_q$ value, as a simple means of dealing with the corrected $\phi'$ values noted earlier. The $M$ set of partial factors is not used to modify the profile values of tan $\delta_r$ or $\phi'$ as they are derived from in situ tests. The model pile approach based on the mean and minimum values of each SPT profile is not considered here.

### 4.3.3 Piles with open ends driven into coarse-grained soils

It was noted in Section 4.3.1 that it is frequently necessary to drive piles supporting offshore petroleum production platforms to considerable depth below the seabed in order to obtain the required resistance to uplift by shaft friction. Driving tubular piles with open ends is usually necessary to achieve this. Driving is relatively easy, even through dense soils, because with each blow of the hammer, the overall pile diameter increases slightly, thereby pushing the soil away from the shaft. When the hammer is operating with a rapid succession of blows, the soil does not return to full contact with the pile. A partial gap is found around each side of the pile wall allowing the pile to slip down. Flexure of the pile in the stick-up length above seabed also reduces resistance to penetration.

At some stage during driving, a plug of soil tends to form at the pile toe after which the plug is carried down with the pile. At this stage, the base resistance increases sharply from that provided by the net cross-sectional area of the pile shoe to some proportion (not 100%) of the gross cross-sectional area.

The stage when a soil plug forms is uncertain; it may form and then yield as denser soil layers are penetrated. It was noted in Section 2.2.4 that 1067 mm steel tube piles showed little indication of a plug moving down with the pile when they were driven to a depth of 22.6 m through loose becoming medium-dense to dense silty sands and gravels in Cromarty Firth. No plugging, even at great penetration depths, may occur in uncemented or weakly cemented calcareous soils. Dutt et al.[4.32] described experiences when driving 1.55 m diameter steel piles with open ends into carbonate soils derived from coral detritus. The piles fell freely to a depth of 21 m below seabed when tapped by a hammer with an 18 tonne ram. At 73 m, the driving resistance was only 15 blows/0.3 m.

It should not be assumed that a solidly plugged pile will mobilise the same base resistance as one with a closed end. In order to mobilise the full resistance developed in friction on the inside face, the relative pile–soil movement at the top of the plug must be of the order of 0.5%–1% of the pile diameter. Thus, with a large-diameter pile and a long plug, a considerable settlement at the toe will be needed to mobilise a total pile resistance equivalent to that of a closed-end pile. Another uncertain factor is the ability of the soil plug to achieve sufficient resistance to yielding by arching of the plug across the pile interior. Research has shown that the arching capacity is related principally to the pile diameter. Clearly, it is not related to the soil density because the soil forming the plug is compacted by the pile driving. The estimated ultimate bearing resistances of sand-plugged piles obtained from published and unpublished sources have been plotted against the pile diameters by Hight et al.[4.33] Approximate upper and lower limits of the plotted points are shown in Figure 4.16. In most cases, the piles were driven into dense or very dense soils, and the test evidence pointed clearly to failure within the plug and not to yielding of the soil beneath the pile toe.

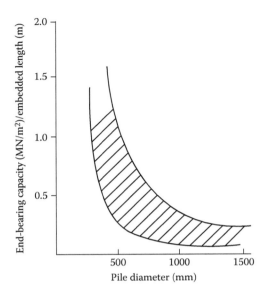

*Figure 4.16* Reduction in end-bearing capacity of open-end piles driven into sand due to increase in diameter. (After Hight, D.W. et al., Evidence for scale effects in the end-bearing capacity of open-end piles in sand, *Proceedings of the 28th Offshore Technology Conference*, Houston, TX, OTC 7975, pp. 181–192, 1996.)

## 4.3.4 Driven and cast-in-place piles in coarse-grained soils

Both the base resistance and shaft friction of driven and cast-in-place piles can be calculated in the same way as described for driven piles in the preceding Section. The installation of driven and cast-in-place types does not loosen the soil beneath the base in any way, and if there is some loosening of the soil around the shaft as the driving tube is pulled out, the original state of density is restored, if not exceeded, as the concrete is rammed or vibrated into place while pulling out the tube. Loosening around the shaft must be allowed for if no positive means are provided for this operation. The provision of an enlarged base adds considerably to the end-bearing resistance of these piles in loose to medium-dense sands and gravels. The gain is not so marked where the base is formed in dense soils, since the enlargement will not greatly exceed the shaft diameter and, in any case, full utilisation of the end-bearing resistance may not be possible because of the need to keep the compressive stress on the pile shaft within design limits.

## 4.3.5 Bored and cast-in-place piles in coarse-grained soils

If drilling for the piles is undertaken by baler (see Section 3.3.7) or by grabbing underwater, there is considerable loosening of the soil beneath the pile toe as the soil is drawn or slumps towards these tools. This causes a marked reduction in end-bearing resistance and shaft friction, since both these components must then be calculated on the basis of a low relative density ($\phi = 28°–30°$). Only if the piles are drilled by power auger or reverse-circulation methods in conjunction with a stabilising slurry or by drilling underwater followed by a base-grouting technique as described in Section 3.3.9 can the end-bearing resistance be calculated on the angle of shearing resistance of the undisturbed soil. However, the effects of entrapping slurry beneath the pile toe[3.12] must be considered. If routine base cleaning is not effective, then the appropriate reduction in resistance should be made. Alternatively, loading tests should be made to prove that the bentonite technique will give a satisfactory end-bearing resistance. Fleming and Sliwinski[4.16] suggest that the shaft friction on bored piles, as calculated from a coefficient of friction and the effective horizontal pressure, should be reduced by 10%–30% if a bentonite slurry is used for drilling in a sand.

The effects of loosening of the soil by conventional drilling techniques on the interface shaft friction and base resistances of a bored pile in a dense sand are well illustrated by the comparative loading tests shown in Figure 4.17. Bored piles having a nominal shaft diameter of 483 mm and a driven precast concrete shell pile with a shaft diameter of 508 mm were installed through peat and loose fine sand into dense sand. The bored piles with toe levels at 4.6 and 9.1 m failed at 220 and 350 kN respectively, while the single precast concrete pile which was only 4 m long carried a 750 kN test load with negligible settlement.

Design by calculation under EC7 procedures is as described in Section 4.3.2, with $N_q$ and $\tan \delta_r$ in Equations 4.13 and 4.14 respectively, being obtained from $\phi'$ values based on SPT or CPT relationships. Judgement is necessary to estimate the reduction in $\phi'$ caused by the pile drilling. Values of $K_s$ are obtained from Table 4.7 with the assumption that $K_0$ represents the loosening of the sand.

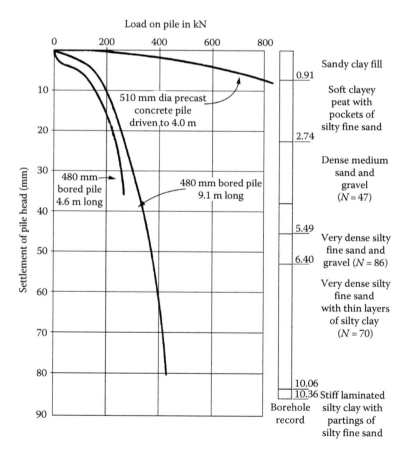

Figure 4.17 Comparison of compressive resistance of driven piles and bored and cast-in-place piles in dense to very dense coarse soils.

## 4.3.6 Use of in situ tests to predict the ultimate resistance of piles in coarse-grained soils

It has been noted that the major component of the ultimate resistance of piles in dense coarse soils is the base resistance. However, Figures 4.13 and 4.14 show that the values of $N_q$ are very sensitive to the values of the angle of shearing resistance of the soils. These values are obtained from in situ tests made in boreholes, and if the boring method has loosened the soil, which can happen if incorrect techniques are used (see Section 11.1.4), then the base resistance of any form of driven pile is grossly underestimated. It is very unlikely that the boring method will compact the soil, and thus any overestimation of the shearing resistance is unlikely.

A reliable method of predicting the shaft friction and base resistance of driven and driven and cast-in-place piles is to make static cone penetration tests at the site investigation stage (CPTM or CPTU, see Section 11.1.4). This equipment produces curves of cone penetration resistance with depth (Figure 4.18). The Bustamante and Gianeselli[4.19] empirical factors noted earlier to determine the end-bearing resistance of bored piles from cone-resistance values must be used with considerable caution in sands because of the loosening of the soil caused by drilling.

Extensive experience with pile predictions based on the cone penetrometer in the Netherlands has produced a set of design rules which have been summarised by Meigh[4.34].

Although engineers in the Netherlands and others elsewhere assess shaft friction values on the measured local sleeve friction ($f_s$), the established empirical correlations between unit friction and cone resistance ($q_c$) are to be preferred. This is because the cone-resistance values are more sensitive to variations in soil density than the sleeve friction and identification of the soil type from the ratio of $q_c$ to $f_s$ is not always clear-cut. Empirical relationships of

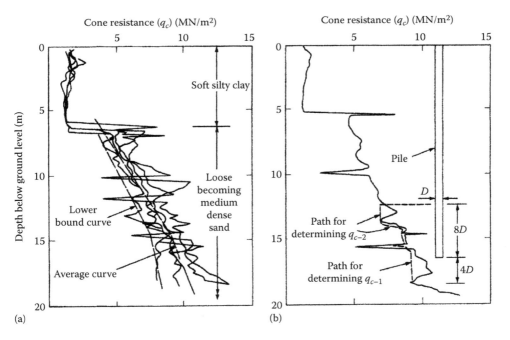

Figure 4.18 Use of static cone penetration tests (CPTs) to obtain design values of average cone resistance ($\bar{q}_c$) in coarse soils. (a) Determining $\bar{q}_c$ from average and lower bound $q_c$/depth curves and (b) Method used in the Netherlands for obtaining base resistance.

*Table 4.9* Relationships between pile shaft friction and cone resistance

| Pile type | Ultimate unit shaft friction |
|---|---|
| Timber | $0.012\,q_c$ |
| Precast concrete | $0.012\,q_c$ |
| Precast concrete enlarged base[a] | $0.018\,q_c$ |
| Steel displacement | $0.012\,q_c$ |
| Open-ended steel tube[b] | $0.008\,q_c$ |
| Open-ended steel tube driven into fine to medium sand | $0.0033\,q_c$ |

Source: After Meigh, A.C., *Cone Penetration Testing*, CIRIA-Butterworth, London, UK, 1987.

[a] Applicable only to piles driven in dense groups; otherwise, use 0.003 where the shaft size is less than the enlarged base.
[b] Also applicable to H-section piles.

pile friction to cone resistance are shown in Table 4.9 and are applicable to piles under static axial compression loading. A limiting value of 0.12 MN/m² is proposed for the ultimate shaft friction.

The end-bearing resistance of piles is calculated from the relationship:

$$q_{ub} = \bar{q}_c \tag{4.15}$$

where $\bar{q}_c$ is the average cone resistance within the zone influenced by stresses imposed by the toe of the pile. This average value can be obtained by plotting the variation of $q_c$ against depth for all tests made within a given area. An average curve is then drawn through the plots either visually or using a computer-based statistical method (Figure 4.18). It is a good practice to draw a lower bound line through the lower cone-resistance values, ignoring sharp peak depressions provided that these are not clay bands in a sand deposit, hence the need to correlate $q_c$ with the soil stratification. The average curve can then be applied to determine the design resistances. Where obvious differences in CPT profiles are present over the site, there is a good case for using the EC7 model procedure for calculating design resistances, as demonstrated by Bauduin[4.35]. Different calibration factors may have to be introduced to account for differences between $q_c$ values from an electric cone and a mechanical cone and when considering cyclic compression loading to allow for the degradation of siliceous sand (see Section 6.2.2).

The method generally used in the Netherlands is to take the average cone resistance $\bar{q}_{c-1}$ over a depth of up to four pile diameters below the pile toe and the average $\bar{q}_{c-2}$ eight pile diameters above the toe as described by Meigh[4.34].

The ultimate base resistance is then

$$q_{ub} = \frac{\bar{q}_{c-1} + \bar{q}_{c-2}}{2} \tag{4.16}$$

The shape of the cone-resistance diagram is studied before selecting the range of depth below the pile to obtain $\bar{q}_{c-1}$. Where the $q_c$ increases continuously to a depth of 4D below the toe, the average value of $\bar{q}_{c-1}$ is obtained only over a depth of 0.7D. If there is a sudden decrease in resistance between 0.7D and 4D, the lowest value in this range should be selected for $\bar{q}_{c-1}$ (Figure 4.18b). To obtain $\bar{q}_{c-2}$, the diagram is followed in an upward direction, and the envelope is drawn only over those values which are decreasing or remain constant at the value at the pile toe. Minor peak depressions are again ignored

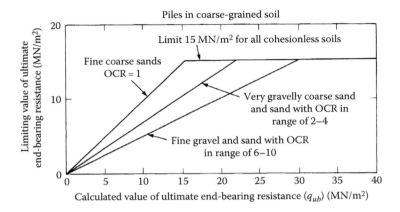

Figure 4.19 Limiting values of pile end-bearing resistance for solid-end piles. (After Te Kamp, W.C., Sondern end funderingen op palen in zand, *Fugro Sounding Symposium*, Utrecht, the Netherlands, 1977.)

provided that they do not represent clay bands; values of $q_c$ higher than 30 MN/m² are disregarded over the 4D–8D range.

An upper limit is placed on the value of the base resistance obtained by either of the methods shown in Figure 4.18. Upper limiting values depend on the particle-size distribution and over-consolidation ratio and are shown in Figure 4.19 after Te Kamp[4.36].

The relationship $q_b = \bar{q}_c$ in Equation 4.15 is valid for piles up to about 500 mm in diameter or breadth, provided that a pile head displacement of 10% of the diameter is taken as the criterion of failure. The reduction of the $q_b/q_c$ ratio with increase in diameter is discussed in Section 4.3.7.

A further factor must be considered when calculating pile shaft friction and end-bearing resistance from CPT data. This is the effect of changes in overburden pressure on the $q_c$ (and also local friction) values at any given level. Changes in overburden pressure can result from excavation, scour of a river or seabed or the loading of the ground surface by placing fill. The direct relationship between $q_c$ and overburden pressure is evident from Figure 4.11. Taking the case of a normally consolidated sand when the vertical effective stress is reduced by excavation, the ratio of the horizontal stress to the vertical stress is also reduced, but not in the same proportion depending on the degree of unloading. The effects are most marked at shallow depths.

Small reductions in overburden pressure cause only elastic movements in the assembly of soil particles. Larger reductions cause plastic yielding of the assembly and a proportionate reduction of horizontal pressures. Broug[4.37] has shown that the threshold value for the change from elastic to elasto-plastic behaviour of the soil assembly occurs when the degree of unloading becomes less than 0.4.

The effect of unloading on cone-resistance values was shown by De Gijt and Brassinga[4.38]. Figure 4.20 shows $q_c$/depth plots before and after dredging to a depth of 30 m in the normally consolidated alluvial sands of the River Maas in connection with an extension to the Euroterminal in the Netherlands. Large reductions in overburden pressure within the zone 10 m below the new harbour bed caused the reduction in cone resistance. The difference between the observed new cone resistance and the mean line predicted by Broug[4.37] did not exceed 5%.

The effects are most marked where the soil deposits contain weak particles such as micaceous or carbonate sands. Broug[4.37] described field tests and laboratory experiments on sands containing 2%–5% of micaceous particles. These studies were made in connection

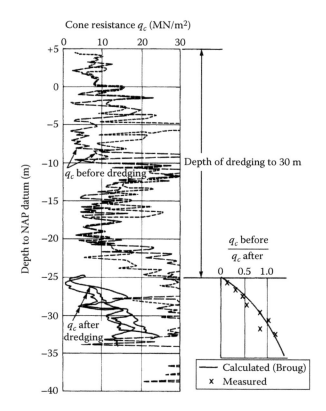

*Figure 4.20* Cone resistance versus depth before and after dredging sand. (After De Gijt, J.G. and Brassinga, H.E., *Land Water*, 1.2, 21, January 1992.)

with the design of piled foundations for the Jamuna River Bridge in Bangladesh where scour depths of 30–35 m occur at times of major floods[4.42].

The static cone penetration test, which measures the resistance of the 'undisturbed' soil, is used as a measure of the resistance to penetration of a pile into a soil which has been compacted by the pile driving. Heijnen[4.39] measured the cone resistance of a loose to medium-dense silty fine sand before and after installing driven and cast-in-place piles. The increase in resistance at various distances from the 1 m diameter enlarged base caused by the pile driving was as follows:

| Distance from pile axis (m) | Increase in static cone resistance (%) |
| --- | --- |
| 1 | 50–100 |
| 2 | About 33 |
| 3.5 | Negligible |

In spite of the considerable increase in resistance close to the pile base, the ultimate resistance of the latter was in fact accurately predicted by the cone-resistance value of the undisturbed soil by using Equation 4.15. This indicates that the effect of compaction both in driven and driven and cast-in-place piles is already allowed for when using this equation.

Field trials to correlate the static cone resistance with pile loading tests are necessary in any locality where there is no previous experience to establish the relationship between the two. In the absence of such tests, the ratio $q_b/q_c$ should be taken as 0.5. The pile head settlement at the applied load is then unlikely to exceed 10 mm for piles of base widths up to about 500 mm. Further reductions in $q_b/q_c$ values may be needed for high effective over-burden stresses. Bustamante and Gianeselli[4.19] propose a reduction of 0.4 for driven piles in very compact sand and gravel with a $q_c$ resistance >12 MN/m². For larger base widths, it is desirable to check that pile head settlements resulting from the design end-bearing pressure are within tolerable limits. Pile head settlements can be calculated using the methods described in Section 4.6.

### 4.3.7  Tubular steel piles driven to deep penetration into clays and sands

The principal users of large tubular steel piles are the offshore petroleum industry, and recently, these piles have found increasing use as monopile foundations for off-shore wind power generators. Guidance for engineers designing offshore piling has been available for many years in the regularly updated recommendations of API in RP2A WSD[4.15]. (Note ISO 19902 has superseded the load and resistance factor design [LRFD] version of API RP2A.) Their recommendations for the shaft friction of piles in clay generally followed the $\alpha c_u$ relationship of Semple and Rigden[4.5]. Equation 4.13 was used for piles in sands with the Brinch Hansen factors of $N_q$ for calculating base capacity. Chow[4.40] found that the API recommendations for piles in sand were over-conservative for short piles with $L/B$ ratios up to 30 and for dense sands with relative densities of 60% or more.

Research work undertaken at Imperial College, London, on the axial capacity of steel tube piles has been referred to briefly in the preceding sections. The initial work has been extended with analysis of further test data and has been published in book form by Jardine et al.[4.41]. The design procedures which have evolved have become known as the ICP method, and while the following comments cover some of the salient points of research behind the method, the reader is referred to the full ICP text for the applications. The reliability of the method depends on continuous CPT/CPTU in situ testing and, for clays, good-quality undisturbed samples using piston samplers and thin-walled tubes followed by sophisticated laboratory testing using oedometer and shear ring apparatus. It is intended that the method be used to predict pile capacities that may be mobilised during slow ML tests conducted 10 days after driving.

The ICP method for piles driven into *clays* is based on effective stresses and takes into account the effects on the interface shaft resistance of the radial displacement of the clay and the gross displacement of the clay beneath the base. To determine shaft resistance, the ICP method calculates the local shear stress at failure on the interface after equalisation of pore pressure changes brought about by the pile driving. The calculations are made for a succession of layers over the embedded length of the shaft. They are then integrated to give the total shaft resistance from the following equation:

$$Q_s = \pi D \int \tau_f dZ \qquad (4.17)$$

The peak local interface shear stress $\tau_f$ is obtained from the following equation:

$$\tau_f = \left(\frac{K_f}{K_c}\right)\sigma'_{rc} \tan \delta_f \tag{4.18}$$

where

$K_f$ is the coefficient of radial effective stress for shaft at failure $= \sigma'_{rf}/\sigma'_{vo}$
$K_c$ is the coefficient of earth pressure at rest $= \sigma'_{rc}/\sigma'_{vo}$
$\sigma'_{rc}$ is the equalised radial effective stress $= K_c\sigma'_{vo}$
$\delta_f$ is the operational interface angle of frictional failure
$K_c$ is obtained from the equation

$$K_c = \left[2.2 + 0.016\,\mathrm{YSR} - 0.870\,\Delta I_{vy}\right]\mathrm{YSR}^{0.42}\left(\frac{h}{R}\right)^{-0.20} \tag{4.19}$$

where

$I_{vy}$ is the relative void index at yield $= \log_{10}S_t$
YSR is the yield stress ratio or apparent over-consolidation ratio
$S_t$ is the clay sensitivity
$h$ is the height of soil layer above pile toe
$R$ is the pile radius
$K_f/K_c = 0.8$

An alternative to Equation 4.19 which is marginally less conservative is

$$K = \left[2 - 0.625\,\Delta I_{vo}\right]\mathrm{YSR}^{0.42}\left(\frac{h}{R}\right)^{-0.20} \tag{4.20}$$

where

$I_{vo}$ is the relative void index
YSR, $\Delta I_{vy}$ and $\Delta I_{vo}$ are obtained either from oedometer tests in the laboratory on good-quality undisturbed samples or from a relationship with consolidated anisotropic undrained triaxial compression tests or by estimation from CPT or field vane tests

The clay sensitivity is determined by dividing the peak intact unconsolidated undrained shear strength by its remoulded undrained shear strength.

The operational interface angle of friction at failure $\delta_f$ lies between the peak effective shear stress angle and its ultimate or long-strain value. The actual value used in Equation 4.18 depends on the soil type, prior shearing history and the clay-to-steel interface properties. It is influenced by local slip at the interface when the blow of the hammer drives the pile downwards and at rebound when the hammer is raised at the end of the stroke. A further influence is progressive failure when the interface shear stress near the ground surface is at the ultimate state, but near the toe, the relative pile–soil movement may be insufficient to reach the peak stress value.

The conditions at the interface can be simulated by determining $\delta_f$ in a ring shear apparatus where the remoulded clay is sheared against an annular ring fabricated from the same material and having the same roughness as the surface of the pile. Details of the apparatus and the testing technique are given in the IC publication.

For calculating the shaft capacity of open-end piles in clay, an equivalent radius $R^*$ is substituted for $R$ in the $h/R$ term where

$$R^* = \left(R_{outer}^2 - R_{inner}^2\right)^{0.5} \tag{4.21}$$

and $h/R^*$ is not less than 8.

Dealing with the base resistance of closed-end piles in clay, the ICP method does not accept the widely used practice of calculating the ultimate resistance from $Q_b = N_c c_u A_b$ where the bearing capacity factor $N_c$ is assumed to be equal to 9. The database of instrumented pile tests used in the IC research showed a wide variation in $N_c$ which was found to be higher than 9 in all the tests analysed. However, the results did demonstrate a close correlation with the results of static cone penetration tests and led to a recommendation to adopt the following relationships:

$$q_b = 0.8q_c \text{ for undrained loading} \tag{4.22}$$

and

$$q_b = 1.3q_c \text{ for drained loading} \tag{4.23}$$

The cone resistance $q_c$ is obtained from CPT's by averaging the readings over a distance of 1.5 pile diameters above and below the toe.

For open-end piles, plugging of the pile toe with clay is defined as the stage when the plug is carried down by the pile during driving. This is deemed to occur when $[D_{inner}/D_{CPT} + 0.45q_c]/P_a$ is less than 36. The cone diameter $D_{CPT}$ is 0.038 m and the normalised atmospheric pressure $P_a$ is 100 kN/m². 

Fully plugged piles as defined above develop half the base resistance calculated by Equations 4.22 and 4.23 for undrained and drained loadings respectively, after a pile head displacement of $D/10$.

The base resistance of an unplugged open-end pile is calculated on the annular area of steel only. The IC proposed 1.6 increase in the value of $Q_b$ for drained loading when $q_b$ is taken as the average $q_c$ at founding depth would seem to be optimistic when compared with Equation 4.23.

Jardine et al.[4.41] recommend safety factors of 1.3–1.6 for the shaft resistance in compression for offshore foundations where uniform settlement of the structure is not critical and the design is based on allowable stress methods.

The ICP method of design for tubular piles in *sands* is a simple one based on CPTs. No other field work or special laboratory testing is required where correlations are available, such as the Chow[4.40] data for the shear modulus. The method is wholly empirical based on small-diameter un-instrumented loading tests and experience. It is justified by the assumption that the penetration of the sleeved cone simulates the displacement of the soil by a closed-end or fully plugged pile.

The expression for the shaft resistance is calculated by the following sequence of equations:

$$\text{Unit shaft resistance} = \tau_f = \sigma'_{rf} \tan\delta_f \tag{4.24}$$

$$\text{Radial effective stress at point of shaft failure} = \sigma'_{rf} = \sigma'_{rc} + \Delta\sigma'_{rd} \tag{4.25}$$

$$\text{Equalised radial effective stress} = \sigma'_{rc} = 0.029q_c \left(\frac{\sigma'_{vo}}{P_a}\right)^{0.13} \left(\frac{h}{R}\right)^{-0.38} \tag{4.26}$$

$$\text{Distant increase in local radial effective stress} = \Delta\sigma'_{rd} = \frac{2G\delta_f}{R} \tag{4.27}$$

where
  $\delta_f = \delta_{cr}$ is the interface angle of friction at failure
  $R$ is the pile radius
  $G$ is the operational shear modulus

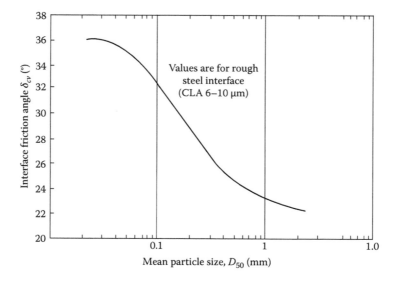

*Figure 4.21* Relationship between interface friction angle and mean particle size of a silica sand. (Based on Jardine, R. et al., *ICP Design Methods for Driven Piles in Sands and Clays*, Thomas Telford, London, UK, 2005.)

In Equation 4.24, $\delta_f$ can be obtained either by constant-volume shear box tests in the laboratory or by relating it to the pile roughness and particle size of the sand (Figure 4.21). The equalised radial stress in Equation 4.26 implies that the elevated pore pressures around the shaft caused by pile driving have dissipated. The term $P_a$ is the atmospheric pressure which is taken as 100 kN/m². Because of the difficulty in calculating or measuring the high radial stresses near the pile toe, $h/R$ is limited to 8.

The shear modulus G in Equation 4.27 can be measured in the field using a pressuremeter (Section 11.1.4) or a seismic cone penetrometer or obtained by correlation with CPT data using the relationship established by Chow[4.40]:

$$G = q_c(A + B\eta - C^2)^{-1} \tag{4.28}$$

and

$$\eta = q_c \sqrt{P_a \sigma'_{vo}} \tag{4.29}$$

The term $\delta_f$ in Equation 4.27 is twice the average roughness $R_{cla}$ of the pile surface which is the average height of the peaks and troughs above and below the centre line. For lightly rusted steel, $\Delta r$ is 0.02 mm. $\Delta\sigma'_{rd}$ is inversely proportional to the pile radius and tends to zero for large-diameter piles.

In Equation 4.28,

$A = 0.0203$

$B = 0.00125$

$C = 1.216 \times 10^{-6}$

Piles driven with open ends develop a lower shaft resistance than closed-end piles because of their smaller volume displacement when a solid plug is not carried down during driving. The open unplugged end is allowed for by adopting an equivalent pile radius $R^*$ (see Equation 4.21). Equation 4.26 becomes

$$\sigma'_{rc} = 0.029q_c \left( \frac{\sigma'_{vo}}{P_a} \right)^{0.13} \left( \frac{h}{R^*} \right)^{-0.38} \qquad (4.30)$$

To use the ICP method, the embedded shaft length is divided into a number of short sections of thickness $h$ depending on the layering of the soil and the variation with depth of the CPT readings. A mean line is drawn through the plotted $q_c$ values over the depths of the identified soil layers. A line somewhat higher than the mean is drawn when the ICP method is used to estimate pile driveability when the shaft resistance must not be underestimated.

From a database of pile tests in calcareous sands, Jardine et al.[4.41] stated that the ICP method was viable in these materials and recommended that the submerged density should be taken as 7.5 kN/m³ for calculating $\sigma'_{vo}$ and the interface angle $\delta_f$ as 25°. The third term in Equation 4.25 is omitted ($\sigma'_{rf} = \sigma'_{rc}$) and Equation 4.26 for open-end piles is modified to become $\sigma'_{rc} = 72(\sigma'_{vo}/P_a)^{0.84}(h/R^*)^{-0.35}$. For closed-end piles, $R$ is substituted for $R^*$.

The ICP method uses CPT data to calculate the base resistance. For closed-end piles, the equation is

$$q_b = q_c \left[ 1 - 0.5 \log \left( \frac{D}{D_{CPT}} \right) \right] \qquad (4.31)$$

where
$q_c$ is the cone resistance averaged over 1.5 pile diameters above and below the toe
$D$ is the pile diameter
$D_{CPT}$ is the cone diameter

The equation is valid provided that the variations in $q_c$ are not extreme and the depth intervals between peaks and troughs of the $q_c$ values are not greater than $D/2$. If these conditions are not met, a $q_c$ value below the mean should be adopted. A lower limit for $q_b$ of $0.3q_c$ is suggested for piles having diameters greater than 0.9 m.

A rigid basal plug within an open-end pile is assumed to develop if the inner diameter in metres is less than $0.02 (D_r - 30)$ where the relative density $D_r$ is expressed as a percentage. Also $D_{inner}/D_{CPT}$ should be less than $0.083q_c/P_a$ and the absolute atmospheric pressure $P_a$ is taken as 100 kN/m².

If the preceding criteria are satisfied, the fully plugged pile is stated to develop a base resistance of 50% of that of a closed-end pile after the head has settled by one-tenth of the diameter. A lower limit of $q_b$ is that it should not be less than that of the unplugged pile and should not be less than $0.15q_c$ for piles having diameters greater than 0.9 m.

The base resistance of unplugged piles is taken as $0.5q_c$ multiplied by the net cross-sectional area of the pile at the toe, where $q_c$ is the cone resistance at toe level. No contribution is allowed from the inner wall shaft friction. For a solid-end pile, $q_b$ at the toe is determined from Equation 4.31.

IC assessed the reliability of their method for piles in sands by comparing the predications of shaft capacity with those of the 1993 version of the API method. The ultimate resistance

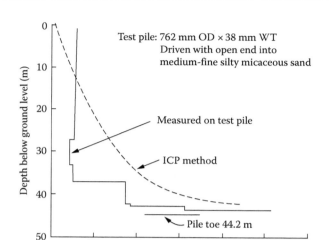

*Figure 4.22* Comparison of measured and calculated interface shear stress on the shaft of a steel tube pile driven into sand.

calculated by the ICP method compared well with measured results, but using the same criteria, the API calculations indicated that they over-predict soil resistance for large-diameter piles.

The ICP method was used to compare the calculated distribution of interface shear stress at failure with stresses measured over the shaft depth of a well-instrumented 762 mm outside diameter (OD) pile driven with an open end to a depth of 44 m into medium-fine silty micaceous sand in Bangladesh. The test was made as part of the trial piling for the foundations of the Jamuna River Bridge at Sirajganj[4.42,4.43] as described in Section 9.6.2. The observed and calculated distributions of stress are compared in Figure 4.22. It will be noted that the ICP method considerably overestimated the short-term measured stresses but appear to correspond with the marked increases in bearing capacity with time as noted in Section 4.3.8. A study of the shaft friction measurements made on two 762 mm trial piles showed that the distribution of interface shear stress could be represented by the relationship $\tau_f = 0.009(h/d)^{-0.5}q_f$ in compression and $0.003(h/d)^{-0.5}q_c$ in tension.

A simplified ICP method has been included in the API RP2 GEO/ISO-19901 commentary of 2011[4.44] and is one of four CPT-based methods considered for the axial capacity of piles in sand. Knudsen et al.[4.45] used parametric studies and pile test data to compare the standard API[4.15] recommended practice with the four new alternatives and found that all the CPT methods should be used with caution for piles larger than 1000 mm.

White and Bolton[4.46] reanalysed the IC database for closed-end piles on the basis that instead of the criterion of failure being the load causing a settlement of 10% of the diameter, they assumed that plunging settlement occurred, that is beyond point D in Figure 4.1. They also made allowance for only partial embedment of some piles into the bearing stratum and the presence in some piles of a weaker layer below base level. They found a mean of $q_b = 0.9q_c$ with no trend towards a reduction of $q_b$ with increase in pile diameter. They suggested that a reduction factor to obtain the ultimate bearing capacity of a closed-end pile in sand should be linked to partial embedment and partial mobilisation rather than to absolute diameter. This suggestion would appear to be part of the methodology of research based on analysis of test pile failures rather than criteria to be adopted at the design stage of piled foundations.

It was generally assumed in past years that no allowance should be made for significant changes in the bearing capacity of piles driven into coarse soils with time after installation. Neither increases nor decreases in capacity were considered although the *set-up* or temporary increase in driving resistance about 24 h after driving was well known. The long-term effects had not been given serious study. However, the research work at Imperial College described earlier did include some long-term tension tests on piles at Dunkirk reported by Jardine and Standing[4.47]. Six 465 mm OD × 19 m long and one 465 mm OD × 10 m long steel tube piles were tested in tension at ages between 10 days and about 6 months. A progressive increase in resistance of about 150% was recorded. All the tests were 'first time', that is, none of the piles were tested a second time. The increased tension capacity at Dunkirk was attributed mainly to relaxation through creep of circumferential arching around the pile shaft leading to increase in radial effective stress.

The 762 mm OD × 44 m long test pile at the Jamuna Bridge site was referred to above[4.42]. There was an increase in tension capacity of about 270% on retest after the initial test made a few days after driving into medium-dense silty micaceous sand. Precast concrete piles on the same site showed a progressive increase of about 200% in compression at various ages up to 80 days after driving. The ultimate resistances were estimated from dynamic tests and graphical analysis of loading tests not taken to failure.

The procedure for calculating pile resistances driven into sand using CPT results as described in Section 4.3.6 and above is wholly empirical. EC7 currently treats CPT methods of calculating resistances the same as for other *ground tests* and requires that the method adopted should have been established from pile loading tests, as required when using soil strength parameters. EC7 offers no comments on the various procedures using CPT results, but it is assumed that the model pile method would apply for base and shaft resistances determined from each CPT profile. Jardine et al.[4.41] do not offer any recommendations for applying EC7 procedures to their design methods. Merritt et al.[4.48] describe the design of piled tripod foundations for the Borkum West II offshore wind farm in the German North Sea based on German Eurocode factors with the ICP procedures. They point out that the high-quality ground investigation was the key to the reliable application of the method.

(BS EN) ISO 19902:2007, which has replaced the LRFD version of API RP2A, deals with the design of offshore platforms. Clause 17 covers the detailed design of piles, giving equations for the adhesion factor $\alpha$ in fine-grained soils and $\beta$ and $N_q$ factors for coarse-grained soils, summarised in Table 17.4.1. It places limits of 3 MN/m² for end bearing in medium-dense sand and silt and 12 MN/m² in very dense sand; the equivalent limits for shaft friction are 67 and 115 kN/m², unless other values are justified by performance data. In cohesive soils, unit end bearing 'shall be computed using $q = 9 c_u$'. Software from Ensoft Inc., APILE Plus5 Offshore, (see Appendix C) features both the ICP and the API methods to compute the axial capacity of driven piles.

## 4.3.8 Time effects for piles in coarse-grained soils

Notwithstanding the comments in the previous section on increases in tension resistance, the engineer should be aware of a possible *reduction* in capacity where piles are driven into fine sands and silts. Peck et al.[4.24] stated that 'If the fine sand or silt is dense, it may be highly resistant to penetration of piles because of the tendency for dilatancy and the development of negative pore pressures during the shearing displacements associated with insertion of the piles. Analysis of the driving records by means of the wave equation may indicate high dynamic capacity but instead of freeze, large relaxations may occur'.

An example of this phenomenon was provided by the experiences of driving large-diameter tubular steel piles into dense sandy clayey silts for the foundations of the new Galata Bridge in Istanbul[4.49]. The relaxation in capacity of the 2 m OD piles in terms of blows per 250 mm

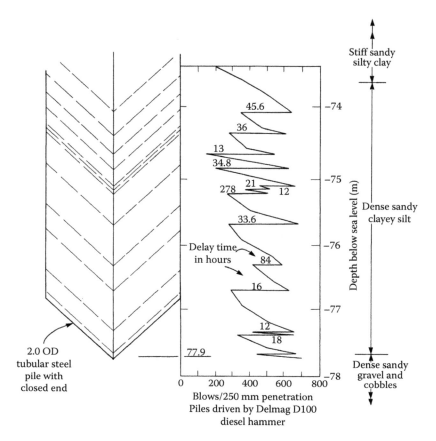

*Figure 4.23* Driving resistance over final 4.5 m of penetration for 2.0 m tubular steel pile showing reduction in driving resistance after various delay periods, New Galata Bridge, Istanbul.

penetration is shown in Figure 4.23. The magnitude of the reduction in driving resistance was not related to the period of time between cessation and resumption of driving. It is likely that most of the reduction occurred within a period of 24 h after completing a stage of driving. The widely varying time periods shown in Figure 4.23 were due to the operational movements of the piling barge from one pile location or group to another.

Correlation of blow count figures with tests made with the dynamic pile analyser (Section 7.3) showed a markedly smaller reduction in dynamic soil resistance than indicated by the reduction in blow count after the delay period.

These experiences emphasise the need to make re-driving tests after a minimum period of 24 h has elapsed after completing the initial drive. Loading tests should not be made on piles in sands until at least 7 days after driving. Where piles are driven into laminated fine sands, silts and clays, special preliminary trial piling should be undertaken to investigate time effects on driving resistance. These trials should include tests with the pile driving analyser.

## 4.4 PILES IN SOILS INTERMEDIATE BETWEEN SANDS AND CLAYS

Where piles are installed in sandy clays or clayey sands which are sufficiently permeable to allow dissipation of excess pore pressure caused by application of load to the pile, the base and shaft resistance can be calculated for the case of drained loading using Equations 4.13

and 4.14. The angle of shearing resistance used for obtaining the bearing capacity factor $N_q$ should be the effective angle $\phi'$ obtained from unconsolidated drained triaxial compression tests. In a uniform soil deposit, Equation 4.13 gives a linear relationship for the increase of base resistance with depth. Therefore, the base resistance should not exceed the peak value of 11 MN/m² unless pile loading tests show that higher ultimate values can be obtained. The effective overburden pressure, $\sigma'_{vo}$, in Equations 4.13 and 4.14 is the total overburden pressure minus the pore-water pressure at the pile toe level. It is important to distinguish between uniform *c–$\phi$ soils* and layered *c* and $\phi$ soils, as sometimes the layering is not detected in a poorly executed soil investigation.

## 4.5 PILES IN LAYERED FINE- AND COARSE-GRAINED SOILS

It will be appreciated from Sections 4.2 and 4.3 that piles in fine-grained soils have a relatively high shaft friction and a low end-bearing resistance and in coarse soils, the reverse is the case. Therefore, when piles are installed in layered soils, the location of the pile toe is of great importance. The first essential is to obtain a reliable picture of the depth and lateral extent of the soil layers. This can be done by making in situ tests with static or dynamic cone test equipment (see Section 11.1.4), correlated by an adequate number of boreholes. If it is desired to utilise the potentially high end-bearing resistance provided by a dense sand or gravel layer, the variation in thickness of the layer should be determined, and its continuity across the site should be reliably established. The bearing stratum should not be in the form of isolated lenses or pockets of varying thickness and lateral extent.

Where driven or driven and cast-in-place piles are to be installed, problems can arise when piles are driven to an arbitrary *set* to a level close to the base of the bearing stratum, with the consequent risk of a breakthrough to the underlying weaker clay layer when the piles are subjected to their applied load (Figure 4.24a). In this respect, the driven and cast-in-place pile with an enlarged base is advantageous, as the bulb can be hammered out close to the top of the bearing stratum (Figure 4.24b). Alternatively, the enlarged base in Figure 4.24b could be achieved using the vibratory concrete column process (Section 2.3.7) for lightly loaded situations. The end-bearing resistance can be calculated conservatively

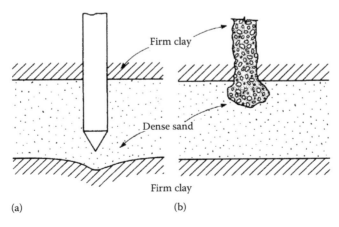

(a)                              (b)

*Figure 4.24* Pile driven to end bearing into relatively thin dense soil layer. (a) Driven pile. (b) Driven and cast-in-place pile.

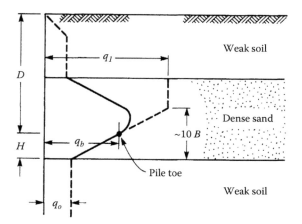

Figure 4.25 End-bearing resistance of piles in layered soils.

on the assumption that the pile always terminates within or just above the clay layer, that is by basing the resistance on that provided by the latter layer. This is the only possible solution for sites where the soils are thinly bedded, and there is no marked change in driving resistance through the various layers. However, this solution can be uneconomical for sites where a dense sand layer has been adequately explored to establish its thickness and continuity. A method of calculating the base resistance of a pile located in a thick stiff or dense layer underlain by a weak stratum has been established by Meyerhof[4.50]. In Figure 4.25, guidance for the unit base resistance of the pile is conservatively given by the following equation:

$$q_b = q_o + \frac{q_1 - q_o}{10B} H \le q_1 \qquad\qquad (4.32)$$

where

$q_o$ is the ultimate base resistance in the lower weak layer
$q_1$ is the ultimate base resistance in the upper stiff or dense stratum
$H$ is the distance from the pile toe to the base of the upper layer ($H$ should be >1 m)
$B$ is the width of the pile at the toe

When applying the effective stress $\beta$ method to calculate shaft resistance as Equation 4.12, $\beta$ should be between 0.05 and 0.1 for driven piles and between 0.5 and 0.8 for bored piles with an upper limit of 100 kN/m². 

Figure 4.26 shows the record of pile driving at British Coal's bulk-handling plant at Immingham, where a layer of fairly dense sandy gravel was shown to exist at a depth of about 14.6 m below ground level. The thickness of the gravel varied between 0.75 and 1.5 m, and it lay between thick deposits of firm to stiff boulder clay. The end-bearing resistance in the gravel of the 508 mm diameter driven and cast-in-place piles was more than 3000 kN as derived from loading tests to obtain separate evaluations of shaft friction and base resistance. It was calculated that if the toe of the pile reached a level at which it was nearly breaking through to the underlying clay, the end-bearing resistance would then fall to 1000 kN. This proved inadequate and further driving was necessary

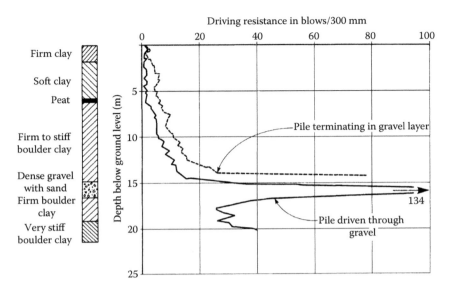

*Figure 4.26* Resistance to driven and cast-in-place piles provided by a thin layer of dense sand and gravel at Immingham.

to mobilise additional shaft friction. The following rules were adopted to ensure that the required pile resistance was achieved:

1. When the driving resistance in the gravel increased rapidly from 20 mm per blow to 5 mm per blow for a complete 300 mm of driving, it was judged that the pile was properly seated in the gravel stratum.
2. The pile was then required to be driven a further 75 mm without any reduction in the driving resistance.
3. If the resistance was not maintained at 5 mm per blow, it was judged that the gravel layer was thin at that point and the pile was liable to break through to the clay. Therefore, the pile had to be driven further to a total penetration of 20 m, which was about 3–4 m below the base of the gravel, to obtain the required additional frictional resistance.

The effects of driving piles in groups onto a resistant layer underlain by a weaker compressible layer must be considered in relation to the settlement of the group. This aspect is discussed in Chapter 5.

## 4.6 SETTLEMENT OF THE SINGLE PILE AT THE APPLIED LOAD FOR PILES IN SOIL

The uncertainties in the calculation of pile capacities using allowable stress design noted in Section 4.1.3 have been traditionally covered by the application of a global safety factor to the ultimate resistances. If the safety factor was greater than 2.5, then from the load/settlement curves obtained from a large number of loading tests in a variety of soil types, both on displacement and non-displacement piles, the settlement under the applied load will not exceed 10 mm for piles of small to medium diameter (up to 600 mm). This is reassuring and avoids the necessity of attempting to calculate settlements on individual piles that are based on the compressibility of the soils. A settlement at the applied load not exceeding 10 mm is satisfactory for most building and civil engineering structures provided that the group settlement is not excessive.

EC7 is silent on the assessment of settlement as such but requires that SLSs be determined by calculating the design values of the effects of actions $E_d$ and comparing them with $C_d$, the 'limiting design value of the effect of an action'. Clause 7.6.4.1 states that where piles are bearing on medium-dense to dense soils, the safety requirements for ULS design are normally sufficient to prevent an SLS in the supported structure. Thus, the combination of EC7 ULS partial factors (e.g. $\gamma_G \times \gamma_{Rd} \times \gamma_b$ or $\gamma_s$) will produce an equivalent global factor of safety between 2 and 3 for single piles (depending on which DA1 combination is used and the variable action applied) and is therefore satisfactory for limiting settlement to 10 mm. As noted in Table 4.5, the lower $R4$ set of partial factors can be used for ULS calculations under certain conditions – 'if settlement is predicted by means no less reliable than load tests...'. However, such reductions (especially if combined with a lower model factor) would not give the same degree of confidence against settlement without further verification from load tests in a range of soils.

For piles larger than 600 mm in diameter, the problem of the settlement of the individual pile under the applied load becomes increasingly severe with the increase in diameter, requiring a separate evaluation of the shaft friction and base load. The load/settlement relationships for the two components of shaft friction and base resistance and for the total resistance of a large-diameter pile in a stiff clay are shown in Figure 4.27. The maximum shaft resistance is mobilised at a settlement of only 10 mm, but the base resistance requires a settlement of nearly 150 mm for it to become fully mobilised. At this stage, the pile has reached the point of ultimate resistance at a failure load of 4.2 MN. A global safety factor of 2 on this condition gives an applied load of 2.1 MN, under which the settlement of the pile will be nearly 5 mm. This is well within the settlement which can be tolerated by ordinary

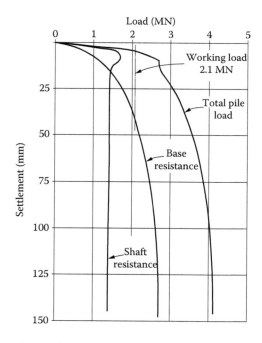

Figure 4.27 Load/settlement relationships for large-diameter bored piles in stiff clay.

building structures. The full shaft resistance will have been mobilised at the applied load, but only 22% of the ultimate base resistance will have been brought into play. This complies with EC7 Clause 2.4.8(4) alternative method for verifying SLS in that a 'sufficiently low fraction of ground resistance is mobilised to keep deformations within the required serviceability limits' (without defining 'sufficiently low'). For economy in pile design, the settlement at the applied load should approach the limit which is acceptable to the structural designer, and this usually involves mobilising the full shaft resistance.

The Oasys PILE program (see Appendix C) applies EC7 rules to traditional empirical methods to determine the capacity of a single axially loaded pile. Settlement analysis in this program is based on the Mattes and Poulos elastic model influence factors as reported by Poulos and Davis[4.31].

Burland et al.[4.51] presented a simple stability criterion for bored piles in clay using global safety factors to produce the expressions $\left(\frac{1}{2}Q_p\right)$ and $\left(Q_s + \frac{1}{3}Q_b\right)$, where the pile has an overall factor of 2; the shaft, a factor of unity; and the base, 3. The allowable load on the pile is the lesser of the two calculations, with $\left(\frac{1}{2}Q_p\right)$ being nearly always dominant for straight-sided piles and for long piles with comparatively small under-reams, whereas $\left(Q_s + \frac{1}{3}Q_b\right)$ often controls piles with large under-reamed bases. However, satisfaction of these criteria does not necessarily mean that the settlement at applied load will be tolerable. Full-scale pile loading tests are necessary where experience of similar piles in similar conditions is not available, but for large diameters, these can be expensive. Loading tests on large piles are more helpful when designing 'ductile piles' (Section 5.2.1). Instrumentation can be provided to determine the relative proportions of load carried in friction on the shaft and transmitted to the base and hence to determine the degree of settlement needed to mobilise peak friction (e.g. at a pile head settlement of about 10 mm in Figure 4.27) and to determine whether or not the lower 'long-strain' value of shaft friction is operating when load distribution between piles in a group takes place.

A more economical procedure is to estimate values from the results of loading tests made on circular plates at the bottom of the pile boreholes or in trial shafts. Burland et al.[4.51] plotted the settlement of test plates divided by the plate diameter $(\rho_i/B)$ against the plate-bearing pressure divided by the ultimate bearing capacity for the soil beneath the plate (i.e. $q/q_f$) and obtained a curve of the type shown in Figure 4.28. If the safety factor on the end-bearing load is greater than 3, the expression for this curve is

$$\frac{\rho_i}{B} = \frac{K \times q}{q_f} \tag{4.33}$$

For piles in London Clay, $K$ in Equation 4.33 has usually been found to lie between 0.01 and 0.02. If no plate bearing tests are made, the adoption of the higher value provides a conservative estimate of settlement. When plate bearing tests are made to failure, the curve can be plotted, and provided that the base safety factor is greater than 3, the settlement of the pile base $\rho_i$ can be obtained for any desired value of $B$.

The procedure used to estimate the settlement of a circular pile is as follows:

1. Obtain $q_f$ from the failure load given by the plate-bearing test.
2. Check $q_f$ against the value obtained by multiplying the shearing strength by the appropriate bearing capacity factor $N_c$, that is, $q_f$ should equal $N_c \times c_{ub}$.
3. Knowing $q_f$, calculate the end-bearing resistance $Q_b$ of the pile from $Q_b = A_b \times q_f$.
4. Obtain the safe end-bearing load on the pile from $W_b = Q_b/F$, where $F$ is a safety factor greater than 3.

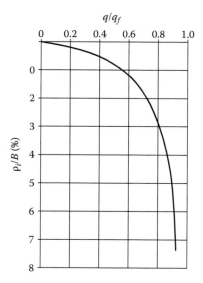

*Figure 4.28* Elastic settlements of bored piles in London Clay at Moorfields. (After Burland, J.B. et al., The behaviour and design of large diameter bored piles in stiff clay, *Proceedings of the Symposium on Large Bored Piles*, Institution of Civil Engineers and Reinforced Concrete Association, London, UK, pp. 51–71, 1966.)

5. Obtain $q$ from $q = W_b / \frac{1}{4} \pi B^2$ and hence determine $q/q_f$.
6. From a curve of the type shown in Figure 4.28, read off $\rho_i/B$ for the value of $q/q_f$ and hence obtain $\rho_i$ (the settlement of the pile base).

Merely increasing the size of the base by providing an under-ream will not reduce the base settlement, and if the settlement is excessive, it should be reduced by one or more of the following measures:

1. Reduce the applied load on the pile.
2. Reduce the load on the base by increasing the shaft resistance, that is by increasing the shaft diameter.
3. Increase the length of the shaft to mobilise greater shaft friction and to take the base down to deeper and less-compressible soil.

Having estimated the settlement of the individual pile using the above-mentioned procedure, it is still necessary to consider the settlement of the pile group as a whole (see Chapter 5).

From their analyses of a large number of load/settlement curves, Weltman and Healy[4.8] established a simple relationship for the settlement of straight-shaft bored and cast-in-place piles in glacial till. The relationship given below assumed a pile diameter not greater than 600 mm, an assumed stress on the pile shaft of about 3 MN/m² in compression, a length-to-diameter ratio of 10 or more, and stiff to hard glacial till with undrained shear strengths in excess of 100 kN/m². The pile head settlement is given by

$$\rho = \frac{l_m}{4} \text{ in mm} \tag{4.34}$$

where $l_m$ is the length of embedment in glacial till in metres.

This is somewhat counter-intuitive showing that longer piles settle more and is essentially the result of the elastic compression stress allowed for by Weltman and Healy and the fact that longer piles will be more heavily loaded. Precast concrete piles and some types of cast-in-place piles are designed to carry applied loads with concrete stress much higher than 3 MN/m². In such cases, the settlement should be calculated from Equation 4.34 assuming a stress of 3 MN/m². The settlement should then be increased pro rata to the actual stress in the concrete.

The above-mentioned methods of Burland et al. and Weltman and Healy were developed specifically for piling in London Clay and glacial till respectively, and were based on the results of field loading tests made at a standard rate of loading as given in SPERW[2.5] (Section 11.4) using the maintained loading procedure. More generally the pile settlements can be calculated if the load carried by shaft friction and the load transferred to the base at the applied load can be reliably estimated. The pile head settlement is then given by the sum of the elastic shortening of the shaft (likely to be small in relation to the overall settlement) and the compression of the soil beneath the base as follows:

$$\rho = \frac{(W_s + 2W_b)L}{2A_sE_p} + \frac{\pi}{4} \cdot \frac{W_b}{A_b} \cdot \frac{B(1-v^2)I_p}{E_b} \tag{4.35}$$

where
   $W_s$ and $W_b$ are the loads on the pile shaft and base, respectively
   $L$ is the shaft length
   $A_s$ and $A_b$ are the cross-sectional area of the shaft and base, respectively
   $E_p$ is the elastic modulus of the pile material
   $B$ is the pile width
   $v$ is the Poisson's ratio of the soil
   $I_p$ is the influence factor related to the ratio of $L/R$
   $E_b$ is the deformation modulus of the soil beneath the pile base

For a Poisson's ratio of 0–0.25 and $L/B > 5$, $I_p$ is taken as 0.5 when the last term approximates to 0.5 $W_b/(BE_b)$. Values of $E_b$ are obtained from plate loading tests at pile base level or from empirical relationships with the results of laboratory or in situ soil tests given in Sections 5.2 and 5.3. The value of $E_b$ for bored piles in coarse soils should correspond to the loose state unless the original in situ density can be maintained by drilling under bentonite or restored by base grouting.

The first term in Equation 4.35 implies that load transfer from pile to soil increases linearly over the depth of the shaft. It is clear from Figure 4.22 that the increase is not linear for a deeply penetrating pile. Section 4.9 comments on the use of computers to simulate the load transfer for wide variations in soil stratification and in cross-sectional dimensions of a pile: the *soil–pile interaction* concept. The basic programs represent an elastic continuum model. A pile carrying an axial compression load is modelled as a system of rigid elements connected by springs and the soil resistance by external non-linear springs (Figure 4.29). The load at the pile head is resisted by frictional forces on each element. The resulting displacement of each of these is obtained from Mindlin's equation for the displacement due to a point load in a semi-infinite mass. The load/deformation behaviour is represented in the form of a *t–z* curve (Figure 4.29). A similar *q–z* curve is produced for the settlement of the pile base.

The concept of modelling a pile as a system of rigid elements and springs for the purpose of determining the stresses in a pile body caused by driving is described in Section 7.3.

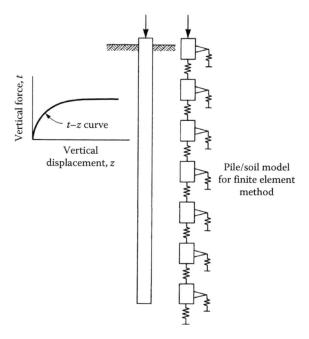

*Figure 4.29* t–z curve for deformation of a pile under vertical axial loading.

It was noted at the beginning of this Section that the adoption of global safety factors in conjunction with allowable stress methods of calculating pile bearing capacity can obviate the necessity of calculating applied load/settlement of small-diameter piles. However, there is not the same mass of experience relating settlements to the design value of actions ($F_d$) obtained by EC7 methods using the partial factors ($\gamma_G$ and $\gamma_Q$). Hence, it is advisable to check that the design pile resistance does not endanger the SLS of the supported structure. Equation 4.35 can be used for this check. A material factor of unity should be adopted for the design value of $E_b$.

An analytical expression for the calculation of the load/settlement curve which is amenable to spreadsheet methods is given in Section 4.9.1. The CEMSET® program (Appendix C) can also be used to obtain estimates of settlement.

## 4.7 PILES BEARING ON ROCK

### 4.7.1 Driven piles

For maximum economy in the cross-sectional area of a pile, it is desirable to drive the pile to virtual refusal on a strong rock stratum, thereby developing its maximum carrying capacity. Piles driven in this manner are regarded as wholly end bearing; friction on the shaft is not considered to contribute to the support of the pile. The depth of penetration required to reach virtual refusal depends on the thickness of any weak or heavily broken material overlying sound rock. If a pile can be driven to near refusal on to a strong 'intact' rock, the actions on the pile are governed by the design resistance of the pile material at the point of minimum cross section; that is the pile is regarded as a short column supported against buckling by the surrounding soil. Where piles are driven through water or through very soft clays and silts of fluid consistency, then buckling as a long strut must be considered (see Section 7.5).

When steel piles are adopted, applied loads based on the steel design resistance at ULS may result in concentrations of very high loading on the rock beneath the toe of the pile. The ability of the rock to sustain the applied loading without yielding depends partly on the compressive strength of the rock and partly on the frequency and inclination of fissures and joints in the rock mass and whether these discontinuities are tightly closed or are open or filled with weathered material. Very high toe loads can be sustained if the rock is strong, with closed joints either in a horizontal plane or inclined at only a shallow angle to the horizontal. If the horizontal or near-horizontal joints are wide, there will be some yielding of the rock mass below the pile toe, but the amount of movement will not necessarily be large since the zone of rock influenced by a pile of slender cross section does not extend very deeply below toe level. However, the temptation to continue the hard driving of slender-section piles to ensure full refusal conditions must be avoided. This is because brittle rocks may be split by the toe of the pile, thus considerably reducing the base resistance. The splitting may continue as the pile is driven down, thus requiring very deep penetration to regain the original resistance.

Where bedding planes are steeply inclined with open transverse joints, there is little resistance to the downward sliding of a block of rock beneath the toe, and the movement will continue until the open joints have become closed or until the rock mass becomes crushed and locked together. This movement and crushing will take place as the pile is driven down, as indicated by a progressive tightening-up in driving resistance. Thus, there should be no appreciable additional settlement when the applied load is applied. However, there may be some deterioration in the end-bearing value if the piles are driven in closely spaced groups at varying toe levels. For this reason, it is desirable to undertake re-driving tests whenever piles are driven to an end bearing into a heavily jointed or steeply dipping rock formation. If the re-driving tests indicate a deterioration in resistance, then loading tests must be made to ensure that the settlement under the applied load is not excessive. Soil heave may also lift piles off their end bearing on a hard rock, particularly if there has been little penetration to anchor the pile into the rock stratum. Observations of the movement of the heads of piles driven in groups, together with re-driving tests, indicate the occurrence of pile lifting due to soil heave. Methods of eliminating or minimising the heave are described in Section 5.7.

Steel tubes driven with open ends or H-section piles are helpful in achieving the penetration of layers of weak or broken rock to reach virtual refusal on a hard unweathered stratum. However, the penetration of such piles causes shattering and disruption of the weak layers to the extent that the shaft friction may be seriously reduced or virtually eliminated. This causes a high concentration of load on the relatively small area of rock beneath the steel cross section. While the concentration of load may be satisfactory for a strong intact rock, it may be excessive for a strong but closely jointed rock mass. The concentration of load can be reduced by welding stiffening rings or plates to the pile toe or, in the case of weak and heavily broken rocks, by adopting winged piles (Figure 2.18).

The H-section pile is particularly economical for structures on land where the shaft is wholly buried in the soil and thus not susceptible to significant loss of cross-sectional area due to corrosion. To achieve the maximum potential bearing capacity, it is desirable to drive the H-pile in conjunction with a pile driving analyser (Section 7.3) to determine its ultimate resistance and hence the design load, verified if necessary by pile loading tests. The *ArcelorMittal Piling Handbook*[4.52] gives guidance on the ultimate load capacity of H-section piles in S235, S275 and S355 steel grades alongside a table with examples of compressive strength of strong and weak rocks.

The methods given below for calculating the pile bearing resistance assume that this is the sum of the shaft and base resistance. Both of these components are based on correlations between pile loading tests and the results of field tests in rock formations or laboratory tests on core specimens.

Where the joints are spaced widely, that is at 600 mm or more apart, or where the joints are tightly closed and remain closed after pile driving, the base resistance may be calculated from the following equation:

$$q_b = 2N_\phi q_{uc} \qquad (4.36)$$

where $q_{uc}$ is the uniaxial compressive strength of the rock and the bearing capacity factor

$$N_\phi = \tan^2\left(45° + \phi/2\right) \qquad (4.37)$$

For (strong) sandstone, which typically has $\phi$ values between 40° and 45°, end bearing at failure is stated by Pells and Turner[4.53] to be between 9 and 12 times $q_{uc}$. As this laboratory assessment of $q_{uc}$ is likely to be considerably less than the in situ strength, a reasonable characteristic value in this case would be $3q_{uc}$ to $4.5q_{uc}$. The variations in $N_\phi$ caused by joints in the rock mass are demonstrated by the comparisons in Table 4.10 of observations of the ultimate base resistance of driven and bored piles terminated in weak mudstones, siltstones and sandstones with the corresponding $N_\phi$ values calculated from Equation 4.37. For these rocks, the $\phi$ values as recommended by Wyllie[4.54] are in the range of 27°–34° giving $N_\phi$ values from 2.7 to 3.4.

It will be noted that the back-calculated $N_\phi$ values in Table 4.10 are considerably lower than the range of 2.7–3.4 established for rocks with widely spaced and tight joints. The reduction is most probably due to the jointing characteristics of the rock formation in which the tests were made. A measure of the joint spacing is the rock quality designation (RQD) determined as described in Section 11.1.4. Kulhawy and Goodman[4.55,4.56] showed that the ultimate base resistance $(q_{ub})$ can be related to the RQD of the rock mass as shown in Table 4.11.

Table 4.10 Observed ultimate base resistance values of piles terminated in weak mudstones, siltstones and sandstones

| Description of rock | Pile type | Plate or pile diameter (mm) | Observed bearing pressure at failure (MN/m²) | Calculated $N_\phi$ |
|---|---|---|---|---|
| Mudstone/siltstone moderately weak | Bored | 900 | 5.6 | 0.25 |
| Mudstone, highly to moderately weathered weak | Plate test | 457 | 9.2 | 1.25 |
| Cretaceous mudstone, weak, weathered, clayey | Bored | 670 | 6.8 | 3.0 |
| Weak carbonate siltstone/ sandstone (coral detrital limestone) | Driven | 762 | 5.11 | 1.5 |
| Calcareous sandstone weak | Driven tube | 200 | 3.0 | 1.2 |
| Sandstone, weak to moderately weak | Driven | 275 | 19[a] | 1.75 |

[a] From dynamic pile test.

*Table 4.11* Ultimate base resistance of piles related to the uniaxial compression strength of the intact rock and the RQD of the rock mass

| RQD (%) | $q_{ub}$ | $c$ | $\phi°$ |
|---|---|---|---|
| 0–70 | $0.33q_{uc}$ | $0.1q_{uc}$ | 30 |
| 70–100 | $0.33$–$0.8q_{uc}$ | $0.1q_{uc}$ | 30–60 |

Note: RQD values may be biased depending on the orientation of the borehole in relation to the dominant discontinuities.

Where laboratory tests can be made on undisturbed samples of weak rocks to obtain the parameters $c$ and $\phi$, Kulhawy and Goodman state that the ultimate bearing capacity of the jointed rock beneath the pile toe can be obtained from the following equation:

$$q_{ub} = cN_c + \gamma DN_q + \gamma \frac{BN_\gamma}{2} \qquad (4.38)$$

where
$c$ is the undrained shearing resistance
$B$ is the base width
$D$ is the base depth below the rock surface
$\gamma$ is the effective density of the rock mass
$N_c$, $N_q$ and $N_\gamma$ are the bearing capacity factors related to $\phi$ as shown in Figure 4.30

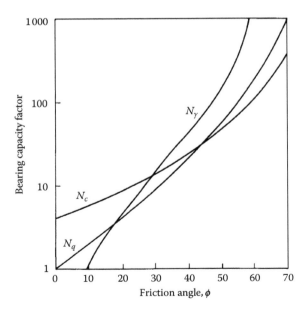

*Figure 4.30* Wedge bearing capacity factors for foundations on rock. (Reprinted from Pells, P.J.N. and Turner, R.M., End bearing on rock with particular reference to sandstone, *Proceedings of the International Conference on Structural Foundations on Rock*, Sydney, New South Wales, Australia, Vol. 1, pp. 181–190, 1980.)

Equation 4.38 represents wedge failure conditions beneath a strip foundation and should not be confused with Terzaghi's equation for spread foundations. Because Equation 4.38 is for strip loading, the value of $cN_c$ should be multiplied by a factor of 1.25 for a square pile or 1.2 for a circular pile base. Also the term $\gamma BN_\gamma/2$ should be corrected by the factors 0.8 or 0.7 for square or circular bases respectively. The term $\gamma BN_\gamma/2$ is small compared with $cN_c$ and is often neglected.

Where it is difficult to obtain satisfactory samples for laboratory testing to determine $c$ or $\phi$, the relationship of these parameters to the uniaxial compression strength and RQD of the rock as shown in Table 4.11 can be used. The $q_{uc}$ values are determined from tests on core specimens of the intact rock to obtain its point load strength (Section 11.1.4).

It is important to note that to mobilise the maximum base resistance from Equation 4.38, the settlement of the pile toe is likely to be of the order of 20% of its diameter. The database of test results produced by Zhang and Einstein[4.57] shows that the end-bearing capacity mobilised at a toe settlement of 10% of the pile diameter can be estimated from

$$\frac{p_b}{P_a} \sim 15\left(\frac{q_{uc}}{P_a}\right)^{0.5} \tag{4.39}$$

where $P_a$ is the normalised atmospheric pressure. While this review shows the wide scatter of sidewall resistance and the uniaxial compressive strength, they summarise the ultimate unit shaft resistance in smooth rock sockets as $0.4q_{uc}^{0.5}$ and in rough sockets as $0.8q_{uc}^{0.5}$ (cf. Equation 4.42 ).

Driving a closed-end pile into low- to medium-density *chalk* causes blocks of the rock to be pushed aside. Crushed and remoulded chalk flows from beneath the toe, and the cellular structure of the rock is broken down releasing water trapped in the cells to form a slurry. This flows into fissures and causes an increase in pore pressure which considerably weakens the shaft resistance, although it is possible that drainage from the fissures will eventually relieve the excess pore pressure, thereby increasing the shaft resistance.

Very little penetration is likely to be achieved when attempting to drive large closed-end piles into a high-density chalk formation with closed joints, but penetration is possible with open-end or H-section plies. As a result of these effects, Equations 4.36 and 4.38 cannot be used to calculate base resistance. From the results of a number of plate and pile loading tests, Lord et al. in CIRIA Report 574[4.58] recommend that the base resistance should be related to the SPT N-values (Section 11.1.4). The report gives the relationship for driven precast piles as

$$\text{Base resistance} = q_{ub} = 300\,N\ \text{kN/m}^2 \tag{4.40}$$

where $N$ is the SPT resistance in blows/300 mm. A lower bound is of the order of $200\,N\,\text{kN/m}^2$.

No correction should be made to the N-values for overburden pressure when using Equation 4.40. The use of this equation is subject to the stress at the base of the pile not exceeding 600–800 kN/m² for low- to medium-density chalk and 1000–1800 kN/m² for medium- to high-density chalk. Report 574 gives recommendations for the allowable pile load using different factors of safety on the ultimate shaft and bearing capacities of the pile. Application of EC7 partial resistance factors is considered in Section 4.7.5.

Dynamic testing (Section 7.3) of preliminary or working piles is frequently used to determine end-bearing resistance on chalk. CIRIA Report 574 states that instrumented dynamic tests using the CAPWAP® program (see Appendix C) can give a good estimate of end-bearing resistance provided that the hammer blow displaces the toe at least 6 mm during the test. Definitions of the density grades of chalk and their characteristics are given in Appendix A.

*Granite* rocks are widely distributed in Hong Kong, where the fresh rock is blanketed by varying thicknesses of weathered rock in the form of a porous mass of quartz particles in a clayey matrix of decomposed feldspar and biotite. The Geotechnical Engineering Office (GEO) of the Hong Kong government[4.59] recommends that piles should be driven to refusal in a fresh to moderately decomposed or partially weathered granite having a rock content greater than 50%. For these conditions, the load on the pile is governed by the design stress on the material forming the pile. CAPWAP® analysis is recommended to determine pile resistance. In the case of open or clay-filled joints, the yielding of the pile at the toe should be calculated using the drained elastic modulus of the rock. The GEO publication gives an $E'_v$ value of 3.5–5.5 N (MN/m²), where $N$ is the SPT value. It is pointed out that $N$ may be increased by compaction during pile driving.

The shaft friction developed on piles driven into *weak weathered* rocks cannot always be calculated from the results of laboratory tests on rock cores. It depends on such factors as the formation of an enlarged hole around the pile, the slurrying and degradation of rocks, the reduction in friction due to shattering of the rock by driving adjacent piles, and the presence of groundwater. In the case of brittle coarse-grained rocks such as sandstones, igneous rocks and some limestones, it can be assumed that pile driving shatters the rock around the pile shaft to the texture of a loose to medium-dense sand. The characteristic shaft resistance can then be calculated from Equation 4.14 using the appropriate values of $K_s$ and $\delta$. Where rocks such as mudstones and siltstones weather to a clayey consistency making it possible to obtain undisturbed samples from boreholes, the weathered rock can be treated as a clay and the shaft friction calculated from the methods described in Section 4.2.1.

The effects of degradation of weakly cemented carbonate soils caused by pile driving have been discussed in Section 4.3.3. Similar effects occur in carbonate rocks such as detrital coral limestones, resulting in very deep penetration of piles without any significant increase in driving resistance. An example of the low driving resistance provided by weak coral limestone to the penetration of closed-end tubular steel piles at a coastal site in Saudi Arabia is shown in Figure 4.31.

Beake and Sutcliffe[4.60] observed ultimate unit shaft resistances of 170 and 300 kN/m² from tension tests on 1067 and 914 mm OD tubular steel piles driven to 4.2 and 4.55 m with open ends in weak carbonate siltstones and sandstones in the Arabian Gulf. The mean compression strengths of the rocks were 3.2 and 4.7 MN/m². The above-mentioned shaft resistances were 0.04–0.10 of the mean uniaxial compression strength of the rock.

Although a relationship was established between the base resistance and SPT $N$-values of piles driven into chalk as noted earlier, no meaningful relationship could be found with shaft resistance. The CIRIA recommendations[4.58] in Table 4.12 are the best possible estimates derived from pile loading tests. An upper limiting value of unit shaft friction for high-strength, high-density chalk is 150 kN/m². The CIRIA Report recommends that whenever possible a preliminary trial pile should be tested to verify the design. It should be noted that dissipation of excess pore pressure caused by pile driving can increase the shaft resistance of piles in chalk. Therefore, as long a delay as possible, at least 28 days, should be allowed between driving and load testing. Some other observed values of the shaft resistance of piles in weak rocks are shown in Table 4.13.

### 4.7.2 Driven and cast-in-place piles

Driven and cast-in-place piles terminated on strong rock can be regarded as end bearing. The actions on the pile are governed by the stress on the pile shaft at the point of minimum cross section. Where these piles are driven into weak or weathered rocks, they should be regarded as partly friction and partly end-bearing piles.

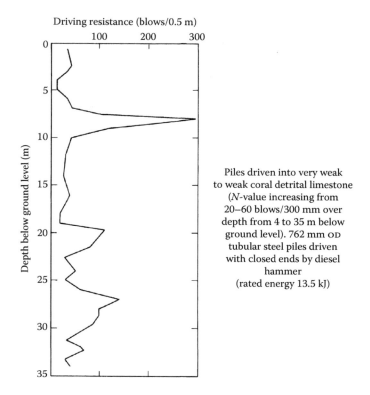

Driving resistance (blows/0.5 m)

Piles driven into very weak
to weak coral detrital limestone
(N-value increasing from
20–60 blows/300 mm over
depth from 4 to 35 m below
ground level). 762 mm OD
tubular steel piles driven
with closed ends by diesel
hammer
(rated energy 13.5 kJ)

*Figure 4.31* Low resistance to driving of tubular steel piles provided by weak coral limestone.

*Table 4.12* CIRIA recommendations for the shaft resistance of displacement piles driven into chalk

| Chalk classification | Type of pile | Ultimate unit shaft resistance (kN/m²) |
|---|---|---|
| Low- to medium-density, open joints | Small displacement | 20 |
| | Small displacement, H-sections | 10 |
| | Large displacement, preformed | 30 |
| High-density, closed joints | Small displacement, open-end tubular | 120 |
| | Large displacement, preformed in predrilled holes | (100) verify by load testing |

CIRIA Report 574[4.58] recommends that the base resistance of driven and cast-in-place piles in chalk should be taken as $250 N$ kN/m² where $N$ is the SPT $N$-value. A lower bound should be $200 N$ kN/m² with the recommendation to make a preliminary test pile whenever possible. For calculating the unit shaft resistance as Equation 4.14, the effective overburden pressure should be multiplied by a factor of 0.8 where $\sigma'_{vo}$ is less than 100 kN/m². If $\sigma'_{vo}$ is greater than 100 kN/m², the design should be confirmed by a loading test.

### 4.7.3  Bored and cast-in-place piles

Where these piles are installed by drilling through soft overburden onto a strong rock, the piles can be regarded as end-bearing elements, and their capacity is determined by the design stress on the pile shaft at the point of minimum cross section. Bored piles drilled down for some depth into weak or weathered rocks and terminated within these rocks act partly as

*Table 4.13* Observed ultimate shaft friction values for piles driven into weak and weathered rocks

| Pile type | Rock description | Ultimate unit shaft friction (kN/m$^2$) | Reference |
|-----------|------------------|------------------------------------------|-----------|
| H-section | Moderately strong slightly weathered slaty mudstone | 28[a] | (4.61) |
| H-section | Moderately strong slightly weathered slaty mudstone | 158[b] | (4.61) |
| Steel tube | Very weak coral detrital limestone (carbonate sandstone/siltstone) | 45 | Unpublished |
| Steel tube | Faintly to moderately weathered moderately strong to strong mudstone | 127 | Unpublished |
| Steel tube | Weak calcareous sandstone | 45 | Unpublished |
| Precast concrete | Very weak closely fissured argillaceous siltstone (Mercia Mudstone) | 130 | (4.62) |

[a] Penetration 1.25 m.
[b] Penetration 2.2 m.

friction and partly as end-bearing piles. The shaft resistance in the overburden is usually neglected. Wyllie[4.54] gives a detailed account of the factors governing the development of shaft friction over the depth of the rock socket (also known as drilled piers). The factors which govern the bearing capacity and settlement of the pile are summarised as the following:

1. The length-to-diameter ratio of the socket
2. The strength and elastic modulus of the rock around and beneath the socket
3. The condition of the sidewalls, that is roughness and the presence of drill cuttings or bentonite slurry
4. Condition of the base of the drilled hole with respect to removal of drill cuttings and other loose debris
5. Layering of the rock with seams of differing strength and moduli
6. Settlement of the pile in relation to the elastic limit of the sidewall strength
7. Creep of the material at the rock–concrete interface resulting in increasing settlement with time

The effect of the length/diameter ratio of the socket is shown in Figure 4.32 for the condition of the rock having a higher elastic modulus than the concrete[4.63]. It will be seen that if it is desired to utilise base resistance as well as socket friction, the socket length should be less than four pile diameters (For example, in a 1m diameter socket 1m deep, 55% of the applied load will be carried on the base). The high interface stress over the upper part of the socket will be noted.

The condition of the sidewalls is an important factor. In a weak rock such as chalk, clayey shale, or clayey weathered marl, the action of the drilling tools is to cause softening and slurrying of the walls of the borehole, and in the most adverse case, the shaft friction corresponds to that typical of a smooth borehole in a soft clay. In stronger and fragmented rocks, the slurrying does not take place to the same extent, and there is a tendency towards the enlargement of the drill hole, resulting in better keying of the concrete to the rock. If the pile borehole is drilled through soft clay, this soil may be carried down by the drilling tools to fill the cavities and smear the sides of the rock socket. This behaviour can be avoided to some extent by inserting a casing and sealing it into the rockhead before continuing the drilling to form the rock socket, but the interior of the casing is likely to be heavily smeared with clay which will be carried down by the drilling tools into the rock socket. Wyllie[4.54] suggests

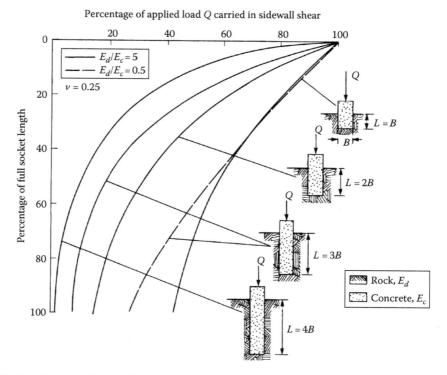

Figure 4.32 Distribution of sidewall shear stress in relation to socket length and modulus ratio. (After Osterberg, J.O. and Gill, S.A., Load transfer mechanisms for piers socketed in hard soils or rock, *Proceedings of the Ninth Canadian Symposium on Rock Mechanics*, Montreal, Quebec, Canada, pp. 235–262, 1973.)

that if bentonite is used as a drilling fluid, the rock socket shaft friction should be reduced to 25% of that of a clean socket unless tests can be made to verify the actual friction which is developed.

It is evident that the keying of the shaft concrete to the rock and hence the strength of the concrete to rock bond is dependent on the strength of the rock. Correlations between the uniaxial compression strength of the rock and rock socket bond stress have been established by Williams and Pells[4.64], Horvath[4.65], and Rosenberg and Journeaux[4.66]. The bond stress, $f_s$, is related to the average uniaxial compression strength, $\bar{q}_{uc}$, by the following equation:

$$f_s = \alpha\beta\bar{q}_{uc} \tag{4.41}$$

where
   $\alpha$ is the reduction factor relating to $\bar{q}_{uc}$ as shown in Figure 4.33
   $\beta$ is the correction factor related to the discontinuity spacing in the rock mass as shown in Figure 4.34

The curve of Williams and Pells[4.64] in Figure 4.33 is higher than the other two, but the $\beta$ factor is unity in all cases for the Horvath and the Rosenberg and Journeaux curves. It should also be noted that the factors for all three curves do not allow for smearing of the rock socket caused by clay overburden dragged into the socket or degradation of the rock.

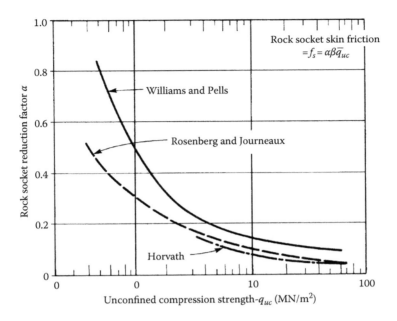

Figure 4.33 Reduction factors for rock socket shaft friction.

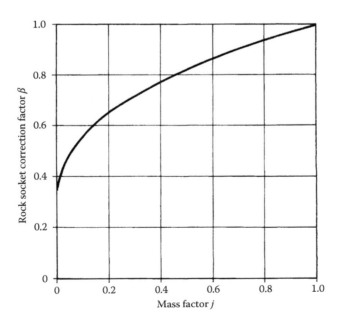

Figure 4.34 Reduction factors for discontinuities in rock mass. (After Williams, A.F. and Pells, P.J.N., *Can. Geotech. J.*, 18, 502, 1981.)

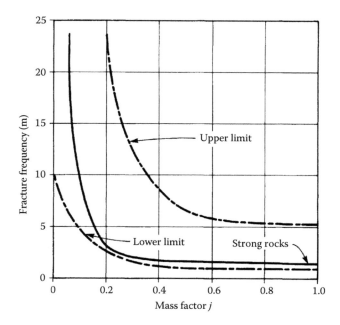

*Figure 4.35* Mass factor value. (After Hobbs, N.B., Review paper – Rocks, *Proceedings of the Conference on Settlement of Structures*, British Geotechnical Society, Pentech Press, pp. 579–610, 1975.)

The $\beta$ factor is related to the mass factor, $j$, which is the ratio of the elastic modulus of the rock mass to that of the intact rock as shown in Figure 4.35. If the mass factor is not known from loading tests or seismic velocity measurements, it can be obtained approximately from the relationships with the RQD or the discontinuity spacing quoted by Hobbs[4.67] as follows:

| RQD (%) | Fracture frequency per metre | Mass factor j |
|---|---|---|
| 0–25 | >15 | 0.2 |
| 25–50 | 15–8 | 0.2 |
| 50–75 | 8–5 | 0.2–0.5 |
| 75–90 | 5–1 | 0.5–0.8 |
| 90–100 | 1 | 0.8–1 |

As a result of later research, Horvath et al. [4.68] derived the following equation for calculating the socket shaft friction of large-diameter piles in mudstones and shales:

$$\text{Unit shaft friction} = f_s = b\sqrt{\sigma'_{ucw}} \qquad (4.42)$$

where

$\sigma'_{ucw}$ is the uniaxial compression strength of the weaker material (concrete or rock)
$f_s$ and $\sigma'_{ucw}$ are expressed in MN/m²
$b$ is given as 0.2–0.3

Alrifai[4.69] provides comparisons of the various methods of calculating shaft friction for rock sockets in carbonate sandstone in Dubai and concludes that, in these conditions, unit shaft friction based on the Horvath et al. Equation 4.42 is the closest to the observed ultimate pile capacity.

The shaft friction can be increased in weak or friable rocks by grooving the socket. Horvath et al.[4.68] described experiments in mudstones using a toothed attachment to a rotary auger. They showed that $f_s$ was related to the depth of the groove by the following equation:

$$\frac{f_s}{\sigma'_{ucw}} = 0.8(\text{RF})^{0.45} \tag{4.43}$$

where
$\sigma'_{ucw}$ is the rock strength defined previously
RF is a roughness factor given as

$$\text{RF} = \frac{\overline{\Delta}_r}{r_s} \times \frac{L_t}{L_s} \tag{4.44}$$

where
$\overline{\Delta}_r$ is the average height of asperities
$r_s$ is the nominal socket radius
$L_t$ is the total travel distance along the grooved profile
$L_s$ is the nominal socket depth

$\overline{\Delta}_r$ is further defined as the radial distance from a socket profile to the surface of an imaginary cylinder which would fit into the grooved socket. There may be practical difficulties in measuring the depth of the groove achieved by the rotary tool, particularly where direct visual or underwater television methods of inspection are used in muddy water.

Chandler and Forster in CIRIA Report 570[4.70] recommend that the shaft friction of bored piles in very weak mudstones can be calculated in the same way as piles in stiff clay using either effective stress methods (Equations 4.9 and 4.12) or undrained shear strengths (Equation 4.10). However, the report points out the difficulty in obtaining satisfactory samples in weak weathered mudstones with the result that the $c_u$ values are likely to be low and hence the calculated shaft friction will be over-conservative. When effective stress methods are used, $c'$ should be taken as zero to allow for softening, and a remoulded value of $\phi'$ of 36° should be assumed. Laboratory tests gave $K_0$ values of 1.5–1.6. Report 570 provides values for the adhesion factor $\alpha$ in Equation 4.10 and $\beta$ in effective stress Equation 4.12 for the weathering grades of mudstone at various sites shown in Table 4.14.

When installing CFA piles in Mercia Mudstone, care must be taken to avoid 'overflighting' (see Sections 2.4.2 and 3.4.7) resulting in remoulding of the sides of the pile shaft.

Table 4.14  $\alpha$ and $\beta$ values of weak mudstones related to weathering grades

| Grade | $\alpha$ | $\beta$ |
|---|---|---|
| IV–III various sites | 0.45 | — |
| IV–II Leicester | 0.45 | 0.5 |
| II Kilroot | 0.3 | 1.71 |
| IV–III Antrim | 0.3 | 0.86 |
| III Berkeley | 0.31–0.44 | 0.86–1.06 |
| IV–II Derby | 0.45 | — |
| IV–II Cardiff | 0.375 | — |

The remoulded layer of clay-enriched material can exceed 50 mm and will significantly affect the $\alpha$ and $\beta$ design parameters. Variations in the measurement of $c_u$ in Mercia Mudstone are not uncommon and can result in conservative estimates for shaft friction.

The end-bearing resistance of bored piles in weak rocks depends to a great extent on drilling techniques. The use of percussive drilling tools can result in the formation of a very soft sludge at the bottom of the drill hole which, apart from weakening the base resistance, makes it difficult to identify the true character of the rock at the design founding level. The use of powerful mechanical augers of the type described in Section 3.3 has eliminated most of the rock identification problems associated with percussion drilling. While SPTs or CPTs can be used to assess rock quality at base level, the examination and testing of cores taken from boreholes at the site investigation stage is preferable, with later correlation by examining drill cuttings from the pile boreholes. This is particularly necessary in thinly bedded strata where weak rocks are interbedded with stronger layers. In such cases, the end-bearing resistance should be governed by the strength of the weak layers, irrespective of the strength of the material on which the pile is terminated.

In the case of sandstones which have been completely weathered to a soil-like consistency, base resistance can be obtained from SPTs and CPTs with calculations from the test results as described for bored piles in coarse-grained soils in Section 4.3.6.

Rotary coring and skilled drilling techniques can provide good-quality undisturbed samples in completely weathered mudstones, siltstones and shales. Shear strength tests can then be made and base resistance calculated as described in Section 4.2.3. In the case of moderately weathered mudstones, siltstones and shales, uniaxial compression tests are made on rock cores, or in the case of poor core recovery, point load tests (Section 11.1.4) are made to obtain the compression strength. The base resistance is then calculated using the relationship with $q_{uc}$ and RQD as shown in Table 4.11. Alternatively, the parameters $c$ and $\phi$ can be obtained from this table and used in conjunction with Equation 4.38.

In the absence of compression strength data, published relationships between the weathering grade, undrained shear strength and elastic properties of the preceding weak rocks can be used to determine the base resistance from Equation 4.38. Gannon et al. in CIRIA Report 181[4.71] give these properties as shown in Table 4.15; note the fracture frequency in chalk is different from those in Figure 4.35.

High values of base resistance resulting from the calculations described earlier should be adopted with caution because of the risk of excessive base settlement. This can be of

Table 4.15 Relationships between weathering grades, undrained shear strength and elastic properties of weak rocks

| Weathering grade | Clay content % | Undrained shear strength ($c_u$, kN/m²) | Shear modulus (G, MN/m²) | Young's modulus (E, MN/m²) |
|---|---|---|---|---|
| V–VI | | 250 | 80 | 115 |
| IV | | 850 | 100 | 230 |
| III | 10 | 1330 | 350 | 820 |
| III | 15 | 1270 | 265 | 615 |
| III | 20 | 1230 | 210 | 490 |
| III | 25 | 1150 | 175 | 405 |
| III | 30 | 1090 | 150 | 350 |
| I and II | | 1450 | 1270 | 2830 |

Source: Seedhouse, R.L. and Sanders, R.L., Investigations for cooling tower foundations in Mercia Mudstone at Ratcliffe-on-Soar, Nottinghamshire, *Proceedings of the Conference on Engineering Geology of Weak Rock*, A.A. Balkema, Rotterdam, the Netherlands, Special Publication No. 8, pp. 465–472, 1993.

the order of 20% of the pile width at the toe which is required to mobilise the ultimate base resistance. Equation 4.40 should be considered in these circumstances. Significant shaft settlement could break down the bond between the rock and concrete, thus weakening the total pile resistance in cases where the design requires the load to be shared between the shaft and the base. A reduction in shaft resistance of 30%–40% of the peak value has been observed where shear displacements of the rock socket of little more than 15 mm have occurred. It may also be difficult to remove soft or loose debris from the whole base area at the time of final clean-out before placing the concrete.

Because of the porous cellular nature of chalk and the consequent breakdown and softening of the material under the action of drilling tools (similar to that described in Section 4.7.1), conventional methods of calculating the base resistance and rock socket shaft friction cannot be used for bored piles in chalk. CIRIA Report 574 states that these two components of bearing capacity are best determined from relationships with the SPT $N$-values uncorrected for overburden pressure. These give a rough indication of the weathering grade to supplement the classification based on examination of rock cores and exposures in the field. CIRIA recommendations for bored and CFA piles are that $q_b$ should be taken as 200 $N$ KN/m² (Report 574) and $q_s$ as $\beta \sigma'_{vo}$ (Report PR86, Lord et al.[4.73]).

Where the average effective overburden pressure, $\sigma'_{vo}$, is less than 400 kN/m² (based on final ground levels and omitting the contribution from made ground and fill), the calculated shaft friction must be confirmed by load testing. In high-density Grade A chalk, the pile may be treated as a rock socket and the shaft friction taken as 0.1 times the uniaxial compressive strength. Report 574 makes a distinction between made ground and fill. The former is regarded as an accumulation of debris resulting from the *activities of man*, whereas fill is purposefully placed.

The shaft factor $\beta$ should be based on SPT $N$-values. For low values of $N$ ($\leq$10) or a cone resistance $q_c$ between 2 and 4 MN/m², $\beta$ should be taken as 0.45. Reports on loading tests on bored piles and driven tubular steel piles in high-strength chalk since PR86 frequently indicate higher shaft friction resistance than expected from this approach. $\beta$ is therefore more usually taken as 0.8, for medium-dense chalk with $N > 10$ and in the absence of flints. CPT values in chalk are not sufficiently reliable for the calculation of base resistance.

However, there is continued uncertainty over shaft friction in chalk. The above-mentioned shaft resistance implies that $\sigma'_{vo}$ and $\sigma'_h$ are directly proportional in chalk, which is questionable. Hence, the recommendations in Reports 574 and PR86 for load testing at some stage as a means of confirming load capacity and achieving economy in design are still important. It is pointed out that a single test made to 3 times the applied load is a much better aid to judgement than two tests to 1.5 times the applied load.

There is a clear distinction between the shaft friction available in the weak rocks (<3 MN/m²) as considered by Williams and Pells[4.64], where friction can be attributed to roughness of the bore and that in strong rock, such as Carboniferous limestone, intact sandstone and igneous rocks. Here, the substantial base resistance can usually only be mobilised by invoking sidewall slip in a straight smooth-sided socket, resulting in possible brittle failure of the rock–pile bond. If the initial drilling produces a degree of roughness, which is difficult to assess even with current devices such as ultrasonic probes, then peak average shear strength will be mobilised at small displacement. Over time, this will produce a plastic load transfer from the walls to the base without causing the shear resistance to fall as noted by Rosenberg and Journeaux[4.66]; their reduction factors in Figure 4.33 are likely to be more representative for shaft friction estimation. There is a case for limiting the contribution of the length available for shaft resistance in a strong rock socket to twice the shaft diameter to ensure effective load distribution.

*Table 4.16* Presumed safe vertical bearing stress for foundations on horizontal ground in Hong Kong

| Category | Weathering grade | Total core recovery (%) | Uniaxial compression strength (MN/m²) | Equivalent point load index strength (MN/m²) | Presumed bearing stress (MN/m²) |
|---|---|---|---|---|---|
| I(a) | I | 100 | 75 | 3 | 10 |
| I(b) | ≥II | 95 | 50 | 2 | 7.5 |
| I(c) | ≥III | 85 | 25 | 1 | 5 |
| I(d) | ≥IV | 50 | — | — | 3 |

Notes: Category I(a), fresh strong to very strong rock. Category I(b), fresh to slightly decomposed strong rock. Category I(c), slightly to moderately decomposed, moderately strong rock. Category I(d): moderately decomposed, moderately strong rock to moderately weak rock.

The practice in Hong Kong for granites and volcanic rocks is to relate the base bearing pressure for bored piles to the weathering grade of the decomposed material. The recommendations of the Government Geotechnical Office[4.59] are shown in Table 4.16. The rock socket shaft friction in weak to moderately weak and strong to moderately strong granites should be determined from correlation with the uniaxial compression strength of sedimentary rocks using the method of Horvath et al.[4.68]. Ng et al.[4.74] point out that observations made in loading tests in granites suggest that the value for $b$ in Equation 4.42 of 0.2 is appropriate. Completely weathered granite should be treated as a soil.

Displacement of piles due to lateral load and moments at the pile head are unlikely to cause deformation in the rock socket, especially where the pile is installed through overburden and the socket is short. However, in situ data and load test observations are not readily available. Piles for integral bridge abutments which are subject to lateral loads from thermal movements and movement of embankments are considered in Section 9.5.

## 4.7.4 Settlement of the single pile at the applied load for piles in rocks

The effects of load transfer from shaft to base of piles on the pile head settlements have been discussed by Wyllie[4.54]. Because of the relatively short penetration into rocks which is needed to mobilise the required total pile resistance, the simpler methods of determining pile head settlement described in Section 4.6 are suitable in most cases. For piles having base diameters up to 600 mm, the settlement at the applied load should not exceed 10 mm if the EC7 partial factors for ULS have been applied.

The settlement of large-diameter piles can be calculated from Equation 4.35. The modulus of deformation of the rock below the pile toe can be obtained from plate bearing tests or PMTs or from empirical relationships developed between the modulus, the weathering grade and the unconfined compression strength of the rock given in Table 4.15 and Section 5.5.

These relationships are not applicable to high-porosity chalk or weathered silty mudstone (Mercia Mudstone). The relationships given in Section 5.5 assume fairly low stress levels. Therefore, calculated values based on the unconfined compression strength of the rock should take into account the high-bearing pressures beneath the base of piles.

In CIRIA Report 574[4.58], the deformation modulus of chalk is related to the weathering grade and SPT $N$-values. For Grade A chalk where the $N$-value is greater than 25, the deformation modulus is 100–300 MN/m². For Grades B, C and D with $N$-values less than 25, the modulus is 25–100 MN/m².

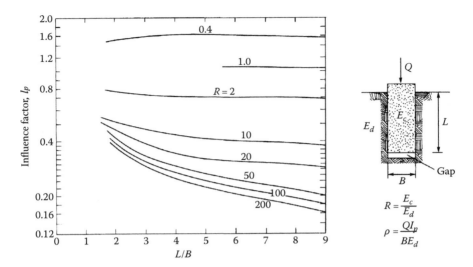

*Figure 4.36* Elastic settlement influence factors for rock socket shaft friction on piles. (After Pells, P.J.N. and Turner, R.M., *Can. Geotech. J.*, 16, 481, 1979; Courtesy of Research Journals, National Research Council, Ottawa, Ontario, Canada.)

Pells and Turner[4.75] have derived influence factors for calculating the settlement of a bored pile where the load is carried by rock socket shaft friction only using the following equation:

$$\text{Settlement} = \rho = \frac{QI_p}{BE_d} \tag{4.45}$$

where
   $Q$ is the total load carried by the pile head
   $I_p$ is the influence factor
   $B$ is the diameter of the socket
   $E_d$ is the deformation modulus of the rock mass surrounding the shaft

The influence factors of Pells and Turner are shown in Figure 4.36. Where the rock sockets are recessed below the ground surface or where a layer of soil or very weak rock overlies competent rock, a reduction factor is applied to Equation 4.45. Values of the reduction factor are shown in Figure 4.37.

Fleming et al.[4.23] describe a method for obtaining the load/settlement relationship in soils (Section 4.9.1) which can be applied to sockets in weak rock with the resistance of the shaft and base treated separately. CIRIA Report 181[4.71] gives further examples of the performance of rock-socket piles. Computer programs, such as ROCKET, are available which are capable of including parameters (generally for weaker rocks) which cannot be considered in the empirical methods (see Appendix C). The American Transportation Research Board has produced a synthesis of information on the design of rock-socketed shafts under axial and lateral loading which has been used in the Oasys ALP program for strong rock.

## 4.7.5 Eurocode recommendations for piles in rock

EC7 makes no specific recommendations for the design of piles carrying axial compression loads in rock. The design methods described in Sections 4.7.1 through 4.7.3 are based either

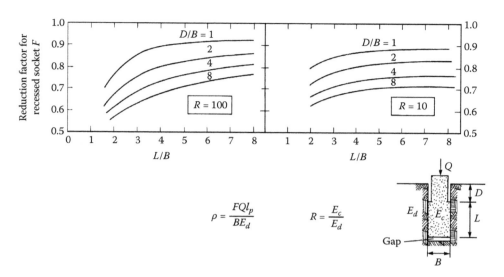

*Figure 4.37* Reduction factors for calculation of settlement of recessed sockets. (After Pells, P.J.N. and Turner, R.M., *Can. Geotech. J.*, 16, 481, 1979; Courtesy of Research Journals, National Research Council, Ottawa, Ontario, Canada.)

on relationships with uniaxial compression strengths or by correlation with SPT $N$-values and are compatible with EC7 procedures. Where the calculations are based on SPT tests, the calibration factor of 1.05 could be applied to the $N$-value, but the effect is small in relation to the other empirical design factors.

The model pile procedure can be used to calculate design total pile resistances based on a series of static or dynamic pile tests using the correlation factors based on the number of tests as in the NA tables. The relevant partial resistance factors as described in Section 4.1.4 are used to obtain the design resistances for ULS calculations. It can be inferred from EC7 Clause 7.6.4.2 that when a pile toe is seated on intact rock, 'the partial safety factors for the ultimate limit state conditions are normally sufficient to satisfy serviceability limit state conditions'.

The selection of the characteristic value of the uniaxial compressive strength of the rock to calculate end-bearing resistance, $R_{bk}$, requires judgement by the designer, and the guidance given by the 'presumed safe bearing stress' for allowable stress calculations should be considered for preliminary design.

## 4.8 PILES IN FILL: NEGATIVE SKIN FRICTION

### 4.8.1 Estimating negative skin friction

Piles are frequently required for supporting structures that are sited in areas of deep fill. The piles are taken through the fill to a suitable bearing stratum in the underlying natural soil or rock. No support for compressive loads from shaft friction can be assumed over the length of the pile shaft through the fill. This is because of the downward movement of the fill as it compresses under its own weight or under the weight of further soil or surcharge placed over the fill area. *Negative skin friction* is the shear stress acting downwards along the pile shaft due to the downward soil movement relative to the pile. The downward movement results in *dragload*, the load transferred to the pile, which must be structurally designed to resist this additional load. *Downdrag* is the downward movement (*drag settlement*) of the pile due to the dragload. A *neutral point* exists where there is equilibrium between the downward permanent actions plus the dragload and the upward acting positive shaft resistance plus the mobilised pile toe resistance. The neutral point is also the point at which the relevant movement between the pile and soil can be considered zero. Where fill is placed over a compressible natural soil, the latter consolidates and moves downwards relative to the pile. Thus, the negative skin friction occurs over the length of the shaft within the natural soil as well as within the fill.

Calculation of the magnitude of the negative skin friction is a complex problem which depends on the following factors:

1. The relative movement between the fill and the pile shaft
2. The relative movement between any underlying compressible soil and the pile shaft
3. The elastic compression of the pile under the applied load
4. The rate of consolidation of the compressible layers

The simplest case is fill that is placed over a relatively incompressible rock with piles driven to refusal in the rock. The toe of the pile does not yield under the combined applied load and downdrag forces. Thus, the negative skin friction on the upper part of the pile shaft is equal to the fully mobilised value. Near the base of the fill, its downward movement may be insufficiently large to mobilise the full skin friction, and immediately above rockhead, the fill will not settle at all relative to the pile shaft. Thus, negative skin friction cannot occur at this point. The distribution of negative skin friction on the shaft of the unloaded pile is shown in Figure 4.38a. If a heavy load is now applied to the pile shaft, the shaft compresses elastically and the head of the pile moves downwards relative to the fill. The upper part of the fill now acts in support of the pile although this contribution is neglected in calculating

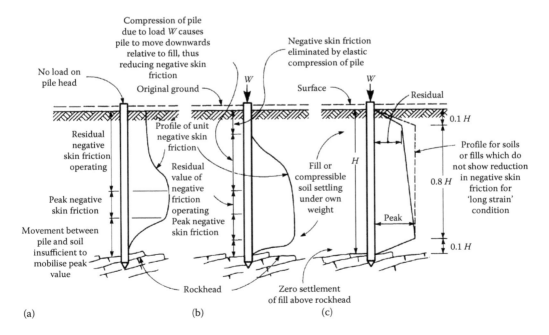

*Figure 4.38* Distribution of negative skin friction on piles terminated on relatively incompressible stratum. (a) No load on pile head. (b) Compressive load on pile head. (c) Design curve for loaded pile.

the pile resistance. The distribution of negative skin friction on the shaft of the loaded pile is shown in Figure 4.38b. Where the fill has been placed at a relatively short period of time before installing the piles, continuing consolidation of the material will again cause it to slip downwards relative to the pile shaft, thus reactivating the downdrag force.

The simplified profile of negative skin friction for a loaded pile on an incompressible stratum is shown in Figure 4.38c. This diagram can be used to calculate the magnitude of the dragload. The peak values for coarse soils and fill material are calculated by the method described in Section 4.3.

In the case where negative skin friction is developed in clays, the rate of loading must be considered. It was noted in Section 4.2.4 that the capacity of a clay to support a pile in shaft friction is substantially reduced if the load is applied to the pile at a very slow rate. The same consideration applies to negative skin friction, but in this case, it works advantageously in reducing the magnitude of the dragload. In most cases of negative skin friction in clays, the relative movement between the soil which causes downdrag and the pile takes place at a very slow rate. The movement is due to the consolidation of the clay under its own weight or under imposed loading, and this process is very slow compared with the rate of application of the applied load to the pile.

Meyerhof[4.50] advises that the negative skin friction on piles driven into soft to firm clays should be calculated in terms of effective stress from the following equation:

$$\tau_{s\ neg} = \beta\sigma'_{vo} \tag{4.46}$$

Values of the negative skin friction factor, $\beta$, which allow for reduction of the effective angle of friction with increasing depth to the residual value $\delta_r$ are shown in Figure 4.39.

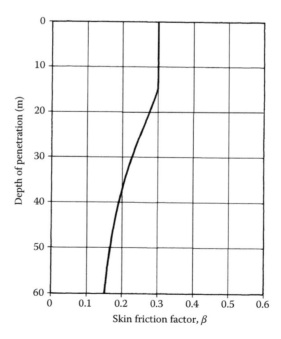

Figure 4.39 Negative skin friction factors for piles driven into soft to firm clays. (After Meyerhof, G.G., Bearing capacity and settlement of pile foundations, Proceedings of the American Society of Civil Engineers, GT3, pp. 197–228, 1976.)

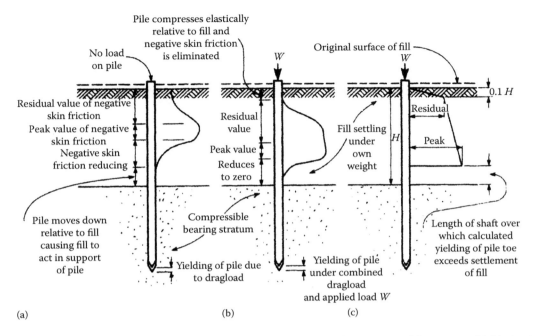

*Figure 4.40* Distribution of negative skin friction on piles terminated in compressible stratum. (a) No load on pile head. (b) Compressive load on pile head. (c) Design curve for loaded pile.

Taking the case of a pile bearing on a compressible stratum, where yielding of the pile toe occurs under the dragload and the subsequently applied load, the downward movement of the pile relative to the lower part of the fill may then be quite large and such that negative skin friction is not developed over an appreciable proportion of the length of the shaft within the fill. Over the upper part of the shaft, the fill moves downwards relative to the pile shaft to an extent such that the negative skin friction operates, whereas in the middle portion of the pile shaft, the small relative movement between the fill and the pile may be insufficient to mobilise the peak skin friction as a downdrag force. The distribution for the unloaded pile is shown in Figure 4.40a.

When the design load is applied to the head of the pile, elastic shortening of the pile occurs, but since the load is limited by the bearing characteristics of the soil at the pile toe, the movement may not be large enough to eliminate the dragload. The distribution of negative friction is then shown in Figure 4.40b. The diagram in Figure 4.40c can be used for design purposes, with the peak value calculated as described in Section 4.3 for coarse soils and fill and by using Equation 4.46 and Figure 4.39 for soft to firm clays.

It may be seen from Figure 4.40a–c that at no time does the maximum skin friction operate as a dragload over the full length of the pile shaft. It is not suggested that these simplified profiles of distribution of negative skin friction represent the actual conditions in all cases where it occurs, since so much depends on the stage reached in the consolidation of the fill and the compression of the natural soil beneath the fill. The time interval between the installation of the pile and the application of the load is also significant. In old fill which has become fully consolidated under its own weight and where it is not proposed to impose surcharge loading, the negative skin friction may be neglected, but shaft friction within the fill layer should not be allowed to help support the pile. In the case of recently placed fill, it may settle by a substantial amount over a long period of years. The fill may also be causing consolidation and settlement of the natural soil, within which the pile obtains its bearing.

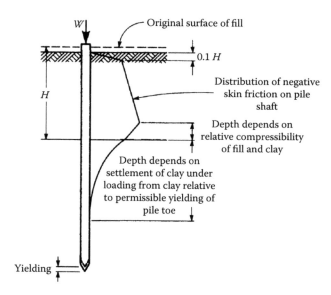

*Figure 4.41* Distribution of negative skin friction on pile driven through recent fill into compressible clay stratum.

The case of recent fill placed over a compressible soil which becomes stiffer and less compressible with depth is shown in Figure 4.41.

Modelling the load transfer by downdrag from fill and the underlying compressible soil and the distribution of resistance in positive shaft friction can be undertaken by using a soil–pile interaction analysis (see Section 4.9.5). The basic $t$–$z$ curve outputs give a more accurate estimate in separate or combined form of the distribution of axial forces over the depth of the pile shaft from the compression load applied to the pile head and the shear stress on the pile surface from the dragload than is possible from semi-empirical diagrams such as shown in Figures 4.38, 4.40 and 4.41. In particular, the $t$–$z$ curves indicate the depth $H$ in Figure 4.41, that is the depth to the *neutral point* at which the shear stress changes from negative, caused by downdrag, to positive, acting in support of the pile.

It is good practice to ignore the contribution to the support provided by friction over the length of a pile in soft clay, where the pile is driven through a soft layer to less-compressible soil. This is because of the dragload on the pile shaft caused by heave and reconsolidation of the soft clay. The same effect occurs if a pile is driven into a stiff clay, but the stiff clay continues to act in support of the pile if yielding at the toe is permitted.

Very large dragload can occur on long piles. In some circumstances, they may exceed the load applied to the head of the pile. Fellenius[4.76] measured the progressive increase in negative skin friction on two precast concrete piles driven through 40 m of soft compressible clay and 15 m of less-compressible silt and sand. Reconsolidation of the soft clay disturbed by pile driving contributed 300 kN to the dragload over a period of 5 months. Thereafter, regional settlement caused a slow increase in negative skin friction at a rate of 150 kN per year. Seventeen months after pile driving, a load of 440 kN was added to each pile, followed by an additional load of 360 kN a year later. Both these loads caused yielding of the pile at the toe to such an extent that all negative skin friction was eliminated, but when the settlement of the pile ceased under the applied load, the continuing regional settlement caused negative skin friction to develop again on the pile shaft. Thus, with a yielding pile toe, the amount of negative skin friction which can be developed depends entirely on the downward movement of the pile toe relative to the settlement of the soil or fill causing the dragload. If the dragload is caused only by the reconsolidation of the heaved soil and if the pile can be permitted to yield by an amount greater than the settlement of the ground surface due to this reconsolidation, then negative friction need not be provided for. If, however, the negative skin friction is due to the consolidation of recent fill under its own weight or to the weight of additional fill, then the movement of the ground surface will be greater than the permissible yielding of the pile toe. Negative skin friction must then be taken into account, the distribution being as shown in Figure 4.40c or Figure 4.41. It follows that negative skin friction will not reduce the ultimate geotechnical capacity of the pile. Geotechnical failure means that the pile plunges through the soil and therefore negative skin friction is not present.

Much greater dragloads occur with piles driven onto a relatively unyielding stratum. Johannessen and Bjerrum[4.77] measured the development of negative skin friction on a steel pile driven through 53 m of soft clay to rock. Sand fill was placed to a thickness of 10 m on the seabed around the pile. The resulting consolidation of the clay produced a settlement of 1.2 m at the original seabed level and a dragload of about 1500 kN at the pile toe. It was estimated that the stress in the steel near the toe could have been about 190 N/mm$^2$, which probably caused the pile to punch into the rock, so relieving some of the dragload. The average unit negative skin friction within the soft clay was equal to 100% of the undrained shearing strength of the clay.

In seismic susceptible areas, the consolidation of soils which have been subject to liquefaction during an earthquake event can produce downdrag on piles and pile caps. In soft fine-grained soils, the depth affected by liquefaction may be greater or less than the depth of compressible soil indicated from the ground investigation. The calculation of the neutral point using unfactored load and resistances is critical as discussed by Fellenius and Siegel[4.78].

### 4.8.2 Partial factors for negative skin friction

Safety factors for piles subjected to negative skin friction required careful consideration when using allowable stress design to arrive at the total allowable pile load. The negative skin friction would be conservatively estimated and deducted from the ultimate pile capacity before deciding the value of a global safety factor, usually 2.5.

The EC7 recommendations for the design of piles subjected to downdrag are much more onerous than the treatment previously applied in that the resulting axial dragload is now treated as a permanent unfavourable action in Table 4.1. This is classed as a geotechnical action in Clause 7.3.2.1(3)P which can be calculated either by a pile–soil interaction analysis (Method (a)) or as an upper-bound force exerted on the pile shaft (Method (b)). As noted in Section 4.1.4, Method (a) is the more effective of the two, particularly in determining the depth to the neutral point. It is evident that if Method (b) is used, the depth $H$ over which the upper-bound force is assumed to act is critical. If the depth is overestimated, application of the action factor $\gamma_G$ of 1.35 in Table 4.17 set A1 will further exaggerate the dragload. Worked Example 4.9 at the end of this chapter demonstrates that extra depth of pile may be needed to cope with the additional dragload action.

There is some inconsistency in the current application of the EC7 partial factors when dealing with negative skin friction, and a review by the CEN technical committee, TC250, is in hand. For structural design, the pile must be capable of supporting the factored applied actions including the dragload. The partial factors in Table 4.17 are provided by Frank et al.[1.5]; other designers use the action factor for the unfavourable dragload, and some omit the model factor $\gamma_{Rd}$ when determining the design resistance from the ground test profile. The application of the M2 partial factors $\gamma_\phi$ and $\gamma_{cu}$ is not required for axially

*Table 4.17* Partial factor sets for a pile axially loaded at the head and subjected to downdrag on the shaft

| Design approach | Structural action $\gamma_G$ | Geotechnical action | | | Resistance to compression $\gamma_s$ or $\gamma_\phi$ |
| | | Shear strength parameter $\gamma_\phi$ | Load $\gamma_G$ | | |
|---|---|---|---|---|---|
| DA1, combination 1 | A1 (1.35) | M1 (1.0) | A1 (1.35) | | R1 (1.0) |
| DA1, combination 2 | A2 (1.0) | M2 (1.25)[a] | A2 (1.0) | | R4 (1.3) |

[a] Applied as a partial action factor, not as a material factor.

loaded piles or when using effective stress calculations, for example, Meyerhof's equation 4.46, which is not directly related to the angle of shearing resistance of the soil. In most fine-grained soils, it is preferable to use the actual characteristic undrained strength directly.

The use of Method (a) requires, as a first step, a settlement analysis to determine the settlement of the fill and underlying compressible soil. Clause 7.3.2.2(5)P requires the design value of the ground settlement analysis to take account of weight densities of the material (M1 and M2 density factors in EC7 are unity and are omitted from NA tables).

When calculating downdrag on the shafts of uncased bored and cast-in-place piles, the possibility of enlargement of the pile cross section due to overbreak should be considered as well as 'waisting' in the supporting soil layer. Clause 2.3.4.2 of EC2 does not consider the possibility of enlargement, but the reductions in diameter given in Table 4.6 in Section 4.1.4 may be used when assessing the concrete design resistance of bored piles.

EC7 points out (Clause 7.3.2.2(7)) that downdrag and transient loading need not normally be considered to act simultaneously in load combinations.

Poulos[4.79] presents a relatively simple design approach which includes limit state factors and serviceability considerations which can be adapted to EC7 rules. He considers the portion of the pile that lies in the 'stable' (non-settling) soil zone (i.e. the ground profile below the neutral point) and takes the resistance of the shaft plus base in this zone into account. By designing this length of pile with a lower factor of safety (1.25 is suggested for shaft friction and end-bearing piles), it is shown that settlement due to the combined effects of applied load and dragload can be limited. The pile settlement reaches a limiting value and does not continue to increase even if the ground continues to settle. Downdrag is further considered in Section 4.9.5.

## 4.8.3 Minimising negative skin friction

The effects of downdrag can be minimised by employing slender piles (e.g. H-sections or precast concrete piles), but more positive measures may be desirable to reduce the magnitude of the dragload. In the case of bored piles, this can be done by placing in situ concrete only in the lower part of the pile within the bearing stratum and using a precast concrete element surrounded by a bentonite slurry within the fill. The use of double casing over the length of pile subject to downdrag is effective provided that the pile is not subjected to lateral load or buckling action. Dragload on precast concrete or steel tubular piles can be reduced by coating the portion of the shaft within the fill with soft bitumen, but there is risk of the coating suffering damage during driving.

Claessen and Horvat[4.80] describe the coating of 380 × 450 mm precast concrete piles with a 10 mm layer of bitumen having a penetration of 40–50 mm at 25°C. The skin friction on the 24 m piles was reduced to 750 kN compared with 1600–1700 kN for the uncoated piles. A 10 mm layer is difficult to apply at the high temperature required, and there is a significant risk that it will spall during pile driving. If bitumen with a penetration capability of

80–100 mm is used at temperatures up to 180°C, the layer can be reduced to 2–3 mm and still be effective in reducing downdrag.

Shell Composites Ltd. markets its Bitumen Compound SL[4.81] for coating bearing piles to form a slip layer. The compound will also adhere to steel, but the pile surface should be cleaned and primed with a compatible solvent primer. Penetration up to 70 mm is claimed for concrete. The bitumen slip layers should not be applied over the length of the shaft which receives support from skin friction, and Claessen and Horvat recommend that a length at the lower end of ten times the diameter or width of the pile should remain uncoated if the full end-bearing resistance is to be mobilised.

Negative skin friction is the most important consideration where piles are installed in groups. The overall settlement of pile groups in fill may be analysed empirically as described in Section 5.5, and comments on the application of soil–pile interactions using computer programs are given in Section 4.9.5.

The above-mentioned measures to minimise negative skin friction can be quite costly. In most cases, it will be found more economical to increase the penetration of the pile into the bearing stratum, thereby increasing its capacity to carry the combined loading.

## 4.9 SOIL–PILE INTERACTION

The *empirical* methods for the calculation of single pile and pile group settlements which are detailed in this text have been proved to be reliable for relatively simple structures and, with the adoption of a high factor of safety, for more complex structures. Spreadsheet calculations using the equations given will allow a limited range of parameters to be studied in the design process. If the soil and the pile are both behaving in a linear elastic mode, then approximate *analytical* solutions can be applied to an axially loaded pile, and if the large computer power now available to run predictive finite element methods (FEMs) and boundary element methods (BEMs) is applied, accurate solutions can be achieved. A key feature of the analytical approach is the consideration of the interaction of Young's modulus of the pile and Young's modulus of the soil (the *pile stiffness ratio* or the *modular ratio* $E_p/E_s$), together with other factors such as the variation of $E_s$ and the shear modulus of the soil ($G$) with depth, the variation of $E_p$ with age, and the pile compressibility, which are described as the soil–pile interaction.

In a weak soil, most of the load applied to the pile at the surface will be transferred to the pile tip, with little load transferred to the ground around the pile, whereas in a stiff soil, the load transferred to the surrounding ground through shear stresses on the shaft decreases the load on the pile with depth, and the settlement of the tip will be less than at the surface. The length of pile will dictate the load transfer – a short pile will take more load at the base and therefore settle more, whereas in a long pile, under similar load, little load will reach the pile tip. If there is a large relative movement between the pile and the soil (such as in a loose sand), there may be a reduction the shaft resistance from a peak value to a residual value. This degradation in shaft friction can now be taken into account for the design of complex foundations. The shear stresses in the soil surrounding a pile shaft reduce exponentially with distance from the pile, but this is difficult to model in a pile group, and a linear decay may be adopted in considering group effects of the stress changes.

The soil–pile–structure interaction analysis is used to predict the distribution of loads from the structure due to deformations of the soil and structure so that distress is not caused to the structural frame, the claddings and foundations. In designing the constitutive soil models for these interaction analyses, it is essential to have reliable soil parameters and layered profiles in order to determine the appropriate soil stiffness. For example, stiffness determined by small strain laboratory tests or in situ dilatometer tests may be more suited to elastic analysis than the subjective secant estimation from a static pile test. Also non-linear soil–pile interaction in pile groups which are highly loaded can have a significant influence on the load/settlement of the group depending on the method used for determining the soil stiffness.

EC7 encourages numerical analysis but makes only limited reference to the application of soil–pile interaction and gives no guidance on when or how it should be considered for design, except that a piled raft has to be accorded Geotechnical Category 3 status. The application of the standard partial factors in EC7 to complex numerical models of piles and pile groups can produce anomalous answers which the designer must understand and resolve. For example, the factoring of stiffness for numerical modelling is an issue which needs to be addressed in EC7 – it may be advisable to investigate the sensitivity of ULS to the variation in soil stiffness. EC2-1-1 is more explicit for the application of soil–structure interaction and in Annex G1 advises that the *relative stiffness*, $K_R$ (as defined, and similar to the Poulos[4.82] model), can be used to determine if the foundation or the structural system should be considered rigid or flexible; $K_R > 0.5$ (or a modular ratio > 500) indicates a rigid structure. In a rigid system or if the ground is very stiff, then 'relative settlements may be ignored and no modification of the loads transmitted from the superstructure is required'. Levels of analysis are given for flexible shallow foundations. However, when dealing with pile caps and piled rafts, the situation is different. Provided that the pile cap does not rotate, then all piles under the cap can be assumed to settle equally, (subject to being of the same dimensions and penetration). EC7 at Annex H gives a range of acceptable foundation rotations to avoid reaching the SLS in the structure. In piled rafts, the more complex interaction between soil, pile and structure will require the use of computer programs to run time-consuming iterative processes to produce an assessment of load distribution and differential settlement.

Space is not available in this text to describe the various FEM and BEM numerical analyses now applied to the solution of soil–pile interactions. The following comments are provided as an introduction to these highly specialised procedures, but it must be recognised that there is no general agreement on appropriate analytical models and boundary conditions for geotechnical design. The outputs from any computer model are affected not only by the parameters selected but by the simplifications made and judgement on the mechanisms which are modelled. It is essential therefore that the designer appreciates and understands the limitations of any software application. The selection of commercial programs listed in Appendix C is for information only; reference must be made to the relevant bureau in respect of a specific application. For a comprehensive insight into finite element analysis as applied to geotechnical engineering, the reader is referred to Potts and Zdravkovic[4.83].

### 4.9.1  Axially loaded single piles

As noted earlier, accurate analysis of the load distribution and settlement of a pile is possible when the pile and soil are both treated as elastic materials. Fleming et al.[4.23] provide a semi-analytical method to determine load/settlement ratios and load distribution which assumes a linear decay of soil stiffness down the pile and a limit to the distance from the pile which is affected by the load transfer to the soil. Figure 4.42 shows a design chart of the method applied to a straight-shaft pile which is under compression, where the following apply:

$\lambda = E_p/G_L$ is the pile-soil stiffness ratio with $G_L$ as the shear modulus at the base

$\rho = \overline{G}/G_L$ is the variation of the shear modulus of the soil with depth.

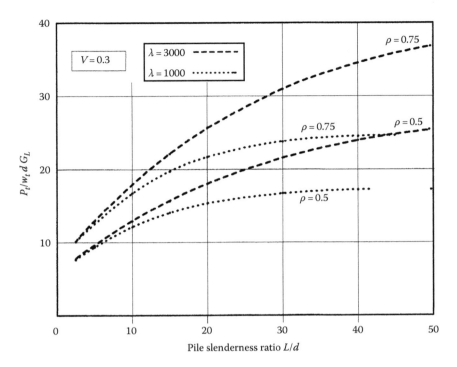

*Figure 4.42* Load/settlement ratios for compressible piles. (After Fleming, W.G.K. et al., *Piling Engineering,* 3rd ed., Taylor & Francis, Abingdon, UK, 2009.)

In the *y-axis* term, the load/settlement ratio is

$$P_t/w_t dG_L$$

where
   $P_t$ is the applied action (load)
   $w_t$ is the settlement at the pile head
   $d$ is the pile diameter
   $P_t/w_t$ is a measure of the pile stiffness ($k_p$)

The method may be applied to under-reamed piles by applying a ratio $\eta = d_b/d$ and to soil where there is an increase in the soil shear modulus from $G_L$ to $G_b$ below the pile tip in an end-bearing pile, with $\xi = G_L/G_b$. (In Figure 4.42, $\eta = \xi = 1$). In assessing the radius of influence of the pile, $\zeta = \ln(2r_m/d)$, the maximum radius $r_m$ is simplified to equal the length of the pile, and the assessment of the pile compressibility, $\mu L$, in Fleming's equation for the load/settlement ratio, depends on $\lambda$, $\zeta$ and the slenderness ratio. Using the chart in Figure 4.42 for a 450 mm diameter concrete pile, $L/d = 30$, a load of 500 kN and $G_L = 30,000$ kN/m² for $\lambda = 22,000/30 = 733$ and $\rho = 0.5$, pile head settlement $w_t$

$$= (500 \times 1000^2)/(14 \times 450 \times 30,000) = 2.6 \text{ mm.}$$

Layered soils can also be considered with the lower layer analysed first using the dimensionless parameters as shown previously with the weaker upper layers superimposed.

The computerised methods for modelling piles under axial loads use three approaches:

1. The elastic methods based on Mindlin's 1936 equations on the effects of subsurface loading in a semi-infinite elastic medium
2. The $t-z$ method
3. FEM

The *elastic method* is noted in Section 4.6, and solutions where the pile is divided into a series of elements each uniformly loaded by skin friction and bearing on a rigid base have been reported by Poulos and Davis[4.31]. It can be used with reasonable confidence in layered soils using an equivalent uniform soil layer with a weighted mean modulus

$$E = \Sigma \frac{E_k \delta_k}{L} \tag{4.47}$$

where
$E_k$ is the elastic modulus of the layer
$\delta_k$ is the thickness of the layer
$L$ is the length of pile

A modified elastic analysis includes the soil–pile interaction when the shear stress at the pile surface reaches a defined failure value and where the load/deflection behaviour is considered to be linear.

The *t–z method* is also noted in Section 4.6 and is widely adopted as a software program where non-linear behaviour at the soil–pile interface has to be considered and there is complex soil stratification. The pile is modelled as a system of rigid elements connected by springs and the soil modelled as external non-linear springs. The internal axial forces in the pile ($t$) are described by *finite difference* expressions in terms of axial displacements ($z$) at equally spaced nodes along the pile. An alternative approach using FEM to produce the $t-z$ curve can accommodate different sizes of pile element and varying pile properties and can be extended to allow for both inelastic behaviour and strength degradation of the soil.

In *finite element analysis*, the modelling of the interface parameters between pile and soil is critical, and while the method has been applied to research applications for many years, it is now a common design tool. Its success in predicting load/deformation behaviour depends on the choice of the size of the interface elements and the stiffness assigned to the soil – as for all soil–pile interaction approaches. However, piles which are subject only to axial loading and analysed with no interface elements can produce adequate results. When considering dragload and lateral load, special interface elements have to be applied. In drained conditions, the interface elements are more significant in determining shaft resistance.

## 4.9.2 Single pile subjected to lateral load

The empirical and semi-analytical design methods described in Section 6.3 are based on subgrade reaction (Terzaghi's[4.84] coefficient, $k$) and $p-y$ curves. The use of computers for elastic analysis has shown that $k$, which is difficult to evaluate, does not deal adequately with pile–soil interaction in assessing resistance to lateral loading. Considerable effort has therefore been made to refine the $p-y$ equations from the results of laterally loaded pile tests. Examples of the construction of $p-y$ curves for cases where the soil yields plastically are given in Section 6.3.5. The examples in the API Code RP2A[4.15] are applicable to piles of less than 1000 mm diameter; caution should be exercised when applying the method to larger piles (say monopiles) as the soil resistance can be over-predicted while underestimating the pile deflection.

The *elastic continuum model*, developed by Poulos[4.82] and Reese and Matlock[6.14], has been enhanced to deal effectively with the soil–pile interaction (see D-PILE Appendix C). The alternative approach adopted by Randolph[4.85] uses FEM to model the pile and soil stiffness to determine ground-level deformation and the critical pile length. The effects of the soil–pile interface elements are critical in determining the horizontal stress in the soil and the displacement of the pile particularly if *gapping* at the back of the pile is likely to occur. The comprehensive BEM analysis in REPUTE allows for linear and non-linear soil models and can handle a wide range of elements in the pile and group.

### 4.9.3  Pile groups

Viggiani et al.[4.14] comment on the division of pile groups and piled rafts into small and large categories: *small* where the raft width to pile length ratio $B/L$ is less than unity and *large* where $B/L$ is greater than unity. In small piled rafts (and small groups with a cap), the raft will generally be stiff and differential settlement is not likely to be a concern. Large rafts are likely to be flexible and the supporting piles contribute significantly to the load transfer to the soil.

Small pile groups where the cap is not in contact with the ground ('free standing') or where the supporting ground is compressible can be analysed accurately if the piles are symmetrically arranged at the corners of a regular shaped cap. In the simple case of a central load $P$ on a rigid pile cap, the load on each pile may be taken as $P/n$ where $n$ is the number of piles and the settlement of the cap may be taken as that for a single pile under this load as applied in Section 7.8. The interaction between piles will be lower where the pile spacing is large resulting in reduction in loads on the peripheral piles. Viggiani et al. provide examples of load sharing for free-standing groups with stiff caps depending on the ratio of pile spacing to pile diameter, for $s/d$ values $\leq 8$.

The various empirical methods which are considered in Chapter 5 do not determine the true load distribution in a pile group but do provide reasonable estimates of the performance of the group. The settlement of the pile group will always be greater than that of a single pile due to the soil–pile–structure interaction, but a relationship between settlement of a single test pile and group settlement can be usefully examined (Dewsbury[4.89]).

The elastic continuum approach can be applied to groups of vertical piles under axial loading to determine the displacement of one pile due to an adjacent pile carrying the same load. The results are expressed as an interaction factor, $\alpha$, defined as

$$\alpha = \frac{\text{Additional settlement caused by adjacent pile}}{\text{Settlement of pile under its own load}} \tag{4.48}$$

The computerised analysis of a pile group requires 3D FEM techniques which simplify the pile group into a segment with an axis of symmetry which will represent the whole group. If there is lateral loading on the group or the cap that has to resist bending moments from the structure, then, as there are fewer axes of symmetry, more piles and elements have to be included.

### 4.9.4  Piled rafts

For small piled rafts in contact with the soil, the load distribution between raft and pile group and the settlement of the group depends essentially on the ratio $B/L$ and on the ratio of the area occupied by the group compared to that of the raft area ($A_g/A_r$, say 0.8–0.9). Viggiani et al.[4.14] have shown that the greater the length of pile, then the average settlement of the raft compared with the settlement of the un-piled raft will be reduced.

For large pile groups supporting a structural raft, the interaction between the piles and between the piles, the soil and the raft requires much more rigorous analysis, such as 3D analysis, necessitating time-consuming iterations of the computer calculations. The load distribution among the piles produces a significant edge effect in the raft and maximises the bending moments in the piles, whether the raft is bearing on the soil or suspended. The raft can no longer be considered rigid, and raft bending moments may be reduced as a result of the load distribution and location of the piles.

Viggiani et al.[4.14] outline the matters which a numerical analysis program for pile groups and piled rafts should consider, particularly when undertaking back analysis and for parametric studies. Such analyses require significant iterations of each pile in the group for the various loading states. To achieve a more flexible approach for use as a design tool, practical simplifications of these requirements can be successfully made by separating the raft and the pile group and applying average interaction factors $\alpha_{rp}$ between the pile and raft[4.23]. The raft and pile group stiffnesses are calculated as normal structural elements with the individual pile stiffness as $k_p = P_p/w_p$ as presented previously and similarly the raft stiffness $k_r = P_r/w_r$. The group stiffness is then approximately $k_g = n\sqrt{k_p}$ where $n$ is the number of piles in the group and rectangular raft stiffness, $k_r$, with length $L$ and width $B$ may be estimated from

$$k_r = \beta_z \sqrt{LB}\,\frac{2G}{1-v} \tag{4.49}$$

where $\beta_z$ is a raft stiffness coefficient $\sim 0.04L/B + 1$. Then the proportion of the load carried by the rigid raft, $P_r$, and the pile group, $P_g$, is given by the approximation

$$\frac{P_r}{P_r + P_g} = \frac{\left(1 - \alpha_{rp}\right)k_r}{k_g + \left(1 - 2\alpha_{rp}\right)k_r} \tag{4.50}$$

The interaction factors are those developed by Randolph[4.85]. As determined by Clancy and Randolph[4.86], $\alpha_{rp}$ tends towards a constant 0.8 for a large piled raft.

This expression does not deal with the differences in load distribution between the peripheral piles and the centre piles under a flexible raft supported by competent soil, where differential settlement must be considered using the appropriate FEM or BEM software.

In large pile groups where $B/L > 1$, the load sharing between raft and piles is affected by the number of piles, the $B/L$ ratio, the pile length-to-diameter ratio $L/d$ and the pile spacing, as well as the $A_g/A_r$ ratio noted earlier. Viggiani et al.[4.14] present the results of numerical analyses showing the effects of these different parameters on the load distribution. The load sharing ranged from 0% to near 100% of the applied load, and the variation in *average* settlement compared with an un-piled raft was around 25%. The differential settlement between the corner of the raft and the centre for a uniform load using the same varied parameters was also studied. This showed that if the relative stiffness of the raft were increased, the average settlement was reduced, but this was not economic as the differential settlement remained high. A conclusion is that for each value of the pile length considered, an optimum number of piles exist to give the maximum reduction in differential settlement. The value of the $A_g/A_r$ ratio can be reduced to between 0.3 and 0.4 in the centre of the raft, thereby reducing differential settlement of the group and bending moments and shear forces in the raft. Padfield and Sharrock[4.87] also investigated the concept of *optimising* the pile support below a raft by applying the so-called settlement-reducing piles (see further examples in Section 5.10).

O'Brien et al.[4.88] provide useful guidelines for piled rafts as an extension of the settlement-reducing piles concept. Firstly, they define two groups which require different design methods: (a) *raft-enhanced* pile groups, where the piles are stiffer than the raft and attract most of the load, and (b) *pile-enhanced* rafts, where the piles will be designed to mobilise all their ultimate capacity under specific columns, with the raft carrying the bulk of the load. The ground conditions appropriate for (a) include competent soils at raft level (e.g. stiff clays, dense sands) and at depth and for (b) deep deposits of homogeneous clays stiff at raft level. In the latter case, there should be consistent 'ductile' behaviour, and therefore the piles should be straight sided. Lateral loads can be better accommodated with a pile-enhanced raft without resort to raking piles by utilising the frictional resistance of the raft–ground contact.

Dewsbury in his comprehensive study[4.89] has tackled the problem of time-consuming iterations for the analysis and design of piled rafts by using a 'modular meshing' technique and ABAQUS software. The model of a theoretical pile group was tested using this technique, against approximate numerical analysis and PIGLET and REPUTE programs (see Appendix C). At relatively low raft loading conditions, agreement between modular meshing, FEM analysis, and the approximate methods for raft settlement and differential settlement was good, although the outputs from the comparative methods were limited. In the case where load had to be distributed between the pile group and the raft, the more rigorous FEM agreed well with the modular meshing, but the results of the approximate methods were more dispersed. He also applied this modular meshing technique to the back analysis of two recently completed piled raft structures and comments on the use of current design standards for considering soil—pile interaction. The results of his numerical analysis showed that on occasions when the relative stiffness of the ground and structure is high, and the effects of the pile—soil—structure interaction are ignored, there is potential for the load distribution and hence the differential movements within the structure to be up to 50% wrong. The impact of such inaccuracies on structural integrity and building finishes are discussed. His work presents a useful approximate method for assessing when to conduct pile—soil—structure interaction analysis, which is in general agreement with the advice in EC2-1-1 Anne4x G. He also comments that, subject to strict criteria, a load test on a single pile can be useful in determining piled raft performance.

The International Society of Soil Mechanics and Ground Engineering (ISSMGE)[4.90] has produced an international guideline for the design and construction of vertically loaded combined pile raft foundations (CPRFs), in accordance with Geotechnical Category 3 of EC7 which takes account of the soil–pile–structure interaction. The computational model proposed for design should simulate the behaviour of a single pile (either from a pile loading test or empirical calculation in similar conditions) and be able to transfer the bearing behaviour of this single pile to the bearing behaviour of the piled raft (as proposed by Dewsbury[4.89], but note precautions in Section 5.1). It must also simulate all relevant interactions which affect the bearing of the piled raft. The guideline shall not be applied to layered soil where the stiffness ratio between the top and the bottom layers is ≤0.1 nor to cases where the *piled raft coefficient* ($\alpha_{rp}$ as defined) is >0.9.

The total characteristic value of the piled raft resistance is given as the sum of the individual characteristic values of the pile resistances and the characteristic value of the raft resistance. A 'sufficient' factor of safety has to be proven for all ULS and SLS combinations of loading in the raft and piles; partial factors to obtain the design actions and resistances are as given in the NA. EC7 Clause 7.1(2) states that the EC7 procedures 'should not be applied to the design of piles intended to act as settlement reducers', but offers no guidance on such designs. The ISSMGE guidance allows for simple cases to be analysed using only the characteristic value of the base resistance of the raft. This includes cases where the piles are of identical length and diameter at constant centres, a rectangular raft, homogeneous soil and the action is concentrated at the centre of the raft.

## 4.9.5 Downdrag

The design of piles subject to downdrag (negative skin friction) and simultaneous vertical axial load has been considered in largely empirical terms in Section 4.8. The elastic solutions to determine dragload originally presented by Poulos and Davis[4.31] for single end-bearing piles have been refined and extended to friction piles and pile groups by the application of complex computer analyses for soil–pile interaction effects. Several authors comment on the comparison between the basic elastic approach and the iterative analyses, and as noted below, there can be considerable variation in downdrag assessment.

Numerical analyses and centrifuge simulations on single piles and pile groups of the effects of *soil slip* at the soil–pile interface have shown that considerable reduction in the dragload can be determined. Parametric analyses using ABAQUS software were undertaken by Lee and Ng[4.91] to study the behaviour of single piles and groups up to 25 piles, with and without considering the soil slip at the pile interface. It was confirmed that for single piles, the development of downdrag was affected by the relative pile–clay stiffness and the relative bearing layer–clay stiffness. For the interface conditions tested, the computed downdrag from the no-slip elastic analysis compared well with the previously published elastic solutions. However, for the no-slip elastic analysis, the predicted downdrag was 8–14 times larger than the *elasto-plastic* slip analysis, in which only limited shear stress is transferred from the consolidating clay to the pile. Similarly, the effective stress $\beta$ method predicted downdrag 2.2–4.2 times larger than that of the slip analyses, leading Lee and Ng to infer that the elasto-plastic method could be considered an economic design tool.

For the 5 × 5 pile group at 2.5 times pile diameter spacing, the maximum downdrag of the centre, inner and corner piles were 63%, 68% and 70% of the maximum downdrag of the single pile slip analyses respectively. This reduction is attributed to the soil–pile interaction within the group – the 'shielding effect' of the outer piles on the inner. The depth of full mobilisation of the interface shear strength (the 'slip length' to the neutral point) depends on the location of a pile in the group. The computed lengths for the centre, inner and corner piles are 25%, 31% and 63% of the 20 m long pile respectively. The slip length of the single pile was 75% of the length. Lee and Ng suggest that this allows for the use of 'sacrificial piles' designed to protect piles in consolidating soils. The study also concludes that the shielding effects in respect of downdrag are likely to be more economical for end-bearing piles and the larger the group, the greater the shielding effect. A centrifuge study into shielding effects is reported by Ng et al.[4.92] and shows similar orders of reduction in dragload and downdrag for the centre piles in a group.

The elasto-plastic solution used by the above-mentioned reporters for soil–pile slip analysis requires considerable computer iterations starting with the simple elastic solution, say as Poulos and Davis[4.31]. It is then required to apply additional external loads from assessed excess soil shear stress until the computed shear stresses along the pile shaft do not exceed the soil limiting values.

### 4.9.6 Rock sockets

Zhang[4.93] provides examples of finite element solutions for axially loaded 'drilled shafts' using linear and non-linear continuum approaches to sidewall slip and non-slip situations. He points out that the FEM results show that the progression of the slip from no slip to full slip takes place over a small interval of displacement. The analyses require sophisticated soil and rock constitutive relations whose parameters can be difficult to obtain and apply in order to produce better economy in design compared with, say, the Pells and Turner approach in Section 4.7.4.

### 4.9.7 Obtaining soil parameters

Several well-known commercial computer programs, which generally follow the established procedures for routine pile design, are mentioned in the text and Appendix C. Designers who wish to apply advanced numerical modelling for innovative solutions should ensure that they have the necessary expertise and that the model is relevant to the problem and ground conditions. It is always useful to undertake a preliminary simple analysis before relying on the results of the sophisticated modelling – which may best be considered as showing trends rather than giving an absolute value. As pointed out by Clayton[4.94], numerical analysis for complex foundations is only part of the design process, and the numerous parameters which are required for some specialised programs are difficult to obtain. In any case, the results are affected more by the simplifications and judgements which have to be applied to the geological model and the difficulty in obtaining representative samples for testing in the laboratory and the simplistic correlation of field tests with soil parameters. Examination of parameters in case histories is useful, but the reviewer should be wary that data quoted may be erroneous.

From the preceding review, it can be concluded that numerical analysis is likely to remain mainly as a research tool until the loads and strength parameters at the soil–pile interface can be determined more accurately and the research is developed into proven commercial applications. In order to adopt these applications, guidelines will be required on appropriate partial factors, model factors, installation factors, and design approach for both ULS and SLS procedures to resolve the different approaches (and different results) to numerical analysis.

## 4.10 LOAD AND RESISTANCE FACTOR DESIGN APPLIED TO PILE DESIGN

This chapter has focussed on design using the partial factor approach in EC7 which is now mandated for many foundation design applications in the United Kingdom. LRFD procedures, which are essentially different from the rational of the Eurocodes, are being applied by many countries to geotechnical design for deep foundations generally and are briefly examined in this Section. The calibration of the load and resistance factors, to account for the uncertainties in the foundation as noted in Section 4.1.3 without relying on the allowable stress global factor of safety, has been developed by a variety of methods. For example, Vardanega et al.[4.95] in their reviews of pile design in London Clay demonstrate the potential for reducing the partial factors for design values of applied load and drained and undrained soil strength. They suggest that such selective reductions, based on statistical procedures (as used in the United States and Australia noted later in the text), would lead to acceptable settlements and a considerable saving on the pile design capacity compared with the current EC7 approach.

In all cases, the basic design philosophy for limit state design, that is where strength or failure conditions are considered, is that the factored strength or resistance must be greater than or equal to the factored load. For the strength limit state, this can be expressed as

$$\phi R_n \geq \sum \eta_i \gamma_i Q_i \qquad (4.51)$$

where
$\phi$ is a statistically based resistance factor
$R_n$ is the nominal (ultimate) resistance or strength of the component or material under consideration
$Q_i$ is a load effect (a force and/or moment)
$\gamma_i$ is a statistically based load factor
$\phi R_n$ is the design resistance, $R_R$
$\Sigma \gamma_i Q_i$ is the summation of all load effects
$\eta_i$ is a *modifier* applied to the load effect

For the service limit state, the expression is

$$\phi \delta_n \geq \sum \eta_i \gamma_i \delta_i \qquad (4.52)$$

where
    $\delta_n$ is the tolerable displacement
    $\delta_i$ is the estimated displacement

This LRFD approach is being widely adopted in the United States, and agencies such as the Federal Highway Administration (FHWA) and the American Institute of Steel Construction (AISC) now require LRFD to be applied to construction works requiring federal funding. The LRFD Bridge Design Specifications[4.96] prepared by the American Association of State Highway and Transport Officials (AASHTO) comprehensively cover all aspects of structural and foundation design. The load and resistance factors for deep foundations are based on extensive research of safety factors in historical databases and the application of reliability theory carried out by the U.S. Transport Research Board as described in NCHRP Report 507. The specification deals with driven piles, drilled shafts and micropiles in separate sections using the earlier prescribed Equations 4.51 and 4.52 and the following general equation for calculating the factored bearing resistance, $R_R$:

$$R_R = \phi R_n = \phi_{stat} R_p + \phi_{stat} R_s \qquad (4.53)$$

where
    $\phi_{stat}$ is the resistance factor for the bearing resistance of a single pile assessed for the shaft and tip separately (based on static analysis)
    $R_p$ is the pile tip resistance, equal to $q_p A_p$
    $R_s$ is the shaft resistance equal to $q_s A_s$ (i.e. as Equations 4.5a and b)

The resistance factors for driven piles are given in Table 10.5.5.2.3-1 of the AASHTO specification for different design methods and soil conditions. The table distinguishes between factors applied to piles analysed by load testing, with factors for $\phi_{dyn}$ (replacing $\phi_{stat}$) ranging from 0.8 for static load tests down to 0.1 for the ENR dynamic pile formula. Factors for static analysis of piles in compression vary from 0.35 to 0.50 depending on the semi-empirical relationship used in clay and sand. For example, the methods of analysis for shaft and end bearing in clay may be by the total stress $\alpha$ adhesion factor (where $q_s = \alpha S_u$) or the $\beta$ factor for effective stress. (Note that large variations in $\alpha$ may be used depending on pile length and $S_u$ which are not normally applied in the United Kingdom.) For shaft resistance and end bearing in sand, $\phi_{stat}$ resistance factors should be obtained from SPT and CPT results. Factors are also provided for block failure in clay, uplift resistance of a single pile and group, and lateral resistance in all soils and rock.

The AASHTO specification implies a preference for using driven piles for bridge foundations. The resistance factors for drilled shafts prescribed in Table 10.5.5.2.4-1 are based on applying total and effective stress methods for shaft and end bearing in sand and clay and range from 0.45 to 0.65. However, these values are to be reduced by 20% when used for the design of a single shaft for a bridge pier, and if high quality procedures are not available during construction, the factors should also be reduced, subject to engineering judgement. Other factors are also provided for block, uplift and lateral resistances. Resistance factors for micropiles in Table 10.5.5.2.5-1 follow similar procedures to those described for driven and drilled shafts.

Load factors for the bridge structure as a whole are prescribed in Tables 3.4.1-1 and 3.4.1-2 of the AASHTO specification and require extensive assessment of limit states under a variety of load combinations. Four groups of limit states are specified: strength, extreme event, service and fatigue. For substructure design, evaluation will generally be limited to performance at *Strength 1* limit state likely to produce the maximum foundation load based on dead loads from the structure and live loads due to normal vehicular traffic without wind load. *Service 1* limit state will define deformation of the foundation being the load combination of normal operation of the bridge and a 90 km/h wind load. For Strength 1, the maximum permanent load factor, $\gamma_p$, for the structural dead load is 1.25 (plus a factor of 1.5 on the wearing surface materials) and for the live load $\gamma_i$ is 1.75; for Service 1, $\gamma_i$ is unity.

Downdrag is considered in some depth in Articles 3.11.8 and 10.7.3.7 and includes downdrag due to liquefaction as an *Extreme Event 1* limit state. The load factors in Table 3.4.1-2 depend on the method used to assess downdrag and range from1.05 to 1.4 (when using an $\alpha$ adhesion factor).

AASHTO has taken a useful step in LRFD by introducing the $\eta_i$ modifier factors which are applied to $\gamma_i$ to account for ductility, redundancy and operational importance. These range from 1.05 to 0.95 and when accumulated should be greater than 0.95 and for the Service 1 state will be unity. The operational importance is based on social, survival and defence requirements and for a seismic structure $\eta_i$ will be 1.05.

Structural resistance of piles is evaluated by comparing the maximum factored stress, $\sigma_{max}$, with the factored unit resistance, $\sigma_r$:

$$\sigma_r = \Sigma\phi\sigma_n \tag{4.54}$$

where
   $\sigma_n$ is the nominal (ultimate) unit structural resistance of the pile material based on the yield strength
   $\phi$ is the resistance factor given in the AASHTO specification for steel (Article 6.5.), concrete (5.5.4) and timber (8.5.2)

Factors for driven H-piles and pipe piles depend on driving conditions.

In the Australian Standard, Piling – Design and Installation[4.97], based on LRFD principles, emphasis is placed on determining the design geotechnical strength $R_{dg}$, from the design ultimate geotechnical strength $R_{d,ug}$, so that

$$R_{dg} = \phi_g R_{d,ug} \tag{4.55}$$

where the geotechnical reduction factor, $\phi_g$, is given by

$$\phi_g = \phi_{g,b} + (\phi_{t,f} - \phi_{g,b})K\phi_{g,b} \tag{4.56}$$

where
   $\phi_{g,b}$ is a basic geotechnical reduction factor
   $\phi_{t,f}$ is the intrinsic test factor
   $K$ is the test benefit factor

The value of $\phi_{g,b}$ depends on the assessed site risk factors and the weighted sum of individual risks multiplied by the risk-weighting factors. Tables are provided to give *individual risk ratings* from 1, very low, to 5, very high; the *basic risk factors* for the site geology are given a weighting of 2, the design risk varies from 1 to 2 and the installation risk varies from 0.5 to 2. The *average risk rating* (ARR) is calculated for the site and $\phi_{g,b}$ determined: for ARR ≤ 1.5 and *high redundancy* $\phi_{g,b} = 0.76$ and for ARR > 4.5 with *low redundancy* $\phi_{g,b} = 0.40$.

In this standard, pile testing is strongly encouraged and will increase the $\phi_g$ factor depending on the type of test: $\phi_{t,f}$ is 0.9 for static load test, 0.85 for Osterberg cell test, 0.8 for dynamic test and 0.75 for Statnamic tests (see Section 11.4). The number of piles tested will also affect the value of $\phi_g$: if no piles are tested, $\phi_{g,b}$ is limited to 0.55, but if 10% or more of piles are tested, $K = 1$. Where $\phi_{g,b} \leq 0.40$, no testing is required, and where $\phi_{g,b} > 0.40$, testing is mandatory. In the absence of tests to verify the ultimate geotechnical strength, tests for serviceability are required for all sites with ARR > 2.5.

From the preceding examples, it can be seen that there are variations in the approach to LRFD which were not so apparent in allowable stress design procedures. The Australian code is unique in providing a guide for the resistance factor and allowing the designer freedom to assess risks associated with soil type and installation methods in order to ensure safe foundations.

## WORKED EXAMPLES

### Example 4.1

A precast concrete pile is required to carry a dead load (*permanent unfavourable*) of 250 kN and an imposed load (*variable unfavourable*) of 115 kN both in compression, together with an uplift load (*variable unfavourable*) of 200 kN. The pile is driven through 6 m of very soft clay into a stiff boulder clay. Determine the required penetration of the 300 × 300 mm pile into the stiff clay to carry the specified loading. Undrained shear strength tests were made on samples from three boreholes as shown in Figure 4.43. Settlements are not critical to the structural design of the jetty. Pile testing on 1% of the working piles is not practical in this case.

The design will be in accordance with the EC7 Clause 7.6.2.3(8) using ground test results to provide the characteristic compressive resistance, using the model factor $\gamma_{Rd} = 1.4$.

Try a 300 mm square pile in concrete grade C40/50 with the toe at –15 m depth.

Actions on the pile in compression are permanent $G_k = 250$ kN and variable $Q_k = 115$ kN.

Using the best-fit line as shown on Figure 4.43 and ignoring any resistance from the shaft in the very soft clay, the undrained shear strength $c_u$ varies from 100 kN/m² at –6 m depth to 217 kN/m² at –15 m:

Unit base resistance = $q_b = N_c c_u$ where $N_c = 9$

Unit shaft resistance = $q_s = \alpha c_u$ where $\alpha = 0.5$

For piles in axial compression from Equations 4.7 and 4.10, characteristic pile resistances for base and shaft are $R_{bk} = (A_b q_{bk})/\gamma_{Rd}$ and $R_{sk} = \Sigma(A_s q_{sk})/\gamma_{Rd}$

$R_{bk} = (0.3^2 \times 9 \times 217)/1.4 = 125.7$ kN

$R_{sk} = 0.3 \times 4 \times 9 \times 0.5 \,(100 + 217)/(2 \times 1.4) = 611.4$ kN

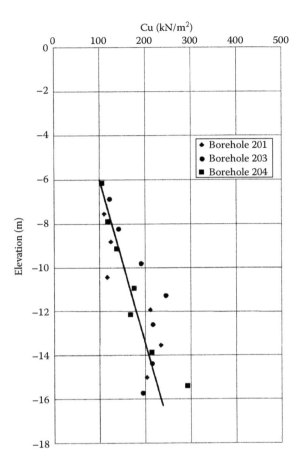

*Figure 4.43* $c_u$ v elevation below GL for Example 4.1 (Courtesy David Beadman.)

*For DA1-1* (sets A1 + M1 + R1 apply), the partial resistance factors for a driven pile as given in Table 4.3 are $\gamma_b = \gamma_s = 1.0$. The partial factors for actions as Table 4.1 are $\gamma_G = 1.35$ and $\gamma_Q = 1.5$.

Design value of actions = $F_d = G_k \gamma_G + Q_k \gamma_Q$

$$= 250 \times 1.35 + 115 \times 1.5 = 510 \text{ kN}$$

Design value of resistances = $R_{cd} = (R_{bk}/\gamma_b + R_{sk}/\gamma_s)$

$$= (125.7/1.0 + 611.4/1.0)$$

$$= 737 \text{ kN} > F_d = 510 \text{ kN and satisfactory}$$

*For DA1-2* (sets A2 + M1 or M2 + R4 apply), the partial resistance factors for a driven pile as given in Table 4.3 are $\gamma_b = 1.7$ and $\gamma_s = 1.5$, assuming no pile tests. The partial factors for actions as Table 4.1 are $\gamma_G = 1.0$ and $\gamma_Q = 1.3$.

Design value of actions = $F_d = G_k \gamma_G + Q_k \gamma_Q$

$$= 250 \times 1.0 + 115 \times 1.3 = 400 \text{ kN}$$

Design value of resistances $= R_{cd} = (R_{bk}/\gamma_b + R_{sk}/\gamma_s)$

$$= (125.7/1.7 + 611.4/1.5)$$

$$= 481 \text{ kN} > F_d = 400 \text{ kN and satisfactory}$$

This example demonstrates that DA1-2 usually defines the pile compressive resistance and is worth considering first.

For piles in tension, only frictional resistance applies:

*For DA1-2* (the critical combination, set A2 + M2 + R4, applies), the partial resistance factor in tension for driven pile is $\gamma_{st} = 2.0$ as Table 4.3, and the partial factor for uplift action is $\gamma_Q = 1.3$. $\gamma_{Rd} = 1.4$ as before.

Design value of actions $= F_d = Q_k\,\gamma_Q$

$$= 200 \times 1.3 = 260 \text{ kN}$$

Design value of resistances $= R_{cd} = (R_{sk}/\gamma_{st})$

$$= (611.4/2.0)$$

$$= 305 \text{ kN} > F_d = 260 \text{ kN and satisfactory}$$

Calculations using the *model pile* procedure are given in Example 4.6 as applied to CPT results from ground tests in accordance with EC7 Clause 7.6.2.3.

## Example 4.2

A steel tubular pile 1.220 m in OD forming part of a berthing structure is required to carry an applied load in compression of 16 MN (*permanent action*) and an uplift of 8 MN (*variable action*). The pile is driven with a closed end into a deep deposit of normally consolidated marine clay. The undrained shearing strength–depth profile of the clay is shown in Figure 4.44. Determine the depth to which the pile must be driven to carry the applied load.

In dealing with problems of this kind, it is a good practice to plot the calculated values of ultimate shaft friction, end bearing and total resistance for various depths of penetration. The required pile length can then be read off from the graph. This is a convenient procedure for a marine structure where the piles may have to carry quite a wide range of loading.

(As an alternative, the design resistances can be calculated for say 0.5 m depth increments using a spreadsheet.)

Outside perimeter of pile $= \pi \times 1.220 = 3.83$ m
Overall base area of pile $= 1/4 \times \pi \times 1.220^2 = 1.169$ m$^2$
From Figure 4.44, at 160 m,

$$\frac{c_u}{\sigma'_{vo}} = \frac{260}{0.65 \times 9.81 \times 160} = 0.25$$

From Figure 4.6a, the adhesion factor, $\alpha_p$, is 1.0 over the full depth.

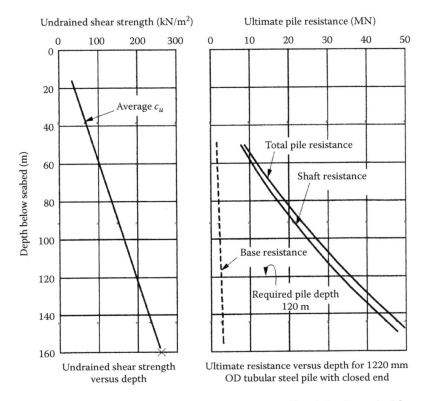

*Figure 4.44* Undrained shear strength and Ultimate pile resistance v Depth for Example 4.2.

At 50 m below the seabed,
Average shearing strength along shaft = 1/2 × 80 = 40 kN/m$^2$
From Figure 4.6b, the length factor $F$ for $L/B$ value of 50/1.22 = 41 is 1.0.
From Equation 4.11, the ultimate shaft resistance on outside of shaft (i.e. no model factor) is

$$\frac{1.0 \times 1.0 \times 40 \times 3.83 \times 50}{1000} = 7.66\,\text{MN}$$

From Equation 4.7, the ultimate end-bearing resistance is

$$\frac{9 \times 80 \times 1.169}{1000} = 0.84\,\text{MN}$$

Thus, the total pile resistance is 8.50 MN.
At 75 m below the seabed,
Average shearing strength along shaft = 1/2 × 120 = 60 kN/m$^2$
Length factor for $L/B$ value of 61 is 0.9
Ultimate shaft resistance on outside of shaft is

$$\frac{1.0 \times 0.9 \times 60 \times 3.83 \times 75}{1000} = 15.51\,\text{MN}$$

The ultimate end-bearing resistance is

$$\frac{9 \times 120 \times 1.169}{1000} = 1.26\,\text{MN}$$

Thus, the total pile resistance is 16.77 MN.

Similarly, the total pile resistances at depths of 100, 125 and 150 m below the seabed are 26.19, 38.01 and 50.78 MN, respectively. The calculated values of pile resistance are plotted in Figure 4.44.

*For DA1 and combination 1* (sets A1 + M1 + R1 apply) and $\gamma_G = 1.35$ in compression,
  Design value of actions = $F_d = 1.35 \times 16 = 21.6$ MN
  Therefore, the design resistance, $R_{cd}$, has to be > $1.4 \times F_d = 1.4 \times 21.6 \sim 31$ MN which, from the graph, requires a penetration of say 113 m (the model factor of 1.4 is now applied as the graph shows ultimate resistances).

*For DA1 and combination 2* (sets A2 + M1 + R4) and $\gamma_G = 1.0$,
  Design value of actions = $F_d = 1.0 \times 16 = 16$ MN
  At 113 m deep, the characteristic resistances are $R_{bk} = 1.9/1.4 = 1.36$ MN and $R_{sk} = 29/1.4 = 20.7$ MN.
  For a driven pile in compression, if the partial factors are $\gamma_s = 1.5$ and $\gamma_b = 1.7$, then

$$R_{cd} = (1.36/1.7 + 20.7/1.5) = 14.6 \text{ MN which is less than } F_d$$

Therefore, the depth must be increased to say 120 m where $R_{cd} = 17.4$ MN which is greater than $F_d$ and satisfactory for both DA1 combinations.

To check the pile in tension, at a depth of 120 m, the ultimate shaft friction is 34 MN.
  With $\gamma_{st} = 2.0$, $\gamma_Q = 1.3$ and model factor = 1.4, $R_{cd} = (34/2.0)/1.4 = 12.1$ MN, which is greater than $F_d = 1.3 \times 8 = 10.4$ MN and satisfactory.

To check the design resistance of the high-tensile steel on the 1.22 m OD steel with a wall thickness of 25 mm, $N_{cRd} = Af_y/\gamma_{M0}$ as in Section 7.10.2, with $\gamma_{M0} = 1.0$ and $f_y = 355$ MN/m$^2$ as stated in Table 3 of EC3-1-1:

$$N_{cRd} = \pi/4 \times (1.22^2 - 1.17^2) \times 355/1.0 = 33.4 \text{ MN, that is } > N_{Ed} = 21.6 \text{ MN for A1 set}$$

Checking the actual stress at the applied load gives the following:

$$\frac{16 \times 10^6}{\pi (1220^2 - 1170^2)/4} = 170\,\text{N/mm}^2$$

This is 48% of the yield strength of high-tensile steel and satisfactory. Subject to driving conditions, the thickness could be reduced to 16 mm over the lower 50 m of the pile.

## Example 4.3

A building column carrying a permanent action of 1100 kN and a variable action of 300 kN is to be supported by a single bored pile installed in firm to stiff fissured London Clay (Figure 4.45). Select suitable pile dimensions and penetration assuming no pile tests are carried out. Calculate the immediate settlement at the applied load.

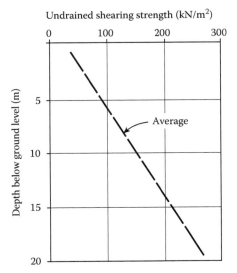

Figure 4.45 Characteristic and average shear strength values from four boreholes for Example 4.3.

Try a 1 m diameter uncased bored pile with 2 m base enlargement at base level of 10 m. The 1 m shaft diameter must be ignored for shaft resistance for 2 m above the top of base enlargement at 9 m deep (as Figure 4.9).

It is assumed that the shear strength/depth relationships in Figure 4.45 were based on an adequate number of boreholes and soil samples and that the straight-line graphs are a cautious estimate (*moderately conservative*) derived from the plotted data. Figure 4.45 is assumed to represent characteristic soil parameters.

The shaft area $A_s$ is $7 \times 1 \times \pi = 22$ m², and the base area $A_b$ is $(\pi \times 2^2/4) = 3.14$ m².

The average unit shearing strength along 7 m pile shaft is $(35 + 111)/2 = 73$ kN/m².

Taking an adhesion factor of 0.45 to allow for possible softening while under-reaming, then from Equation 4.10 with the model factor $\gamma_{Rd} = 1.4$,

Characteristic shaft resistance $R_{sk} = (0.45 \times 73 \times 22.0)/1.4 = 516.4$ kN

Because the clay is fissured, it is desirable to reduce the average shearing strength at pile base level (155 kN/m²) by a factor of 0.75 to obtain the end-bearing resistance. Then from Equation 4.7,

Characteristic end-bearing resistance $R_{bk} = (9 \times (0.75 \times 155) \times 3.14)/1.4 = 2347$ kN.

*For DA1-1 (sets A1 + M1 + R1 apply)*
From Table 4.1, the A1 factors are 1.35 (permanent unfavourable) and 1.5 (variable unfavourable). Therefore, the design value of actions is as follows:

$$F_d = 1.35 \times 1100 + 1.5 \times 300 = 1935 \text{ kN}$$

From Table 4.4, the R1 partial factors are unity:
Pile design resistance $R_{cd} = (516.4/1.0 + 2347/1.0) = 2863$ kN/m² > $F_d$ of 1935 kN

*For DA1-2 (sets A2 + M1 + R4 apply)*
From Table 4.1, the A2 factors are 1.0 (permanent unfavourable) and 1.3 (variable unfavourable). Therefore, the design value of actions is as follows:

$$F_d = 1.0 \times 1100 + 1.3 \times 300 = 1490 \text{ kN}$$

From Table 4.4, the R4 partial factors are $\gamma_s = 1.6$ and $\gamma_b = 2.0$ assuming no pile testing:
Pile design resistance $R_{cd} = (516.4/1.6 + 2347/2.0) = 1496$ kN/m² > $F_d$ of 1490 kN, which is satisfactory for both DA1 combinations.

However, in view of the efficiency of modern pile drilling equipment, it is likely that the cost of under-reaming and concreting to form an enlarged base at 10 m will exceed the cost of drilling the extra 5 m for a 1 m diameter straight-sided uncased pile shaft 14 m long below the cut-off at 1–15 m deep.

The area of 14 m long shaft is $A_s = 44.0$ m², and the base area is $A_b = 0.785$ m².

*For DA1-1,* the design value of actions as before is as follows:

$$F_d = 1.35 \times 1100 + 1.5 \times 300 = 1935 \text{ kN}$$

The average shearing strength along 14 m pile shaft between 1 and 15 m deep (from Figure 4.45) is $(35 + 210)/2 = 122.5$ kN/m².

Taking an adhesion factor of 0.5 over the greater depth, then
Characteristic shaft resistance $R_{sk} = (0.5 \times 122.5 \times 44.0)/1.4 = 1925$ kN
As before, the shear strength at the base will be reduced by a factor of 0.75; thus,
Characteristic end-bearing resistance $R_{bk} = (9 \times (0.75 \times 210) \times 0.785)/1.4 = 795$ kN
From Table 4.4, the R1 and M1 partial factors are unity:
Pile design resistance $R_{cd} = (1925/1.0 + 795/1.0) = 2720$ kN > $F_d$ of 1935 kN

*For DA1-2,* the design value of actions as before is as follows:

$$F_d = 1.0 \times 1100 + 1.3 \times 300 = 1490 \text{ kN}$$

From Table 4.4, the R4 partial factors are $\gamma_s = 1.6$ and $\gamma_b = 2.0$:
Pile design resistance $R_{cd} = (1925/1.6 + 795/2.0) = 1601$ kN > $F_d$ of 1490 kN, which shows that the deeper pile is satisfactory.

Considering the settlement at the total applied load of 1400 kN for straight pile 14 m long (after cut-off), the following applies:

As the ultimate shaft resistance (unfactored) of $1.4 \times 1925 = 2695$ kN exceeds the applied load, the settlement of the pile will be no more than that required to mobilise the ultimate resistance. Hence, settlement of less than 10 mm can be expected.

For long-term loading, using Equation 4.35 (as modified for Poisson's ratio between 0 and 0.25 and $L/B > 5$), assume a deformation modulus at the base of $140c_u$. With $L/B = 14$, only a small proportion of the load will be taken in end bearing, say 12.5% with 87.5% in shear. Partial factors for design actions are in unity for serviceability state.

Then $W_s = 0.85(1100 + 300) = 1225$ kN and $W_b = 175$ kN. The area of the 14 m shaft is 44 m² and, allowing for softening at the pile tip, $c_u = 0.75 \times 210$ kN/m² giving $E_b = 140 \times 0.75 \times 210 = 22{,}050$ kN/m:

$$\rho = \frac{(1225 + 2 \times 210) \times 14{,}000}{2 \times 44.0 \times 30 \times 10^6} + \frac{0.5 \times 175 \times 1{,}000}{1 \times 22{,}050}$$

$$= 0.08 + 3.96 = 4.04 \text{ mm}$$

If the approximate analytical solution given in Section 4.9.1 is applied, then $L/d = 14$, and from Figure 4.42, $P_t/w_t (G_L d) = 15$, giving settlement $w_t = 1400/(15 \times 22.05) = 4.2$ mm (the reduction in Poisson's ratio from 0.3 in the figure to 0.2 used in the example will be negligible).

## Example 4.4

A precast pile 450 mm square forming part of a jetty structure is driven into a medium-dense over-consolidated sand. SPTs made in the sand gave an average value of $N$ of 15 blows/300 mm. The pile is required to carry a compressive load of 250 kN (permanent action) and an uplift load of 180 kN (permanent unfavourable). Determine the required penetration depth of the pile.

The unit shaft friction developed on a pile in sand is rather low, and thus the penetration depth in this case is likely to be governed by the requirements for uplift resistance.

Take a trial penetration depth of 8.5 m below seabed. From Figure 4.10 for $N$ = 15, $\phi$ = 31°. The submerged density of the sand may be taken as 1.2 Mg/m³. For an over-consolidated sand, we can take $K_0$ = 1. Table 4.7 gives $K_s/K_0$ = say 1.5, giving $K_s$ = 1.5. From Table 4.8, take $\delta$ = 0.8 $\phi$ = 0.8 × 31° = 24.8°. Then from Equation 4.14 taking the average effective overburden pressure and applying the model factor $\gamma_{Rd}$ = 1.4,

Characteristic shaft friction resistance $R_{sk}$ = 1.5 × ½ (1.2 × 9.81 × 8.5) × tan 24.8° × (8.5 × 4 × 0.45)/1.4 = 379 kN

*For DA1-1* (sets A1 + M1 + R1 apply) where $\gamma_{st}$ = 1.0 and $\gamma_G$ = 1.35,
Design action in tension $F_{td}$ = 1.35 × 180 = 243 kN
Design friction resistance = $R_{td}$ = (379/1.0) = 379 kN > $F_{td}$ = 243 kN

*For DA1-2* (sets A2 + M1 or M2 + R4 apply) where $\gamma_{st}$ = 2.0 and $\gamma_G$ = 1.0,
Design action in tension $F_{td}$ = 1.0 × 180 = 180 kN
Design friction resistance = $R_{td}$ = (379/2.0) = 189 kN > $F_{td}$ = 180 kN
Hence, an 8.5 m penetration is satisfactory for uplift resistance.
Checking the base resistance using the Berezantsev value of $N_q$ in Equation 4.13, from Figure 4.13 with $\phi$ = 31°, $N_q$ = 20 (for $D/B$ = 19). Thus,
Characteristic base resistance $R_{bk}$ = (20 × (1.2 × 9.81 × 8.5) × 0.45²)/1.4 = 289.3 kN

*For DA1-1* (sets A1 + M1 + R1) where $\gamma_b$ and $\gamma_b$ = 1.0 and $\gamma_G$ = 1.35,
Design action in compression $F_d$ = 1.35 × 250 = 337.5 kN

$R_{cd}$ = (289.3/1.0 + 379/1.0) = 668 kN > $F_d$ = 337.5 kN

*For DA1-2* (sets A2 + M2 + R4) where $\gamma_b$ = 1.7, $\gamma_s$ = 1.5 and $\gamma_G$ = 1.0,
Design action in compression $F_d$ = 1.0 × 250 = 250 kN

$R_{cd}$ = (289.3/1.7 + 379/1.5) = 423 kN > $F_d$ = 250 kN

Hence, an 8.5 m penetration is satisfactory in compression, and the pile penetration is governed by tensile resistance.

## Example 4.5

Isolated piles are required to carry a permanent load (action) of 900 kN on a site where borings and static cone penetration tests recorded the soil profile shown in Figure 4.46. Select a suitable type of pile and determine the required penetration depth to carry the load. Previous tests in the area have shown that the ultimate base resistance of piles driven into the dense sand stratum is equal to the static cone resistance.

The piles will attain their resistance within the sand stratum (12–28 m). Any type of bored and cast-in-place pile will be uneconomical compared with the driven type. A driven and cast-in-place pile is suitable.

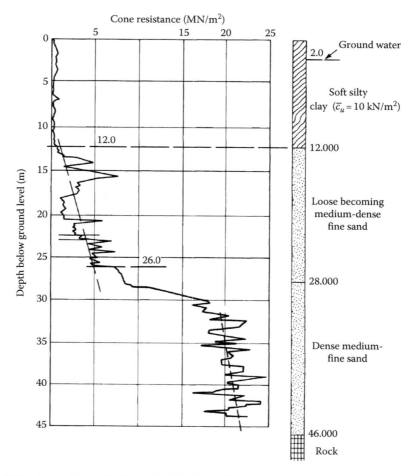

Figure 4.46 CPT values v Depth for Example 4.5, 4.9 and 5.3.

*Driven and cast-in-place pile*
Use concrete grade C25/35 with characteristic strength of $f_{ck}$ = 25 N/mm². From Section 7.10.1,

$$N_{Ed}/N_{Rd} \leq 1.0$$

For a concrete pile that does not have permanent casing, the dimensional factor as in Table 4.6 and the $k_f$ factor of 1.1 as in Section 7.10.1 have to be applied to the STR calculation:

$$N_{Ed} = 900 \times 1.35 = 1215 \text{ kN for a permanent action}$$

$$N_{Rd} = A \times 0.85 \times 25/(1.5 \times 1.1) \text{ with } \gamma_C = 1.5 \times 1.1$$

Hence, the area required is $A$ = 1215 × (1.5 × 1.1) × 1,000/(0.85 × 25) = 94,341 mm², and the diameter required is $d_{nom} = \sqrt{(4 \times 94341/\pi)}$ = 346 mm.

A 20 mm dimensional reduction factor has to be added to $d_{nom}$; hence, try pile shaft diameter of 400 mm.

From Figure 4.46, consider a pile penetration depth of 28 m where the CPT values indicate suitable bearing. From Table 4.9, taking the unit shaft friction, $f_s = 0.012\ q_c$, then

At 12 m, unit shaft friction $= 0.012 \times 1 \times 10^3 = 12\ kN/m^2$
At 26 m, unit shaft friction $= 0.012 \times 5 \times 10^3 = 60\ kN/m^2$
At 28 m, unit shaft friction $= 0.012 \times 8 \times 10^3 = 96\ kN/m^2$

With $\gamma_{Rd} = 1.4$, the characteristic shaft friction is

$$R_{sk} = \frac{(12+60)\times 14+(60+96)\times 2}{2}\times \pi \times 0.40\,/\,1.4 = 592\,kN$$

To obtain base resistance as Equation 4.15, $q_{ub} = \bar{q}_c$.
The resistance at 28 m can be taken as an average over a distance of 8 pile diameters above the toe and 4 diameters below the toe. Over 4 diameters below the toe, $q_{c-1} = (7 + 15) \times 0.5 = 11\ MN/m^2$, and over 8 diameters above the toe, $q_{c-2} = (5 + 8) \times 0.5 = 6.5\ MN/m^2$. Then as in Equation 4.16, the average $q_c$ at 28 m is 8.75 $MN/m^2$. Therefore,
Characteristic base resistance $R_{bk} = (8.75 \times 10^3 \times 0.40^2 \times \pi/4)/1.4 = 785\ kN$

*For DA1-2 (sets A2 + M1+ R4 apply) and assuming no pile testing where $\gamma_G = 1.0$, $\gamma_s = 1.5$ and $\gamma_b = 1.7$ from Table 4.3,*

$F_d = 900 \times 1.0\ kN$

Total pile resistance $R_{cd} = (592/1.5 + 785/1.7) = 856\ kN < F_d = 900\ kN$ which fails
Therefore, increase pile diameter to 450 mm gives the following:
Shaft friction resistance $R_{sk} = 592 \times 0.45/0.40 = 666.4\ kN$
Base resistance $R_{bk} = (8.75 \times 10^3 \times 0.45^2 \times \pi/4)/1.4 = 960\ kN$
Total pile resistance $R_{cd} = (666.4/1.5 + 1392/1.7) = 1008\ kN > F_d = 900\ kN$ and satisfactory

*By inspection, DA1-1 is not critical for the GEO check.*
If pile testing were to be carried out on the 400 mm pile, then the factors for $R_{cd}$ would be $\gamma_s = 1.3$ and $\gamma_b = 1.5$ from Table 4.3 for DA1 (2), giving $R_{cd} = 978\ kN > 900\ kN$ and satisfactory.
Check concrete stress:
The reduction factor 0.95 as in Table 4.6 has to be applied to the diameter of an uncased concrete pile $d = 0.95 \times 450 = 427.5$ mm:

$\sigma_{cd} = F_d/A = 1215/(427.5^2 \times \pi/4) = 8.5\ N/mm^2$

$f_{cd} = 0.85 \times 25/(1.5 \times 1.1) = 12.9\ N/mm^2 > \sigma_{cd}$ and satisfactory

## Example 4.6

Calculate the resistance of a 914 mm OD × 19 mm wall thickness tubular steel pile driven with a closed end to a depth of 17 m below ground level in the soil conditions shown in Figure 4.18a, and compare the capacity with that of an open-end pile driven to the same depth.
The pile characteristics are as follows:

External perimeter = 2.87 m
Internal perimeter = 2.75 m
Gross base area = 0.656 $m^2$

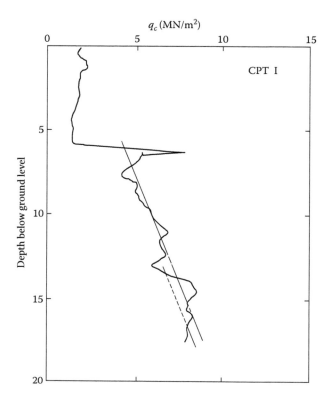

*Figure 4.47* CPT values v Depth for Example 4.6.

For 38 mm shoe $(D_{CPT})$, the net cross-sectional area at toe is 0.1046 m².

The results of the CPT tests are combined as shown in Figure 4.18a. Any shaft friction in the soft clay will be ignored; hence, the length providing frictional resistance is 10.5 m. The coefficient from Table 4.9 is 0.008.

From plots of individual cone readings, shown typically for CPT 1 in Figure 4.47, the average and characteristic cone resistances over the length of the shaft and at the base are as follows:

|  | CPT1 | CPT2 | CPT3 | CPT4 | Average | Characteristic |
|---|---|---|---|---|---|---|
| Shaft, average $q_c$ MN/m² | 6.0 | 7.5 | 8.4 | 8.5 | 7.6 | 7.35 |
| Base, min $q_c$ MN/m² | 8.2 | 9.7 | 10.1 | 10.7 | 9.7 | 9.3 |

The shaft friction $R_{s\,cal}$ is $0.008\,q_{c\,av}A$, and the base resistance $R_{b\,cal}$ is $q_{c\,b}A$.

For the closed-end driven pile, the *model pile* method will be used to check resistances from ground tests:

| CPT | Shaft av $q_c$ | $R_{s\,cal}$ | Base $q_c$ | $R_{b\,cal}$ | $R_{s\,cal} + R_{b\,cal}$ |
|---|---|---|---|---|---|
| 1 | 6.0 | 1.45 | 8.2 | 5.38 | 6.83 |
| 2 | 7.5 | 1.81 | 9.7 | 6.36 | 8.17 |
| 3 | 8.4 | 2.03 | 10.1 | 6.63 | 8.65 |
| 4 | 8.5 | 2.05 | 10.7 | 7.02 | 9.07 |
| $R_{s\,cal}$ (mean) | | 1.83 | $R_{b\,cal}$ (mean) | 6.35 | |
| $R_{s\,cal}$ (min) | | 1.45 | $R_{b\,cal}$ (min) | 5.38 | |

The correlation factors for 4 tests as in Table A.NA.10 are: $\xi_3$ (mean) = 1.38 and $\xi_4$ (min) = 1.29.

*Taking DA1-2* (as likely defining resistance, sets A2 + M1 + R4 apply) assuming no pile tests, preliminary or working, where $\gamma_G$ = 1.0, $\gamma_s$ = 1.5 and $\gamma_b$ = 1.7,

$R_{sd}$ = minimum of $R_{s\,cal}$ (mean) or $R_{s\,cal}$ (min),
   that is 1.83/(1.5 × 1.38) = 0.89 MN or 1.45/(1.5 × 1.29) = 0.75 MN

$R_{bd}$ = minimum of $R_{b\,cal}$ (mean) or $R_{b\,cal}$ (min),
   that is 6.35/(1.7 × 1.38) = 2.71 MN or 5.38/(1.7 × 1.29) = 2.45 MN

Therefore, $R_{cd} = R_{sd\,min} + R_{bd\,min}$ = 0.75 + 2.45 = 3.2 MN.

The design action $F_d = G_k \gamma_G \leq R_{cd}$.

Hence, $F_d \leq 3.20 \times 1.0$ = 3.20 MN as the maximum permanent unfavourable action for the pile.

*Checking DA1-1* (sets A1 + M1 + R1 apply) where $\gamma_s = \gamma_b = 1.0$ and $\gamma_G$ = 1.35,

$R_{sd}$ = minimum of $R_{s\,cal}$ (mean) or $R_{s\,cal}$ (min),
   that is 1.83/(1.0 × 1.38) = 1.33 MN or 1.45/(1.0 × 1.29) = 1.12 MN

$R_{bd}$ = minimum of $R_{b\,cal}$ (mean) or $R_{b\,cal}$ (min),
   that is 6.35/(1.0 × 1.38) = 4.60 MN or 5.38/(1.0 × 1.29) = 4.17 MN

Therefore, $R_{cd} = R_{sd\,min} + R_{bd\,min}$ = 1.12 + 4.17 = 5.29 MN.

$F_d$ = 1.35 × 3.20 = 4.32 MN < $R_{cd}$ = 5.29 MN and satisfactory.

The ultimate resistance of the open-end pile can be calculated by the *ICP method*.

Assuming that the sand has a $D_{50}$ size of 0.3 mm, Figure 4.21 gives a value of 27° for the interface angle of friction, $\delta_{cv}$. From Equation 4.21, the equivalent radius of the open-end pile is $R^*$ = $(0.457^2 - 0.438^2)^{0.5}$ = 0.130 m.

The shear modulus G in Equation 4.28 can be calculated from Figure 5.22 and Equation 6.49. Figure 5.22 gives $E_{50}$ = 30 MN/m². Take Poisson's ratio as 0.2, giving G = 30/2(1 + 0.2) = 12.5 MN/m².

Take the average roughness as 2 × 1 × 10⁻⁵ mm. From Equation 4.27, $\Delta\sigma'_{rd}$ = 2 × 12.5 × 2 × 10⁻⁵/0.457 = 0.001 MN/m².

This is small in relation to $\sigma'_{rc}$ as calculated below and can be neglected.

The 10.5 m of embedment into the sand is divided into 9 by 1.0 m segments and an uppermost segment of 1.5 m (the limiting height to the lowermost segment of 8.0 × 0.13 m = 1.04 m is not exceeded).

Calculating $\sigma'_{rc}$ for the lowermost layer, the effective overburden pressure at the centre of the layer is (8 × 6.50 + 10 × 10.0) = 152 kN/m², and the average $q_c$ is 9.5 MN/m². Take $P_a$ = 100 kN/m².

From Equation 4.26,

$\sigma'_{rc}$ = 0.029 × 9.5 × $(152/100)^{0.13}$ × $(0.5/0.13)^{-0.38}$

   = 0.174 MN/m²

From Equation 4.24,
Unit shaft resistance = $0.174 \times \tan 27° = 0.087$ MN/m$^2$
Shaft resistance on segment = $0.087 \times 2.87 = 0.250$ MN
The resistance of the remaining segments to the top of the sand layer is calculated in the same way as shown in the following table:

| Depth of segment (m bgl) | h (m) | $\left(\dfrac{h}{R^*}\right)^{-0.38}$ | $\left(\dfrac{\sigma'_{vo}}{p_o}\right)^{0.13}$ | $q_c$ (MN/m$^2$) | $\sigma_{rc}$ (MN/m$^2$) | $\tau_f = \sigma_{rc} \tan \delta_{cv}$ (MN/m$^2$) | $Q_s$ (MN) |
|---|---|---|---|---|---|---|---|
| 17–16 | 0.5 | 0.599 | 1.055 | 9.5 | 0.174 | 0.087 | 0.250 |
| 16–15 | 1.5 | 0.395 | 1.047 | 8.9 | 0.107 | 0.053 | 0.152 |
| 15–14 | 2.5 | 0.325 | 1.037 | 8.4 | 0.082 | 0.041 | 0.118 |
| 14–13 | 3.5 | 0.286 | 1.026 | 7.9 | 0.067 | 0.033 | 0.095 |
| 13–12 | 4.5 | 0.260 | 1.015 | 7.4 | 0.057 | 0.028 | 0.080 |
| 12–11 | 5.5 | 0.241 | 1.003 | 6.8 | 0.048 | 0.024 | 0.069 |
| 11–10 | 6.5 | 0.226 | 0.989 | 6.3 | 0.041 | 0.020 | 0.057 |
| 10–9 | 7.5 | 0.214 | 0.974 | 5.8 | 0.035 | 0.017 | 0.049 |
| 9–8 | 8.5 | 0.204 | 0.958 | 5.2 | 0.029 | 0.015 | 0.043 |
| 8–6.5 | 7.25 | 0.217 | 0.935 | 4.4 | 0.026 | 0.013 | 0.037 |
| | | | | | | Total $Q_s$ = | 0.95 MN |

Calculating the base resistance, $q_c$ at base = 9.7 MN/m$^2$ $D_{inner}/D_{CPT} = 0.876/0.038 = 24.3$ which is greater than $0.083 \times 9.7 \times 10^3/100 = 8.05$. Therefore, a rigid basal plug will not develop.

Taking the base resistance of an unplugged pile as $0.5\ q_c$, then $q_b = 0.5 \times 9.7 = 4.85$ MN/m$^2$ and base resistance $Q_b = 4.85(0.914^2 - 0.876^2) \times \pi/4 = 0.26$ MN.

Assuming no frictional contribution from the inner surface of the pile,
Total pile resistance = $0.95 + 0.26 = 1.21$ MN

## Example 4.7

A bored and cast-in-place pile is required to carry an applied load of 9000 kN (permanent unfavourable action) at a site where 4 m of loose sand overlies a weak jointed cemented mudstone. Core drilling into the mudstone showed partly open joints and RQD values increased from an average of 15% at rockhead to 35% at a depth of 10 m. Tests on rock cores gave an average unconfined compression strength of 4.5 MN/m$^2$. Determine the required depth of the pile below rockhead, and calculate the settlement of the pile at the applied load.

The effective diameter of a 1.5 m diameter pile as in Table 4.6 is $1500 - 50 = 1450$ mm for the length of pile in loose sand. Concrete grade will be C25/35, with the characteristic strength $f_{ck}$ of 25 MN/m$^2$ and $\gamma_C \times k_f = 1.5 \times 1.1$ and $\alpha = 0.85$. From Section 7.10, $N_{Ed}/N_{Rd} \leq 1.0$:

$$N_{Ed} = 9 \times \gamma_G = 9 \times 1.35 = 12.15 \text{ MN}$$

$$N_{Rd} = A\,f_{ck}\,\alpha/\gamma_C = (1.45^2 \times \pi/4 \times 25 \times 0.85)/(1.5 \times 1.1) = 21.3 \text{ MN which is satisfactory}$$

Load carried in shaft friction in the loose sand will be negligible.
From Figure 4.33 for $q_{uc} = 4.5$ MN/m$^2$, $\alpha = 0.2$. The mass factor, $j$, for RQD from 15% to 35% is 0.2. Therefore, $\beta$, from Figure 4.34, is 0.65.
From Equation 4.41, unit rock socket shaft friction = $0.2 \times 0.65 \times 4500 = 585$ kN/m$^2$.

Taking a 7 m socket length and a1400 mm rock bit drilling inside temporary casing and model factor $\gamma_{Rd}$ = 1.4,

Characteristic shaft friction resistance $R_{sk}$ = (585 × $\pi$ × 1.4 × 7)/1.4 = 12,864 kN

Because of the open joints in the rock, it will be advisable to assume that the base resistance does not exceed the unconfined compression strength of the rock:

Characteristic base resistance $R_{bk}$ = (4500 × 1.4² × $\pi$/4)/1.4 = 4948 kN

*For DA1-1* (sets A1 + M1 + R1 apply) and using the partial factors for bored piles (Table 4.4) where $\gamma_b$ = $\gamma_s$ = 1.0 and $\gamma_G$ = 1.35,

$$F_d = 1.35 \times 9000 \text{ kN}$$

Design resistance $R_{cd}$ = (12,864 + 4,948) = 17,812 kN = 12,150 kN > $F_d$ and satisfactory

*For DA1-2* (sets A2 + M1 + R4 apply) with $\gamma_b$ = 2.0, $\gamma_s$ = 1.6, assuming no pile tests, and $\gamma_G$ = 1.0,

Design resistance $R_{cd}$ = (12,864/1.6 + 4,948/2.0) = 10,514 kN > $F_d$ = 9,000 kN and satisfactory

Check concrete stress under load:

$$\sigma_{cd} = F_d/A = 12.15/(1.45^2 \times \pi/4) = 7.36 \text{ MN/m}^2$$

$$f_{cd} = f_{ck} \, \alpha \, /\gamma_C \times k_f = 25 \times 0.85)/(1.5 \times 1.1) = 12.9 \text{ MN/m}^2 > \sigma_{cd} \text{ and satisfactory}$$

If, considering the factors noted by Wyllie[4.54] in Section 4.3.7, the rock socket shaft friction were to be only half the calculated value, no load would be transferred to the pile base. Therefore, the pile head settlement will be caused by compression in the rock socket only.

From Section 5.5, the modulus ratio of a cemented mudstone is 150, and for a mass factor of 0.2, the deformation modulus of the rock mass is 0.2 × 150 × 4.5 = 135 MN/m². In Figure 4.36, the modulus ratio $E_c/E_d$ is 20 × 10³/135 = 148, and for L/B = 7/1.5 = 4.7, the influence factor *I* is 0.25. The ratio D/B for a recessed socket is 4/1.5 = 2.7. There, the reduction factor from Figure 4.37 is about 0.8. Hence, from Equation 4.45,

$$\text{Pile head settlement} = \frac{0.8 \times 9 \times 10^3 \times 0.25}{1.5 \times 135} = 9 \text{ mm}$$

Checking the calculated unit shaft friction from Equation 4.42 and taking *b* as 0.25, $f_s$ = 0.25$\sqrt{4.5}$ = 0.53 MN/m² which agrees closely with Equation 4.41.

If the socket is grooved to an average depth of 25 mm over shortened socket length of 5.0 m with the grooves at vertical intervals of 0.75 m, say 6 grooves, then the following applies:

In Equation 4.44, if the $\Delta_r$ = 0.775 – 0.75 = 0.025 m and the total length of travel = $\pi$ × 1.4 × 6 = 26.39 m, then RF = 0.025 × 26.39/(0.75 × 5.0) = 0.18.

From Equation 4.43, the unit shaft friction $f_s$ = 0.8(0.18)^0.45 × 4.5 = 1.66 MN/m².

The characteristic shaft friction resistance on 5 m socket length, with the model factor of 1.4, is $R_{sk}$ = (1.66 × $\pi$ × 1.4 × 5)/1.4 = 26.1 MN.

*For DA1-2* with factors as above, the design resistance $R_{cd}$ = 26.1/1.6 = 16.3 MN > $F_d$ = 9 MN and satisfactory.

Therefore, grooving the socket would theoretically provide a much shorter socket length than the 7 m required for an un-grooved shaft.

## Example 4.8

A tubular steel pile with an OD of 1067 mm is driven with a closed end to near refusal in a moderately strong sandstone (average $q_{uc}$ = 20 MN/m²) overlain by 15 m of soft clay. Core drilling in the rock showed a fracture frequency of 5 joints per metre. Calculate the maximum load (permanent unfavourable action) which can be applied to the pile and the settlement at this load.

Only a small penetration below rockhead will be possible with sandstone of this quality, and the rock will be shattered by the impact. Hence, frictional support both in the soft clay and the rock will be negligible compared with the base resistance.

Pile-driving impact is likely to open joints in the rock; hence, the base resistance should not exceed the unconfined compression strength of the intact rock:

From Section 7.10, $N_{Ed}/N_{Rd} \leq 1.0$

$$N_{Ed} = G_k \times \gamma_G = G_k \times 1.35$$

$$N_{Rd} = Af_y/\gamma_{M0}$$

Using S235 JRH tubular steel pile with wall thickness of 19 mm, characteristic yield strength of 235 N/mm², and $\gamma_{M0}$ = 1.0, then $N_{Rd}$ = [(1067² – 1029²) × 235 × π/4]/1.0 × 10⁶ = 14.7 MN.

Hence, the maximum load based on steel strength $G_k$ = 14.7/1.35 = 10.9 MN.

If the characteristic pile resistance $R_{ck}$ is equal to the base resistance with $\gamma_{Rd}$ = 1.4, then $R_{bk}$ = (1.067² × π/4 × 20)/1.4 = 12.7 MN.

*For DA1-1* (sets A1 + M1 + R1 apply) for a driven pile, if the partial factors are $\gamma_b$ = 1 and $\gamma_G$ = 1.35, then $R_{cd}$ = 12.7/1.0 = 12.7 MN < $F_d$ = 1.35 × 10.9 = 14.7 MN which fails.

*For DA1-2* (sets A2 + M1 + R4), if the partial factors are $\gamma_b$ = 1.7 (no pile testing) and $\gamma_G$ = 1.0, then $R_{cd}$ = 12.7/1.7 = 7.5 MN.

This is the maximum unfavourable action which the pile can resist and will satisfy DA1-1:

$$F_d = 1.35 \times 7.5 = 10.13 \text{ MN} < R_{cd} = 12.7 \text{ MN}$$

Pile-driving impact may increase the fracture frequency from 5 to 10, say, fractures per metre giving a mass factor of 0.2. From Section 5.4, the modulus ratio of sandstone is 300:

$$\text{Deformation modulus} = 0.2 \times 300 \times 20 = 1200 \, \text{MN/m}^2$$

From Equation 4.35,

$$\text{Settlement of pile head} = \frac{7.52 \times 15 \times 1000}{0.0626 \times 2 \times 10^5} + \frac{0.5 \times 7.52 \times 1000}{1.067 \times 1200}$$

$$= 9.0 + 2.9$$

$$= 11.9 \, \text{mm}$$

(Range is likely to be from 10 to 15 mm.)

## Example 4.9

A 5 m layer of hydraulic fill consisting of sand is pumped into place over the ground shown in Figure 4.46. The calculated time/settlement curve for the surface of the hydraulic fill is

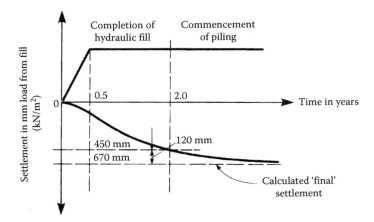

*Figure 4.48* Settlement v Time for Example 4.9.

shown in Figure 4.48. Two years after the completion of filling, a closed-end steel cased pile with an OD of 517 mm is driven to a penetration of 27 m to carry a permanent (unfavourable) load of 900 kN. Calculate the negative friction which is developed on the pile shaft, and assess whether or not any deeper penetration is required to carry the combined load and negative skin friction.

It can be seen from the time/settlement curve that about 120 mm of settlement will take place from the time of driving the pile until the clay beneath the fill layer is fully consolidated. This movement is considerably larger than the compression of the pile head under the applied load (about 10 mm of settlement would be expected under the applied load of 900 kN). Therefore, negative skin friction will be developed over the whole depth of the pile within the hydraulic fill. Considering now the negative skin friction within the soft clay, if it is assumed that downdrag will not occur if the clay settles relatively to the pile by less than 5 mm, then adding the settlement of the pile toe (10 mm at the applied load) negative skin friction will not be developed below the point where the clay settles by less than 15 mm relative to site datum. After pile driving, the full thickness of the clay settles by 120 mm at the surface of the layer. By simple proportion, a settlement of 5 mm occurs at a point 12 × 15/120 = 1.5 m above the base of the layer. This assumes uniform compressibility in the clay, but there is decreasing compressibility with increasing depth such that the settlement decreases to less than 15 mm at a point not less than 2 m above the base of the layer. A closer estimate could be obtained by a $t$–$z$ analysis. However, the above approximate assessment will be adequate for the present case.

Adopting Meyerhof's factor from Figure 4.39 for the negative skin friction and applying Equation 4.46 gives the following:
The unit negative skin friction 2 m above the base of clay layer is

$$0.3\sigma'_{vo} = 0.3 \times 9.81[(5 \times 2) + (2 \times 1.9) + (8 \times 0.9)] = 62 \text{ kN/m}^2$$

The unit negative skin friction at the top of clay stratum is

$$0.3 \times 9.81 \times 5 \times 2 = 29 \text{ kN/m}^2$$

The unit negative skin friction 2 m below the top of clay stratum (at groundwater level) is

$$0.3 \times 9.81 [(5 \times 2) + (2 \times 1.9)] = 41 \text{ kN/m}^2$$

The total negative skin friction in clay is

$$\pi \times 0.517 \, [0.5(29 + 41)2 + 0.5(41 + 62)8] = 783 \text{ kN}$$

Drainage of the fill will produce a medium-dense state of compaction for which $K_0$ is 0.45 and $K_s$ in Equation 4.14 is 0.67 (Table 4.7) and $\delta = 0.7 \times 30 = 21°$ (Table 4.8). Therefore, the additional negative skin friction as in Equation 4.14 is

$$0.67 \times (9.81 \times 2 \times 5) \times 0.5 \times \tan 21° \times \pi \times 0.517 \times 5 = 102 \text{ kN}$$

Hence, the total negative skin friction on pile (dragload) $D_{Gd} = 102 + 783 = 885$ kN.

From Example 4.5, the average CPT resistance at a penetration of 28 m is about 8.75 MN/m², and applying the model factor $\gamma_{Rd} = 1.4$ as the resistances have been calculated from the cautious estimate of CPT results and pile tests is not practical in downdrag conditions; then

Characteristic base resistance at 28 m = $R_{bk} = (\pi/4 \times 0.517^2 \times 8.75 \times 10^3)/1.4 = 1312$ kN

Also from Example 4.5 by proportion,

Characteristic shaft friction resistance from 12 to 28 m = $R_{sk} = 666.4 \times 517/450 = 765.6$ kN

The shaft resistance is fully mobilised over the depth below the base of the fill.

*For DA1-2*, the partial factors are as Table 4.17 for sets *A1 + M1* in downdrag with Table 4.3 *R4* resistance factors:

Structural action, *A1* = 1.0. Downdrag action, *A1* = 1.25. *R4* set, $\gamma_s = 1.5 \, \gamma_b = 1.7$

Design value of actions $F_d = 900 \times 1.0 + 885 \times 1.25 = 2006$ kN

Design resistance $R_{cd} = R_{cs} + R_{cb} = (765.6/1.5 + 1312/1.7) = 1282$ kN $< F_d$ which fails

Therefore, the pile has to be driven a further 4 m to 32 m depth in order to support the dragload:

At 32 m depth, the average $q_c$ from Figure 4.46 and Equation 4.16 is approximately 12 MN/m². The characteristic base resistance at 32 m is $R_{bk} = (\pi/4 \times 0.517^2 \times 12 \times 10^3)/1.4 = 1799$ kN.

At 30 m depth, the cone resistance $q_c = 16$ MN; hence, the unit shaft resistance is $0.012 \times 16 \times 1000 = 192$ kN/m².

At 32 m depth, $q_c = 19$ MN; hence, the unit shaft resistance is $0.012 \times 19 \times 1000 = 228$ kN/m².

The increase in total shaft friction over the extra 2 m to 30 m depth is

$$\frac{(96 + 192)2}{2} \times \pi \times 0.517 = 468 \text{kN}$$

The increase in total shaft friction over extra 2–32 m is

$$\frac{(192 + 228)2}{2} \times \pi \times 0.517 = 682 \text{kN}$$

The characteristic shaft friction resistance from 12 to 32 m is $R_{sk} = 765.6 + (468 + 682)/1.4 = 1587$ kN.

The design resistance is $R_{cd} = R_{cs} + R_{cb} = (1587/1.5 + 1799/1.7) = 2116$ kN $> F_d = 2006$ kN which is therefore satisfactory.

*By inspection, DA1-1* is not a critical.

The preceding calculations show that the penetration of 32 m is needed for the pile to satisfy the downdrag requirements of the ULS using EC7 procedures. Whereas using the allowable stress method for the 28 m deep pile,

Total load on the pile = 855 + 900 = 1785 kN

Ultimate base resistance (from above) = (1.4) × 1312 = 1837 kN

Ultimate shaft resistance = (1.4) × 765.6 = 1072 kN

Factor of safety = (1837 + 1072)/1785 = 1.6 which would be considered satisfactory

The stress on the pile shaft must be checked under the maximum factored action for set *A1* where $\gamma_G$ = 1.35 and $\gamma_G$ downdrag = 1.25; hence,

$$F_d = 900 \times 1.35 + 885 \times 1.25 = 2410 \text{ kN}$$

For a wall thickness of 4.47 mm, the steel area is 7193 mm².

From Section 7.10.2, using S355 JRH steel with characteristic yield strength of 355 N/mm² and $\gamma_{M0}$ = 1.0,

$$\sigma_{cd} = F_{cd}/A_s = 2410/7193 = 335 \text{ N/mm}^2 < f_{cd} = 355/1.0 = 355 \text{ N/mm}^2$$

If the pile is filled with C25/35 grade concrete, the characteristic strength is 25 N/mm² and $\gamma_C$ is 1.5 × $k_f$. As the pile is permanently cased, no dimensional factor is applied:

$$\sigma_{cd} = F_d/A_c = 2410/(\pi/4 \times 0.508^2 \times 1000) = 11.9 \text{ N/mm}^2$$

$$f_{cd} = 0.85 \times 25/(1.5 \times 1.1) = 12.9 \text{ N/mm}^2 > \sigma_{cd}$$

Both pile materials are therefore satisfactory.

If there is concern about long-term corrosion of the steel section in the hydraulic fill, the strength of the concrete filling could be increased so that the whole of the load is carried by the concrete.

## REFERENCES

4.1    Tomlinson, M.J. The adhesion of piles driven in clay soils, *Proceedings of Fifth International Conference, ISSMFE*, London, UK, 1957, Vol. 2, 66–71.

4.2    Tomlinson, M.J. The adhesion of piles in stiff clay, Construction Industry Research and Information Association (CIRIA), Research Report No. 26, London, UK, 1970.

4.3    Bond, A.J. and Jardine, R.J. Effects of installing displacement piles in a high OCR clay, *Geotechnique*, 41 (3), 1991, 341–363.

4.4    Randolph, M.F. and Wroth, C.P. Recent developments in understanding the axial capacity of piles in clay, *Ground Engineering*, 15 (7), 1982, 17–25.

4.5    Semple, R.M. and Rigden, W.J. Shaft capacity of driven pipe piles in clay, *Symposium on Analysis and Design of Pile Foundations*, American Society of Civil Engineers, San Francisco, CA, 1984, 59–79.

4.6    Bown, A.S. and O'Brien, A.S. Shaft friction in London Clay – Modified effective stress approach, *Foundations: Proceedings of Second BGA International Conference on Foundations*, Dundee, UK. M.J. Brown, M.F. Bransby, A.J. Brennan and J.A. Knappett (ed.), BRE Press, 2008, 92–100.

4.7    Rigden, W.J., Pettit, J.J., St. John, H.D. and Poskitt, T.J. Developments in piling for offshore structures, *Proceedings of the Second International Conference on the Behaviour of Offshore Structures*, London, UK, 1979, Vol. 2, 276–296.

4.8 Weltman, A.J. and Healy, P.R. Piling in 'boulder clay' and other glacial tills, Construction Industry Research and Information Association (CIRIA), Report PG5, London, UK, 1978.

4.9 Trenter, N.A. Engineering in glacial tills, Construction Industry Research and Information Association (CIRIA), Report C504, London, UK, 1999.

4.10 Meyerhof, G.G. and Murdock, L.J. An investigation of the bearing capacity of some bored and driven piles in London Clay, *Geotechnique*, 3 (7), 1953, 267–282.

4.11 Whitaker, T. and Cooke, R.W. Bored piles with enlarged bases in London Clay, *Proceedings of Sixth International Conference, ISSMFE*, Montreal, Quebec, Canada, 1965, Vol. 2, pp. 342–346.

4.12 Skempton, A.W. Cast in-situ bored piles in London Clay, *Geotechnique*, 9 (4), 1959, 153–173.

4.13 London District Surveyors Association. Guidance notes for the design of straight shafted bored piles in London Clay, No. 1. LDSA Publications Bromley, London, UK, 2009.

4.14 Viggiani, C., Mandolini, A. and Russo, G. *Pile Design and Pile Foundations*. Spon Press, Abingdon, UK, 2012.

4.15 American Petroleum Institute, RP2A-WSD. *Recommended Practice for Planning, Designing and Constructing Fixed Offshore Platforms – Working Stress Design*, 21st ed. Washington, DC, 2000 (reaffirmed 2010).

4.16 Fleming, W.G.K. and Sliwinski, Z. The use and influence of bentonite in bored pile construction, Construction Industry Research and Information Association (CIRIA), Report PG3, London, UK, 1977.

4.17 Kraft, L.M. and Lyons, C.G. Ultimate axial capacity of grouted piles, *Proceedings of the Sixth Annual Offshore Technology Conference*, Houston, TX, 1974, pp. 485–499.

4.18 Jones, D.A. and Turner, M.J. Load tests on post-grouted micropiles in London Clay, *Ground Engineering*, September 1980, 47–53.

4.19 Bustamante, M. and Gianeselli, L. Pile bearing capacity prediction by means of static penetration CPT, *Proceedings of the Second European Symposium on Penetration Testing, ESOPT2*, Amsterdam, the Netherlands, 1982, pp. 493–500.

4.20 Bjerrum, L. Problems of soil mechanics and construction in soft clay, *Proceedings of the Eighth International Conference, ISSMFE*, Moscow, Russia, 1973, Vol. 3, pp. 150–157.

4.21 Taylor, P.T. Age effect on shaft resistance and effect of loading rate on load distribution of bored piles, PhD thesis, University of Sheffield, Sheffield, UK, October 1966.

4.22 Doherty, P. and Gavin, K. The aged reloading response of piles in clay, *36th Annual Conference on Deep Foundations, Boston, Massachusetts*, Deep Foundations Institute, Hawthorne, NJ, 2011, pp. 67–73.

4.23 Fleming, W.G.K., Weltman, A.J., Randolph, M.F. and Elson, W.K. *Piling Engineering*, 3rd ed. Taylor & Francis, Abingdon, UK, 2009.

4.24 Peck, R.B., Hanson, W.E. and Thornburn, T.H. *Foundation Engineering*, 2nd ed. John Wiley, New York, 1974.

4.25 Durgonoglu, H.T. and Mitchell, J.K. Static penetration resistance of soils, *Proceedings of the Conference on In-Situ Measurement of Soil Properties*, American Society of Civil Engineers, Raleigh, NC, 1975, Vol. 1, pp. 151–188.

4.26 Vesic, A.S. *Design of Pile Foundations, NCHRP Synthesis 42*. Transportation Research Board, Washington, DC, 1977.

4.27 Berezantsev, V.G. et al. Load bearing capacity and deformation of piled foundations, *Proceedings of the Fifth International Conference, ISSMFE*, Paris, France, 1961, Vol. 2, pp. 11–12.

4.28 Bolton, M.D. The strength and dilatancy of sands, *Geotechnique*, 36 (1), 1986, 65–78.

4.29 Kulhawy, F.H. Limiting tip and side resistance, fact or fallacy, *Proceedings of the Symposium on Analysis and Design of Pile Foundations*, American Society of Civil Engineers, San Francisco, CA, 1984, pp. 80–98.

4.30 Cheng, Y.M. $N_q$ factor for pile foundations by Berezantsev, *Geotechnique*, 54 (2), 2005, 149–150.

4.31 Poulos, H.G. and Davis, E.H. *Pile Foundation Analysis and Design*. John Wiley, New York, 1980.

4.32 Dutt, R.N., Moore, J.E. and Rees, T.C. Behaviour of piles in granular carbonate sediments from offshore Philippines, *Proceedings of the Offshore Technology Conference*, Houston, TX, Paper No. OTC 4849, 1985, pp. 73–82.

4.33  Hight, D.W., Lawrence, D.M., Farquhar, G.B. and Potts, D.M. Evidence for scale effects in the end-bearing capacity of open-end piles in sand, *Proceedings of the 28th Offshore Technology Conference*, Houston, TX, OTC 7975, 1996, pp. 181–192.

4.34  Meigh, A.C. *Cone Penetration Testing*. CIRIA-Butterworth, London, UK, 1987.

4.35  Bauduin, C.M. Design of axially loaded piles according to Eurocode 7, *Proceeding of the Ninth International Conference on Piling and Deep Foundations (DFI202)*, Nice, Presses de l'EPNC, Paris, France, 2002, pp. 301–312.

4.36  Te Kamp, W.C. Sondern end funderingen op palen in zand, *Fugro Sounding Symposium*, Utrecht, the Netherlands, 1977.

4.37  Broug, N.W.A. The effect of vertical unloading on cone resistance $q_c$, a theoretical analysis and a practical confirmation, *Proceedings First International Geotechnical Seminar on Deep Foundations on Bored and Auger Piles*, J. Van Impe (ed.), Balkema, Rotterdam, the Netherlands, 1988, pp. 523–530.

4.38  De Gijt, J.G. and Brassinga, H.E. Ontgraving beïnvlodet de conusweerstand, *Land and Water*, 1.2, January 1992, 21–25.

4.39  Heijnen, W.J. Tests on Frankipiles at Zwolle, the Netherlands, *La technique des Travaux*, No. 345, January/February 1974, pp. I–XIX.

4.40  Chow, F.C. Investigations into displacement pile behaviour for offshore foundations, PhD thesis, Imperial College, London, UK, 1997.

4.41  Jardine, R., Chow, F.C., Overy, R. and Standing, J. *ICP Design Methods for Driven Piles in Sands and Clays*. Thomas Telford, London, UK, 2005.

4.42  Tappin, R.G.R., van Duivandijk and Haque, M. The design and construction of Jamuna Bridge, Bangladesh, *Proceedings of the Institution of Civil Engineers*, 1998, Vol. 126, pp. 162–180.

4.43  Tomlinson, M.J. Discussion, British Geotechnical Society Meeting, 1996, *Ground Engineering*, 29 (10), 1996, 31–33.

4.44  American Petroleum Institute (API). *Geotechnical and Foundation Design Considerations*, 1st ed. ANSI/API RP2GEO/ISO19901-4:2003, Washington, DC, 2011.

4.45  Knudsen, S., Langford, T., Lacasse, S. and Aas P.M. Axial capacity of offshore piles driven in sand using four CPT-based methods. Offshore site investigation and geotechnics. Integrated geotechnologies – present and future, *Proceedings of the Seventh International Conference*, Society for Underwater Technology, London, UK, 2012, pp. 449–457.

4.46  White, D.J. and Bolton, M.D. Comparing CPT and pile base resistance in sand, *Proceedings of the Institution of Civil Engineers, Geotechnical Engineering*, 158 (GE1), 2005, 3–14.

4.47  Jardine, R.J. and Standing, J.R. Pile load testing performed for HSE cyclic loading study at Dunkirk, France, Offshore Technology Report OTO 2000 007. Vol. 2, Health and Safety Executive, London, UK, 2000.

4.48  Merritt, A.S., Schroeder, F.C., Jardine, R.J., Stuyts, B., Cathie, D. and Cleverly, W. Development of pile design methodology for an offshore wind farm in the North Sea. Offshore site investigation and geotechnics. Integrated geotechnologies – present and future, *Proceedings of the Seventh International Conference*, Society for Underwater Technology, London, UK, 2012, pp.439–447.

4.49  Togrol, E., Aydinoglu, N., Tugcu, E.K. and Bekaroglu, O. Design and Construction of large piles, *Proceedings of the 12th International Conference on Soil Mechanics*, Rio de Janeiro, Brazil, 1992, Vol. 2, pp. 1067–1072.

4.50  Meyerhof, G.G. Bearing capacity and settlement of pile foundations, *Proceedings of the American Society of Civil Engineers*, GT3, 1976, 197–228.

4.51  Burland, J.B., Butler, F.G. and Duncan, P. The behaviour and design of large diameter bored piles in stiff clay, *Proceedings of the Symposium on Large Bored Piles*, Institution of Civil Engineers and Reinforced Concrete Association, London, UK, 1966, pp. 51–71.

4.52  *ArcelorMittal Piling Handbook*, 8th ed. ArcelorMittal Group, 2005.

4.53  Pells, P.J.N. and Turner, R.M. End bearing on rock with particular reference to sandstone, *Proceedings of the International Conference on Structural Foundations on Rock*, Sydney, New South Wales, Australia, 1980, Vol. 1, pp. 181–190.

4.54  Wyllie, D.C. *Foundations on Rock*, 1st ed. E & FN Spon, London, UK, 1991.

4.55 Kulhawy, F.H. and Goodman, R.E. Design of foundations on discontinuous rock, *Proceedings of the International Conference on Structural Foundations on rock*, Sydney, New South Wales, Australia, 1980, Vol. 1, pp. 209–220.

4.56 Kulhawy, F.H. and Goodman, R.E. Foundations in rock, Chapter 15, *Ground Engineering Reference Book*, F.G. Bell (ed.), Butterworth, London, UK, 1987.

4.57 Zhang, L. and Einstein, H.H. End-bearing capacity of drilled shafts in rock, *Journal of the Geotechnical Engineering Division, American Society of Civil Engineers*, 124 (GT7), 1998, 574–584.

4.58 Lord, J.A., Clayton, C.R.I. and Mortimore, R.N. Engineering in chalk, Construction Industry Research and Information Association (CIRIA), Report No 574, London, UK, 2002.

4.59 Foundation design and Construction, Geo Publication No 1/2006, Geotechnical Engineering Office, Civil Engineering and Development Department, Government of Hong Kong.

4.60 Beake, R.H. and Sutcliffe, G. Pipe pile driveability in the carbonate rocks of the Southern Arabian Gulf, *Proceedings of the International Conference on Structural Foundations on Rock*, Balkema, Rotterdam, the Netherlands, 1980, Vol. 1, pp. 133–149.

4.61 George, A.B., Sherrell, F.W. and Tomlinson, M.J. The behaviour of steel H-piles in slaty mudstone, *Geotechnique*, 26 (1), 1976, 95–104.

4.62 Leach, B. and Mallard, D.J. The design and installation of precast concrete piles in the Keuper Marl of the Severn Estuary, *Proceedings of the Conference on Recent Developments in the Design and Construction of Piles*, Institution of Civil Engineers, London, UK, 1980, pp. 33–43.

4.63 Osterberg, J.O. and Gill, S.A. Load transfer mechanisms for piers socketed in hard soils or rock, *Proceedings of the Ninth Canadian Symposium on Rock Mechanics*, Montreal, Quebec, Canada, 1973, pp. 235–262.

4.64 Williams, A.F. and Pells, P.J.N. Side resistance rock sockets in sandstone, mudstone and shale, *Canadian Geotechnical Journal*, 18, 1981, 502–513.

4.65 Horvath, R.G. *Field Load Test Data on Concrete-to-Rock Bond Strength for Drilled Pier Foundations*. University of Toronto, Toronto, Ontario, Canada, publication 78–07, 1978.

4.66 Rosenberg, P. and Journeaux, N.L. Friction and end bearing tests on bedrock for high capacity socket design, *Canadian Geotechnical Journal*, 13, 1976, 324–333.

4.67 Hobbs, N.B. Review paper – Rocks, *Proceedings of the Conference on Settlement of Structures*, British Geotechnical Society, Pentech Press, 1975, pp. 579–610.

4.68 Horvath, R.G., Kenney, T.C. and Kozicki, P. Methods of improving the performance of drilled piles in weak rock, *Canadian Geotechnical Journal*, 20, 1983, 758–772.

4.69 Alrifai, L. Rock socket piles at Mall of the Emirates, Dubai, *Proceedings of the Institution of Civil Engineers, Geotechnical Engineering*, 160 (GE2), 2007, 105–120.

4.70 Chandler, R.J. and Forster, A. Engineering in Mercia mudstone, Construction Industry Research and Information Association (CIRIA), Report No. 570, London, UK, 2001.

4.71 Gannon, J.A., Masterton, G.G.T., Wallace, W.A. and Muir-Wood, D. Piled foundations in weak rock, Construction Industry Research and Information Association, (CIRIA), Report No. 181, London, UK, 1999/2004.

4.72 Seedhouse, R.L. and Sanders, R.L. Investigations for cooling tower foundations in Mercia mudstone at Ratcliffe-on-Soar, Nottinghamshire, *Proceedings of the Conference on Engineering Geology of Weak Rock*, A.A. Balkema, Rotterdam, the Netherlands, Special Publication No. 8, 1993, pp. 465–472.

4.73 Lord, J.A., Hayward, T. and Clayton, C.R.I. Shaft friction of CFA piles in chalk, Construction Industry Research and Information Association (CIRIA), Report No. PR86, London, UK, 2003.

4.74 Ng, C.W.W., Simons, N. and Menzies, B. *A Short Course in Soil-Structure Engineering of Deep Foundations, Excavations, and Tunnels*. Thomas Telford, London, UK, 2004, pp. 86.

4.75 Pells, P.J.N. and Turner, R.M. Elastic solutions for design and analysis of rock socketed piles, *Canadian Geotechnical Journal*, 16, 1979, 481–487.

4.76 Fellenius, B.H. Down drag on piles in clay due to negative skin friction, *Canadian Geotechnical Journal*, 9 (4), 1972, 323–337.

4.77 Johannessen, I.J. and Bjerrum, L. Measurement of the compression of a steel pile to rock due to settlement of the surrounding clay, *Proceedings of the Sixth International Conference, ISSMFE*, Montreal, Quebec, Canada, 1965, Vol. 2, pp. 261–264.

4.78 Fellenius, B.H. and Siegel T.C. Pile drag load and downdrag in a liquefaction event, *Journal of Geotechnical Engineering and Geoenvironmental Engineering, American Society of Civil Engineers*, 134 (9), 2008, 1412–1417.

4.79 Poulos, H.G. A practical design approach for piles with negative friction, *Proceedings of the Institution of Civil Engineers, Geotechnical Engineering*, 161 (GE1), 2008, 19–27.

4.80 Claessen, A.I.M. and Horvat, E. Reducing negative skin friction with bitumen slip layers, *Journal of the Geotechnical Engineering Division, American Society of Civil Engineers*, 100 (GT8), August 1974, 925–944.

4.81 *The Shell Bitumen Industrial Handbook*. Thomas Telford, London, UK, 1995.

4.82 Poulos, H.G. Behaviour of laterally loaded piles – single piles, *Proceedings of the American Society of Civil Engineers*, 97(SM5), 1971.

4.83 Potts, D.M. and Zdravkovic, L. *Finite Element Analysis in Geotechnical Engineering*, Vol. 1-Theory and Vol. 2-Application. Thomas Telford, London, UK, 1999.

4.84 Terzaghi, K. Evaluation of coefficients of subgrade reaction, *Geotechnique*, 5 (4), 1955, 297–326.

4.85 Randolph, M.F. The response of flexible piles to lateral loading, *Geotechnique*, 31 (2), 1981, 247–249.

4.86 Clancy, P. and Randolph, M.F. An approximate analysis procedure for piled raft foundations, *International Journal of Numerical and Analytical Methods in Geomechanics*, 17 (12), 1993, 849–869.

4.87 Padfield, C.J. and Sharrock, M.J. Settlement of structures on clay soils, Construction Industry Research and Information Association (CIRIA), Special Publication 27, London, UK, 1983.

4.88 O'Brien, A.S., Burland, J.B. and Chapman, T. Rafts and piled rafts, in *ICE Manual of Geotechnical Engineering*, J. Burland, T. Chapman, H. Skinner and M. Brown (ed.), Thomas Telford, London, UK, 2012.

4.89 Dewsbury, J.J. Numerical modelling of soil-pile-structure interaction, PhD thesis, University of Southampton, Southampton, UK, 2012.

4.90 International Society for Soil Mechanics and Ground Engineering *ISSMGE, CPRF—Combined piled raft foundations guidelines*. Deep foundations Commitee TC210. Draft 2012.

4.91 Lee, C.J. and Ng, C.W.W. Development of downdrag on piles and pile groups in consolidating soil, *Journal of Geotechnical and Geoenvironmental Engineering, American Society of Civil Engineers*, 130 (9), 2004, 905–914.

4.92 Ng, C.W.W., Poulos, H.G., Chan, V.S.H., Lam, S.S.Y. and Chan, G.C.Y. Effects of tip location and shielding on piles in consolidating ground, *Journal of Geotechnical and Geoenvironmental Engineering, American Society of Civil Engineers*, 134 (9), 2008, 1245–1250.

4.93 Zhang, L. *Drilled Shafts in Rock: Analysis and Design*. Taylor & Francis, London, UK, 2004.

4.94 Clayton, C.R.I. Obtaining parameters for numerical analysis, *BGA Symposium of Eurocode 7 – Today and Tomorrow*, Cambridge, UK, March 2011.

4.95 Vardanega, P.J., Williamson, M.G. and Bolton M.D. Bored pile design in stiff clay I and II, *Proceedings of the Institution of Civil Engineers, Geotechnical Engineering*, 165 (GE4), 2012, 213–246.

4.96 American Association of State Highway and Transport Officials. *AASHTO, LRFD Bridge Design Specifications, Customary US Units*. American Association of State Highway and Transport Officials, Washington, DC, 2012.

4.97 *Piling – Design and Installation. AS 2159 2009*. Standards Australia Limited, Sydney, New South Wales, Australia.

# Chapter 5

# Pile groups under compressive loading

## 5.1 GROUP ACTION IN PILED FOUNDATIONS

The supporting capacity of a group of vertically loaded piles can, in many situations, be considerably less than the sum of the capacities of the individual piles comprising the group. In all cases, the elastic and consolidation settlements of the group are greater than those of a single pile carrying the same applied load as that on each pile within the group. This is because the zone of soil or rock which is stressed by the entire group extends to a much greater width and depth than the zone beneath the single pile (Figure 5.1). Even when a pile group is bearing on rock, the elastic deformation of the body of rock within the stressed zone can be quite appreciable if the piles are loaded to their maximum safe capacity.

Group action in piled foundations has resulted in many recorded cases of failure or excessive settlement, even though loading tests made on a single pile have indicated satisfactory performance. A typical case of foundation failure is the single pile driven to a satisfactory set in a compact or stiff soil layer underlain by soft compressible clay. The latter formation is not stressed to any significant extent when the single pile is loaded (Figure 5.2a), but when the load from the superstructure is applied to the whole group, the stressed zone extends down into the soft clay. Excessive settlement or complete general shear failure of the group can then occur (Figure 5.2b).

The allowable loading on pile groups is sometimes determined by the so-called efficiency formulae, in which the efficiency of the group is defined as the ratio of the average load per pile when failure of the complete group occurs to the load at failure of a single comparable pile. The various efficiency ratios are based simply on experience without any relationship to soil mechanics principles. It is preferable to base design methods on the assumption that the pile group behaves as a block foundation with a degree of flexibility which depends on the rigidity of the capping system and the superimposed structure. By treating the foundation in this manner, normal soil mechanics practice can be followed in the calculations to determine the ultimate bearing capacity and settlement. Load transfer in shaft friction from the pile shaft to the surrounding soil is allowed for by assuming that the load is spread from the shafts of friction piles at an angle of 1 in 4 from the vertical. Three cases of load transfer are shown in Figure 5.3a through c.

An important point to note in the application of soil mechanics methods to the design of pile groups is that, whereas in the case of the single pile the installation method has a very significant effect on the selection of design parameters for shaft friction and end bearing, the installation procedure is of lesser importance when considering group behaviour. This is because the zone of disturbance of the soil occurs only within a radius of a few pile

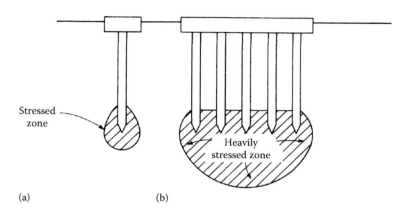

Figure 5.1 Comparison of stressed zones beneath single pile and pile group: (a) single pile; (b) pile group.

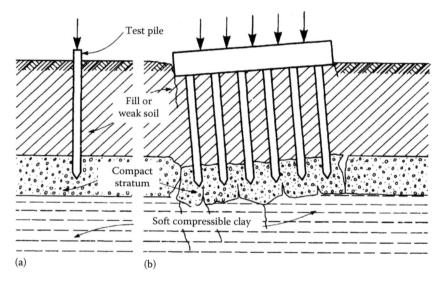

Figure 5.2 Shear failure of pile group: (a) test load on single isolated pile when soft clay is not stressed significantly; (b) load applied to group of piles when soft clay is stressed heavily.

diameters around and beneath the individual pile, whereas the soil is significantly stressed to a depth to or greater than the width of the group (Figure 5.1). The greater part of this zone is well below the ground which has been disturbed by the pile installation.

Section 4.9 outlines how computer programs have been established to model pile–soil interaction behaviour from which the settlement of pile groups and the loads on individual piles within the group can be determined. Some of the current programs which have been developed and are commercially available are given in Appendix C.

In the DEFPIG program by Poulos, soil behaviour is modelled on the basis of the theory of elasticity using interaction factors. Poulos[5.1] states, 'Despite the gross simplification which this model involves when applied to real soil, it provides a useful basis for the prediction of pile behaviour provided that appropriate elastic parameters are selected for the soil. A significant advantage of using an elastic model for soil is that it provides a rational means of

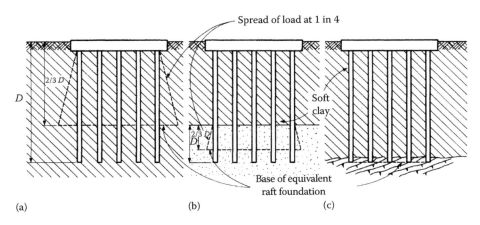

*Figure 5.3* Load transfer to soil from pile group: (a) group of piles supported predominantly by shaft friction; (b) group of piles driven through soft clay to combined shaft friction and end bearing in stratum of dense granular soil; (c) group of piles supported in end bearing on hard rock stratum.

analysis of pile groups and evaluation of immediate and final movement of a pile. In determining immediate movements, the undrained elastic parameters of the soil are used in the theory, whereas for final movements the drained parameters are used'. Poulos also provides comparisons of the predicted settlements for other programs which indicate that although settlement of single piles was predicted accurately, the group settlement was over-predicted, probably due to overestimation of the interaction effects. The interaction factors depend on the geometry, stiffness and spacing of the piles and the assessment of the elastic modulus of the soil between them. In a later paper dealing with complex vertical and lateral loading on a large pile group, Poulos[6.28] outlines the requirements for overall stability and serviceability analysis. He states that apart from the 3D finite element packages such as PLAXIS 3D, many of the current programs fall short of a number of critical aspects, particularly in their ability to include soil–raft contact and raft flexibility. However, the advances in mesh generation for finite element analysis as described by Dewsbury[4.89] are allowing more rigorous 3D models of rafts and piles to be made.

In view of the difficulties of obtaining representative values of the undrained and drained deformation parameters (particularly the latter) from field or laboratory testing of soils and rock, it is considered that the *equivalent raft method* is sufficiently reliable for most day-to-day settlement predictions. It is widely used to determine settlement either for preliminary design purposes or to check the output of computer programs. Whichever software is applied, it needs to be explicit as to how the soil is modelled and how the soil below the group is simulated, requiring careful selection of soil testing procedures and the resulting design parameters. The ability of software to assess load redistribution within the group, deformations, bending moments under lateral load and the use of raking piles clearly facilitates economic design. Also the use of computers allows rapid iterations to be made to study the effects of varying basic parameters such as pile diameter, length and spacing. In addition to the guidance in Sections 4.9.3 and 4.9.4, the pile group *aspect ratio* ($R = (ns/L)^{0.5}$, where $n$ is the number of piles in group, $s$ the spacing and $L$ the length, provides a useful indication of how the group will perform and which software is best suited to design. When $R$ is small (<2), most of the load will be taken in end bearing influencing compressible layers below the pile; when large (>4), the load is resisted by shaft friction on long piles.

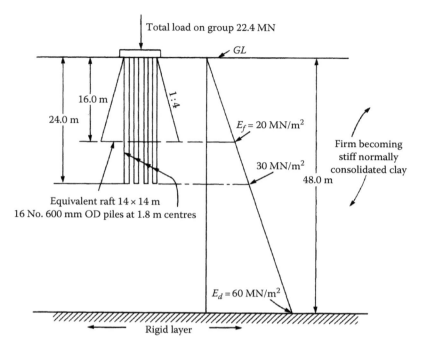

*Figure 5.4* Pile group settlement by equivalent raft method.

In most practical problems, piles are taken down to a stratum of relatively low compressibility and the resulting total and differential settlements are quite small such that an error of ±50% due to deficiencies in theory or unrepresentative deformation parameters need not necessarily be detrimental to the structure carried by the pile group.

As an example of the relative accuracy of the methods, Figure 5.4 shows a 4×4 pile group where the piles spaced at 3 diameters centre-to-centre are taken down to a depth of 24 m into a firm becoming stiff normally consolidated clay where the undrained shear strength and compressibility vary linearly with depth. The group settlements calculated by the equivalent raft method (Figure 5.19) used the influence factors of Butler (Figure 5.17).

The comparative group settlements were

DEFPIG 42 mm
PGROUP 31 mm
Equivalent raft 30 mm

The principal problems concerned with pile groups are constructional effects such as ground heave, the interference of closely spaced piles which have deviated from line during driving (see Section 3.4.13), and the possibilities of damage to adjacent structures and services. It is, of course, necessary to calculate the total and differential settlements of pile groups and overall piled areas to ensure that these are within limits acceptable to the design of the superstructure. The criteria of relative deflections, angular distortion and horizontal strain which can be tolerated by structures of various types have been reviewed by Burland and Wroth[5.2].

When checking group settlement calculations to verify compliance with serviceability limit criteria, EC7 recommends a partial factor of 1.0 for actions and ground properties unless otherwise specified.

## 5.2 PILE GROUPS IN FINE-GRAINED SOILS

### 5.2.1 Ultimate bearing resistance

Burland[5.3] has stated his strong opinion that specifying authorities' requirement for each pile in a group to be designed to carry an applied load which has a global safety factor on its ultimate bearing capacity can result in grossly uneconomic foundation design. This is because it ignores the capability of a raft to redistribute loads from the superstructure on to the piles forming the group. Redistribution of loading can be permitted provided that

1. The raft has sufficient flexibility (ductility) to perform this function without failure as a structural unit
2. The superstructure has sufficient flexibility to accommodate any resulting movements in the raft
3. The pile group has adequate resistance against failure or excessive settlement when considered as an *equivalent block foundation*
4. Account is taken of the effects of ground heave or subsidence of the mass of soil encompassed by the pile group during the construction stage (Section 5.7)

Burland recommends that redistribution should be effected by permitting piles carrying the heavier loading to mobilise their ultimate resistance in shaft friction, thereby yielding and transferring some of their load to surrounding piles within the group. This concept of *ductile foundations* where load sharing is designed between raft and piles and from pile to pile is discussed further in Section 5.10.

In all cases where piles are designed to transmit loading as a group terminating in a clay or sand stratum, whether or not some of the piles are permitted to yield, it is essential to consider the risks of general shear failure or excessive total and differential settlement of the equivalent block foundation taking the form shown in Figure 5.5.

The bearing resistance (ultimate limit state [ULS]) of the block foundation as shown in Figure 5.5 can be calculated by using the Brinch Hansen general equation[5.4]. This was referred to in Section 4.3 with reference to the bearing capacity factor $N_q$ in Equation 4.13. The complete Brinch Hansen equation as applied to a shallow spread foundation embedded in soil with a level ground surface is

$$q_u = cN_c s_c d_c i_c b_c + p_o N_q s_q d_q i_q b_q + 0.5\, \gamma B N_\gamma s_\gamma d_\gamma i_\gamma b_\gamma \qquad (5.1)$$

where
   $c$ is the cohesion intercept of soil
   $N_c$, $N_q$ and $N_\gamma$ are bearing capacity factors
   $s_c$, $s_q$ and $s_\gamma$ are shape factors
   $d_c$, $d_q$ and $d_\gamma$ are depth factors
   $i_c$, $i_q$ and $i_\gamma$ are load inclination factors
   $b_c$, $b_q$ and $b_\gamma$ are base inclination factors
   $\gamma$ is the density of the soil
   $p_o$ is the pressure of the overburden soil at foundation level

For undrained conditions ($\phi = 0°$), the second term of the equation is omitted and $c_u$ is substituted for $c$. For drained conditions, $c'$ (the cohesion intercept in terms of effective stress) is used instead of $c$. Values of the factors in Equation 5.1 are shown in Figures 5.6 through 5.10.

The equation in similar form is given in EC7 Annex D for drained bearing resistance, but as this is essentially an expression for shallow spread foundations ($D$ not greater than $B$),

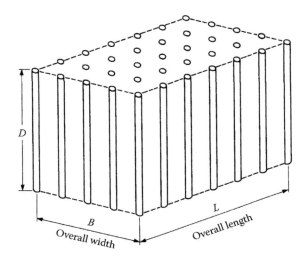

*Figure 5.5* Pile group acting as block foundation.

several factors are omitted which are critical to the safe design of the group as a block foundation (see Section 5.4).

Values of the shape factors $s_c$ and $s_\gamma$ for centrally applied vertical loading are obtained from Figure 5.7 and $s_q$ from the equation

$$s_q = \frac{s_c(s_c - 1)}{N_q} \tag{5.2}$$

Inclined loading is considered in relation to the effective breadth $B'$ and the effective length $L'$ of the equivalent block foundation. The plan dimensions of the block, as derived by Meyerhof[5.5], are shown in Figure 5.8. Thus, for loading in the direction of the breadth,

$$B' = B - 2e_x \tag{5.3a}$$

where $e_x$ is the eccentricity of loading in relation to the centroid of the base.

Similarly,

$$L' = L - 2e_y \tag{5.3b}$$

The shape factors, $s$, are modified for inclined loading by the equations

$$s_{CB} = 1 + 0.2i_{CB}B'/L' \tag{5.4}$$

$$s_{CL} = 1 + 0.2i_{CL}L'/B' \tag{5.5}$$

$$s_{qB} = 1 + \sin\phi\, i_{qB}B'/L' \tag{5.6}$$

$$s_{qL} = 1 + \sin\phi\, i_{qL}L'/B' \tag{5.7}$$

$$s_{\gamma B} = 1 - 0.4i_{\gamma B}B'/L' \tag{5.8}$$

$$s_{\gamma L} = 1 - 0.4i_{\gamma L}L'/B' \tag{5.9}$$

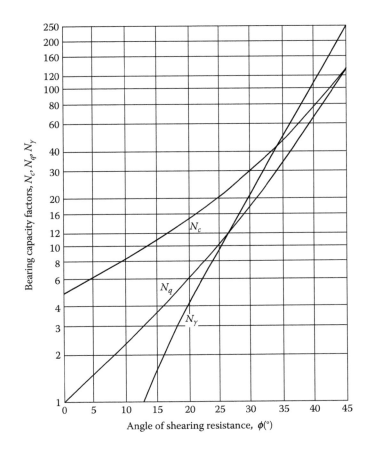

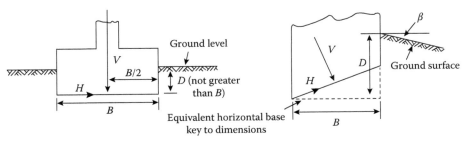

*Figure 5.6* Bearing capacity factors $N_c$, $N_q$ and $N_\gamma$. (After Hansen, J.B., A general formula for bearing capacity, Danish Geotechnical Institute, Bulletin No. 11, 1961; also, A revised and extended formula for bearing capacity, Danish Geotechnical Institute, Bulletin No. 28, 1968, and Code of Practice for Foundation Engineering, Danish Geotechnical Institute, Bulletin No. 32, 1978.)

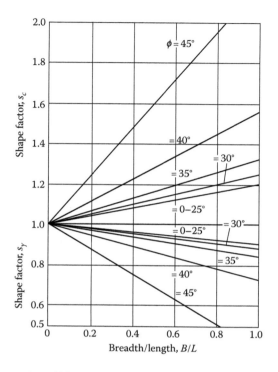

*Figure 5.7* Shape factors $s_c$ and $s_y$. (After Hansen, J.B., A general formula for bearing capacity, Danish Geotechnical Institute, Bulletin No. II, 1961; also, A revised and extended formula for bearing capacity, Danish Geotechnical Institute, Bulletin No. 28, 1968, and Code of Practice for Foundation Engineering, Danish Geotechnical Institute, Bulletin No. 32, 1978.)

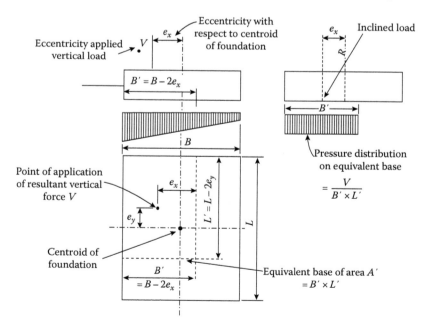

*Figure 5.8* Transformation of eccentrically loaded foundation to equivalent rectangular area carrying uniformly distributed load.

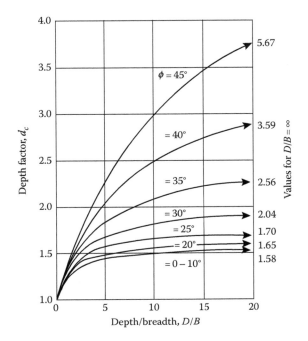

*Figure 5.9* Depth factor $d_c$. (After Hansen, J.B., A general formula for bearing capacity, Danish Geotechnical Institute, Bulletin No. 11, 1961; also, A revised and extended formula for bearing capacity, Danish Geotechnical Institute, Bulletin No. 28, 1968, and Code of Practice for Foundation Engineering, Danish Geotechnical Institute, Bulletin No. 32, 1978.)

Where $B'$ is less than $L'$, approximate values of the shape factors for centrally applied vertical loading which are sufficiently accurate for most practical purposes are as follows:

| Shape of base | $s_c$ | $s_q$ | $s_\gamma$ |
|---|---|---|---|
| Continuous strip | 1.0 | 1.0 | 1.0 |
| Rectangle | 1 + 0.2 B/L | 1 + 0.2 B/L | 1 − 0.4 B/L |
| Square | 1.3 | 1.2 | 0.8 |
| Circle (diameter B) | 1.3 | 1.2 | 0.6 |

Values of the depth factor $d_c$ are obtained from Figure 5.9. The values on the right-hand side of the figure are for $D$ = infinity. $d_q$ is obtained from

$$d_q = \frac{d_c - 1}{N_q} \tag{5.10}$$

The depth factor $d_\gamma$ can be taken as unity in all cases, and also when $\phi = 0°$, $d_q = 1.0$. Where $\phi$ is greater than $25°$, $d_q$ can be taken as equal to $d_c$. A simplified value of $d_c$ and $d_q$ where $\phi$ is less than $25°$ is $1 + 0.35\ D/B$. The use of the depth factors assumes that the soil above foundation level is not significantly weaker in shear strength than that of the soil below this level. However, in the case of pile groups, the piles are usually taken down through weak soils into stronger material, when either the depth factors should not be used or the depth $D$ should be taken as the penetration depth of the piles into the bearing stratum. Values of the

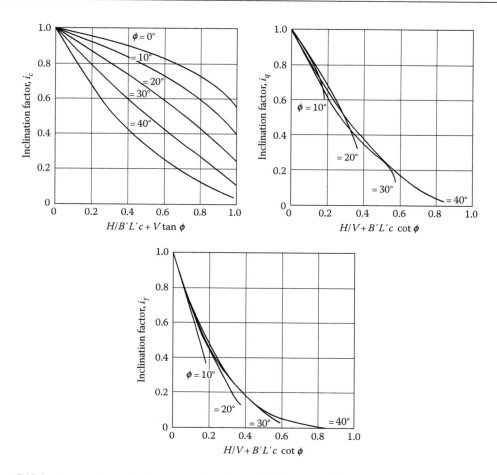

*Figure 5.10* Inclination factors $i_c$, $i_q$ and $i_\gamma$. (After Hansen, J.B., A general formula for bearing capacity, Danish Geotechnical Institute, Bulletin No. 11, 1961; also, A revised and extended formula for bearing capacity, Danish Geotechnical Institute, Bulletin No. 28, 1968, and Code of Practice for Foundation Engineering, Danish Geotechnical Institute, Bulletin No. 32, 1978.)

load inclination factors $i_c$, $i_q$ and $i_\gamma$ are shown in Figure 5.10 in relation to $\phi$ and the effective breadth $B'$ and length $L'$ of the foundation. Simplified values where the horizontal load $H$ is not greater than $V \tan \delta + cB'L'$ and where $c$ and $\delta$ are the parameters for cohesion and friction respectively, of the soil beneath the base are given by the following equations:

$$i_c = 1 - \frac{H}{2cB'L'} \tag{5.11}$$

$$i_q = 1 - \frac{1.5H}{V} \tag{5.12}$$

$$i_\gamma = i_q^2 \tag{5.13}$$

Equation 5.13 is strictly applicable only for $c = 0$ and $\phi = 30°$ but Brinch Hansen advises that it can be used for other values of $\phi$.

The base of an equivalent block foundation, that is, pile toe level, is usually horizontal, but where piles are terminated on a sloping bearing stratum, the base of the block can be treated as horizontal at a depth equal to that of the lowest edge and bounded by vertical planes through the other three edges (Figure 5.6). The base factors $b_c$, $b_q$ and $b_\gamma$ are unity for a horizontal base.

It is evident from the foregoing account of the application of the Brinch Hansen equation that it is not readily adaptable from its original use in the design of relatively shallow spread foundations to deep pile groups subjected to high levels of transverse loading. In such cases, it is preferable to use a computer program which can simulate interaction between the piles and the surrounding soil and can give a visual display of the extent of any overstressed zones in the soil below the group (see Section 4.9). Further aspects of group behaviour under transverse loading are discussed in Section 6.4.

Equation 5.1 ignores friction on the sides of the block foundation. The contribution of side shear is only a small proportion of the total where piles are taken down through a weak soil into a stronger stratum (but see Section 5.4). In cases of marginal stability, side shear resistance can be calculated as the shear resistance on a soil–soil interface on the sides of the group.

Where piles are installed in relatively small numbers, there is a possibility of excessive base settlement if two or more piles deviate from line and come into near or close contact at the toe and the toe loads are concentrated over a small area. While failure would not occur if end-bearing resistance was adequate, the settlement would be higher than that which would occur when the piles were at their design spacing. This would lead to differential settlement between the piles in the group. A safeguard against this occurrence is the adoption of a centre-to-centre spacing of piles in clay of at least three pile diameters, with a minimum of 1 m. The recommendations for friction piles are that the spacing should not be less than the perimeter of the pile or for circular piles three times the diameter. Closer spacing can be adopted for piles carrying their load mainly in end bearing, but the space between adjacent piles must not be less than their least width. Special consideration must be given to the spacing of piles with enlarged bases, including a study of interaction of stresses and the effect of construction tolerances. Where adjacent piles in a group have to be bored within 4 h, BS EN 1536 states the centre-to-centre distance must be greater than four times the diameter with a minimum of 2 m. The optimum spacing of piles can depend on the ULS due to tensile resistance failure and the uplift resistance of the block of soil containing the piles.

## 5.2.2 Settlement

The first step in the settlement analysis is to determine the vertical stress distribution below the base of the equivalent raft or block foundation (Figure 5.3) using the curves shown in Figure 5.11, where the stress at any depth $z$ below this level is related to its length/breath ratio. The curves assume that the foundation is rigid, but it is sufficiently accurate to assume that the superstructure, pile cap, piles and soil surrounding them have the required degree of rigidity.

The second step is to determine the depth of soil over which the stresses transmitted by the block foundation are significant. This is usually taken as the depth at which the vertical stress resulting from the net pressure at foundation level has decreased to 20% of the net overburden pressure at that level (Figure 5.12). A deeper level should be considered for soft highly compressible alluvial clays and peats.

The third step is to calculate the settlement of the foundation which takes place in two phases. The first is immediate settlement ($\rho_i$) caused by elastic compression of the soil without dissipation of pore pressure. It is followed by consolidation settlement ($\rho_c$) which takes place over the period of pore pressure dissipation at a rate which depends upon the permeability of the soil. There is also the possibility of very long-term secondary settlement ($\rho_\infty$)

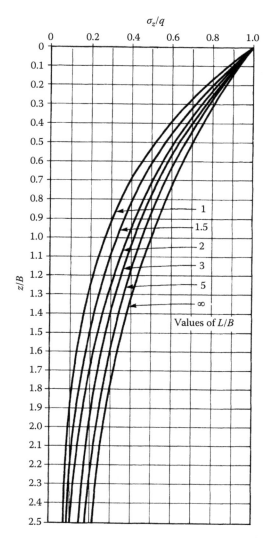

*Figure 5.11* Calculation of mean vertical stress ($\sigma_z$) at depth z beneath rectangular area $a \cdot b$ on surface loaded at uniform pressure q.

or creep of the soil. In the case of the very soft soils referred to in the previous paragraph, secondary settlement could be a significant proportion of the total.

The net immediate settlement of foundations on clays is calculated from the equation

$$\rho_i = \frac{q_n \times B \times (1 - \upsilon^2) \times I_p}{E_u} \tag{5.14}$$

where
  $q_n$ is the net foundation pressure
  $B$ is the foundation width
  $\upsilon$ is Poisson's ratio
  $E_u$ is the undrained deformation modulus
  $I_p$ is the influence factor

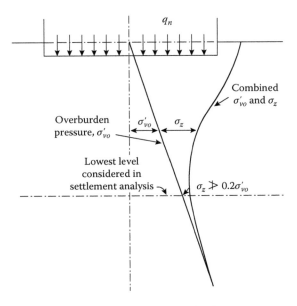

*Figure 5.12* Vertical pressure and stress distribution for deep clay layer.

$E_u$ (or for drained conditions designated $E'_v$) can be obtained by one or more of the following methods:

1. From the stress/strain curves established in the field by plate bearing tests
2. From drained triaxial compression tests on good-quality samples (to obtain $E'_v$)
3. From oedometer tests to obtain the modulus of volume compressibility ($m_v$), when $E'_v$ is the reciprocal of $m_v$
4. From relationships with the shear modulus ($G$) obtained in the field by pressuremeter tests:
   $E_u = 2G(1 + \nu_u)$ and $E'_v = 2G(1 + \nu')$, where $\nu_u$ and $\nu'$ are the undrained and drained values of Poisson's ratio, respectively

With regard to method (1), a typical stress/strain curve obtained by a plate-bearing test in undrained conditions is shown in Figure 5.13. Purely elastic behaviour occurs only at low stress levels (line AB in Figure 5.13). Adoption of a modulus of elasticity (Young's modulus) corresponding to AB could result in underestimating the settlement. The usual procedure is to draw a secant AC to the curve corresponding to a compressive stress equal to the net foundation pressure at the base of the equivalent block foundation. More conservatively, the secant AD can be drawn at a compressive stress of 1.5 times or some other suitable multiple of the foundation pressure. The deformation modulus $E_u = q/x$ for the particular condition.

As an alternative to direct determination of $E_u$ from field tests, it can be obtained from a relationship with the undrained shear strength $c_u$, the plasticity index and over-consolidation ratio of the clay established by Jamiolkowski et al.[5.6] (Figure 5.14). The latter value is derived from oedometer tests or from a knowledge of the geological history of the deposit.[5.7] These tests are used to calculate the long-term consolidation settlement of the foundation as described in the succeeding texts. Knowing the oedometer settlement ($\rho_{oed}$) provides another way of determining the immediate, consolidation and final settlements using the following relationships established by Burland et al.[5.8]

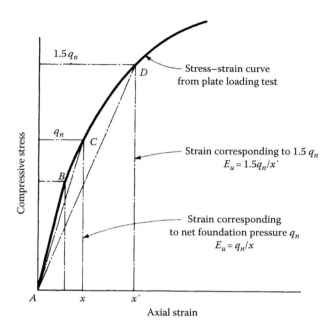

Figure 5.13 Determining deformation modulus $E_u$ from stress/strain curve.

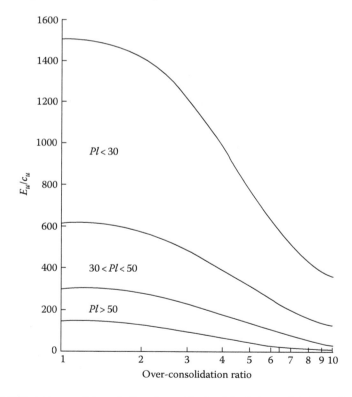

Figure 5.14 Relationship between $E_u/c_u$ ratio for clays with plasticity index and degree of over-consolidation. (After Jamiolkowski, M. et al., Design parameters for soft clays, *Proceedings of the 7th European Conference on Soil Mechanics*, Brighton, UK, pp. 21–57, 1979.)

For stiff over-consolidated clays:

$$\text{Immediate settlement} = \rho_i = 0.5 \text{ to } 0.6\rho_{oed} \tag{5.15}$$

$$\text{Consolidation settlement} = \rho_c = 0.4 \text{ to } 0.5\rho_{oed} \tag{5.16}$$

$$\text{Final settlement} = \rho_{oed} \tag{5.17}$$

For soft normally consolidated clays:

$$\text{Immediate settlement} = \rho_i = 0.1\rho_{oed} \tag{5.18}$$

$$\text{Consolidation settlement} = \rho_c = \rho_{oed} \tag{5.19}$$

$$\text{Final settlement} = \rho_{oed} \tag{5.20}$$

The $E_u/c_u$ ratio is also strain dependent showing a reduction in the ratio with increasing strain. Jardine et al.[5.9] showed this effect in London Clay from the results of undrained triaxial tests on good-quality samples (Figure 5.15). Normally loaded foundations, including pile groups, usually exhibit a strain of 0.01%–0.1%, which validates the frequently used relationship $E_u = 400c_u$ for the deformation modulus of intact blue London Clay.

Marsland[5.10] obtained $E_u/c_u$ ratios equal to 348 for an upper glacial till and 540 for a laminated glacial clay at Redcar, North Yorks.

The influence factor $I_p$ in Equation 5.14 is obtained from Steinbrenner's curves (Figure 5.16) using the method developed by Terzaghi[5.11]. Values of $F_1$ and $F_2$ in Figure 5.16 are related to Poisson's ratio ($\nu$) of the foundation soil. For a ratio of 0.5, $I_p = F_1$. When the ratio is zero, $I_p = F_1 + F_2$. Some values of Poisson's ratio are shown in Table 5.1.

When using the curves in Figure 5.16 to calculate the immediate settlement of a flexible pile group, the square or rectangular area in Figure 5.5 is divided into four equal rectangles. Equation 5.14 then gives the settlement at the corner of each rectangle. The settlement at the centre is then equal to four times the corner settlement. In the case of a rigid pile group such as a group with a rigid cap or supporting a rigid superstructure, the settlement at the centre

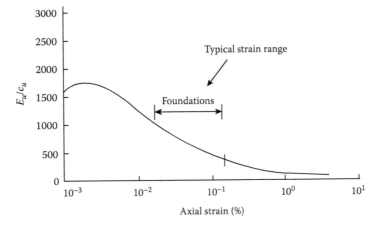

Figure 5.15 Relationship between $E_u/c_u$ and axial strain. (After Jardine, R. et al., Field and laboratory measurements of soil stiffness, *Proceedings of the 11th International Conference on Soil Mechanics*, San Francisco, CA, Vol. 2, pp. 511–514, 1985.)

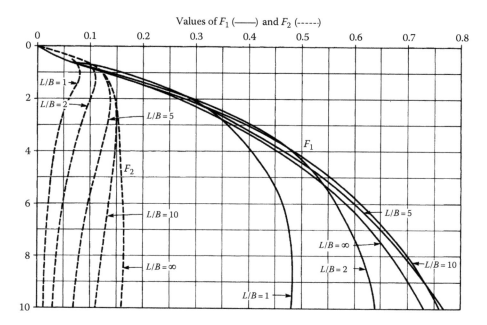

*Figure 5.16* Values of Steinbrenner's influence factor $I_p$ (for $\nu$ of 0.5, $I_p = F_1$, for $\nu = 0$, $I_p = F_1 + F_2$). Note: When using this diagram to calculate at the centre of a rectangular area, take $B$ as half the foundation width to obtain $H/B$ and $L/B$.

*Table 5.1* Poisson's ratio for various soils and rocks

| | |
|---|---|
| Clays (undrained) | 0.5 |
| Clays (stiff, drained) | 0.1–0.3 |
| Silt | 0.3 |
| Sands | 0.1–0.3 |
| Rocks | 0.2 |

of the longest edge (twice the corner settlement) is obtained and the average settlement of the group is obtained from the equation

$$\rho_{average} = \frac{(\rho_{centre} + \rho_{corner} + \rho_{centre\ long\ edge})}{3} \tag{5.20a}$$

These calculations can be performed by computer using a program such as PDISP from Oasys (see Appendix C).

The curves in Figure 5.16 assume that $E_u$ is constant with depth. Calculations based on a constant value can overestimate the settlement. Usually, the deformation modulus in soils and rocks increases with depth. For materials with a linear increase, Butler[5.12] developed a method based on the research of Brown and Gibson[5.13], for calculating settlements where $E_u$ or $E_v'$ increases linearly with depth through a layer of finite thickness. The value of the modulus at any depth $z$ below the base of the equivalent block foundation is given by the equation

$$E = E_f(1 + kz/B) \tag{5.21}$$

where $E_f$ is the modulus at the base of the equivalent foundation as Figure 5.17.

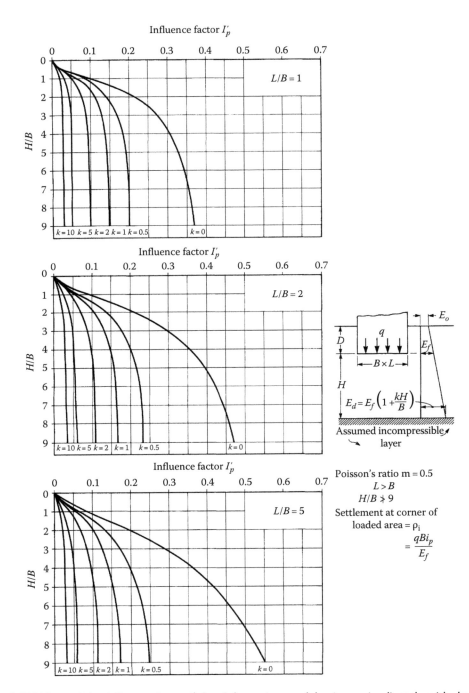

*Figure 5.17* Values of the influence factor $I'_p$ for deformation modulus increasing linearly with depth and modular ratio of 0.5 for normally consolidated days. (After Butler, F.G., General Report and state-of-the-art review, Session 3, *Proceedings of the Conference on Settlement of Structures*, Cambridge, UK, 1974, Pentech Press, London, UK, pp. 531–578, 1975.)

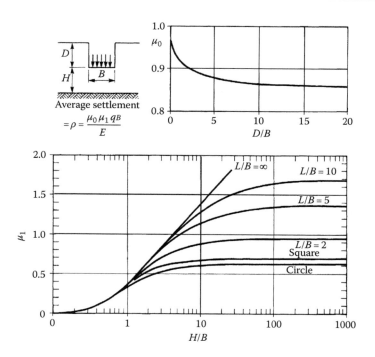

*Figure 5.18* Influence factors for calculating immediate settlements of flexible foundations of width *B* at depth *D* below ground surface. (After Christian, J.T. and Carrier, W.D., *Can. Geotech. J.*, 15, 123, 1978.)

To obtain $k$, values of $E_u$ or $E'_v$ obtained by one or more of the methods listed earlier are plotted against depth and a straight is drawn through the plotted points. The value of $k$ is then obtained using Figure 5.17 which also shows the values of the influence factor $I_p$. The curves in this figure are based on normally consolidated clays having a Poisson's ratio of 0.5 and are appropriate to a compressible layer of thickness not greater than nine times the breadth of the foundation. For a rigid pile group, the immediate settlement as calculated for a flexible pile group is multiplied by a factor of 0.8 to obtain the average settlement of the rigid group, and a depth factor is applied using the curves in Figure 5.18.

Where a piled foundation consists of a number of small clusters of piles or individual piles connected by ground beams or a flexible ground floor slab, the foundation arrangement can be considered as flexible.

When making a settlement analysis for a pile group underlain by layered soil strata with different but progressively increasing modulus values with depth, the strata are divided into a number of representative horizontal layers. An average modulus value is assigned to each layer. The dimensions $L$ and $B$ in Figure 5.18 are determined for each layer on the assumption that the vertical stress is spread to the surface of each layer at an angle of 30° from the edges of the equivalent raft or block foundation (Figure 5.19). The total settlement of the piled foundation is then the sum of the average settlements calculated for each layer.

The procedure in Equation 5.21 is referred to in EC7, Annex F, as the 'stress/strain' method. The other procedure described in Annex F is the 'adjusted elasticity' method. A typical example of the latter is the use of the Christian and Carrier[5.14] influence factors shown in Figure 5.18. These give the average settlement of the pile group from the equation

$$\text{Average settlement} = \rho_i = \mu_1 \mu_0 q_n B / E_u \qquad (5.22)$$

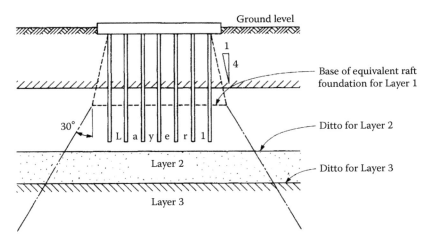

*Figure 5.19* Load distribution beneath pile group in layered soil formation.

In the previous equation, Poisson's ratio is taken as 0.5. The influence factors $\mu_1$ and $\mu_0$ are related to the depth and the length/breadth ratio of the equivalent block foundation and the thickness of the compressible layer as shown in Figure 5.19. $E_u$ is obtained by means of one or more of the methods listed earlier.

The *consolidation* settlement $\rho_c$ is calculated from the results of oedometer tests made on clay samples in the laboratory. The curves for the pressure/voids ratio obtained from these tests are used to establish the coefficient of volume compressibility $m_v$.

In hard glacial tills or weak highly weathered rock, it may be difficult to obtain satisfactory undisturbed samples for oedometer tests. If the results of standard penetration tests (SPTs) are available, values of $m_v$ (and also $c_u$) can be obtained from empirical relationships established by Stroud[5.7] shown in Figure 5.20.

Having obtained a representative value of $m_v$ for each soil layer stressed by the pile group, the *oedometer settlement* $\rho_{oed}$ for this layer at the centre of the loaded area is calculated from the equation

$$\rho_{oed} = \mu_d m_v \times \sigma_z \times H \tag{5.23}$$

where
  $\mu_d$ is a depth factor
  $\sigma_z$ is the average effective vertical stress imposed on the soil layer due to the net foundation pressure $q_n$ at the base of the equivalent raft foundation
  $H$ is the thickness of the soil layer

The depth factor $\mu_d$ is obtained from Fox's correction curves[5.15] shown in Figure 5.21. To obtain the average vertical stress $\sigma_z$ at the centre of each soil layer, the coefficients in Figure 5.11 should be used. The oedometer settlement must now be corrected to obtain the field value of the consolidation settlement. The correction is made by applying a *geological factor* $\mu_g$ to the oedometer settlement, where

$$\rho_c = \mu_g \times \rho_{oed} \tag{5.24}$$

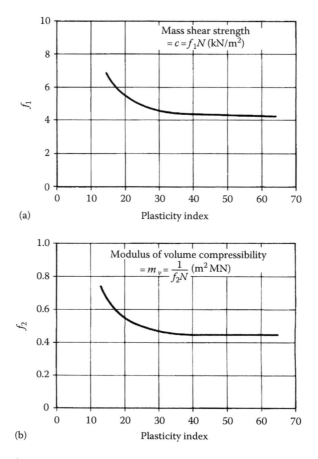

*Figure 5.20* Relationship between mass shear strength, modulus of volume compressibility, plasticity index and SPT N-values. (a) N-value versus undrained shear strength; (b) N-value versus modulus of volume compressibility. (After Stroud, M.A., The standard penetration test in insensitive clays, *Proceedings of the European Symposium on Penetration Testing*, Stockholm, Sweden, Vol. 2, pp. 367–375, 1975.)

Published values of $\mu_g$ have been based on comparisons of the settlement of actual structures with computations made from laboratory oedometer tests. Values established by Skempton and Bjerrum[5.16] are shown in Table 5.2.

The total settlement of the pile group is then the sum of the immediate and consolidation settlements calculated for each separate layer. A typical case is a gradual decrease in compressibility with depth. In such a case, the stressed zone beneath the pile group is divided into a number of separate horizontal layers, the value of $m_v$ for each layer being obtained by plotting $m_v$ against the depth as determined from the laboratory oedometer tests. The base of the lowermost layer is taken as the level at which the vertical stress has decreased to $q_n/10$. The depth factor $\mu_d$ is applied to the sum of the consolidation settlements calculated for each layer. It is not applied to the immediate settlement if the latter has been calculated from the factors in Figure 5.18.

Another method of estimating the total settlement of a structure on an over-consolidated clay is to use Equation 5.14, making the substitution of a deformation modulus obtained for loading under drained conditions. This modulus is designated by the term $E'_v$, which is

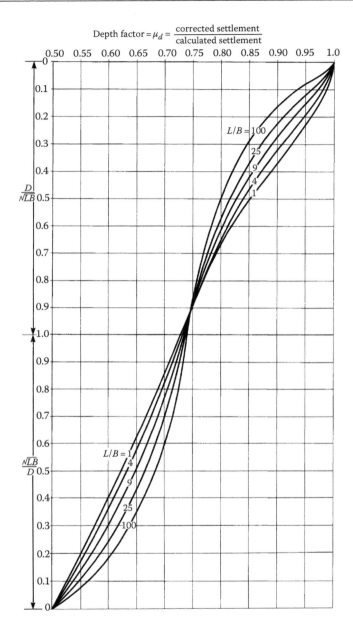

Depth factor $= \mu_d = \dfrac{\text{corrected settlement}}{\text{calculated settlement}}$

*Figure 5.21* Depth factor $\mu_d$ for calculating oedometer settlements. (After Fox, E.N., The mean elastic settlement of a uniformly-loaded area at a depth below the ground surface, *Proceedings of the Second International Conference, ISSMFE*, Rotterdam, the Netherlands, Vol. I, pp. 129–132, 1948.)

substituted for $E_u$ in the equation. It is approximately equal to $1/m_v$. The equation implies a homogeneous and elastic material and thus it is not strictly valid when used to calculate consolidation settlements. However, when applied to over-consolidated clays for which the settlements are relatively small, the method has been found by experience to give reasonably reliable predictions. Success in using the method depends on the collection of sufficient data correlating the observed settlements of structures with the determinations of $E_v'$ from plate loading tests and laboratory tests on good undisturbed samples of clay. Butler[5.12] in

*Table 5.2* Value of geological factor $\mu_g$

| Type of clay | $\mu_g$ value |
|---|---|
| Very sensitive clays (soft alluvial, estuarine and marine clays) | 1.0–1.2 |
| Normally consolidated clays | 0.7–1.0 |
| Over-consolidated clays (London Clay, Weald, Kimmeridge, Oxford and Lias Clays) | 0.5–0.7 |
| Heavily over-consolidated clays (unweathered glacial till, Mercia Mudstone) | 0.2–0.5 |

his review of the settlement of structures on over-consolidated clays has related $E'_v$ to the undrained cohesion $c_u$ and arrived at the relationship $E'_v = 130c$ for London Clay.

Various correlations between the soil modulus and the undrained shear strength of clays for piles with a length to diameter ratio equal to or greater than 15 are shown in Figure 5.22. In commenting on these data, Poulos[5.1] stated that they should be taken as representing values of the undrained modulus. He commented on the wide spread of the data suggesting that this could be due to differences in the method of measuring $c_u$ and the soil modulus, differences in the level of loading at which the modulus was measured, and differences between the type and over-consolidation ratio of the various clays. The size of the sample used to determine parameters is also critical. Where the undrained shear strength increases linearly with depth, Equation 5.21 can be used to obtain $E'_v$ and hence the total settlements from Figure 5.17. From an extensive review of published and unpublished data, Burland and Kalra[5.17] established the relationship for London Clay:

$E'_v = 7.5 + 3.9z$ (MN/m²), where $z$ is the depth in metres below ground level.

Generally, it is preferable to consider immediate and consolidation settlements separately. This properly takes into account time effects and the geological history of the site. Provided

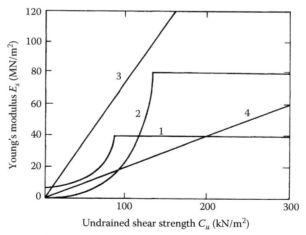

| Curve | Pile type | Reference |
|---|---|---|
| 1 | Driven | Poulos (1972) |
| 2 | Bored | Poulos (1972) |
| 3 | Driven ($E_s = 750C_u$) | Aschenbrenner and Disen (1984) |
| 4 | Bored (lower bound) $E_s = 200C_u$ | Callanan and Kulhawy (1985) |

*Figure 5.22* Correlations for soil modulus for piles in clay. (After Callanan and Kulhawy, for references see Poulos, H.G., *Geotechnique*, 39, 365, 1989.)

that a sufficient number of good undisturbed samples have been obtained at the site investigation stage, the prediction of consolidation settlements from oedometer tests made in
the laboratory has been found to lead to reasonably accurate results. The adoption of the
method based on the total settlement deformation modulus depends on the collection of
adequate observational data, first regarding the relationship between the undrained shearing strength and the deformation modulus and secondly regarding the actual settlement of
structures from which the relationships can be checked. Any attempt to obtain a deformation modulus from triaxial compression tests in the laboratory is likely to result in serious
error. The modulus is best obtained from the $E_u/c_u$ and $E_u/E_v'$ relationships, which must be
established from well-conducted plate bearing tests and field observations of settlement.

The steps in making a settlement analysis of a pile group in, or transmitting stress to, a
fine-grained soil can be summarised as follows:

1. For the required length of pile, and form of pile bearing (i.e. friction pile or end-
   bearing pile), draw the equivalent flexible raft foundation represented by the group (see
   Figure 5.3).
2. From the results of field or laboratory tests, assign values to $E_u$ and $m_v$ for each soil
   layer significantly stressed by the equivalent raft.
3. Calculate the immediate settlement of $\rho_i$ of each soil layer using Equation 5.22, and
   assuming a spread of load of 30° from the vertical, obtain $q_n$ at the surface of each
   layer (Figure 5.19). Alternatively calculate on the assumption of a linearly increasing
   modulus.
4. Calculate the consolidation settlement $\rho_c$ for each soil layer from Equations 5.23 and
   5.24, using Figure 5.11 to obtain the vertical stress at the centre of each layer.
5. Apply a rigidity factor to obtain the average settlement for a rigid pile group.

The consolidation settlement calculated as described earlier is the final settlement after a
period of some months or years after the completion of loading. It is rarely necessary to
calculate the movement at intermediate times, that is, to establish the time/settlement curve,
since in most cases the movement is virtually complete after a period of a very few years and
it is the final settlement which is the main interest of the structural engineer. If time effects
are of significance, however, the procedure for obtaining the time/settlement curve can be
obtained from standard works of reference on soil mechanics.

Morton and Au[5.18] provide detailed case histories of settlement rates over a period of
6 years for three high-rise blocks in London with pile group foundations in London Clay.
Also described are the settlements of five similar-size buildings supported on thick rafts on
Woolwich and Reading Beds. The maximum settlement of the piled structures was approximately 30 mm and for the rafts 100 mm under gross applied pressures of between 209
and 244 kN/m²; settlements at the end of construction for both types of foundation were
around 60% of the maximum observed. The majority of the settlement in the piled blocks
had occurred in the first 3 years. Distortion for both types was within safe limits for all the
structures.

## 5.3 PILE GROUPS IN COARSE-GRAINED SOILS

### 5.3.1 Estimating settlements from standard penetration tests

Where piles are driven in groups to near refusal into a dense sand or gravel, it is unlikely that
there will be sufficient yielding of individual piles under applied load to permit redistribution of superstructure loading to surrounding piles as described in Section 5.2.1. Sufficient

yielding to allow redistribution may occur where bored pile groups are terminated in sand or where piles are driven to a set predetermined from loading tests to allow a specified amount of settlement under applied loads.

Provided that the individual pile has adequate resistance against failure under compressive loading, there can be no risk of the block failure of a pile group terminated in and applying stress to a coarse soil. The end-bearing capacity due to the overburden pressure in Equation 5.1 will now be more significant. As in the case of piles terminated in a clay, there is a risk of differential settlement between adjacent piles in small groups if the toe loads of a small group become concentrated in a small area when the piles deviate from their intended line. The best safeguard against this occurrence is to adopt a reasonably wide spacing between the piles. Methods of checking the deviation of piles caused by the installation method are described in Chapter 11.

The immediate settlement of the pile group due to *elastic* deformation of the coarse soil beneath the equivalent flexible raft foundation must be calculated. Equation 5.22 is applicable to this case and the deformation modulus $E'_v$ is substituted for $E_u$ as obtained from plate loading tests in trial pits, or from standard penetration, pressuremeter, or Camkometer tests, made in boreholes. Schultze and Sherif[5.19] used case histories to establish a method for predicting foundation settlements from the results of SPTs using the equation

$$\rho = \frac{s \times p}{N^{0.87}\left(1 + 0.4D/B\right)} \tag{5.25}$$

where
  $s$ is a settlement coefficient
  $p$ is the applied stress at foundation level
  $N$ is the average SPT $N$-value over a depth of $2B$ below foundation level or $d_s$ if the depth of cohesion-less soil is less than $2B$
  $D$ and $B$ are the foundation depth and width, respectively

Values of the coefficient $s$ and $d_s$ are obtained from Figure 5.23.

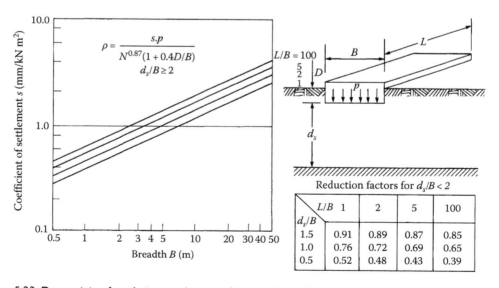

Figure 5.23 Determining foundation settlements from results of SPTs. (After Schultze, E. and Sherif, G., Predictions of settlements from evaluated settlement observations for sand, *Proceedings of the 8th International Conference*, ISSMFE, Moscow, Russia, Vol. I, pp. 225, 1973.)

Burland and Burbidge[5.20] have developed an empirical relationship between SPT $N$-values and a term they have called the foundation subgrade compressibility, $a_f$. This term is used in the equations

$$I_c = \frac{a_f}{B^{0.7}}$$

(5.26)

and

$$a_f = \frac{\Delta_{pi}}{\Delta_q} \text{ (in mm/kN/m}^2\text{)}$$

(5.27)

where

$I_c$ is a compressibility index
$B$ is the foundation width
$\Delta_{pi}$ is the immediate settlement in mm
$\Delta_q$ is the increment of foundation pressure in kN/m$^2$

$I_c$ and $a_f$ are related to the SPT results shown in Figure 5.24 for normally consolidated coarse-grained soils. In very fine and silty sands below the water table where $N$ is greater than 15, the Terzaghi and Peck correction factor should be applied, giving

$$N \text{ (corrected)} = 15 + 0.5(N - 15)$$

(5.28)

Where the material is gravel or sandy gravel, Burland and Burbidge recommend a correction:

$$N \text{ (corrected)} = 1.25 N$$

(5.29)

It should be noted that the $I_c$ values in Figure 5.24 are based on the average $N$-values over the depth of influence, $z_I$, of the foundation pressure. The depth of influence is related to the width of the loaded breadth $B$ in Figure 5.25 for cases where $N$ increases or is constant with depth. Where $N$ shows consistent decrease with depth, $z_f$ is taken as equal to $2B$ or the base of the compressive layer, whichever is the lesser. The average $N$ in Figure 5.24 is the arithmetic mean of the $N$-values over the depth of influence. Clayton also comments in CIRIA Report 143[11.9] on the need to apply corrections to $N$-values in different soils depending on the parameter to be assessed.

In a normally consolidated sand, the immediate average settlement, $\rho_i$, corresponding to the average net applied pressure, $q'$, is given by

$$\rho_i = q' \times B^{0.7} \times I_c \text{ (in mm)}$$

(5.30)

In an over-consolidated sand or for loading at the base of an excavation for which the maximum previous overburden pressure was $\sigma_{vo}$ and where $q'$ is greater than $\sigma_{vo}$, the immediate settlement is given by

$$\rho_i = \left( q' - \frac{2}{3}\sigma_{vo} \right) B^{0.7} \times I_c \text{ (in mm)}$$

(5.31a)

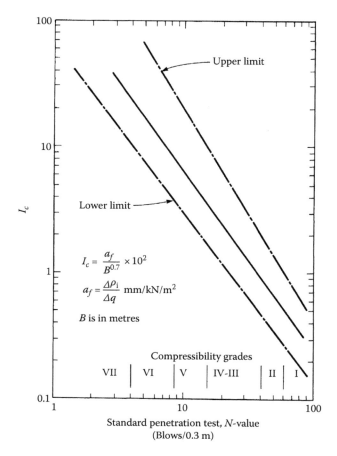

*Figure 5.24* Relationship between compressibility index and average *N*-value over depth of influence. (After Burland, J.B. and Burbidge, M.C., *Proc. Inst. Civil Eng.*, 78, 1325, 1985.)

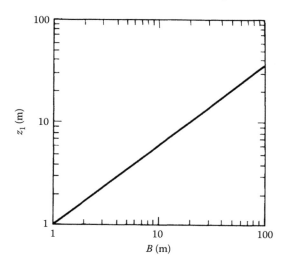

*Figure 5.25* Relationship between breadth of loaded area and depth of influence $z_I$. (After Burland, J.B. and Burbidge, M.C., *Proc. Inst. Civil Eng.*, 78, 1325, 1985.)

Where $q'$ is less than $\sigma_{vo}$, Equation 5.31a becomes

$$\rho_i = q' \times B^{0.7} \times \frac{I_c}{3} \text{ (in mm)} \tag{5.31b}$$

In the case of pile groups, the width $B$ is the width at the base of the equivalent raft as shown in Figure 5.3. The Burland and Burbidge method was developed essentially for shallow foundations and correlations with published settlement records given in their paper were mainly confined to foundations where their depth was not greater than their width. They state that the depth to width ratio did not influence the settlements to any significant degree and hence a depth factor of the type shown in Figure 5.18 should not be applied. However, a correction should be applied to allow for the foundation shape and for the thickness of the compressible layer beneath the foundation where this is less than the depth of influence, $z_I$.

The correction factors are

$$\text{Shape factor} = f_s = \left( \frac{1.25 L/B}{L/B + 0.25} \right)^2 \tag{5.32a}$$

$$\text{Thickness factor} = f_l = \frac{H_s}{z_I} \left( 2 - \frac{H_s}{z_I} \right) \tag{5.32b}$$

where
> $L$ is the length of the loaded area $(L > B)$
> $B$ is the width of the loaded area
> $H_s$ is the thickness of the compressible layer $(H_s < z_I)$

Burland and Burbidge state that most settlements on coarse-grained soils are time dependent, that is they show a long-term creep settlement and a further time correction factor is applied using the equation

$$f_t = \frac{\rho_t}{\rho_i} = \left( 1 + R_3 + R \log \frac{t}{3} \right) \tag{5.33}$$

where
> $t$ is equal to or greater than 3 years
> $R_3$ is the proportion of the immediate settlement which takes place in the loaded area
> $R$ is the creep ratio expressed as the proportion of the immediate settlement that takes place per log cycle of time

Burland and Burbidge give conservative values of $R$ and $R_3$ as 0.2 and 0.3 respectively, for static loading and 0.8 and 0.7, respectively for fluctuating loads.

Summarising all the previous corrections, the *average consolidation settlement* is given by

$$\rho_c = f_s f_l f_t \left[ \left( q' - \frac{2}{3} \sigma'_{vo} \right) \times B^{0.7} \times I_c \right] \text{ (in mm)} \tag{5.34}$$

The wide range of $I_c$ values between the upper and lower limit shown in Figure 5.24 can cause difficulty in obtaining a reasonably close estimate of pile group settlements, particularly where the group is underlain by medium-dense sands. For example, the average $I_c$ value for a sand with an N-value of 10 is 6 compared with upper and lower limit values of 20 and 3 respectively, giving an upper limit of settlement of three times that calculated from the average curve. However, in most cases, piles are taken down to dense sands to obtain the maximum end-bearing resistance, where the settlement calculated from the upper limit curve is likely to be relatively small.

## 5.3.2 Estimating settlements from static cone penetration tests

Where total and differential settlements are shown to be large and critical to the superstructure design, it is desirable to make static cone penetration tests, CPTs (Section 11.1.4), from which the soil modulus values can be derived, and then to use the Steinbrenner (Figure 5.16) or Christian and Carrier (Figure 5.18) charts to obtain the group settlement. Relationships between the cone-resistance $(q_c)$ values and the drained Young's modulus for normally consolidated quartz sands from several researchers are shown in Figure 5.26. The $E_{25}$ and $E_{50}$ values represent the secant drained modulus at a stress level of 25% and 50% respectively, of the failure stress. In a general review of the application of cone penetration testing to foundation design, Meigh[5.23] stated that the $E_{25}$ values are appropriate for most foundation problems but the $E_{50}$ values may be more relevant to calculating settlements of the single pile.

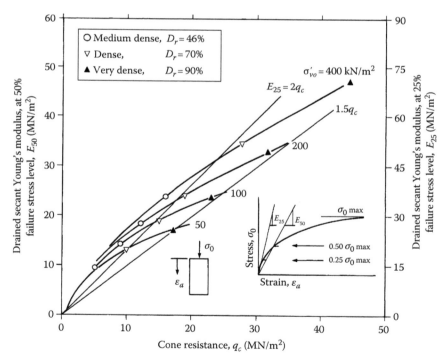

*Figure 5.26* Drained deformation modulus values $(E_d)$ for uncemented normally consolidated quartz sands in relation to cone resistance. (After Meigh, A.C., *Cone Penetration Testing*, CIRIA-Butterworth, London, UK, 1987; Robertson, P.K. and Campanella, R.G., *Can. Geotech. J.*, 20, 718, 1983; Baldi, G. et al., Cone resistance of dry medium sand, *Proceedings of the 10th International Conference*, ISSMFE, Stockholm, Sweden, Vol. 2, pp. 427–432, 1981.)

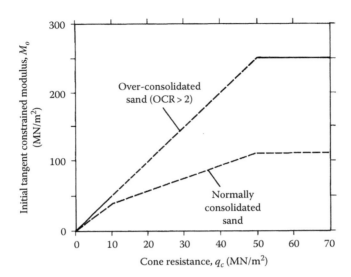

*Figure 5.27* Initial tangent constrained modulus for normally consolidated and over-consolidated sand related to cone resistance. (After Lunne, T. and Christoffersen, H.P., Interpretation of cone penetration data for offshore sands, *Proceedings of the Offshore Technology Conference* 15, Houston, TX, Vol. I, pp. 181–192, 1983.)

The $E$ values in Figure 5.26 greatly overestimate settlements in over-consolidated sands. Lunne and Christoffersen[5.24] established a relationship between initial tangent constrained modulus (the reciprocal of the modulus of volume compressibility $m_v$) and $q_c$ for normally and over-consolidated sands as shown in Figure 5.27.

Another method of estimating the settlements of pile groups in coarse-grained soils based on static CPT values has been developed by Schmertmann[5.25] and Schmertmann et al.[5.26] Their basic equation for the settlement of a loaded area is

$$r = C_1 C_2 \Delta_p \sum_0^{2B} \frac{I_z}{E_v'} \Delta_z \tag{5.35}$$

where
  $C_1$ is a depth correction factor (see below)
  $C_2$ is a creep factor (see below)
  $\Delta_p$ is the net increase of load on the soil at the base of the foundation due to the applied loading
  $B$ is the width of the loaded area
  $I_z$ is the vertical-strain influence factor (see Figure 5.28)
  $E_v'$ is the deformation modulus
  $\Delta_z$ is the thickness of the soil layer

The value of the depth correction factor is given by

$$C_1 = 1 - 0.5 \left( \frac{\sigma_{vo}'}{\Delta_p} \right) \tag{5.36}$$

where $\sigma_{vo}'$ is the effective overburden pressure at foundation level (i.e. at the base of the equivalent raft).

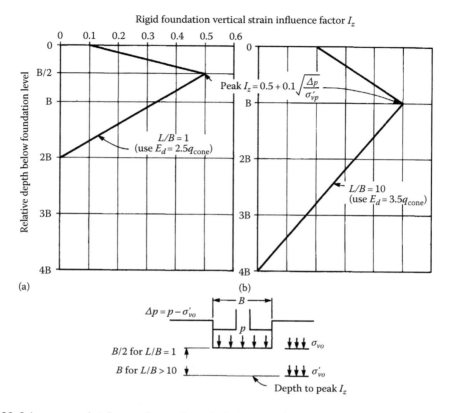

*Figure 5.28* Schmertmann's influence factors for calculating immediate settlements of foundations on sands. (a) for L/B = I and (b) for L/B = I0 (After Schmertmann, J.H. et al., *Proc. Am. Soc. Civil Eng.*, GT8, 1131, 1978.)

Schmertmann[5.25] states that while the settlement of foundations on coarse-grained soils is usually regarded as immediate, that is the settlement is complete within a short time after the completion of the application of load, observations have frequently shown long-continuing secondary settlement or creep. He gives the value of the creep factor as

$$C_2 = 1 + 0.2 \log_{10} \left( \frac{\text{time in years}}{0.1} \right) \tag{5.37}$$

Schmertmann et al.[5.26] have established an improved curve for obtaining the vertical-strain influence factor based on elastic half-space theory where the factor $I_z$ is related to the foundation width, as shown in Figure 5.28.

The vertical-strain influence factor is obtained from one of the two curves shown in Figure 5.28. For square pile groups (axisymmetric loading), the curve in Figure 5.28a should be used. For long pile groups (the plane strain case) where the length is more than 10 times the breadth, use the curve in Figure 5.28b. Values for rectangular foundations for $L/B$ of less than 10 can be obtained by interpolation.

The deformation modulus for square and long pile groups in normally consolidated sands is obtained by multiplying the static cone resistance, $q_c$, by a factor of 2.5 and 3.5 respectively.

The deformation modulus applicable for a stress increase of $\Delta_p$ above the effective overburden pressure, $\sigma'_{vo}$, is given by the equation

$$E'_v = E\sqrt{\frac{\sigma'_{vo} + \left(\Delta_p/2\right)}{\sigma'_{vo}}} \qquad (5.38)$$

Where SPTs only are available, the static cone resistance ($q_c$ in MN/m²) can be obtained by multiplying the SPT $N$-values (in blows/300 mm) by an empirical factor for which Schmertmann suggests the following values:

| | |
|---|---|
| Silts, sandy silts and slightly cohesive silty sands | $q_c = 0.2N$ |
| Clean fine to medium sands, slightly silty sands | $q_c = 0.35N$ |
| Coarse sands and sands with a little gravel | $q_c = 0.5N$ |
| Sandy gravel and gravels | $q_c = 0.6N$ |

Where static cone-resistance data are available, the relationships in Figures 5.26 or 5.27 can be used to obtain values of for substitution in Equation 5.35.

The procedure for estimating settlements by the Schmertmann method is first to divide the static cone-resistance diagram into layers of approximately equal or representative values of $q_c$ in a manner shown in Figure 5.29. The base of the equivalent raft representing the pile group is then drawn to scale on this diagram and the influence curve is superimposed

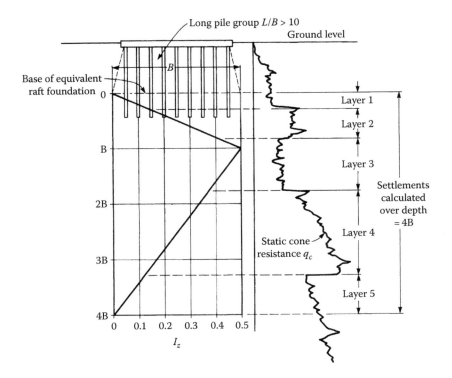

Figure 5.29 Establishing the vertical strain from static CPTs.

beneath the base of the raft. The settlements in each layer resulting from the loading $\varDelta_p$ at the base of the equivalent raft are then calculated using the values of $E_v'$ and $I_z$ appropriate to each of the representative layers. The sum of these settlements is corrected for depth and creep from Equations 5.36 through 5.38. The various steps in the calculation are made in tabular form as illustrated in Example 5.3. The computer program GEO5/Pile CPT is based on the application of the Schmertmann method (see Appendix C).

Where piles are terminated in a coarse soil stratum underlain by compressible clay, the settlements within the zone of clay stressed by the pile group are calculated by the methods described in Section 5.2.2. The form of load distribution to be used in this analysis to obtain the dimensions of the equivalent raft on the surface of the clay layer is shown in Figure 5.19.

## 5.4 EUROCODE 7 RECOMMENDATIONS FOR PILE GROUPS

Clause 7.6.2.1 of EC7 requires the stability of a pile group to be considered both in relation to the risk of failure of an individual pile in the group and to the failure of the group considered as an equivalent block foundation. Subclause (4) states that the block foundation can be considered to act as a single large-diameter pile. However, no guidance is given as to relationship between the diameter and depth of this pile to the shape, base area and depth of the group. If it is assumed that the plan area of the large-diameter pile is equal to the gross area of the group, then in the case of square (or rectangular) groups, the resulting bearing calculations could give an over-conservative value of the design load. Also it is reasonable to assume that the shaft friction of the *equivalent pile* should be calculated on the basis of a soil–soil interface using the undisturbed shear strength of the surrounding soil. Whereas when calculating the shaft friction on an individual pile, the installation method has an important influence on the resistance of a pile–soil interface. Where a group of piles is driven into a clay, the surrounding soil is strengthened by expulsion of pore water, and a sand is strengthened by densification. Conversely, drilling for a group of bored piles could cause weakening of a clay due to relaxation of a fissured structure or drilling in sand could result in loss of resistance in friction.

If, as an alternative to the large-diameter pile assumption, the pile group is treated as an equivalent block foundation as in Section 5.1, the partial factors for actions and material properties are the same as used for piled foundations (Tables 4.1 through 4.5 in Section 4.1.4). The base resistance factor for spread foundations, $\gamma_{Rv}$, and the factor for sliding, $\gamma_{Rh}$, are both unity for Set 1 in the National Annex (NA). There are no R4 resistance factors for spread foundations. Annex D of EC7 provides two (*sample*) equations for the calculation of the bearing resistance of a spread foundation, which could be applied to the equivalent block. In undrained conditions, equation D1 adds a surcharge pressure which will overestimate the base resistance for large $D/B$ ratios, and in equation D2 for drained conditions, the depth factor is omitted which, when the $D/B$ ratio is large, will underestimate the base resistance. The NA (NA 3.3) recognises the potential anomalies and allows for the use of alternative approaches. The general Brinch Hansen[5.4] Equation 5.1 deals with this critical point, and while there are several expressions available for the depth, shape and inclination factors, including examples determined by finite element analysis, it is considered that the consistent approach of Brinch Hansen provides a reasonable empirical solution to the preferred equivalent block method. A global safety factor of 2.5 was used with Equation 5.1 to calculate the allowable bearing pressure, and to satisfy EC7 procedures, a model factor, $\gamma_{Rd}$, will be needed to obtain the characteristic resistance with an appropriate $\gamma_b$ resistance factor to give the design resistance.

No consensus exists among engineers at present as to whether the equivalent pile or equivalent block is best suited to the assessment of group bearing capacity using EC7. Development of numerical analytical modelling will assist in resolving the issue.

Clause 7.6.4.2(2)P states that the assessment of settlement of pile groups should take into account the settlement of the individual piles as well as that of the group, but it does not make it clear whether the settlement analysis should assume that the group acts as an equivalent large-diameter pile or as a block foundation. Presumably, the latter is the case, for which Clause 6.6.2(6), considering the settlement of spread foundations, requires the depth of the compressible soil layer to be taken normally as the depth at which the effective vertical stress due to the foundation load is 20% of that of the effective overburden stress. In many cases, this may be roughly estimated as one to two times the foundation width or less for lightly loaded foundation rafts. In the case of pile groups, it is assumed that this is to be the depth below the base of the equivalent rafts shown in Figure 5.3. An aspect ratio of $R > 4$ may indicate that the equivalent raft method is best suited for determining group settlement and $R < 2$ for the equivalent pile method. Again, as there is no consensus on the approach, it may be feasible to use one scheme for immediate settlement and the other for consolidation depending on the soil profile. Comments on the analytical methods for determining load distribution and settlement in pile groups are given in Section 4.9.

## 5.5 PILE GROUPS TERMINATING IN ROCK

The stability of a pile group bearing on a rock formation is governed by that of the individual pile. For example, one or more of the piles might yield due to the presence of a pocket of weathered rock beneath the toe. There is no risk of block failure unless the piles are terminated on a sloping rock formation, when sliding on a weak clay-filled bedding plane might occur if the bedding is unfavourably inclined to the direction of loading (Figure 5.30). The possibility of such occurrences must be studied in the light of the information available on the geology of the site.

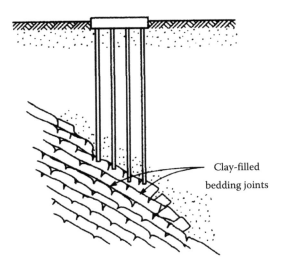

Figure 5.30 Instability of pile group bearing on sloping rock surface.

The settlement of a pile group may be of significance if the piles are heavily loaded. Immediate settlements can be calculated as described in Section 5.2.2, and Equations 5.14 and 5.22 are applicable where the deformation modulus for the rock mass $E_d$ is reasonably constant with depth.

It is possible to obtain a rough estimate of the deformation modulus of a jointed rock mass from empirical relationships with the unconfined compression strength of the intact rock, using the equation previously recorded in BS 8004 $E_d = j \times M_r \times q_c$ where $j$ is the mass factor (see Section 4.7.3 for values) and $M_r$ is the ratio of the elastic modulus of the intact rock to its unconfined compression strength. The following values for $M_r$ were quoted:

|  |  | Values for $M_r$ |
|---|---|---|
| Group 1 | Pure limestones and dolomites | 600 |
|  | Carbonate sandstones of low porosity |  |
| Group 2 | Igneous | 300 |
|  | Oolitic and marly limestones |  |
|  | Well-cemented sandstones |  |
|  | Indurated carbonate mudstones |  |
|  | Metamorphic rocks including slates and schists (flat cleavage/foliation) |  |
| Group 3 | Very marly limestones | 150 |
|  | Poorly cemented sandstones |  |
|  | Cemented mudstones and shales |  |
|  | Slates and schists (steep cleavage/foliation) |  |
| Group 4 | Uncemented mudstones and shales | 75 |

The conservative values mentioned earlier apply to constant $E_d$ with depth and to a thick rock layer; for more general application, see Meigh[5.23]. Chalk and Mercia Mudstone (Keuper Marl) are excluded from the above-mentioned groups. Some observed values of $E_d$ for chalk are given in Table 5.3 and for Mercia Mudstone in Table 5.4.

It is likely that weathered rocks will show an increase in $E_d$ with depth as the state of weathering decreases from complete at rockhead to the unweathered condition. If it is possible to draw a straight line through the increasing values, the influence factors in Figure 5.17 can be used in conjunction with Equation 5.21 to obtain the settlement at the centre of the loaded area. These curves were established by Butler[5.12] for a Poisson's ratio

Table 5.3 Values of deformation modulus of chalk

| Density | Grade | Yield stress (MN/m²) | Ultimate bearing capacity (MN/m²) | Secant modulus at applied stress of 200 kN/m² (MN/m²) | Yield modulus (MN/m²) |
|---|---|---|---|---|---|
| Medium/high | A | — | 16 | 1500–3000 | — |
|  | B | 0.3–0.5 | 4.0–7.7 | 1500–2000 | 35–80 |
|  | C | 0.3–0.5 | 4.0–7.7 | 300–1500 | 35–80 |
| Low | B and C | 0.25–0.5 | 1.5–2.0 | 200–700 | 15–35 |
| (Low) | $D_c$ | 0.25–0.5 | — | 200 | 20–30 |
|  | $D_m$ | — | — | 6 | — |

Source: Lord, J.A. et al., Engineering in chalk, Construction Industry Research and Information Association, Report No 574, 2002.

*Table 5.4* Values of deformation modulus of Mercia Mudstone
(Keuper Marl) at low stress levels

| Zone | Deformation modulus (MN/m²) |
|------|------|
| I | 26–250 |
| II | 9–70 |
| III | 2–48 |
| IV | 2–13 |

Source: Chandler, R.J. and Davis, A.G., Further work on the engineering
properties of Keuper Marl, Construction Industry Research and Information
Association (CIRIA), Report 47, 1973.

of 0.5, but most rock formations have lower ratios. Meigh[5.23] stated that Poisson's ratio of Triassic rocks is about 0.1–0.3.

Meigh[5.23] derived curves for the influence factors shown in Figure 5.31 for various values of the constant $k$ in Equation 5.39 where

$$k = \frac{(E_d - E_f)}{E_f} \cdot \frac{B}{H} \tag{5.39}$$

for a Poisson's ratio of 0.2 and where $E_f$ is the modulus at foundation depth as mentioned previously.

He applied further corrections to the calculation of the settlement at the *corner* of the foundation where

$$\text{Settlement at corner} = \rho_i = \frac{q_n B I_p'}{E_f} \tag{5.40}$$

(as shown in Figure 5.17).

The corrected settlement is given by

$$\rho_c(\text{corrected}) = \frac{q_n B I_p'}{E_f} \times F_B \times F_D \tag{5.41}$$

where
$F_B$ is the correction factor for roughness of base (Figure 5.32)
$F_D$ is the correction factor for depth of embedment (Figure 5.33)

The equivalent raft is assumed to have a rough base and is divided into four equal rectangles and the settlement computed for the corner of each rectangle from Equation 5.41. The settlement at the centre of the pile group is then four times the corner settlement.

## 5.6 PILE GROUPS IN FILLED GROUND

The problem of negative skin friction or downdrag on the shafts of isolated piles embedded in fill was discussed in Sections 4.8 and 4.9. This downdrag is caused by the consolidation of the fill under its own weight or under the weight of additional imposed fill. If the fill is underlain by a compressible clay, the consolidation of the clay under the weight of the fill

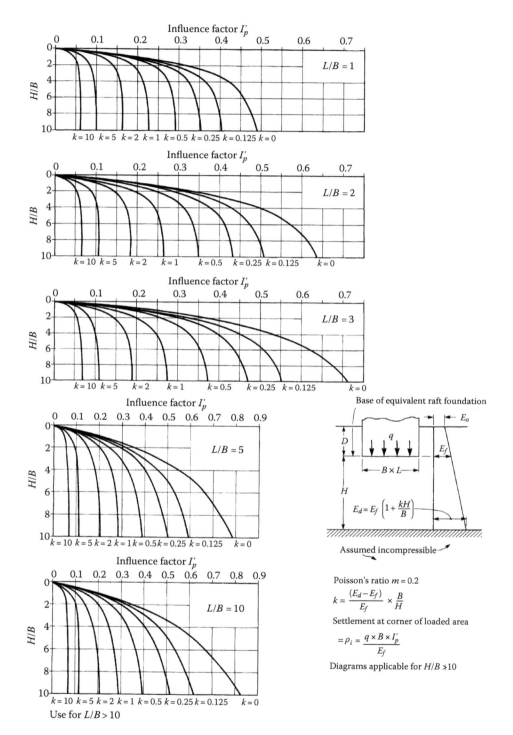

*Figure 5.31* Values of influence factor for deformation modulus increasing linearly with depth and modular ratio of 0.2 in rock. (After Meigh, A.C., *Geotechnique*, 26, 393, 1976.)

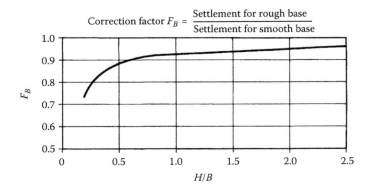

Figure 5.32 Correction factors for roughness of base of foundation. (After Meigh, A.C., *Geotechnique*, **26**, 393, 1976.)

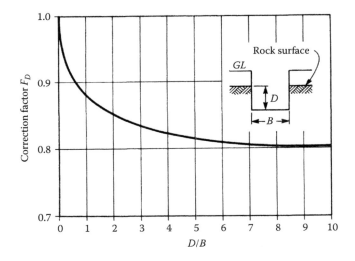

Figure 5.33 Correction factors for depth of embedment of foundation below surface of rock. (After Meigh, A.C., *Geotechnique*, **26**, 393, 1976.)

also causes negative skin friction in the portion of the shaft within this clay. Negative skin friction also occurs on piles installed in groups but the addition to the applied load on each of the piles in the group is not necessarily more severe than that calculated for the isolated pile. The basis for calculating the negative skin friction as described in Section 4.8.1 is that the ultimate skin friction on the pile shaft is assumed to act on that length of pile over which the fill and any underlying compressible clay move downwards relative to the shaft. The magnitude of this skin friction cannot increase as a result of grouping the piles at close centres, and the total negative skin friction acting on the group cannot exceed the total weight of fill enclosed by the piles. Thus, in Figure 5.34a,

$$\text{Total load on pile group} = \text{applied load} + (B \times L \times \gamma'D') \qquad (5.42)$$

where
   $\gamma'$ is the unit weight of fill
   $D'$ is the depth over which the fill is moving downwards relative to the piles

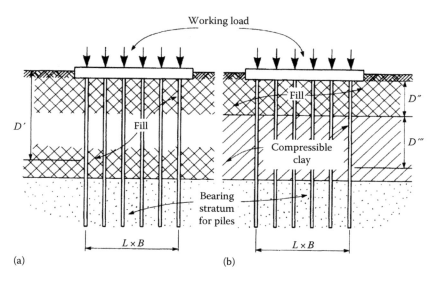

*Figure 5.34* Negative skin friction on pile groups in filled ground: (a) fill overlying relatively incompressible bearing stratum and (b) fill placed on compressible clay layer.

where the fill is underlain by a compressible clay, as in Figure 5.34b,

$$\text{Total load on pile group} = \text{applied load} + B \times L(\gamma'D' + \gamma''D'') \tag{5.43}$$

where

$D'$ is the total thickness of fill
$\gamma''$ is the unit weight of compressible clay
$D''$ is the thickness of compressible clay moving downwards relative to the piles

It should also be noted that the negative skin friction acting on the piles in the group does not increase the settlement of the group caused by the applied load on the piles. If the filling has been in place for a long period of years, any underlying compressible soil will have been fully consolidated and the only additional load on the compressible soil causing settlement of the group is that from the applied load on the piles. However, if the fill is to be placed only a short time before driving the piles, then any compressible soil below the fill will consolidate. The amount of this consolidation can be calculated separately and added to the settlement caused by the applied load on the piles. The negative skin friction on the piles is not included in the applied load for the latter analysis.

EC7 gives no specific guidance for the design of pile groups carrying compression loading in filled ground. As in the case of the single pile calculation, the load distribution on individual piles in the group is best undertaken by an interaction analysis as discussed in Section 4.9. It is evident that treatment of the group as a single large-diameter pile as proposed in Clause 7.6.2.1(4) for the determination of group stability is not valid for application to an interaction analysis.

Clause 7.3.2.2(5)P requires account to be taken of the weight density of materials in a settlement analysis for piles in filled ground. As noted in the case of the single pile, the partial factors for weight density are omitted from the NA.

## 5.7 EFFECTS ON PILE GROUPS OF INSTALLATION METHODS

When piles are driven in groups into clay, the mass of soil within the ground heaves and also expands laterally, the volume of this expansive movement being approximately equal to the volume occupied by the piles. High pore pressures are developed in the soil mass, but in the course of a few days or weeks, these pore pressures dissipate and the heaving directly caused by pore pressure subsides. In soft clays, the subsidence of the heaved soil can cause negative skin friction to develop. It is not usual to add this negative skin friction to the applied load since it is of relatively short duration, but its effect can be allowed for by ignoring any support provided in shaft friction to the portion of the pile shaft within the soft clay. Methods of calculating the surface heave within a pile group have been discussed by Hagerty and Peck[5.28]. Chow and Teh[5.29] have established a theoretical model relating the pile head heave/diameter ratio to the pile spacing/diameter ratio for a range of length/diameter ratios in soft, firm and stiff clays.

It is not good practice to terminate pile groups within a soft clay since the reconsolidation of the heaved and remoulded soil can result in the substantial settlement of a pile group, and neighbouring structures can be affected. It may be seen from Figure 5.35 that there is little difference between the extent of the stressed zone around and beneath a surface raft and a group of short friction piles. The soil beneath the raft is not disturbed during construction and hence the settlement of the raft may be much less than that of a pile group carrying the same overall loading. This was illustrated by Bjerrum[5.30], who compared the settlement of buildings erected on the two types of foundation construction on the deep soft and sensitive clays of Drammen near Oslo.

A building where the gross loading of 65 kN/m² was reduced by excavation for a basement to a net loading of 25 kN/m² was supported on 300 timber friction piles 23 m long.

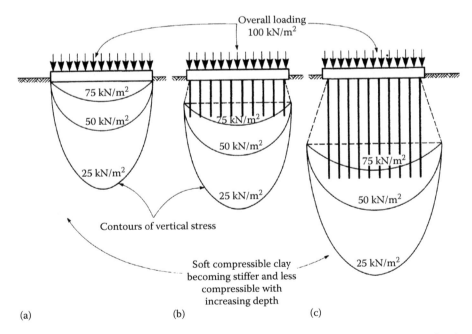

Figure 5.35 Comparison of stress distribution beneath shallow raft foundation and beneath pile groups: (a) shallow raft, (b) short friction piles and (c) long friction piles.

In 10 years, the building had settled by 110 mm and the surrounding ground surface had settled by 80 mm. A nearby building with a gross loading of 55 kN/m² had a fully compensated un-piled foundation, that is, the weight of the soil removed in excavating for the basement balanced the superstructure and substructure giving a net intensity of loading of zero on the soil. Nearly 30 mm of heave occurred in the base of excavation and thus the settlement of the building was limited to the reconsolidation of the heaved soil. The net settlement 9 years after completing the building was only 5 mm.

Lateral movement of a clay soil and the development of high pore pressures can damage structures or buried services close to a pile group. Adams and Hanna[5.31] measured the pore pressures developed within the centre of a large group of driven piles at Pickering Nuclear Power Station, Ontario. The horizontal ground strains were also measured at various radial distances from the centre. The group consisted of 750 piles driven within a circle about 46 m in diameter. Steel H-section piles were selected to give a minimum of displacement of the 15 m of firm to very stiff and dense glacial till, through which the piles were driven to reach bedrock. From measurements of the change in the distance between adjacent surface markers, it was calculated that the horizontal earth pressure at a point 1.5 m from the edge of the group was 84 kN/m², while at 18.8 m from the edge, the calculated pressure was only 1 kN/m². Earth pressure cells mounted behind a retaining wall 9 m from the group showed no increase in earth pressure due to the pile driving. Very high pore pressures were developed at the centre of the piled area, the increase being 138 kN/m² at a depth of 6 m, dissipating to 41 kN/m², 80 days after completing driving of the instrumented pile, when all pile driving in the group had been completed.

The average ground heave of 114 mm measured over the piled area represented a volume of soil displacement greater than the volume of steel piles which had been driven into the soil, for which the theoretical ground heave was 108 mm.

Substantial heave accompanied by the lifting of piles already driven can occur with large displacement piles. Brzezinski et al.[5.32] made measurements of the heave of 270 driven and cast-in-place piles in a group supporting a 14-storey building in Quebec. The piles had a shaft diameter of 406 mm and the bases were expanded by driving. The piles were driven through 6.7–11 m of stiff clay to a very dense glacial till. Precautions against uplift were taken by providing a permanent casing to the piles and the concrete was not placed in the shafts until the pile bases had been re-driven by tapping with a drop hammer to the extent necessary to overcome the effects of uplift. The measured heave of a cross section of the piled area is shown in Figure 5.36. It was found that the soil heave caused the permanent casing to become detached from the bases, as much as 300 mm of separation being observed. Heave effects were not observed if the piles were driven at a spacing wider than 12 diameters. This agrees with the curves established by Chow and Teh[5.29] which show a pile head heave of only about 1 mm for a spacing of 12 diameters.

Similar effects were observed by Cole[5.33]. At three sites, the heave was negligible at pile spacings wider than 8–10 diameters. Cole observed that uplift was more a function of the pile diameter and spacing than of the soil type or pile length. Where piles carry their load mainly in end bearing, the effect of uplift is most damaging to their performance, and on all sites where soil displacement is liable to cause uplift, precautions must be taken as described in Section 5.8. Heave is not necessarily detrimental where piles are carried by shaft friction in firm to stiff clays in which there will be no appreciable subsidence of the heaved soil to cause negative skin friction to develop on the pile shaft. On a site where a 12-storey block of flats was supported by driven and cast-in-place piles installed in 5 m of firm London Clay to terminate at the base of a 4 m layer of stiff London Clay, about 0.5 m of heave was observed in the ground surface after 70 piles had been driven within the 24 × 20 m area of the block. A pile was tested in an area where 220 mm of heave had occurred.

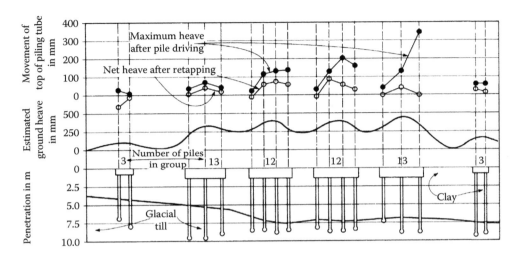

*Figure 5.36* Observations of heave due to pile driving in clay. (After Brzezinski, L. et al., *Can. Geotech. J.*, 10, 246, 1973.)

The settlement at 1300 kN (i.e. twice the applied load) was 23 mm, while the settlement at the applied load was only 2.5 mm.

Heaving and the development of high pore pressures do not occur when bored and cast-in-place piles are installed in groups. However, general subsidence around the piled area can be caused by the 'draw' or relaxation of the ground during boring. In soft sensitive clays, the bottom of a pile borehole can heave up due to 'piping', with a considerable loss of ground. These effects can be minimised by keeping the pile borehole full of water or bentonite slurry during drilling and by placing the concrete within a casing which is only withdrawn after all concrete placing is completed.

Detrimental effects from heave are not usually experienced when driving piles in groups in coarse soils. A loose soil is densified, potentially requiring imported filling to make up the subsided ground surface within and around the group. Adjacent structures may be damaged if they are within the area of subsidence. A problem can arise when the first piles to be installed drive easily through a loose sand but, as more piles are driven, the sand becomes denser thus preventing the full penetration of all the remaining piles. This problem can be avoided by paying attention to the order of driving, as described in Section 5.8.

Subsidence due to the loss of ground within and around a group in a coarse soil can be quite severe when bored and cast-in-place piles are installed, particularly when 'shelling' is used as the boring method (see Section 3.3.7). The subsidence can be very much reduced, if not entirely eliminated, by the use of rotary drilling with the assistance of a bentonite slurry (see Section 3.3.8).

## 5.8 PRECAUTIONS AGAINST HEAVE EFFECTS IN PILE GROUPS

It will have been noted from Section 5.7 that the principal problems with soil heave and the uplift of piles occur when large displacement piles are driven into clay. In coarse soils, the problems can be overcome to a great extent by using small displacement piles such as H-sections or open-ended steel tubes. To adopt a spacing between piles of 10 or more diameters is not usually practical if pile group dimensions are to be kept within economical limits. Pre-boring the pile shaft is not always effective unless the pre-bored hole is taken down to the pile base, in which case the shaft friction will be substantially reduced if not entirely eliminated. Jetting piles is only effective in a coarse soil and the problems associated with this method are described in Section 3.1.9. The most effective method is to re-drive any risen piles, after driving all the piles in a cluster that are separated from adjacent piles by at least 12 diameters has been completed. Re-driving friction piles in clay can result in reduced resistance in the short term.

In the case of driven and cast-in-place piles, a permanent casing should be used and the re-driving of the risen casing and pile base should be effected by tapping the permanent casing with a 3-tonne hammer, as described by Brzezinski et al.[5.32] Alternatively, the *multitube* method described by Cole[5.33] can be used. This consists of providing sufficient lengths of withdrawable casing to enable all the piling tubes to be driven to their full depth and all the pile bases to be formed before the pile shafts in any given cluster are concreted. An individual cluster dealt with in this way must be separated from a neighbouring cluster by a sufficient distance to prevent the uplift of neighbouring piles or to reduce this to an acceptable amount. On the three sites described by Cole, it was found possible to drive piles to within 6.5 diameters of adjacent clusters without causing an uplift of more than 3 mm to the latter. This movement was not regarded as detrimental to the load/settlement behaviour. Cole stated that, although the multitube system required eight driving tubes to each piling rig, the cost did not exceed that of an additional 2 m on each pile.

It is possible to re-drive risen driven and cast-in-place piles using a 3- to 4-tonne hammer with a drop not exceeding 1.5 m. The head of the pile should be protected by casting on a 0.6 m capping cube in rapid-hardening cement concrete.

Cole[5.33] stated that the order of driving piles did not affect the incidence of risen piles but it did change the degree of uplift on any given pile in a group. Generally, the aim should be to work progressively outwards or across a group and in the case of an elongated group from end to end or from the middle outwards in both directions. This procedure is particularly important when driving piles in coarse soils. If piles are driven from the perimeter towards the centre of a group, a coarse-grained soil will tighten up so much due to ground vibrations that it will be found impossible to drive the interior piles.

It is desirable to adopt systematic monitoring of the behaviour of all piles installed in groups by taking check levels on the pile heads, by carrying out re-driving tests and by making loading tests on working piles selected at random from within the groups. Loading tests undertaken on isolated piles before the main pile driving commences give no indication of the possible detrimental effects of heave. Lateral movements should also be monitored as necessary.

## 5.9 PILE GROUPS BENEATH BASEMENTS

Basements may be required beneath a building for their functional purpose, for example as an underground car park or for storage. The provision of a basement can be advantageous in reducing the loading which is applied to the soil by the building. For example, if a basement is constructed in an excavation 7 m deep, the soil at foundation level is relieved of a pressure equivalent to 7 m of overburden, and the gross loading imposed by the building is reduced by this amount of pressure relief. It is thus possible to relieve completely the net loading on the soil. An approximate guide to the required depth of excavation is the fact that a multi-storey dwelling block in reinforced concrete with brick and concrete external walls, light-weight concrete partition walls and plastered finishes weighs about 12.5 kN/m² per storey. This loading is inclusive of 100% of the permanent load and 60% of the variable load. Thus, a 20-storey building would weigh 250 kN/m² at ground level, requiring a basement to be

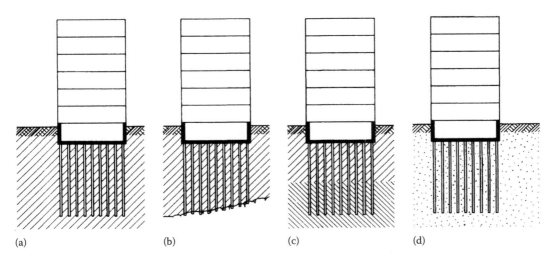

(a)                    (b)                    (c)                    (d)

*Figure 5.37* Piled basements in various ground conditions: (a) wholly in compressible clay, (b) compressible clay over bedrock, (c) soft clay over stiff clay and (d) loose sand becoming denser with depth.

excavated to a depth of about 20 m to balance the loading (assuming the groundwater level to be 3 m below ground level and taking the submerged density of the soil below water level).

Deep basement excavations in soft compressible soils can cause considerable constructional problems due to heave, instability and the settlement of the surrounding ground surface. Because of this, it may be desirable to adopt only a partial relief of loading by excavating a basement to a moderate depth and then carrying the net loading on piles taken down to soil having a lesser compressibility.

In all cases where piles are installed to support structures, it is necessary to consider the effects of soil swelling and heave on the transfer of load from the basement floor slab to the piles. Four cases can be considered as described in the following texts and shown in Figure 5.37.

## 5.9.1 Piles wholly in compressible clay

In the case shown in Figure 5.37a, the soil initially heaves due to swelling consequent on excavating the foundation, and further heave results from pile driving. The heaved soil is then trimmed off to the correct level and the basement slab concreted. If the concreting is undertaken within a few days or a week after the pile driving, there is a tendency for the heaved soil to slump down, particularly in a soft clay which developed high pore pressures. A space may tend to open between the underside of the concrete and the soil surface. When the superstructure is erected, the piles will carry their applied load, and if correctly designed, they will settle to an acceptable degree. This will in turn cause the basement slab to settle but pressure will not develop on its underside because the soil within and beneath the settling piles will move down with them. Thus, the maximum pressure on the underside of the basement slab is due to the soil swelling at an early stage before partial slumping of the heaved soil takes place and before the piles carry any of their designed loading. The uplift pressure on the basement slab will be greater if bored piles are used since no heaving of the soil is caused by installing the piles, and if the basement slab is completed and attached to the piles soon after completing the excavation, the swelling pressures on the underside of the slab will cause tension to be developed in the piles. This is particularly liable to happen where bored piles are installed from the ground surface before the excavation for the basement commences. Concreting of the pile shaft is terminated at the level of the underside of the basement slab and the construction of the basement slab usually takes place immediately after the completion of excavation and before any heave of the excavation can take place to relieve the swelling pressure. Generally, in any piled basement where bored piles are installed wholly in compressible clay, the basement slab should be designed to withstand an uplift pressure equal at least to one-half of the permanent and variable load of the superstructure. Alternatively, a void can be provided beneath the basement slab by means of collapsible cardboard or plastics formers. The piles can be designed to be anchored against uplift or they can be sleeved over the zone of swelling. Anchoring the piles against uplift by increasing the shaft length to increase shaft friction below the swelling zone is often the most economical solution to the problem. Where void formers made of cardboard or plastics are used to eliminate swelling pressure beneath the basement slab, there is a risk of biodegradation of the organic materials causing an accumulation of methane gas in the void. Venting the underside of the slab can be difficult and costly.

Providing an increased shaft length can be made more economical than sleeving the pile shaft within the swelling zone. Fleming and Powderham[5.34] recommended that where piles are reinforced to restrain uplift the friction forces should not be underestimated and they suggest that if the forces are estimated conservatively it would be appropriate to reduce the load factors on the steel, perhaps to about 1.1.

Hydrostatic pressure will, of course, act on the basement slab in water-bearing soil. The piles must be designed to carry the net full weight of the structure (i.e. the total weight less the weight of soil and soil water excavated from the basement).

When installing piles for 'top-down' construction as shown in Figure 2.33, with the steel stanchion plunged into the bored pile, particular care is required to establish the position of the pile borehole and maintain verticality in drilling. If this is not done, there could be considerable error in the position of the pile head, leading to eccentric loading on the pile and off-plumb column. Taking the case of a 3-storey basement with an overall depth from ground surface to pile head level (beneath the lowest floor slab) of 15 m and applying the tolerances noted in Section 3.4.13, the pile could be critically displaced from its design position. The ICE SPERW[2.5] tolerances would result in an out of position of 275 mm and the BS EN 1536 tolerance would be 400 mm. Specifications for plunge column alignment are therefore much more stringent, leading to the use of large-diameter piles (2 m and above) and verticality limits up to 1 in 400.

### 5.9.2  Piles driven through compressible clay to bedrock

In the case shown in Figure 5.37b, soil swelling takes place at the base of the excavation followed by heave if driven piles are employed. As before, the heaved soil tends to slump away from the underside of the basement slab if the latter is concreted soon after pile driving. Any gap which might form will be permanent since the piles will not settle except due to a very small elastic shortening of the shaft. If bored piles are adopted, with a long delay between concreting the base slab and applying the superstructure loading to the piles, the pressure of the underside of the slab due to long-term soil swelling might be sufficient to cause the piles to lift from their seating on the rock. The remedy then is to provide a void beneath the slab and to anchor the piles to rock or to sleeve them through the swelling zone.

### 5.9.3  Piles driven through soft clay into stiff clay

The case shown in Figure 5.37c, intermediate between the first two. There is a continuing tendency for the heaved soft clay to settle away from the underside of the basement slab, because the settlement of the piles taking their bearing in the stiff clay is less than that caused by the reconsolidation of the heaved and disturbed soft clay. Uplift pressure occurs on the underside of the base slab if bored piles are used, and a design value equal at least to one-half of the combined permanent and variable load of the superstructure should be considered. Alternatively, the effects of heave should be eliminated as described earlier.

### 5.9.4 Piles driven into loose sand

In the case shown in Figure 5.37d, it is presumed that the piles are driven through loose sand to an end bearing in deeper and denser sand. The slight heave of the soil caused by excavating the basement is an instantaneous elastic movement. No heave occurs because either pile driving causes some settlement of the ground surface due to densification or a loss of ground results due to pile boring. When the superstructure load is applied to the piles, they compress but the soil follows the pile movement, and any soil pressures developed on the underside of the basement slab are relatively small. Hydrostatic pressure occurs in a water-bearing soil.

In all cases when designing piled basements, the full applied load should be considered as acting on the piles and, in the case of piles bearing on rock or coarse-grained soils of low compressibility, the load on the underside of the basement slab can be limited to that caused by the soil pressure (i.e. the overburden pressure measured from the ground surface around the basement) and hydrostatic pressure. Sometimes, a tall building is constructed close to a low-rise podium (Figure 5.38) and both structures are provided with a piled basement. Piling beneath the podium is required to reduce differential movement between the heavily loaded tower block and the podium. Uplift of the latter may occur if the weight of the superstructure is less than that of the soil removed in excavating for the basement. In such

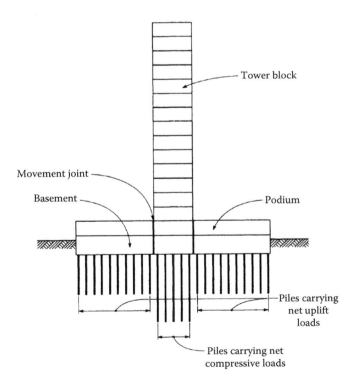

*Figure 5.38* Tower block and podium supported by piled basement.

a case, the piles must be anchored below the zone of soil swelling and designed to take or eliminate tension. The pressure on the underside of the podium basement slab will be equal to the swelling pressure exerted by the soil unless a void former is used to eliminate the pressure. A vertical movement joint passing completely through the basement and super-structure should be provided between the tower and podium to allow freedom of movement.

Measurements of the relative loads carried by the piles and the underside of the slab of a piled basement raft were described by Hooper[5.35]. The measurements were made during and subsequently to the construction of the 31-storey building of the Hyde Park Cavalry Barracks in London. The 90 m high building was constructed on the piled raft 8.8 m below ground level. The 51 bored and cast-in-place piles supporting the raft had a shaft diameter of 910 mm and an enlarged base 2400 mm in diameter (Figure 5.39a).

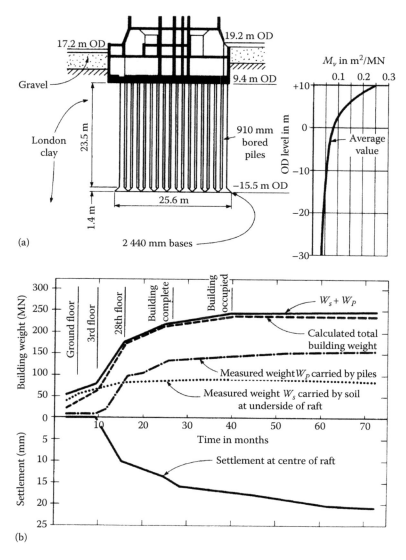

Figure 5.39 Piled raft foundations for Hyde Park Cavalry Barracks, London: (a) foundation arrangements and soil characteristics; (b) distribution of loading between raft and piles.

The piles were installed by drilling from ground level and concreting the shaft up to raft level before commencing the bulk excavation.

The weight of the building (including imposed load but excluding wind load) was calculated to be 228 MN. The weight of soil removed when excavating through gravel on to the stiff London Clay at raft level was 107 MN, giving a net load to be transferred by the raft and piles to the London Clay of 121 MN or a net bearing pressure at raft level of 196 kN/m². 

Load cells were installed in three of the piles to measure the load transferred from the raft to the pile shaft, and three earth pressure cells were placed between the raft and the soil to measure the contact pressures developed at this interface. Settlements of the raft at various points were also measured by means of levelling points installed at ground level.

The observations of pile loadings and contact pressures were used to estimate the proportion of the total load carried by the piles and the basement raft from the initial stages of construction up to 3 years after completing the building. The results of these calculations are shown in Figure 5.39b and are compared with the calculated total weight of the building at the various stages of construction. Hooper[5.35] estimated that at the end of construction, 60% of the building load was carried by the piles and 40% by the underside of the raft. In the post-construction period, there was a continuing trend towards the slow transfer of more load to the piles, about 6% of the total downward structural load being transferred to the piles in the 3-year period.

## 5.10 OPTIMISATION OF PILE GROUPS TO REDUCE DIFFERENTIAL SETTLEMENTS IN CLAY

Cooke et al.[5.36] measured the proportion of load shared between the piles and raft and also the distribution of load to selected piles in different parts of a 43.3 m by 19.2 m piled raft supporting a 16-storey building in London Clay at Stonebridge Park. There were 351 piles in the group with a diameter of 0.45 m and a length of 13 m. The piles were uniformly spaced on a 1.6 m square grid. The overall loading on the pile group was about 200 kN/m².

At the end of construction, the piles carried 78% of the total building load, the remainder being carried by the raft. The distribution of the load to selected piles near the centre, at the edges, and at the corners of the group is shown in Figure 5.40. It will be seen that the loads carried by the corner and edge piles were much higher than those on the centre piles. The loading was distributed in the ratio 2.2:1.4:1 for the corner, edge and centre respectively.

Advantages can be taken of the load sharing between raft and piles and between various piles in a group to optimise the load sharing whereby differential settlement is minimised and economies obtained in the design of the structural frame and in the penetration depth and/or diameter of the piles (Section 5.3). The procedure in optimisation is described by Padfield and Sharrock[4.87]. Central piles are influenced by a larger number of adjacent piles than those at the edges. Hence, they settle to a greater extent and produce the characteristic dished settlement. Therefore, if longer stiffer piles are provided at the centre, they will attract a higher proportion of the load. The outer piles are shorter and thus less stiff and will yield and settle more, thus reducing the differential settlement across the group. The alternative method of varying the settlement response to load is to vary the cross-sectional dimensions. The centre piles are made long with straight shafts and mobilise the whole of their bearing capacity in shaft friction at a settlement of between 10 and 15 mm. The shorter outer piles can be provided with enlarged bases which require a greater settlement to mobilise the total ultimate bearing capacity (see Section 4.6). An example of this is given by Burland and Kalra[5.17]. Viggiani et al.[4.14] carried out an exercise using a finite element

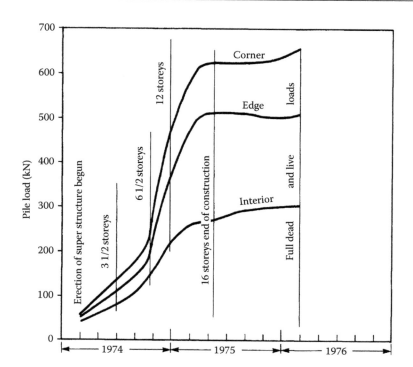

Figure 5.40 Load distribution on piled raft in London Clay. (After Cooke, R.W. et al., *Proc. Inst. Civil Eng.*, 7, 433, 1981.)

program based on plate theory to show that by concentrating piles in the centre of the Stonebridge Park raft, and as the raft itself had sufficient bearing capacity to support the uniformly distributed load, the number of piles could be optimised with only a marginal increase in differential settlement. Padfield and Sharrock also demonstrated an alternative design for this site where the number of piles could be reduced to 40 placed under the central 30% of the raft at 3.2 m spacing. Wind loading will affect the need for peripheral piles to accommodate the lateral actions.

Randolph[1.1] pointed out that where the ratio of the width of a pile group to the pile length is greater than unity, the pile cap contributes significantly to the load transfer from the superstructure to the soil. Hence, the stiffness of a piled raft where the piles are arranged to cover the whole foundation area will be similar to that of the raft structure without the piles. Thus, by concentrating the piles in the central area and using shorter piles (or no piles) around the edges, the bending moments due to dishing of the raft are considerably reduced. In the case of a uniformly-loaded foundation area, analyses show that piles of length greater than 70% of the foundation width situated over the central 25%–40% of the raft area are required (see also Section 4.9.4). Hence, instead of conventionally spreading the piles uniformly over the whole foundation area, as little as 30%–50% of the cumulative length of all the piles is needed.

Load distribution between the piles is achieved through the continuous pile cap which must be designed to be stiff enough to achieve this. With perfect optimisation, differential settlement can be reduced to zero. The analysis to achieve optimisation is complex and is best resolved by interactive analyses using iterative computer models as discussed in Sections 4.9.4 and 5.4. It is also necessary to check that the stress is not excessive on the shafts of the central piles which are designed to carry a high proportion of the load.

## WORKED EXAMPLES

### Example 5.1

Bored piles 500 mm in diameter drilled to a depth of 13.9 m below ground level into a firm to stiff clay are arranged in a group consisting of 10 rows each of seven piles at 1100 mm centres, each carrying a permanent load of 250 kN and a variable load of 110 kN. From the results of tests on samples from three boreholes, the characteristic undrained shear strength of the clay increases from 60 kN/m² at 1.5 m below ground surface to 110 kN/m² at the base of the pile group. The strength of the clay at pile toe level is 80 kN/m². Profiles of the undrained deformation modulus $E_u$ and the coefficient of compressibility $m_v$ are shown in Figure 5.41. Determine the overall stability and settlement of the pile group.

The first step is to calculate the characteristic resistance of the individual bored pile under the design actions so that $R_{cd} \geq F_d$, from Equation 4.7 with $N_c = 9$ and Equation 4.10 with $\alpha = 0.5$ on the characteristic strength and the model factor $\gamma_{Rd} = 1.4$ assuming no pile testing:

$$R_{ck} = R_{bk} + R_{sk} = (9 \times 80 \times \pi/4 \times 0.5^2)/1.4 \ + (0.5 \times (60+110)/2 \times 0.5 \times \pi \times 12.4)/1.4$$

$$= (101 + 591) \text{ kN}$$

The *alternative* EC7 procedure will be used:

For *DA1 combination 2* from Table 4.4, R4 factors are $\gamma_b = 2.0$ and $\gamma_s = 1.6$. From Table 4.1, the A2 permanent action factor is $\gamma_G = 1.0$ and variable $\gamma_Q = 1.3$; hence,

$$F_d = 1.0 \times 250 + 1.3 \times 110 = 393 \text{ kN}$$

$$R_{cd} = (101/2.0 + 591/1.6) = 419 \text{ kN} > F_d \text{ and satisfactory}$$
$$\text{(DA1 combination 1 will also be satisfactory by inspection)}$$

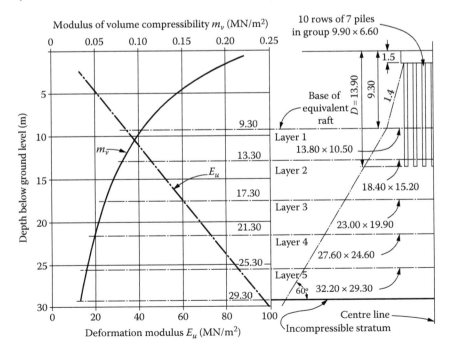

*Figure 5.41* Profiles of the undrained deformation modulus $E_u$ and the coefficient of compressibility $m_v$.

Because of the increasing strength of the clay below toe level, block failure of the group should not occur. However, to comply with EC7 Clause 7.6.2.1, it will be assumed that the pile group acts as a single large-diameter pile to determine the stability of the group. For the arrangement of the piles shown in Figure 5.41, the overall dimensions of the pile group are $9 \times 1.1 = 9.9$ m, $6 \times 1.1 = 6.6$, and 12.4 m deep at the bearing stratum (13.9 m – 1.5 m cap). The diameter of the *equivalent pile* is $(9.9 \times 6.6 \times 4/\pi)^{0.5} = 9.12$ m and the base area $A_b = 65.34$ m². The adhesion factor of 0.5 will be applied.

From Equation 4.7,

$$R_{bk} = (9 \times 110 \times 65.34)/1.4 = 46{,}205 \text{ kN (no material factor used for axially loaded piles)}$$

From Equation 4.10,

$$R_{sk} = (0.5 \times (60 + 110)/2 \times 9.12 \times \pi \times 12.4)/1.4 = 10{,}785 \text{ kN}$$

*For DAI combination 2* and using the resistance factors for bored piles,

$$R_{cd} = (46{,}205/2.0 + 10{,}785/1.6) = 29{,}843 \text{ kN} > F_d = 70 \times 393$$

$$= 27{,}510 \text{ kN and satisfactory}$$

(DA1 combination 1 will also be satisfactory by inspection)

As an alternative, and with no transverse loading, the Brinch Hansen procedure in Sections 5.1 and 5.2.1 can be applied to the equivalent block of $9.9 \times 6.6$ m.

For calculating the ultimate base resistance from Equation 5.1,

$$R_b = A_b \left( c_u N_c s_c d_c i_c b_c + p_0 N_q s_q d_q i_q b_q + 0.5 \gamma B N_\gamma s_\gamma d_\gamma i_\gamma b_\gamma \right)$$

$N_c$ from Figure 5.6 is 5.14 (the classic value for a shallow foundation on clay in undrained shear, i.e. $\pi + 2$); $s_c$ is 1.3; for $D/B = 2.1$ and $\phi = 0°$, $d_c = 1.3$ (Figure 5.9); $i_c$ is 1.0 for a centrally applied vertical load; and $b_c$ is 1.0. The second term is zero for $\phi = 0°$ and in the third term $N_\gamma = 1.0$, $s_\gamma = 0.95$, and $d_\gamma = i_\gamma = b_\gamma = 1.0$. Applying the M2 material factor, $\gamma_{cu} = 1.4$ from Table 4.2, the characteristic shear strength is 78.6 kN/m² and

$$R_b = (9.9 \times 6.6) \ (78.6 \times 5.14 \times 1.3 \times 1.3 \times 1.0 \times 1.0 + 0.5 \times 18 \times 6.6 \times 1.0 \times 0.95)$$

$$= 39{,}596 + 3{,}687 = 43{,}283 \text{ kN}, \qquad \text{hence } R_{bk} = 43{,}283/1.4 = 30{,}916 \text{ kN}$$

*For DA1 combination 2* and applying the spread foundation base factor, $\gamma_{Rv} = 1.0$ to the block, $\gamma_G = 1.0$, and $\gamma_Q = 1.3$ as before:

$$F_d = 27{,}510 \text{ kN}$$

$$R_{cd} = (30{,}916/1.0) = 30{,}916 \text{ kN} > F_d \text{ and satisfactory}$$

(DA1–1 will also be satisfactory by inspection)

*Settlement of pile group*

As the resistance is partly from shaft friction, take the spread of the load shown in Figure 5.3a:

Depth to centre of equivalent raft $= \frac{2}{3} \times 13.90 = 9.3$ m

Dimensions of equivalent raft $= 6.60 + \left(\frac{1}{4} \times 7.80 \times 2\right) = 10.5$ m

$$\text{and } 9.90 + \left(\frac{1}{4} \times 7.80 \times 2\right) = 13.8 \text{ m}$$

Unfactored pressure at level of equivalent raft (note that for SLS as EC7 Clause 2.4.8(2) and the NA the partial factors are taken as unity):

$$q_n = \frac{70 \times (250 + 110)}{10.5 \times 13.8} = 174 \text{ kN/m}^2$$

The settlements are calculated over the zone of soil down to the level of the incompressible stratum, that is, at a depth of 20 m below the base of the equivalent raft. It is convenient to divide the soil into five 4 m layers commencing at 9.30 m and extending to 29.30 m. The immediate and consolidation settlements are then calculated for each layer.

*Immediate settlement in Layer 1*
From Figure 5.41, average $E_u = 39$ MN/m². From Figure 5.18, for $H/B = 4/10.5 = 0.38$ and $L/B = 13.8/10.5 = 1.3$, $= \mu_1 = 0.15$, and for $D/B = 9.3/10.5 = 0.9$ and $L/B = 1.3$, $= \mu_0 = 0.93$. Therefore, from Equation 5.22,

$$\text{Immediate settlement} = \rho_i = \frac{0.15 \times 0.93 \times 174 \times 10.5 \times 1000}{39 \times 1000} = 6.5 \text{ mm}$$

The settlements in the underlying four layers are calculated in a similar manner, the calculations for all five layers being tabulated thus:

| Layer | B (m) | L (m) | $q_n$ (kN/m²) | $\mu_1$ | $\mu_0$ | $E_u$ (MN/mv²) | $\rho_i$ (mm) |
|---|---|---|---|---|---|---|---|
| 1 | 10.5 | 13.8 | 174 | 0.15 | 0.93 | 39 | 6.5 |
| 2 | 15.1 | 18.4 | 90 | 0.06 | 0.93 | 52 | 1.5 |
| 3 | 19.7 | 23.0 | 55 | 0.03 | 0.92 | 64 | 0.5 |
| 4 | 24.3 | 27.6 | 37 | 0.02 | 0.92 | 76 | 0.2 |
| 5 | 28.9 | 32.2 | 27 | 0.01 | 0.93 | 88 | 0.1 |
| Total immediate settlement | | | | | | | 8.8 |

The immediate settlement can be checked from Equation 5.21 because the deformation modulus increases linearly with depth. At the level of equivalent raft, $E_u$ is 32 MN/m² and at 20 m below this level, it is 97 MN/m². Therefore, from Equation 5.21,

$$97 = 32(1 + 20k/10.5)$$

$$k = 1.1$$

Dividing equivalent raft into four rectangles, each 6.9 × 5.25 m. From Figure 5.18 for $L/B = 6.9/5.25 = 1.3$, $H/B = 20/5.25 = 3.8$, and $k = 1.1$, $I_p'$ is 0.13. From equation in Figure 5.17,

$$\text{Settlement at corner of rectangle} = \frac{174 \times 5.25 \times 0.13 \times 1000}{32 \times 1000} = 3.7 \text{ mm}$$

Settlement at centre of equivalent raft $= 4 \times 3.7 = 14.8$ mm.

*Oedometer settlement for Layer 1*

Depth to centre of layer = 9.3 + 2.0 = 11.3 mm

From Figure 5.11 with $L/B$ = 13.8/10.5 = 1.3 and $z/B$ = 2/10.5 = 0.19, stress at the centre of layer = 0.8 × 174 kN/m². From Figure 5.41, average $m_v$ at centre of layer = 0.09 MN/m². Therefore, oedometer settlement from Equation 5.23

$$\rho_{oed} = \frac{0.09 \times 0.80 \times 174 \times 4 \times 1000}{1000} = 50.1 \text{ mm}$$

The oedometer settlements for all five layers are calculated in a similar manner and are tabulated in the following.

| Layer | Depth to centre of layer (m) | z (m) | z/B | $\sigma_z$ | $m_v$ (MN/m²) | $\rho_{oed}$ (mm) |
|-------|------------------------------|-------|------|------------|---------------|-------------------|
| 1 | 11.3 | 2 | 0.19 | 0.80 × 174 | 0.09 | 50.1 |
| 2 | 15.3 | 6 | 0.57 | 0.51 × 174 | 0.07 | 24.8 |
| 3 | 19.3 | 10 | 0.95 | 0.33 × 174 | 0.05 | 11.5 |
| 4 | 23.3 | 14 | 0.33 | 0.22 × 174 | 0.04 | 6.1 |
| 5 | 27.3 | 18 | 0.71 | 0.15 × 174 | 0.04 | 4.2 |
| Total oedometer settlement | | | | | | 96.7 |

From Figure 5.21, the depth factor $\mu_d$ for $D/\sqrt{LB}$ = $9.30/\sqrt{13.8 \times 10.5}$ = 0.77 is 0.78, and for London Clay, the geological factor $\mu_g$ is about 0.5. Therefore,

Corrected consolidation settlement = $\rho_c$ = $0.5 \times 0.78 \times 96.7$ = 37.7 mm.

Total settlement of pile group = $\rho_i + \rho_c$ = $8.8 + 37.7$ = 46.5 mm.

In practice, a settlement between 30 and 60 mm would be expected.

## Example 5.2

Part of the jetty structure referred to in Example 4.4 carries bulk-handling equipment with a permanent vertical action of 3 MN and variable action of 3 MN. Design a suitable pile group to carry this equipment and calculate the settlement under the permanent and variable loading.

It has been calculated in Example 4.4 that a 450 × 450 mm precast concrete pile driven to 8.5 m below the seabed was needed to resist the uplift load of 180 kN. The compressive load of 250 kN was adequately resisted at this penetration. For uniformity in design and construction, it is desirable to adopt a pile of the same dimensions to carry the bulk-handling plant. However, it is possible to reduce the depth of piles to 7 m as there is no requirement to resist uplift. A group of 42 piles arranged in seven rows of six piles should be satisfactory.

Spacing the piles at centres equal to three times the width, the dimensions of the group are 6 × 1.35 = 8.10 m by 5 × 1.35 = 6.75 m. A suitable pile cap in the form of a thick slab would be 10.5 × 9.0 × 1.25 m deep. Take a depth of water of 12 m and a height of 4 m from water level to the underside of the pile cap.

The weight of the pile group above seabed level (with concrete weight density 2.5 tonne/m³) is as follows:

$$= 9.81\left([10.5 \times 9.0 \times 1.25 \times 2.5] + \left\{42 \times 0.45^2\left[(12 \times 1.5) + (4 \times 2.5)\right]\right\}\right) = 5233 \text{ kN}$$

*Check resistance of single pile*
From Example 4.4, characteristic shaft friction resistance for a 7 m penetration by comparison is

$R_{sk}$ = 379 × 7/8.5 = 312 kN and characteristic end bearing is

$R_{bk}$ = 289.3 × 7/8.5 = 238 kN

For *DA1 combination 2* (driven pile), $\gamma_b$ = 1.7, $\gamma_s$ = 1.5 and $\gamma_{Rd}$ = 1.4, and for actions $\gamma_G$ = 1.0 and $\gamma_Q$ = 1.3,
  Permanent action = (3000 + 5233)/42 = 196 kN/pile
  Variable action = 3000/42 = 71 kN/pile

$F_d$ = 196 × 1.0 + 71 × 1.3 = 288 kN

$R_{cd}$ = (238/1.7 + 312/1.5) = 348 kN > $F_d$ and satisfactory

For *DA1 combination 1* (driven pile), $\gamma_b$ = 1.0, $\gamma_s$ = 1.0 and $\gamma_{Rd}$ = 1.4, and for actions $\gamma_G$ = 1.35 and $\gamma_Q$ = 1.5,

$F_d$ = 1.35 × 196 + 71 × 1.5 = 372 kN

$R_{cd}$ = (238/1.0 + 312/1.0) = 550 kN > $F_d$ and satisfactory

*Check settlement of pile group*
Because the piles are driven into a uniform sand carrying their load partly in skin friction and partly in end bearing, the distribution of load shown in Figure 5.3a applies:

Depth below seabed to equivalent raft = $\frac{2}{3} \times 7$ = 4.67 m

Thus, the dimensions of the equivalent raft are

$L = 8.1 + \left(\frac{1}{4} \times 2 \times 4.67\right) = 10.4$ m

$B = 6.75 + \left(\frac{1}{4} \times 2 \times 4.67\right) = 9.1$ m

In calculating settlements, it is only necessary to consider the unfactored actions from the bulk-handling plant. The piles and pile cap settle immediately as they are constructed and the pile cap is finished to a level surface:

Pressure on sand below raft = $\dfrac{6 \times 1000}{10.4 \times 9.1}$ = 63 kN/m²

At level of raft, effective overburden pressure = 1.2 × 9.81 × 4.67 = 55 kN/m²

From Figure 5.24 for an SPT N-value of 15 blows/300 mm, $I_c$ is 4 × 10⁻².

Assume for the purposes of illustration that the previous overburden pressure was 75 kN/m². Then from Equation 5.31b, the immediate settlement for an effective pressure increase, $p$, of 63 kN/m² is

$$\rho_i = 63 \times 9.1^{0.7} \times \frac{4 \times 10^{-2}}{3} = 3.9 \text{ mm}$$

From Figure 5.25, the depth of influence $z_I$ for $B$ of 9.1 m is 5 m. This is less than the thickness of the compressible layer. Hence, the thickness factor, $f_s$, is unity. From Equation 5.32a,

$$\text{Shape factor, } f_s = \left( \frac{1.25 \times 10.4/9.1}{10.4/9.1 + 0.25} \right)^2 = 1.05$$

The time factor for settlement at 30 years and static loading condition from Equation 5.33 is

$$f_t = 1 + 0.3 + 0.2 \log \frac{30}{3} = 1.5$$

Therefore, from Equation 5.34, consolidation settlement = 1.05 × 1.0 × 1.5 × 3.9 = 6.1 mm. The imposed loading would be intermittent in operation.

Checking from Equation 5.25, for $d_s$ greater than $2B$ and $L/B = 1.1$, Figure 5.23 gives $s = 1.1$:

$$\text{Immediate settlement} = \frac{1.1 \times 63}{15^{0.87} \left( 1 + 0.4 \times \frac{4.67}{9.1} \right)} = 5 \text{ mm}$$

Therefore, the pile group would be expected to settle between 5 and 10 mm under the permanent and variable loads from the bulk-handling equipment.

## Example 5.3

The driven and cast-in-place piles in Example 4.5 each carry a permanent action of 900 kN and are arranged in a group of 20 rows of 15 piles spaced at 1.60 m centres in both directions. Calculate the settlement of the pile group using the static cone-resistance diagram in Figure 4.46. Length of pile group = 19 × 1.6 = 30.4 m. Width of pile group = 14 × 1.6 = 22.4 m. The transfer of load from the piles to the soft clay in skin friction is relatively small, and therefore the distribution of load shown in Figure 5.3b applies.

Depth to equivalent raft foundation = $\frac{2}{3} \times 15 = 10$ m below the surface of the sand stratum or 22 m below ground level, as shown in Figure 5.42:

$$\text{Length of equivalent raft } L = 30.4 + \left( 2 \times 10 \times \tfrac{1}{4} \right) = 35.4 \text{ m}$$

$$\text{Width of equivalent raft } B = 22.4 + \left( 2 \times 10 \times \tfrac{1}{4} \right) = 27.4 \text{ m}$$

$$\text{Pressure on soil beneath raft} = \frac{270 \times 1000}{35.4 \times 27.4} = 278 \text{ kN/m}^2$$

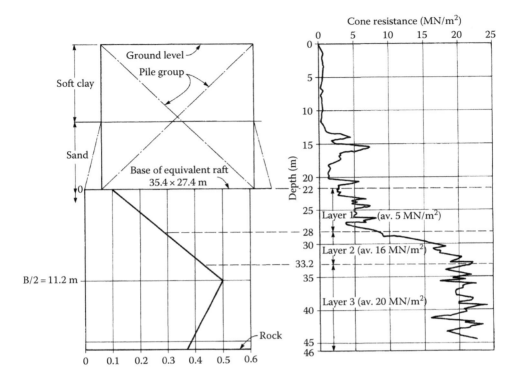

*Figure 5.42* Cone resistance and factors for Example 5.3.

The settlement can be calculated by the Schmertmann method. It is convenient to divide the cone-resistance diagram shown in Figure 4.46 into three layers between the base of the equivalent raft and rockhead. The subdivision of these layers and the superimposition of the Schmertmann curves beneath the base of the raft are shown in Figure 5.42. The settlement is calculated over a period of 25 years. For SLS calculations, the partial factors are unity.

From Figures 5.28 and 5.42, the values for $I_z$ and $E_d$ are as follows.

For $I_z$:

| Layer | For L/B = 1 | For L/B = 10 |
|---|---|---|
| 1 | 0.20 | 0.24 |
| 2 | 0.36 | 0.3 |
| 3a | 0.46 | 0.39 |
| 3b | 0.4 | 0.39 |

For $E_d$:

| Layer | For L/B = 1 | For L/B = 10 |
|---|---|---|
| 1 | 5 × 2.5 = 12.5 MN/m² | 5 × 3.5 = 17.5 MN/m² |
| 2 | 16 × 2.5 = 40 MN/m² | 16 × 3.5 = 56 MN/m² |
| 3 | 20 × 2.5 = 50 MN/m² | 20 × 3.5 = 70 MN/m² |

$q_c$ factors as Figure 5.28.

For axisymmetric loading ($L/B = 1$) from Equation 5.35, uncorrected settlements are given by

$$\text{Layer}\,1 = \frac{278 \times 0.20 \times 6 \times 1000}{12.5 \times 1000} = 27 \text{ mm}$$

$$\text{Layer}\,2 = \frac{278 \times 0.36 \times 5.2 \times 1000}{40 \times 1000} = 13 \text{ mm}$$

$$\text{Layer}\,3a = \frac{278 \times 0.46 \times 2.5 \times 1000}{50 \times 1000} = 6 \text{ mm}$$

$$\text{Layer}\,3b = \frac{278 \times 0.4 \times 10.3 \times 1000}{50 \times 1000} = 23 \text{ mm}$$

$$\text{Total} = 69 \text{ mm}$$

Similarly, for $L/B > 10$, the uncorrected settlements are

$$\text{Layer}\,1 = 19 \text{ mm}$$

$$\text{Layer}\,2 = 9 \text{ mm}$$

$$\text{Layer}\,3 = 21 \text{ mm}$$

$$\text{Total} = 49 \text{ mm}$$

By interpolation, the settlement for $L/B = 1.3$ is 66 mm.
Effective overburden pressure at base of raft

$$= 9.81[(2 \times 1.9) + (10 \times 0.9) + (10 \times 0.9)] = 214 \text{ kN/m}^2$$

From Equation 5.36, $C_1 = 1 - 0.5 \times \dfrac{214}{278} = 0.62$

From Equation 5.37, $C_2 = 1 + 0.2 lg \dfrac{25}{0.1} = 1.48$

Corrected settlement at 25 years = $0.62 \times 1.48 \times 66 = 61$ mm, say, between 50 and 70 mm.

## Example 5.4

Nuclear reactors and their containment structures and ancillary units weighing 900 MN are to be constructed on a base 70 × 32 m sited on 8 m loose to medium-dense sand overlying a moderately strong sandstone. Rotary cored boreholes showed that below a thin zone of weak weathered rock, the RQD value of the sandstone was 85% and the average unconfined compression strength was 14 MN/m². For this loading, a piled foundation is required using 1.5 m diameter bored piles taken 2 m below weak weathered rock on to the moderately strong sandstone. Calculate the concrete stress and settlement of a group of 84 piles arranged in 14 rows of six piles each at 5 m centres in both directions.
Use class C25/30 concrete with $\gamma_C = 1.5 \times 1.1$:
Design concrete compressive strength = $0.85 \times 25/(1.5 \times 1.1) = 12.9$ MN/m²

Actual stress on 1.5 m piles allowing for reduction of 50 mm in diameter (as Table 4.6) at a design action with $\gamma_G = 1.35$ of $F_d = 900 \times 1.35 = 1215$ MN

$$\sigma = 1215/(\pi/4 \times 1.45^2 \times 84) = 8.7 \text{ MN/m}^2 \text{ and satisfactory}$$

Length of pile group $L = 13 \times 5 = 65$ m
Width of pile group $B = 5 \times 5 = 25$ m
The transfer of load in skin friction to the sand is relatively small and the piles can be regarded as end bearing on the rock. The base of the equivalent raft will be as shown in Figure 5.3c.

Overall loading at base of raft (SLS partial factors are unity) = $900/(65 \times 25) = 0.55$ MN/m²

From Section 4.7.3 for RQD of 85%, mass factor = 0.7, and from Section 5.5, the modulus ratio of a well-cemented sandstone is 300 and deformation modulus of sandstone = $E_d = 0.7 \times 300 \times 14 = 2940$ MN/m², say, 3000 MN/m².
From Figure 5.18 with $H/B = \infty$ and $L/B = 65/25 = 2.6$, $\mu_1 = 1.1$, and with $D/B = (8 + 2)/25 = 0.4$ and $L/B = 2.6$, $\mu_0 = 0.95$. From Equation 5.22,

$$\text{Settlement of foundation} = \frac{1.1 \times 0.95 \times 0.55 \times 25 \times 1000}{3000} = 5 \text{ mm}$$

## Example 5.5

A site, where the ground conditions consist of 5.5 m of soft organic silty clay overlying 35 m of stiff to very stiff over-consolidated clay followed by rock, is reclaimed by placing and compacting 4 m of sand fill covering the entire site area. Six months after completing the reclamation, a 12-storey building imposing an overall permanent load of 160 kN/m² on a ground floor area of 48 m by 21 m is to be constructed on the site. The average undrained shearing strength of the stiff clay stratum is 90 kN/m² at the surface of the stratum, increasing to 430 kN/m² at rockhead. Measurements of the deformation modulus and modulus of volume compressibility show a linear variation, with average values at the top and bottom of the stiff clay stratum as follows:
At top: $E_u = 40$ MN/m², $m_v = 0.8$ m²/MN
At bottom: $E_u = 120$ MN/m², $m_v = 0.04$ m²/MN
Design suitable piled foundations and estimate the settlement of the completed building.
Because of the heavy loading, it is economical to provide large-diameter bored and cast-in-place piled foundations. A suitable arrangement consists of fourteen rows of six piles (Figure 5.43). Trial-and-adjustment calculations show that a pile diameter of 1200 mm is suitable. The pile spacing must be a minimum of 3 diameters, giving a spacing of at least 3.6 m.
Two different approaches to determining the effect of negative skin friction will be used to calculate the size of piles and the block and determine the resulting settlement: the first is the traditional allowable stress method, then it is checked against current EC7 recommendations.
Adopt a spacing of say 3.75 m in both directions. Thus the dimensions of the pile group are $5 \times 3.75 = 18.75$ m and $13 \times 3.75 = 48.75$ m.

$$\text{Average permanent action carried by piles} = \frac{48 \times 21 \times 160}{14 \times 6} = 1920 \text{ kN/pile}$$

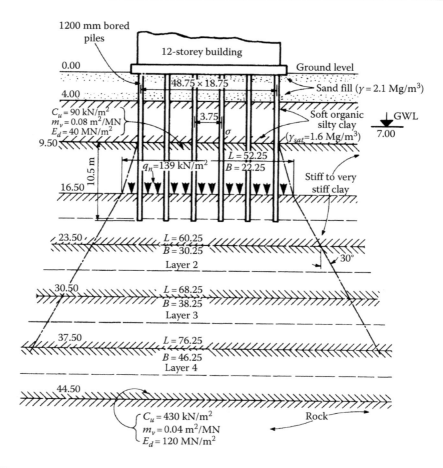

*Figure 5.43* Pile group and stratification for Example 5.5.

The central two rows of piles carry higher loads than the outer two rows on each side. A likely loading for the centre rows is 2200 kN per pile. The required penetration of the piles will be calculated on this loading. The exterior piles will be taken to the same depth but adopting a reduced diameter as required by the lesser loading.

The piles carry negative skin friction due to the consolidation of the soft clay under the imposed loading of the sand fill. At 6 months, settlement of the soft clay will be continuing at a very slow rate and it is appropriate to use Figure 4.39 (Meyerhof) to calculate the negative skin friction in this layer.

Unit negative skin friction at top of layer

$$= 0.30\sigma'_{vo} = 0.30 \times 9.81 \times 2.1 \times 4 = 24.7 \text{ kN/m}^2$$

Unit negative skin friction at groundwater level (see Figure 5.43)

$$= 0.30 \times 9.81[(2.1 \times 4) + (1.6 \times 3)] = 38.8 \text{ kN/m}^2$$

Unit negative skin friction at bottom of layer

$$= 0.30 \times 9.81[(2.1 \times 4) + (1.6 \times 3) + (0.6 \times 2.5)] = 43.3 \text{ kN/m}^2$$

Therefore, total negative skin friction in soft clay

$$= \pi \times 1.2 \left[\tfrac{1}{2}(24.7 + 38.8) \times 3 + \tfrac{1}{2}(28.8 + 43.3) \times 2.5\right] = 746 \text{ kN}$$

Because the pile will settle due to yielding of the stiff clay when the full load is applied, the pile will move downwards relative to the lower part of the soft clay. Thus, negative skin friction will be developed only over about 80% of the length within the soft clay. Thus, approximate total negative skin friction in soft clay = 0.8 × 746 = 597 kN.

The negative skin friction in the sand can be calculated using the coefficients for $K_s$ in Table 4.7. Although the compacted sand fill is dense, it will be loosened by pile boring to give a coefficient $K_s$ of 1 and a $\phi$ value of 30°. From Equation 4.14 using average overburden pressure (but ignoring the $\gamma_{Rd}$ factor for allowable stress application),

Negative skin friction on pile in sand fill

$$= 0.5 \times 1 \times 9.81 \times 2.1 \times 4 \times \tan 30° \times \pi \times 1.2 \times 4 = 359 \text{ kN}$$

Total negative skin friction on pile = 359 + 597 = 956 kN

Total applied load on piles in centre rows = 956 + 2200 = 3156 kN

The required pile penetration depth is calculated on the basis of the building loading, with a check being made to ensure that the safety factor on the combined building load and negative skin friction is adequate.

Required ultimate pile resistance for overall safety factor of 2 (Section 4.6) = 2 × 2200 = 4400 kN.

Take a trial penetration depth of 10 m into the stiff clay stratum. At the pile base level, $c_{ub}$ = 190 kN/m$^2$ and the average value of $c_u$ on the shaft is 140 kN/m$^2$. Thus,

Ultimate base resistance $= \tfrac{1}{4} \times \pi \times 1.2^2 \times 9 \times 190 = 1935 \text{ kN}$

Load to be carried in skin friction = 4400 − 1935 = 2465 N

The adhesion factor for a straight-sided pile can be taken as 0.45. Therefore, from Equation 4.10 (and again ignoring the $\gamma_{Rd}$ factor for allowable stress application),

Total load to be resisted by the pile shaft = 2465 = 0.45 × 140 × $\pi$ × 1.2 × $l$

from which $l$ = 10.4 m (say 10.5 m) and the trial depth is satisfactory.

Checking the criterion of a safety factor of 3 in end bearing and unity in skin friction, allowable load = $(\tfrac{1}{3} \times 1935) + 2465 = 3110 \text{ kN}$ which roughly equals the building load plus the negative skin friction. Checking the overall safety factor on the combined loading,

Safety factor (1935 + 2465)/3156 = 1.4

This is satisfactory since the negative skin friction on the piles will not contribute to the settlement of the pile group.

The transfer of load from the pile group to the soil will be as shown in Figure 5.3b. The dimensions of the equivalent raft are

$$L = 48.75 + \left(\tfrac{2}{3} \times 10.5 \times 2 \times \tfrac{1}{4}\right) = 52.25 \text{ m}$$

$$B = 18.75 \left(\tfrac{2}{3} \times 10.5 \times 2 \times \tfrac{1}{4}\right) = 22.25 \text{ m}$$

$$\text{Pressure on base of equivalent raft due to building load} = \frac{48 \times 21 \times 160}{52.25 \times 22.25} = 139 \text{ kN/m}^2$$

*Calculating the immediate settlement*
At a level of equivalent raft, $E_u = E_f = 65 \text{ MN/m}^2$
At rockhead, $E_u = 120 \text{ MN/m}^2$

From Equation 5.21, $120 = 65(1 + 28k/22.25)$

$$k = 0.7$$

Divide equivalent raft into four rectangles, each 26.1 × 11.1 m.
From Figure 5.17 for $L/B = 26.1/11.1 = 2.3$, $H/B = 28/11.1 = 2.5$ and $k = 0.7$, $I_p'$ is 0.14.

$$\text{Settlement at corner of rectangle} = \frac{139 \times 11.1 \times 0.14 \times 1000}{65 \times 1000} = 3.3 \text{ mm}$$

Settlement at centre of equivalent raft = 4 × 3.3 = 13.2 mm

*Calculating the consolidation settlement*
To calculate the settlement of the pile group due to the building loads only, the 28 m layer of clay between the equivalent raft and rockhead is divided into four 7 m layers.

*Oedometer settlement in Layer 1*
From Figure 5.11 with and $z/B = 3.5/22.25 = 0.16$ and $L/B = 52.25/22.25 = 2.3$, stress at centre of rectangle = 0.83 × 139 = 118 kN/m². Modulus of volume compressibility = 0.07 m²/MN.

Then from Equation 5.23 the uncorrected settlement $= \dfrac{0.07 \times 118 \times 7 \times 1000}{1000} = 57.8 \text{ mm}$.

The settlements in the remaining layers are calculated similarly and the results for the four layers are tabulated as follows:

| Layer | Depth to centre of layer (m) | z (m) | z/B | $\sigma_z$ (kN/m²) | $m_v$ (MN/m²) | $\rho_{oed}$ (mm) |
|---|---|---|---|---|---|---|
| 1 | 20.00 | 3.5 | 0.16 | 118 | 0.07 | 57.8 |
| 2 | 27.00 | 10.5 | 0.47 | 88 | 0.06 | 37.0 |
| 3 | 34.00 | 17.5 | 0.79 | 64 | 0.05 | 22.4 |
| 4 | 41.00 | 24.5 | 1.10 | 50 | 0.04 | 14.0 |
| Total uncorrected oedometer settlement | | | | | | 131.2 |

The previous summation must be corrected by a depth factor which is given by Figure 5.21, with $D/\sqrt{LB} = 16.5/\sqrt{52.25 \times 22.25} = 0.48$ and $L/B = 2.35$ as $\mu_d = 0.85$.

To obtain the consolidation settlement $\rho_c$, the summation is also multiplied by the geological factor $\mu_g$, which is 0.5 for an over-consolidated clay. Therefore,

Total consolidation settlement = $0.85 \times 0.5 \times 131.2 = 55.8$ mm

Total settlement of pile group due to building load only = $\rho_i + \rho_c = 13.2 + 55.8 = 69.0$ mm

To this figure must be added the consolidation settlement of the stiff clay due to the sand filling. The immediate settlement is not taken into account since this will have taken place before commencing the construction of the building.

Oedometer settlement due to 4 m of sand fill for an average $m_v$ of 0.06 m²/MN in clay layer

$$= \frac{0.06 \times 9.81 \times 2.1 \times 4 \times 1000}{1000} = 4.9 \text{ mm}$$

Correcting for the geological factor as Equation 5.24,

$\rho_c = 0.5 \times 4.9 = 2.4$ mm

A time/settlement calculation would show that about one-third of this settlement would be complete before completing the pile installation. Thus, settlement of 12-storey building due to combined loading from building and sand layer

$$= 69.0 + \left(\tfrac{2}{3} \times 2.4\right) = 70.6 \text{ mm}$$

It will be noted that the negative skin friction on the piles was not added to the loading on the equivalent raft when calculating the settlement of the building. However, it is necessary to check that the individual piles will not settle excessively under the combined building load and negative skin friction.

Maximum load on pile = 3156 kN. If shaft friction on pile is fully developed, the end-bearing load is $3156 - 2465 = 691$ kN, and thus

$$\text{End-bearing pressure} = \frac{691}{\tfrac{1}{4}\pi \times 1.2^2} = 611 \text{ kN/m}^2$$

Ultimate unit base resistance = $9 \times 190 = 1710$ kN/m²

From Equation 4.33, with K = 0.01, $\rho_i = 0.01 \times \dfrac{611}{1710} \times 1200 = 4$ mm

Therefore, individual piles will not settle excessively and the critical factor is the overall settlement of the complete pile group, for which a movement of 50–100 mm over a long period of years is by no means excessive.

### Checking the pile length using the EC7 recommendations
Applying the same penetration into the stiff clay of 10.5 m for the 1.2 m diameter piles as determined above and taking the characteristic strengths at base level of $c_{ub}$ = 192 kN/m² and average value of $c_u$ = 140 kN/m² on the shaft, then from Equation 4.7 and applying model factor $\gamma_{Rd}$ = 1.4,

Characteristic base resistance $R_{bk} = (9 \times 192 \times \pi/4 \times 1.2^2)/1.4 = 1395$ kN

The adhesion factor for a straight-sided pile can be taken as 0.5. Therefore, from Equation 4.10,

Characteristic shaft resistance $R_{sk} = (0.5 \times 140 \times \pi \times 1.2 \times 10.5)/1.4 = 1979$ kN

*Consider DA1 combination 2* as being critical. The previous calculation for negative skin friction in clay and in sand is based on effective stress; hence, an M2 material factor is not required to be applied. $\gamma_G = 1.0$ from Table 4.1 will be used for permanent unfavourable actions due to structural action and downdrag as calculated earlier, and from Table 4.4, $\gamma_b = 2.0$ and $\gamma_s = 1.6$ for bored piles without testing in downdrag conditions.

Total design action = $F_d = (2200 + 597 + 359) \times 1.0 = 3156$ kN

Total design resistance = $R_{cd} = (1395/2.0 + 1979/1.6) = 1934$ kN and fails

Therefore, increase the pile penetration in stiff clay to 16.5 m where $c_{ub} = 250$ kN/m² and $c_u = 170$ kN/m²:

Characteristic base resistance $R_{bk} = (9 \times 250 \times \pi/4 \times 1.2^2)/1.4 = 1818$ kN

Characteristic shaft resistance $R_{sk} = (0.5 \times 170 \times \pi \times 1.2 \times 16.5)/1.4 = 3778$ kN

Total design resistance = $R_{cd} = (1818/2.0 + 3778/1.6) = 3270$ kN > $F_d = 3156$ kN and satisfactory

An increase of approximately 57% in penetration is therefore necessary for the pile block to conform to EC7 factors; the extra depth means that the equivalent raft is lowered and settlement will be less than calculated earlier. This DA1-2 calculation indicates that the current EC7 treatment of negative skin friction is very conservative and a safe design can be achieved using the traditional allowable stress approach.

# REFERENCES

5.1   Poulos, H.G. Pile behaviour – Theory and application, *Geotechnique*, 39 (3), 1989, 365–415.

5.2   Burland, J.B. and Wroth, C.P. General report on Session 5: Allowable and differential settlements of structures, including settlement damage and soil structure interaction, *Proceedings of the Conference on Settlement of Structures*, Cambridge, UK, 1974, Pentech Press, London, UK, 1975, pp. 611–654.

5.3   Burland, J.B. Interaction between structural engineers and geotechnical engineers, *The Structural Engineer*, 84 (8), 2006, 29–37.

5.4   Hansen, J.B. A general formula for bearing capacity, Danish Geotechnical Institute, Bulletin No. 11, 1961; also, A revised and extended formula for bearing capacity, Danish Geotechnical Institute, Bulletin No. 28, 1968, and Code of Practice for Foundation Engineering, Danish Geotechnical Institute, Bulletin No. 32, 1978.

5.5   Meyerhof, G.G. Some recent research on bearing capacity, *Canadian Geotechnical Journal*, 1, 1963, 16–26.

5.6   Jamiolkowski, M. et al. Design parameters for soft clays, *Proceedings of the Seventh European Conference on Soil Mechanics*, Brighton, UK, 1979, pp. 21–57.

5.7   Stroud, M.A. The standard penetration test in insensitive clays, *Proceedings of the European Symposium on Penetration Testing*, Stockholm, Sweden, 1975, Vol. 2, pp. 367–375.

5.8   Burland, J.B., Broms, B.B. and de Mello, V. Behaviour of foundations and structures, *Proceedings of the Ninth International Conference on Soil Mechanics*, Tokyo, Japan, Session 2, 1977.

5.9   Jardine, R., Fourie, A., Maswose, J. and Burland, J.B. Field and laboratory measurements of soil stiffness, *Proceedings of the 11th International Conference on Soil Mechanics*, San Francisco, CA, 1985, Vol. 2, pp. 511–514.

5.10  Marsland, A. In-situ and laboratory tests on glacial clays at Redcar, *Proceedings of a Symposium on Behaviour of Glacial Materials*, Midlands Geotechnical Society, University of Brimingham, UK, 1975, pp. 164–180.

5.11 Terzaghi, K. *Theoretical Soil Mechanics*. John Wiley, New York, 1943, pp. 425.

5.12 Butler, F.G. General Report and state-of the-art review, Session 3, *Proceedings of the Conference on Settlement of Structures*, Cambridge, UK, 1974, Pentech Press, London, UK, 1975, pp. 531–578.

5.13 Brown, P.T. and Gibson, R.E. Rectangular loads on inhomogeneous soil, *Proceedings of the American Society of Civil Engineers*, 99 (SM10), 1973, 917–920.

5.14 Christian, J.T. and Carrier, W.D. Janbu, Bjerrum and Kjaernsli's chart reinterpreted, *Canadian Geotechnical Journal*, 15, 1978, 123–128.

5.15 Fox, E.N. The mean elastic settlement of a uniformly-loaded area at a depth below the ground surface, *Proceedings of the Second International Conference, ISSMFE*, Rotterdam, the Netherlands, 1948, Vol. 1, pp. 129–132.

5.16 Skempton, A.W. and Bjerrum, L. A contribution to the settlement analysis of foundations on clay, *Geotechnique*, 7 (4), 1957, 168–178.

5.17 Burland, J.B. and Kalra, J.C. Queen Elizabeth Conference Centre: Geotechnical aspects, *Proceedings of the Institution of Civil Engineers*, 80 (1), 1986, 1479–1503.

5.18 Morton, K. and Au, E. Settlement observations on eight structures in London, Session 3, *Proceedings of the Conference on Settlement of Structures*, Cambridge, UK, 1974, Pentech Press, London, UK, 1975, pp. 183–203.

5.19 Schultze, E. and Sherif, G. Predictions of settlements from evaluated settlement observations for sand, *Proceedings of the Eighth International Conference, ISSMFE*, Moscow, Russia, 1973, Vol. 1, pp. 225.

5.20 Burland, J.B. and Burbidge, M.C. Settlement of foundations on sand and gravel, *Proceedings of the Institution of Civil Engineers*, 78 (1), 1985, 1325–1337.

5.21 Robertson, P.K. and Campanella, R.G. Interpretation of cone penetration tests, parts 1 and 2, *Canadian Geotechnical Journal*, 20, 1983, 718–745.

5.22 Baldi, G., Belloti, R., Ghionna, V., Jamiolkowski, M. and Pasqualini, E. Cone resistance of dry medium sand, *Proceedings of the 10th International Conference, ISSMFE*, Stockholm, Sweden, 1981, Vol. 2, pp. 427–432.

5.23 Meigh, A.C. The Triassic rocks, with particular reference to predicted and observed performance of some major structures, *Geotechnique*, 26, 1976, 393–451.

5.24 Lunne, T. and Christoffersen, H.P. Interpretation of cone penetration data for offshore sands, *Proceedings of the Offshore Technology Conference 15*, Houston, TX, 1983, Vol. 1, pp. 181–192.

5.25 Schmertmann, J.H. Static cone to compute static settlement over sand, *Journal of the Soil Mechanics and Foundations Division, American Society of Civil Engineers*, 96 (SM3), May 1970, 1011–1043.

5.26 Schmertmann, J.H., Hartman, J.P. and Brown, P.R. Improved strain influence diagrams, *Proceedings of the American Society of Civil Engineers*, GT8, 1978, 1131–1135.

5.27 Chandler, R.J. and Davis, A.G. Further work on the engineering properties of Keuper Marl, Construction Industry Research and Information Association (CIRIA), Report 47, 1973.

5.28 Hagerty, A. and Peck, R.B. Heave and lateral movements due to pile driving, *Journal of the Soil Mechanics and Foundation Division, American Society of Civil Engineers*, (SM11), November 1971, 1513–1532.

5.29 Chow, Y.K. and Teh, C.I. A theoretical study of pile heave, *Geotechnique*, 40 (1), 1990, 1–14.

5.30 Bjerrum, L. Engineering geology of normally-consolidated marine clays as related to the settlement of buildings, *Geotechnique*, 17 (2), 1967, 83–117.

5.31 Adams, J.I. and Hanna, T.H. Ground movements due to pile driving, *Proceedings of the Conference on the Behaviour of Piles*, Institution of Civil Engineers, London, UK, 1970, pp. 127–33.

5.32 Brzezinski, L.S., Shector, L., Macphie, H.L. and Van Der Noot, H.J. An experience with heave of cast-in-situ expanded base piles, *Canadian Geotechnical Journal*, 10 (2), May 1973, 246–260.

5.33 Cole, K.W. Uplift of piles due to driving displacement, *Civil Engineering and Public Works Review*, March 1972, 263–269.

5.34 Fleming, W.G.K. and Powderham, A.J. Soil down-drag and heave on piles, Institution of Civil Engineers, *Ground Engineering Group*, notes for meeting on 25 October 1989.

5.35 Hooper, J.A. Observations on the behaviour of a piled-raft foundation on London Clay, *Proceedings of the Institution of Civil Engineers*, 55 (2), December 1973, 855–877.

5.36 Cooke, R.W., Bryden Smith, D.W., Gooch, M.N. and Sillett, D.F. Some observations on the foundation loading and settlement of a multi-storey building on a piled raft foundation in London Clay, *Proceedings of the Institution of Civil Engineers*, 7 (1), 1981, 433–460.

# Chapter 6

# Design of piled foundations to resist uplift and lateral loading

## 6.1 OCCURRENCE OF UPLIFT AND LATERAL LOADING

Piles are used to resist tension loads for buoyant structures such as dry docks, basements and pumping stations. Where the hydrostatic pressure always exceeds the downward loading, as in the case of some underground tanks and pumping stations, the anchorages are permanently under tension and cable anchors may be preferred to piles. However, in the case of the shipbuilding dock floor in Figure 6.1 for example, the anchorages may be under tension only when the dock is pumped dry before the commencement of shipbuilding. As the loading on the floor from ship construction increases to the stage at which the uplift pressure is exceeded, the anchor piles are required to carry compressive loads. Cable anchors might not then be suitable if the dock floor was underlain by soft or loose soil.

Vertical piles are also used to restrain buildings against uplift caused by the swelling of clay soils. Swelling can occur for example, when mature trees are removed from a building site. The desiccated soil in the root zone of the trees gradually absorbs water from the surrounding clay, and the consequent swelling of the clay, if unrestrained, may amount to an uplift of 50–100 mm of the ground surface, causing severe damage to buildings sited over the root zone. In subtropical countries where there is a wide difference in seasonal climatic conditions, that is a hot dry summer and a cool wet winter, the soil zone affected by seasonal moisture changes can extend to a depth of several metres below the ground surface. In clay soils, these changes cause the ground surface to alternately rise and fall with a differential movement of 50 mm or more. The depth to which these swelling (or alternate swelling and shrinkage) movements can occur usually makes the use of piled foundations taken below the zone of soil movements more economical and technically more suitable than deep strip or pad foundations.

Vertical piles must have a sufficient depth of penetration to resist uplift forces by the development of shaft friction in the soil beneath the zone of soil movements (Figure 6.2). Uplift on bored piles can be reduced by casting the concrete in the upper part of the pile within a smooth polyvinylchloride (PVC) sleeve or by coating a precast concrete or steel tubular pile with soft bitumen (see Section 4.8.3). Uplift can be further reduced by supporting the superstructure clear of the ground surface or by providing a compressible layer beneath pile caps and ground beams (see Figure 7.16). Downdrag on a friction pile should not be included in calculations to resist uplift. Piles in large groups may also be lifted due to ground heave, as described in Section 5.7.

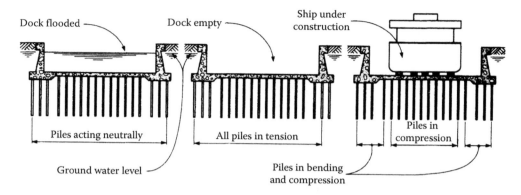

Figure 6.1 Tension/compression piles beneath the floor of shipbuilding dock.

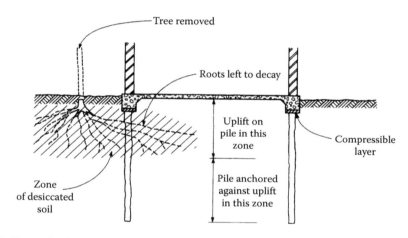

Figure 6.2 Uplift on pile due to swelling of soil after removal of mature tree.

In countries where frost penetrates deeply below the ground surface, frost expansion of the soil can cause uplift on piles, resulting in severe effects in *permafrost* regions, as described in Section 9.4. Floating ice on lakes and rivers can jam between piles in groups causing them to lift when water levels rise or when the ice sheet buckles.

The most frequent situation necessitating design against lateral and uplift forces occurs when the piles are required to restrain forces causing the sliding or overturning of structures. Lateral forces may be imposed by earth pressure (Figure 6.3a), by the wind (Figure 6.3b), by earthquakes or by the traction of braking vehicles (Figure 6.3c). In marine structures, lateral forces are caused by the impact of berthing ships (Figure 6.4), by the pull from mooring ropes and by the pressure of winds, currents, waves and floating ice. A vertical pile generally has a low resistance to lateral loads and, for economy, substantial loadings are designed to be resisted by groups of inclined or raking piles (sometimes referred to as 'batter' piles). Thus, as shown in Figure 6.5, the horizontal force can be resolved into two components, producing an axial compressive force in pile A and a tensile force in pile B.

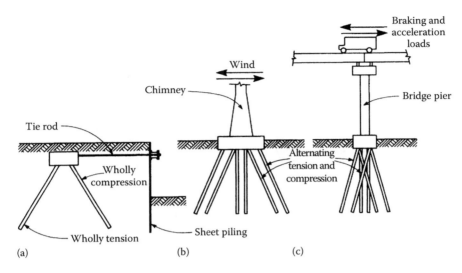

*Figure 6.3* Raking piles to resist overturning forces: (a) piled anchorage to tie rods restraining sheet-piled retaining wall; (b) raking piles to withstand wind forces on chimney; (c) raking piles to withstand traction forces from vehicles on bridge.

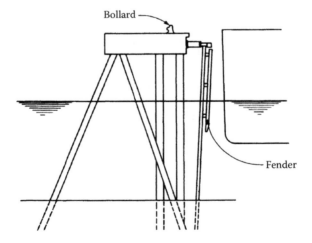

*Figure 6.4* Raking and vertical piles in breasting dolphin.

It is usual to ignore the restraint offered by a ground bearing pile cap; thus, the magnitude of each component is obtained from a simple triangle of forces as shown. Where lateral forces are transient in character, for example for wind loadings, they may be permitted to be carried wholly or partly by the pile cap where this is bearing on the ground (see Section 7.8). If raking piles are installed in fill or compressible soil which is settling under its own weight or under a surcharge pressure, considerable bending stresses can be induced in the piles, requiring a high moment of resistance to withstand the combined axial and bending stresses as discussed in Section 6.4.

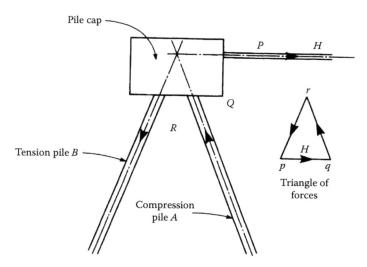

*Figure 6.5* Restraint of horizontal force by raking piles.

## 6.2 UPLIFT RESISTANCE OF PILES

### 6.2.1 General

The simplest method of restraining piles against uplift is to make the pile sufficiently long to take the whole of the uplift load in shaft friction. However, where there is rock beneath a shallow soil overburden, it may not be possible to drive the piles deeply enough to mobilise the required frictional resistance. In such cases, the shaft resistance must be augmented by adding dead weight to the pile to overcome the uplift load or by anchoring the pile to the rock.

Adding dead weight to counteract uplift loading is not usually feasible or economical. The piles may be required to carry alternating uplift and compressive loads, in which case the added dead weight would result in a large increase in the compressive loading. In the case of shipbuilding dock floors (Figure 6.1), dead weight in the form of a thick floor would add considerably to the construction costs, and in piled dolphins (Figure 6.4), the provision of a massive pile cap could make a substantial addition to the load on the compression rakers. Experience has shown that anchors in the form of grouted-in bars, tubes or cables are the most economical means of providing the required uplift resistance for piles taken down to a shallow rock layer.

### 6.2.2 Uplift resistance of friction piles

The resistance of straight-sided piles in shaft friction to statically applied uplift loads is calculated in exactly the same way as the shaft friction on compression piles, and the calculation methods given in Sections 4.2 through 4.5 can be used. However, for cyclic loading, the frictional resistance is influenced by the rate of application of the load and the degree of degradation of the soil particles at the interface with the pile wall. In the short term, the uplift resistance of a bored pile in clay is likely to be equal to its frictional resistance in compression; however, Radhakrishna and Adams[6.1] noted a 50% reduction in the uplift resistance of cylindrical augered footings and a 30%–50% reduction in belled footings in clay when sustained loads were carried over a period of 3–4 months. It was considered that the reduction in uplift was due to a loss of suction beneath the pile base and the dissipation of negative pore pressures set up at the initial loading stage. These authors pointed out that such reductions are unlikely for piles where the depth/width ratio is greater than 5.

The ICP method[4.41] can be used to determine the tension capacity of driven piles. For piles in clay, the method does not differentiate between shaft resistance in compression and tension, that is Equations 4.17 through 4.21 can be used without modification for either type of loading. Conditions are different for piles in sands where the degradation of soil particles at the pile–soil interface has an effect. Also in the case of tubular steel piles, the radial contraction across the diameter under tension loads is a further weakening effect on frictional resistance, particularly for open-ended piles. Accordingly, Equation 4.24 is modified to become

$$\tau_f = (0.8\sigma'_{rc} + \Delta\sigma'_{rd})\tan\delta'_{cr} \qquad (6.1)$$

where $\sigma'_{rc}$ and $\Delta\sigma'_{rd}$ are calculated as described for compression loading in Section 4.3.7. For open-ended piles in tension, $\tau_f$ as calculated by Equation 6.1 is reduced by a factor of 0.9.

Cyclic loading generally results in a weakening of shaft capacity. The reduction can be significant for offshore structures where piles are subjected to repetitive loading from wave action. The degree of reduction depends on the amplitude of shear strain at the pile–soil interface, the susceptibility of the soil grains to attrition, and the number and direction of the load cycles, that is one-way or two-way loading. The amplitude of the shear strain depends in turn on the ratio of the applied load to the ultimate shaft capacity. In clays, the repeated load applications increase the tendency for the soil particles to become realigned in a direction parallel to the pile axis at the interface which may eventually result in residual shear conditions with a correspondingly low value of $\delta_{cr}$. In sands, it is evident that the greater the number of load cycles, the greater the degree of degradation, although the residual silt-sized particles produced by a silica sand will have an appreciable frictional resistance.

As in the case of compression loading, degradation, both in sands and clays, takes place initially at ground level where the amplitude of the tensile strain is a maximum; it then decreases progressively down the shaft but may not reach the pile toe if the applied load is a relatively small proportion of the ultimate shaft capacity.

Jardine et al.[4.41] recommend cyclic shear tests in the laboratory using the site-specific materials as a means of quantifying the reduction in friction capacity. In clays, the interface shear is likely to occur in undrained conditions; accordingly, the laboratory testing programme should provide for simple cyclic undrained shear tests. An alternative to laboratory testing suggested by Jardine et al. is to simulate the relative movement between pile and soil under repetitive loading by finite element or $t$–$z$ analyses (Section 4.6).

EC7 adopts a criterion for avoiding the ultimate limit state for single piles or pile groups in tension by the expression similar to that for compression loading, that is,

$$F_{td} \leq R_{td} \qquad (6.2a)$$

where
    $F_{td}$ is the design value for actions in tension on a pile or pile group
    $R_{td}$ is the design value of resistance in tension of the pile or the foundation

$$F_{td} = (\gamma_G\, T_{Gk} - \gamma_{G\,fav}\, W_{Gk}) + \Sigma\gamma_Q\, T_{Qk} \qquad (6.2b)$$

where $T_{Gk}$ and $T_{Qk}$ are the characteristic permanent and variable tension loads on the pile, respectively. $W_{Gk}$ is the self-weight of the pile – usually ignored. $\gamma_G$, $\gamma_{G\,fav}$ and $\gamma_Q$ are the partial factors for permanent and variable actions as shown for compression piles in Table 4.1.

$$R_{td} = \frac{R_{stk}}{\gamma_{st}}$$

(6.2c)

where

$R_{stk}$ is the characteristic shaft resistance
$\gamma_{st}$ is the partial factor for the particular pile type in tension (in Tables 4.3 through 4.5)

Two modes of failure are to be examined:

1. The pull-out of the pile from the ground mass
2. Uplift of a block of ground containing the piles

For condition (1), the risk of pull-out of a cone of soil adhering to the pile is to be considered. The adverse effects of cyclic loading as described earlier are to be taken into account.

Calculation methods based on ground test results as described in Chapter 4 for compression loading are permitted by EC7 to be used for calculating resistance to tension loading. When using the EC7 *model pile* approach as described in Section 4.1.4, the correlation factors shown in Table A.NA.10 are applied to the results of the calculations to obtain characteristic values ($R_{stk}$). The factors depend on the number of ground test results used to provide the basis for the calculations. The partial factor for shaft resistance, $\gamma_{st}$, depending on the type of pile in Tables 4.3 through 4.5, is then applied to obtain $R_{std}$, plus any resistance from an enlarged pile base to give $R_{tk}$. It will be noted that the factors are generally higher than those for shaft resistance in compression reflecting the potentially more damaging effects of failure of a foundation in uplift. As before, where the preferred best-fit profile of the site ground tests is used for the calculations (the EC7 *alternative* method), then the model factor, $\gamma_{Rd}$, has to be applied to the resistances and not the correlation factors.

EC7 permits the ultimate tensile resistance to be determined by pile loading tests. It is recommended that more than one test should be made, and in the case of a large number of piles, at least 2% should be tested. Correlation factors given in Table A.NA 11 are applied to a series of load test results to obtain the characteristic tension resistance $R_{tk}$.

Where vertical piles are arranged in closely spaced groups, the uplift resistance of the complete group may not be equal to the sum of the resistances of the individual piles. This is because, at ultimate load conditions, the block of soil enclosed by the pile group is lifted. The manner in which the load is transferred from the pile to the soil is complex and depends on the elasticity of the pile, the layering of the soil and the disturbance to the ground caused by installing the pile. A spread of load of one in four from the pile to the soil provides a simplified and conservative estimate of the volume of a coarse-grained soil available to be lifted by the pile group, as shown in Figure 6.6. For simplicity in calculation, the weight of the pile embedded in the ground is assumed to be equal to that of the volume of soil it displaces. If the weight of the block of soil is calculated by using a diagram of the type shown in Figure 6.6, then the safety factor against uplift can be taken as unity, since frictional

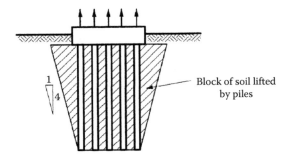

Figure 6.6 Uplift of group of closely spaced piles in coarse-grained soils.

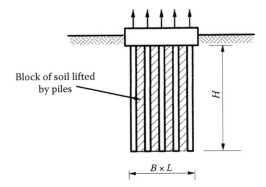

Figure 6.7 Uplift of group of piles in fine-grained soils.

resistance around the periphery of the group is ignored in the calculation. The submerged weight of the soil should be taken below groundwater level.

In the case of fine-grained soils, the uplift resistance of the block of soil in undrained shear enclosed by the pile group in Figure 6.7 is given by the equation

$$Q_u = (2LH + 2BH)\bar{c}_u + W \tag{6.3}$$

where
 $Q_u$ is the total uplift resistance of the pile group
 $L$ and $B$ are the overall length and width of the group, respectively
 $H$ is the depth of the block of soil below pile cap level
 $\bar{c}_u$ is the value of average undisturbed undrained shear strength of the soil around the sides of the group
 $W$ is the combined weight of the block of soil enclosed by the pile group plus the weight of the piles and pile cap

Submerged densities are used for the soil and portion of the structure below groundwater level when calculating $W$. $W$ is designated a favourable permanent action $G_{stbk}$ when calculating the factored design actions.

EC7 (Clause 7.6.3.1) recommends calculating the uplift resistance of a block of soil surrounding the pile group in a manner similar to that described earlier. The design value of the uplift load combined with the uplift force from buoyancy on the underside of the soil block, $V_{dstd}$, is resisted by the design values of the friction on the vertical outer surfaces of the block, $T_d$, and the stabilising forces, $G_{stbd}$, of the mass of soil composing the block, the pile cap or other substructures supported by the piles and the weight of any soil overburden above these structures. The resistances of the piles to pull-out are not included in the stabilising forces but are considered separately since they provide no resistance if failure is by lifting of the mass of soil.

Whichever of the previous methods is used to calculate the combined uplift resistance of a pile group, the resistance must not be greater than that provided by the sum of the shaft friction resistance of the individual piles in the group, taking account of the relevant partial factors.

Because buoyancy is a destabilising factor, EC7 (Clause 2.4.7.4) requires verification of stability by the uplift limit state (UPL) criteria as given by the equation

$$V_{dstd} \leq G_{stbd} + R_d \qquad\qquad (6.4a)$$

where

$$V_{dstd} = G_{dstd} + Q_{dstd} \qquad\qquad (6.4b)$$

and

$V_{dstd}$ is the design value of the permanent destabilising vertical action on the substructure
$G_{stbd}$ is the design value of the permanent stabilising vertical actions
$G_{dstd}$ is the design value of the permanent destabilising actions
$Q_{dstd}$ is the design values of the variable actions
$R_d$ is any additional resistance to uplift

The EC7 partial factors for actions for the ultimate UPL are shown in Table 6.1. For verification of the uplift resistances of the soil surrounding the block, and of the pull-out resistances of the piles in the group, where derived by calculations using soil parameters, the partial factors shown in Table 6.2 are used.

In allowable stress calculations, a safety factor of 2 would be used with Equation 6.3 to allow for the possible weakening of the soil around the pile group caused by the method of installation. For long-term sustained loading, a safety factor of 2.5–3 would be appropriate.

Table 6.1 Partial factors for actions ($\gamma_F$) for UPL verifications (EC7 Table A.NA.15)

| Action | Symbol | Value |
|---|---|---|
| Permanent | | |
| Unfavourable[a] | $\gamma_{Gdst}$ | 1.1 |
| Favourable[b] | $\gamma_{Gstb}$ | 0.9 |
| Variable | | |
| Unfavourable[a] | $\gamma_{Qdst}$ | 1.5 |
| Favourable[b] | $\gamma_{Qstb}$ | 0 |

[a] Destabilising.
[b] Stabilising.

*Table 6.2* Partial factors ($\gamma_M$) for soil parameters and resistances ($\gamma_R$) (EC7 Table A.NA.16)

| Soil parameter | Symbol | Values |
|---|---|---|
| Angle of shearing resistance[a] | $\gamma'_\phi$ | 1.25 |
| Effective cohesion | $\gamma'_c$ | 1.25 |
| Undrained shear strength | $\gamma_{cu}$ | 1.40 |
| Tensile pile resistance ($\gamma_R$)[b] | $\gamma_{st}$ | 2.0 |
| Anchorage resistance ($\gamma_R$) | $\gamma_a$ | 1.40[c] |

[a] This value is applied to tan $\phi'$.
[b] The shaft resistance partial factor in tension $\gamma_{st}$ depends on the type of pile and pile testing (as shown in Tables 4.3 through 4.5).
[c] Larger values of $\gamma_a$ should be used for non-prestressed anchors consistent with those for tension piles (as shown in Tables 4.3 through 4.5).

### 6.2.3 Piles with base enlargements

When bored piles are constructed in clay soils, base enlargements can be formed to anchor the piles against uplift. The enlargements are made by the belling tools described in Section 3.3.1. The size and stability of an enlargement formed in coarse-grained soil is problematical, whether bored with or without a support fluid. Full-scale loading tests are essential to prove the reliability of the bentonite method for any particular site. Reliable predictions cannot be made of the size and shape of base enlargements formed by hammering out a bulb of concrete at the bottom of a driven and cast-in-place pile as described in Section 2.3.2. End enlargements formed on precast concrete or steel piles, although providing a substantial increase in compressive resistance when driven to a dense or hard stratum, do not offer much uplift resistance since a gap of loosened soil is formed around the shaft as the pile is driven down.

In the case of bored piles in fine-grained soils installed using belling tools, resistance to uplift loading provided by the straight-sided portion of the shaft is calculated over the depth *H* in Figure 6.8 minus the overall depth of the under-ream. Failure under short-term loading takes place in undrained shear on the pile to clay interface. The mobilised resistance should

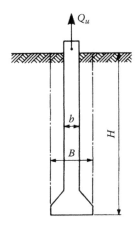

*Figure 6.8* Uplift of single pile with base enlargement in fine-grained soil ($\phi = 0$).

take into account the effects of installation as described in Section 4.2.3. Uplift resistance of the projecting portion of the enlarged base is assumed to be provided by compression resistance of the soil overburden.

Resistance to long-term uplift loading on piles in fine-grained soils is calculated by effective stress methods as described for clayey sands in the following paragraphs.

Meyerhof and Adams[6.2] investigated the uplift resistance of a circular plate embedded in a partly clayey $(c - \phi)$ soil and established the equation

$$Q_u = \pi cBH + s \times 0.5\pi \times \gamma \times B(2D - H)HK_u \tan\phi + W \qquad (6.5)$$

where

$Q_u$ is the ultimate uplift resistance of the plate
$B$ is the diameter of the plate
$H$ is the height of the block of soil lifted by the pile (Figure 6.9)
$c$ is the cohesive strength of the soil
$s$ is a shape factor (see below)
$\gamma$ is the density of the soil (the submerged density being taken below groundwater level)
$D$ is the depth of the plate
$K_u$ is a coefficient obtained from Figure 6.9
$\phi$ is the angle of shearing resistance of the soil
$W$ is the weight of the soil resisting uplift by the plate

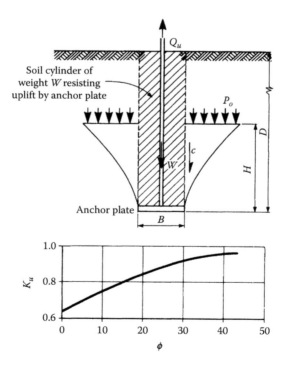

Figure 6.9 Uplift of circular plate in partly clayey $(c - \phi)$ or sandy $(c = 0)$ soil. (After Meyerhof, G.G. and Adams, J.I., Can. Geotech. J., 5(4), 225, 1968.)

If Equation 6.5 is adapted to a pile with an enlarged base, the weight of the pile is taken in conjunction with the weight of the soil when calculating $W$ (i.e. $G_{stbk}$).

It will be noted that for deeply embedded plates or pile enlargements, $H$ does not extend up to ground level and its value can be obtained from tests made by Meyerhof which gave the following results:

| $\phi$ | 20° | 25° | 30° | 35° | 40° | 45° | 48° |
|---|---|---|---|---|---|---|---|
| $H/B$ | 2.5 | 3 | 4 | 5 | 7 | 9 | 11 |

The shape factor $s$ for deep foundations (including piles) is equal to $1 + mH/B$, where $m$ depends on the angle of shearing resistance $\phi$ of the soil. Meyerhof's values of $m$ and the maximum permissible values of the shape factor are as follows:

| $\phi$ | 20° | 25° | 30° | 35° | 40° | 45° | 48° |
|---|---|---|---|---|---|---|---|
| $m$ | 0.05 | 0.1 | 0.15 | 0.25 | 0.35 | 0.50 | 0.60 |
| max. $s$ | 1.12 | 1.30 | 1.60 | 2.25 | 3.45 | 5.50 | 7.60 |

The value of $Q_u$ calculated from Equation 6.5 must not exceed the combined resistance of the enlarged base (considered as a buried deep foundation) and the pile shaft friction. These components are calculated as described in Chapter 4.

The shaft length is taken as the overall depth of the pile, from which the depth of the enlargement and any allowance made for the shrinkage of the soil away from the pile at the ground surface are deducted. Where piles in clay have to carry long-term sustained uplift loading, and the ratio of the depth of these piles to the width of the enlarged base is less than 5, the uplift resistance, as calculated by Equation 6.5 or the methods in Chapter 4, should be reduced by one-half.

## 6.2.4 Anchoring piles to rock

Rock anchors are provided for tension piles when the depth of soil overburden is insufficient to develop the required uplift resistance on the pile in shaft friction. In weak rocks such as chalk or marl, it is possible to drive piles into the rock or to drill holes for bored piles so that the frictional resistance can be obtained on the pile shaft at its contact surface with the rock. However, driving piles into a strong rock achieves only a small penetration and so shatters the rock that no worthwhile frictional resistance can be obtained. The cost of drilling into a strong rock to form a bored pile is not usually economical compared with that of drilling smaller and deeper holes for anchors as described below, although drilling in large-diameter piles to carry ship-berthing forces in marine structures is sometimes undertaken (see Section 8.1).

Anchorages in rock are formed after driving an open-ended tubular pile to seat the toe of the pile into the rock surface. The pile must not be driven too hard at this stage as otherwise the toe will buckle, thus preventing the entry of the cleaning-out tools and the anchor drilling assembly. If a bored pile is to be anchored, the borehole casing is drilled below rock level to seal off the overburden. All the soil within the piling tube is cleaned out by baling, washing or *airlifting*, and drill pipes with centralisers are lowered down to the rock level. The anchor hole is then drilled to the required depth and the cuttings cleaned by reverse-circulation drilling or air lift. The anchor, which can consist of a high-tensile steel bar or a stranded cable, is fed down the hole and grout injected at the base to fill the hole. (Figure 6.10 shows a *doubly*

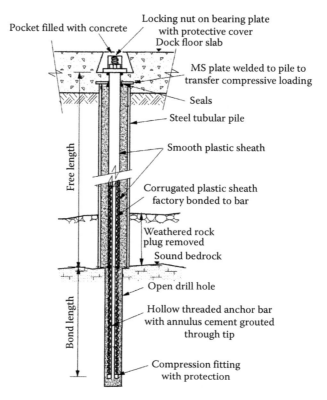

*Figure 6.10* Doubly protected, hollow threaded bar forming stressed tendon in tubular steel pile supporting dock floor.

*protected* bar anchor.) BS 8081 Code of practice for ground anchors provides comprehensive information on the requirements for corrosion protection of anchor bars and strand. The anchors are usually factory fitted with the necessary grout injection tube and plastic sheathing over the bond length (corrugated) and the tendon 'free length.' Where the anchors are stressed, the bar or cable is carried up to the top of the pile or pile cap to which the stress from the anchor is transferred by a stressing head and jack. Care is needed to ensure protection of the anchor head and fittings. In marine piles, the space around the sheath is usually left void to allow for flexing under lateral load.

Unstressed or 'dead' anchors can consist of steel tubes installed by drilling them down into rock. On reaching the required depth, grout is pumped down the drill pipe, through the drill bit, and fills the annulus between the anchor tube and the rock. A sealing plate prevents the grout from entering the space between the anchor tube and the drilling pipe, as shown in Figure 6.11. The grout is allowed to fill the pile to the height necessary to cover the top of the anchor tube, so as to protect it from corrosion and to serve as the medium transferring the uplift load from the pile on to the anchor. Where large uplift loads are carried, the transfer of load is effected by welding mild steel strips onto the interior surface of the pile and the exterior of the anchor tube to act as shear keys, as described in the following Section. The drill bit is left in place at the bottom of the tube where it acts as a compression fitting, but the drilling rods are disconnected at a special back-off coupling.

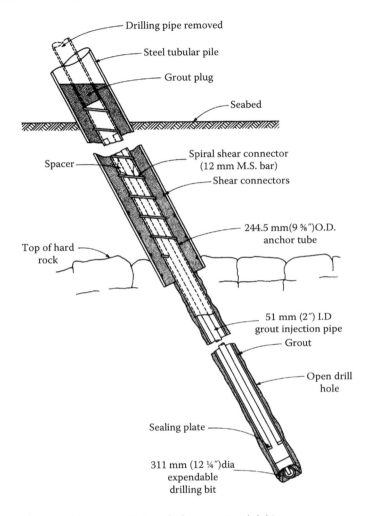

Figure 6.11 Dead anchor in raking steel tubular pile for mooring dolphin.

## 6.2.5 Uplift resistance of drilled-in rock anchors

The resistance to pull-out of anchors drilled and grouted into rock depends on five factors, each of which must be separately evaluated. They are as follows:

1. The stress in the steel forming the anchor
2. The bond stress between the anchor and the grout
3. The bond stress between the grout and the rock
4. The dead weight of the mass of rock and any overlying soil which is lifted by the anchor, if prior failure does not occur due to the preceding three factors
5. The dead weight of the mass of rock and any soil overburden which is lifted by a group of closely spaced anchors

The bond stress between the anchor and the grout depends on the compressive strength of the grout, the amount of keying or roughening given to the steel surface, the diameter of the anchor and the influence of the bottom compression fitting in short anchors. The anchor diameter is of significance since with large-diameter high-capacity anchors, there is an appreciable diminution of diameter caused by the inward radial strain that occurs under the tensile load. This creates a tendency to weaken the bond between the steel and the grout.

Specifications for anchorage materials and grouting cements and recommendations for the bond strength at the grout to tendon interface are given in BS 8081. Clause 6.3.2 gives the following guidance for ultimate bond strength:

| | |
|---|---|
| Plain bar | Not greater than 1 N/mm² |
| Clean strand or deformed bar | Not greater than 2 N/mm² |
| Locally noded strand | Not greater than 3 N/mm² |

The strength of the steel to grout bond given in EC2-1-1 depends on the design compressive strength and hence the design tensile strength of the concrete: $f_{ctd} = \alpha_{ct} f_{ctk\,0.05}/\gamma_C$. The value of $f_{ctk\,0.05}$ is taken from Table 3.1 for the class of grout and $\gamma_C$ is the concrete material factor (see Section 7.10.1). Clause 8.4.2 in EC2 gives factors to be applied to $f_{ctd}$ to determine the bond strength for ribbed reinforcing steel.

Special grouts are formulated for injection into the annulus between an anchor and a tubular pile or between a pile and a surrounding sleeve, to reduce bleeding or shrinkage and increase the rate of strength gain. For example, a grout with a water/cement ratio of about 0.5 and an appropriate plasticiser can attain compressive strengths of the order of 24 N/mm² at 3 days. In these conditions, the annulus should be kept to a minimum to reduce potential shrinkage effects. For marine pile connections, a mix consisting of 100 parts of API Oilwell B cement to 34 parts of *seawater* developing a characteristic cube strength of about 22 N/mm² at 3 days is extensively used on oil platforms (notwithstanding the comments on the use of fresh water for grouts in BS EN ISO 19902 Clause 19.6.1). The transfer of load from a pile to the sleeve can be significantly improved through shear keys formed on the inner surface of the sleeve and outer surface of the pile, in the form of beads of weld metal or welded-on steel strips.

The ultimate grout to steel bond strength on the surface of tubular piles within pile sleeves, either with or without mechanical shear connectors, can be calculated using the equations included in BS EN ISO 19902 given below. Clause 15 of this document is based on the major research programme described in Refs 6.3a through c and was originally produced in the UK Department of Energy Guidance and in HSE Report 2001/016. The equations were refined in a later paper by Harwood et al.[6.4]

The bond strength of the pile–sleeve connection, now described in Clause 15.1 of ISO 19902 as the 'design interface transfer strength' of the grout, $f_d$, is defined as

$$f_d = \frac{f_g k_{red}}{\gamma_{Rg}} \tag{6.6}$$

where
   $k_{red}$ is a reduction factor for movement during grout setting (see below)
   $\gamma_{Rg} = 2.0$, the partial resistance factor

The lesser of the representative interface transfer strength for sliding at the grout–steel interface $f_{g\,sliding}$ (Equation 6.7a) and the representative interface transfer strength for shear for grout matrix failure $f_{g\,shear}$ (Equation 6.7b) are then applied in Equation 6.6:

$$f_{g\,sliding} = C_p\left[2 + 140\left(\frac{h}{s}\right)^{0.8}\right] \times K^{0.6}(f_{cu})^{0.3} \qquad (6.7a)$$

$$f_{g\,shear} = \left[0.75 - 1.4\left(\frac{h}{s}\right)\right](f_{cu})^{0.5} \qquad (6.7b)$$

where
  $f_{cu}$ is the *unconfined cube* compressive strength in N/mm²
  $K$ is the radial stiffness factor (see below)
  $C_p$ is the scale factor for the diameter of the pile (see below)
  $h$ is the minimum shear key outstand in mm
  $s$ is the nominal shear key spacing in mm

The stiffness factor is given by

$$K = \frac{1}{m}\left(\frac{D}{t}\right)_g^{-1} + \left[\left(\frac{D}{t}\right)_p + \left(\frac{D}{t}\right)_s\right]^{-1} \qquad (6.8a)$$

where
  $m$ is the modular ratio of steel to grout (18 in the absence of other data for long term, i.e. 28 days or more)
  $D$ is the outside diameter
  $t$ is the wall thickness

The suffixes $g$, $p$ and $s$ refer to the grout, pile and sleeve, respectively.
The scale factor for $D_p \le 1000$ mm is given by

$$C_p = \left(\frac{D_p}{1000\text{ mm}}\right)^2 - \left(\frac{D_p}{500\text{ mm}}\right) + 2 \qquad (6.8b)$$

The scale factor for $D_p > 1000$ is

$$C_p = 1.0 \qquad (6.8c)$$

The transfer stress at the pile is given by

$$\sigma_a = \frac{P}{\pi D_p L_e} \qquad (6.9)$$

where
  $P$ is the largest force on the connection from factored (design) actions
  $L_e$ is the grouted connection length, dependent on whether the annulus is sealed by a grout plug (length not to be included), an allowance for slump and shear key spacing

The above equations are subject to limitations on grout strength and geometric dimensions:

- Grout strength: $20 \leq f_{cu} \leq 80$ N/mm² (but <20 N/mm² strength and sand–cement grouts may be considered subject to testing)
- $1.5 \leq w/h \leq 3.0$ (where $w$ is the width of shear key)
- $0.0 \leq h/s \leq 0.10$
- $20 \leq D_p/t_p \leq 40$
- $30 \leq D_s/t_s \leq 140$
- $h/D_p \leq 0.012$
- $D_p/s \leq 16$
- $1 \leq L_e/D_p \leq 10$

The limits for $C_p$ and $K$ shall be

- $C_p \leq 1.5$
- $K \leq 0.02$

Shear keys, where needed, must be either continuous hoops or a continuous helix, be in contact with the grout and be present on both pile and sleeve with outstand and spacing the same. For driven piles, the shear keys should cover a sufficient length to ensure contact with the grout after driving. Where helical keys are used, the representative interface transfer strengths in Equations 6.7a and b need to be reduced by a factor of 0.75 and the following additional allowances applied:

- For relative axial movement between tubular steel members of 0.035% of $D_p$, then $k_{red} = 1.0$
- For relative axial movement between tubular steel members of between 0.035% and 0.35% of $D_p$ and for $h/s \leq 0.6$, then $k_{red} = 1.0-0.1(h/s)f_{cu}$

Annex 15.1.5.3 of BS EN ISO 19902 gives guidance on movement allowances. The Standard also requires a strength check (Clause 15.1.6) and a fatigue assessment in certain conditions (Clause 15.1.7). The API recommendations[4.15] for the design of steel pile–sleeve connections are similar to those mentioned above but have different provisos as to their application.

The bond stress between grout and rock depends on the compressive strength of the intact rock, the size and spacing of joints and fissures in the rock, the keying of the rock affected by the drilling bit and the cleanliness of the rock surface obtained by the flushing water. The size of the drill hole and the size of the annular space between the anchor and the wall of the hole are also important. As noted for the pile–sleeve connection, the annulus between anchor and rock should be minimised in order to reduce shrinkage effects and the consequent weakening of the grout to rock bond. Typically, the diameter of the drill hole will be 1.3–2 times the anchor diameter and a 'non-shrink' grout will be used. The smaller the annulus and the shorter the bonded length, the higher is the compressive stress in the grout and hence its ability to lock into the surrounding rock. A compression fitting at the bottom of the anchor will also increase the compressive stress in the grout column. The value of the bond between grout and rock will be small if the rock softens to a slurry under the action of drilling and flushing. This occurs with chalk, weathered marl and weathered clayey shales. Some observed values of bond stress at failure for drill holes of up to 75 mm in diameter are given in Table 6.3.

If the bond stress between the grout and the rock is a critical factor in designing the anchors, the required resistance should be obtained by increasing the length of the anchor rather than

*Table 6.3* Examples of bond stress between grout and rock

| Type of rock | Bond stress between grout and rock at failure (N/mm²) | Reference |
|---|---|---|
| Chalk (Grade I) | 0.21 | Littlejohn[6.5] |
| Chalk (Grade III) | 0.80 | Littlejohn[6.5] |
| Keuper Marl (Zones I and II) | 0.17–0.25 | Littlejohn[6.5] |
| Chalk | 1.0 | Hutchinson[6.6] |
| Weathered shaley slate | 0.27 | Unpublished[6.7] |
| Hard shaley slate | 1.0–1.7 | Unpublished[6.7] |
| Billings shale (Ottawa) | 3.0 | Freeman et al.[6.8] |
| Sandstone | >0.6 | Unpublished[6.7] |

by increasing the diameter of the drill hole, for the reasons already stated. However, in certain conditions, it is possible that the bond stress will not be reduced in direct proportion to the increase in bond length. This is because of the possibility of progressive failure in a hard rock, similar to the sidewall slip in rock-socketed piles (Section 4.7). The maximum stretch in the anchor occurs at the top of the bonded length, and this may cause local bond failure with the rock or the pulling out of a small cone of rock (Figure 6.12a). Progressive failure then extends down to the bottom of the anchor. By limiting the bond length and sheathing the tendon free length within the rock as described in BS 8081 (Figure 6.12b), the pulling out of a cone of rock is prevented and the column of grout is compressed and acts in bond resistance with the rock.

The pull-out resistance of the mass of rock (as shown in Figure 6.12b) is the final criterion for the performance of an individual anchor. The actual shape of the mass of rock lifted depends on the degree of jointing and fissuring of the rock and the inclination of the bedding planes. Various forms of failure are sketched in Figure 6.13. A cone with a half angle

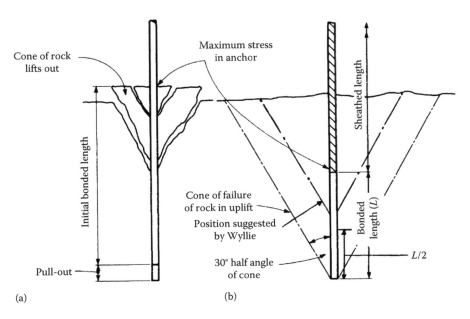

*Figure 6.12* Pull-out of cone of rock: (a) fully bonded anchor; (b) upper part sheathed, lower part bonded.

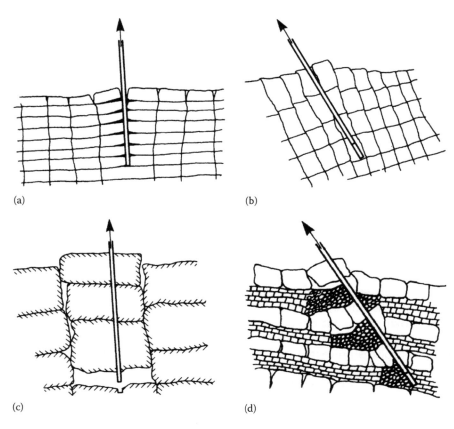

*Figure 6.13* Pull-out failure in rock anchors: (a) horizontally bedded rock (thinly bedded); (b) steeply inclined bedding planes with anchor raked in direction of bedding joints; (c) horizontally bedded rock; (d) alternating thinly and thickly bedded rocks.

of 30° gives a conservative value for the pull-out resistance and represents conditions for a heavily jointed or shattered rock. Wyllie[(4.54)] suggests that the base of the cone should be taken at the midpoint of the bonded length (Figure 6.12b), but this arrangement would not apply for the case of a compression fitting at the bottom of the anchor. Because shear at the interface between the surface of the cone and the surrounding rock is neglected, a factor of unity can be taken on the weight of the rock cone, where the rock is bedded horizontally or at moderate angles from the vertical (Figure 6.13a). Where the bedding planes or other joint systems are steeply inclined, as shown in Figure 6.13b through d, either an increased factor should be allowed or an attempt should be made to calculate the uplift resistance based on the understanding the behaviour of the rock mass. The submerged weight should be taken for rock below groundwater level or below the sea. The uplift resistance of the cylinder or cone of soil overburden above the rock cone can be calculated as described in Section 6.2.3. The dimensions $B$ and $H$ in Equations 6.3 and 6.5 are as shown in Figure 6.14. Shaft friction on the pile above the anchor does not operate to resist uplift for this mode of failure. The mode of failure of a group of anchors, assuming no failure occurs in the bond between grout and steel or grout and rock, is shown in Figure 6.15. The piles and anchors can be splayed out as shown in Figure 6.16 to increase the volume of rock bounded by the group.

The calculation of the volume of rock $V_c$ in a single cone with a half angle of 30° at various angles of inclination $\theta$ to a horizontal rock surface can be performed with the aid of the

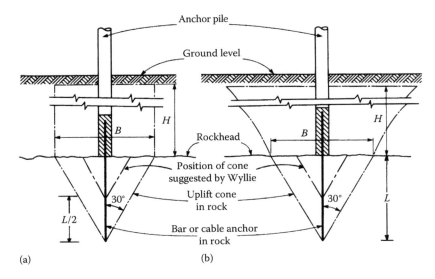

Figure 6.14 Approximate method of calculating ultimate uplift resistance of rock anchors with soil over-burden: (a) clay overburden and (b) granular soil overburden.

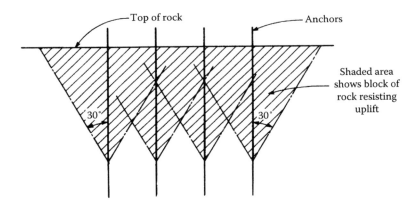

Figure 6.15 Failure condition at group of anchors in rock.

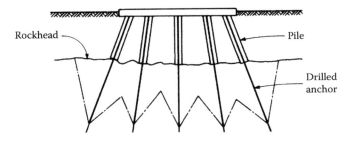

Figure 6.16 Splaying anchors in group to increase uplift resistance.

curve for $V_c/L^3$ in Figure 6.17a. The effect of overlapping cones of rock in groups of vertical or raking anchors can be calculated by reference to Figure 6.17a and b. These charts enable the overlapping volumes $\Delta V_m$ and $\Delta V_n$ to be calculated for a group of anchors arranged on a rectangular grid. They are not applicable to a diagonal (i.e. staggered) pattern. All the anchors in the group are assumed to be arranged at the same angle of inclination to the horizontal and the charts are based on a cone with a 30° half angle. The charts are not valid if the sum of $(P/n)^2$ and $(S/m)^2$ (as defined in the Figure 6.17a and b) is less than 4 when composite overlapping occurs. In such a case, the total volume acting against uplift needs to be estimated from the geometry of the system.

Because of the various uncertainties in the design of rock anchors as described earlier, it is evident that it is desirable to adopt post-tensioned anchors. Every anchor is individually stressed and hence checked for pull-out resistance, at a proof load of 1.5 times the design load. However, it should be noted that the technique of stressing anchors by jacking against the reaction provided by the pile does not check the pull-out resistance of the cone of rock: this is clear from Figure 6.10. The resistance offered by the mass of rock can be tested only by providing a reaction beam with bearers sited beyond the influence of the conjectural rock cone. Tests of this description are very expensive to perform and it is usual to avoid them by adopting conservative assumptions for the dimensions of the cone and applying a safety factor to the calculated weight if required.

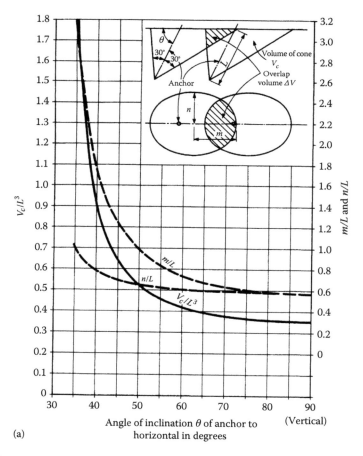

(a)

*Figure 6.17* (a) Chart for calculating volumes of single or overlapping cones with vertical or inclined axes.

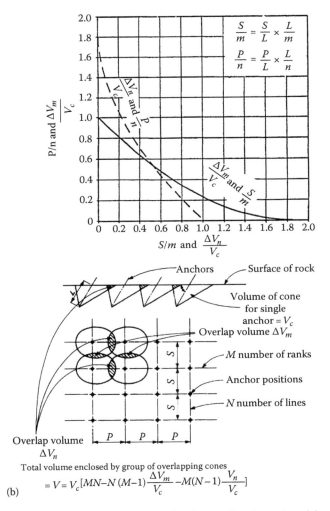

(b)

Total volume enclosed by group of overlapping cones

$$= V = V_c [MN - N(M-1)\frac{\Delta V_m}{V_c} - M(N-1)\frac{V_n}{V_c}]$$

*Figure 6.17* (continued) (b) Charts for calculating total volume of rock enclosed by groups of anchors arranged on rectangular grid pattern.

EC7 Clause 8.1.1(4)P requires tension piles to be designed as described in Section 7 of the code. EC7 is not appropriate for anchorages formed by grouting tendons into drilled holes and requires their design and installation to be in accordance with BS EN 1537.

BS EN 1537 defines temporary anchors as those with a design life of less than 2 years and permanent anchors as those with a design life of 2 years or more. The ultimate limit states to be considered are the same as those listed at the beginning of this Section. In addition, EC7 and BS EN 1537 require design measures to check the following:

1. Structural failure of the anchor head
2. Distortion or corrosion of the anchor head
3. Loss of anchorage force by excessive displacement of the anchor head or by creep and relaxation
4. Failure or excessive distortion on parts of the structure due to the applied anchorage force
5. Interaction of groups of anchorages with the ground and adjoining structures

BS EN 1537 requires construction steel in anchors to be in accordance with EC2 and EC3. Prestressing steel used for tendons is to comply with the information on prestressing tendons now included in EC2-1-1. Cement grouts are to comply with BS EN 445, 446 and 447; resin grouts may be used subject to appropriate tests for the particular application. Admixtures and inert fillers are permitted to be used in grout mixes provided that they do not contain materials liable to cause corrosion of the tendons. Corrosion protection of tendons using plastic sheathing of tendons should be detailed as shown in BS 8081, but temporary anchors need not be sheathed provided that they have protection from corrosion suitable for their design life of 2 years.

Drilling for anchorages is required to be within a deviation limit of not more than 1/30 of the anchor length as Clause 8.1.1 of BS EN 1537. The procedure for making permeability tests in the drilled holes using water and grout to investigate the possibility of grout loss is described.

BS EN 1537 gives detailed information on the procedure for conducting three types of test on an anchorage, including the interpretation of the results, monitoring of behaviour and record keeping. Items such as health and safety and environmental matters are dealt with. The tests are as follows:

1. Investigation test
2. Suitability test
3. Acceptance test

The investigation test is made on expendable anchors to establish the ultimate resistance of the anchor at the grout–ground interface and to determine the characteristics of the anchorage in the working load range.

The suitability test is made to confirm that a particular anchorage system will be adequate for the ground conditions on the project site. In the case of permanent anchorages, the test is made with sheathed tendons and is required to establish acceptable limits of creep or load loss at the proof and lock-off loads. In cases where no investigation tests are made, the suitability test is undertaken on expendable anchors to demonstrate anchorage characteristics and to provide criteria for acceptance tests.

The acceptance test is made at the project construction stage on each working anchor with the following requirements:

1. To demonstrate that the proof load can be sustained
2. To determine the apparent free length
3. To ensure that the lock-off load is at the design load level, excluding friction
4. To determine creep or load loss characteristics at the serviceability limit state where necessary

For the purpose of design verification, characteristic values of anchorage resistance $R_{ak}$ obtained from pull-out tests are divided by the partial factor $\gamma_a$ to determine the design resistance, so that $R_{ad} = R_{ak}/\gamma_a$. Values of $\gamma_a$ are given in EC7 Table A.NA12 where $\gamma_{at}$ for temporary anchors and $\gamma_{ap}$ for permanent anchors are both unity for the R sets. Correlation factors ($\xi$) depending on the number of tests are not provided in the EC7 NA, but it is specified that at least three suitability tests should be made for each distinct condition of ground and structure.

Where $R_{ad}$ is derived by calculation, the design approach DAI as described in Section 4.1.4 needs to be used, with verification of stability against uplift of the structure by application of the UPL partial factors as described in Section 6.2.2 for friction piles. As the model factor for the 'SLS force' noted in EC7 Clause 8.6(4) is not provided in the U.K. NA (Clause A6.6), it may be necessary for the designer to agree a value with the structural engineer and client to 'ensure the resistance of the anchor is sufficiently safe.'

To verify the serviceability limit state of a structure restrained by prestressed anchorages, the tendons are regarded as elastic prestressed springs. Analysis needs to consider the most adverse combinations of minimum and maximum anchorage stiffness and minimum and maximum prestress. To prevent damaging effects of interaction between close-spaced groups of anchors, EC7 and BS EN 1537 require tendons to be spaced at least 1.5 m apart.

## 6.3 SINGLE VERTICAL PILES SUBJECTED TO LATERAL LOADS

The ultimate internal resistance of a vertical pile and the deflection of the pile are complex matters involving the interaction between a structural element and the soil. Taking the case of a vertical pile unrestrained at the head, the lateral loading on the pile head is initially carried by the soil close to the ground surface. At a low loading, the soil compresses elastically but the movement is sufficient to transfer some pressure from the pile to the soil at a greater depth. At a further stage of loading, the soil yields plastically and the pile transfers its load to greater depths. A short rigid pile unrestrained at the top and having a length-to-width ratio of less than 10 to 12 (Figure 6.18a) rotates, and passive resistance develops above the toe on the opposite face to add to the resistance of the soil near the ground surface. Eventually, the rigid pile will fail by rotation when the passive resistance of the soil at the head and toe are exceeded. The short rigid pile restrained at the head by a cap or bracing will fail by translation in a similar manner to an anchor block which fails to restrain the movement of a retaining wall transmitted through a horizontal tied rod (Figure 6.18b).

The failure mechanism of an infinitely long pile is different. Theoretically, the passive resistance to yielding provided by the soil below the yield point can be considered infinite and rotation of the pile cannot occur, the lower part remaining vertical while the upper part deforms to a shape shown in Figure 6.19a. Failure takes place when the pile yields at the point of maximum bending moment, and for the purpose of analysis, a plastic hinge capable of transmitting shear is assumed to develop at this point. In the case of a long pile restrained at the head, high bending stresses develop at the point of restraint, for example just beneath the pile cap, and the pile may yield at this point (Figure 6.19b).

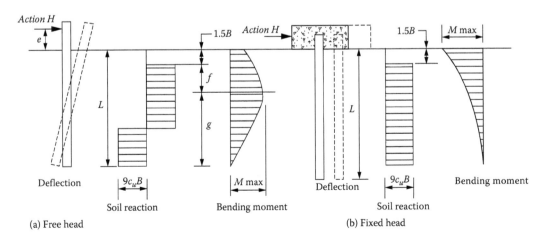

*Figure 6.18* Short vertical pile under horizontal load in fine-grained soil. Soil reactions and bending moments after Broms. (a) Free head pile and (b) Fixed head pile.

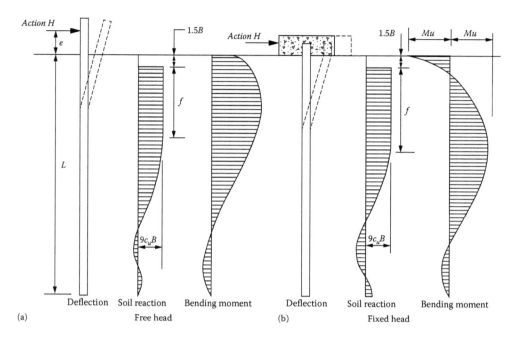

*Figure 6.19* Long vertical pile under horizontal load in fine-grained soil. Soil reactions and bending moments after Broms. (a) Free head pile and (b) Fixed head pile.

The pile head may move horizontally over an appreciable distance before rotation or failure of the soil–pile system occurs. Therefore, having calculated the ultimate resistance and applied the appropriate partial factor, it is still necessary to check that the serviceability limit of the structure supported by the pile is not exceeded.

There are many interrelated factors which govern the behaviour of laterally loaded piles. The dominant one is the pile stiffness, which influences the deflection and determines whether the failure mechanism is one of the rotations of a short rigid element or is due to flexure followed by the failure in bending of a long pile. The type of loading, whether sustained (as in the case of earth pressure transmitted by a retaining wall) or alternating (say, from reciprocating machinery) or pulsating (as from the traffic loading on a bridge pier), influences the behaviour of the soil. External influences such as scouring around piles at seabed level, or the seasonal shrinkage of clay soils away from the upper part of the pile shaft, affect the resistance of the soil at a shallow depth. Scour or *gapping* of stiff clay around the pile at the seabed may also be attributed to cyclic loading. The to-and-fro movement of the pile forces water up the sides of the pile producing turbulence as the gap is closed. Remedial measures are difficult, with pea gravel placement being the most effective. The problem is less likely in coarse sand seabed conditions.

Methods of calculating ultimate resistance and deflection under lateral loads are presented in the following sections of this chapter. No attempt is made to give their complete theoretical basis. Various simplifications have been necessary in order to provide simple solutions to complex problems of soil–structure interaction, and the limitations of the methods are stated where these are particularly relevant. Most practical calculations are processes of trial and adjustment, starting with a very simple approach to obtain an approximate measure of the required stiffness, and embedment depth of the pile. The process can then be elaborated to some degree to narrow the margin of error and to provide the essential data for calculating bending moments, shearing forces and deflections at the applied load. In general, very elaborate calculation processes are not justified, because of the non-homogeneity of most natural soil deposits and the disturbance to the soil caused by installing piles. Some of these significant factors can be reproduced in the constitutive models in commercial computer programs, but several simplifying assumptions have to be made.

EC7 Section 7.7 requires the design of transversely loaded piles to be consistent with the design rules previously described in Chapter 4 for piles under compression loading:

$$F_{trd} \leq R_{trd} \tag{6.10a}$$

$$F_{trd} = \gamma_G H_{Gk} + \Sigma \gamma_Q H_{Qk} \tag{6.10b}$$

where

$F_{trd}$ is the design transverse load

$H_{Gk}$ and $H_{Qk}$ are the characteristic values of the permanent and variable components of the horizontal load, $H_u$ (as below)

$\gamma_G$ and $\gamma_Q$ are the relevant action factors

Failure mechanisms to be considered are failure of a short rigid pile by rotation or translation, where the resistance to $F_{trd}$ is governed by the ground strength, and failure of a long slender pile in bending producing local yield and displacement of the soil near the pile head. The assessment of lateral loading on both driven and bored piles in soft clay needs to take account of adjacent piling or imposed loading, possible adjacent excavation and settlement of nearby structures resulting in lateral soil movement and possible curvature of the pile. The reinforcement cage pushed into the concrete of a CFA pile must be designed to resist the bending and be long enough to ensure anchoring in the pile concrete.

Pile load tests, when undertaken as a means of determining the transverse resistance, are not generally required to be taken to failure, but the magnitude and line of action of the test load should conform to the design requirements. The effects of interaction between piles in groups and fixity at the pile head are required to be considered.

Where transverse resistance is determined by calculation, the method based on the concept of a modulus of horizontal subgrade reaction as described in Section 6.3.1 is permitted. The structural rigidity of the connection of the piles to the pile cap or substructure is to be considered as well as the effects of load reversals and cyclic loading.

For any important foundation structure which has to carry high or sustained lateral loading, it is advisable to make field loading tests on trial piles having at least three different shaft lengths, in order to assess the effects of embedment depth and structural stiffness. For less important structures, or where there is previous experience of pile behaviour to guide the designer, it may be sufficient to make lateral loading tests on pairs of working piles by jacking or pulling them apart. These tests are rapid and economical to perform (see Section 11.4.4) and provide a reliable check that the design requirements have been met.

### 6.3.1 Calculating the ultimate resistance of short rigid piles to lateral loads

The first step is to determine whether the pile will behave as a short rigid unit with resistance governed by ground strength alone or as an infinitely long flexible member dependent on both pile and ground strength. This is done by calculating the stiffness factors $R$ and $T$ for the particular combination of pile and soil. The stiffness factors are governed by the flexural stiffness ($EI$ value) of the pile and the compressibility of the soil. The latter is expressed in terms of a *soil modulus*, which is not constant for any soil type but depends on the width of the pile $B$ and the depth of the particular loaded area of soil being considered. The soil modulus $k$ has been related to Terzaghi's concept of a modulus of horizontal

*Table 6.4* Relationship between modulus of subgrade reaction ($k_1$) and undrained shearing strength of stiff over-consolidated clay

| Consistency | Firm to stiff | Stiff to very stiff | Hard |
|---|---|---|---|
| Undrained shear strength ($c_u$) (kN/m²) | 50–100 | 100–200 | >200 |
| Range of $k_1$ (MN/m³) | 15–30 | 30–60 | >60 |

subgrade reaction[4.84]. In the case of a stiff over-consolidated clay, the soil modulus is generally assumed to be constant with depth. For this case,

$$\text{Stiffness factor } R = \sqrt[4]{\frac{EI}{kB}} \text{ (in units of length)} \qquad (6.11)$$

where
  E is the elastic modulus of the material forming the pile shaft
  I is the moment of inertia of the cross section of the pile shaft

For short rigid piles, it is sufficient to take k in the above equation as equal to the Terzaghi modulus $k_1$, as obtained from load/deflection measurements on a 305 mm square plate. It is related to the undrained shearing strength of the clay, as shown in Table 6.4.

For most normally consolidated clays and for coarse-grained soils, the soil modulus is assumed to increase linearly with depth, for which

$$\text{Stiffness factor } T = \sqrt[5]{\frac{EI}{n_h}} \text{ (in units of length)} \qquad (6.12)$$

where

$$\text{Soil modulus } K = n_h \times \frac{x}{B} \qquad (6.13)$$

and x is the depth below ground level as shown in Figure 6.21.

Values of the coefficient of modulus variation $n_h$ were obtained directly from lateral loading tests on instrumented piles in submerged sand at Mustang Island, Texas. The tests were made for both static and cyclic loading conditions and the values obtained, as quoted by Reese et al.[6.9], were considerably higher than those of Terzaghi. The investigators recommended that the Mustang Island values should be used for pile design and these are shown together with the Terzaghi values in Figure 6.20[6.10].

Other observed values of $n_h$ are as follows:

Soft normally consolidated clays: 350–700 kN/m³
Soft organic silts: 150 kN/m³

Having calculated the stiffness factors R or T using estimates of $n_h$ and k appropriate to ground conditions, the criteria for behaviour as a short rigid pile or as a long elastic pile are related to the embedded length L as follows:

| Pile type | Soil modulus | |
|---|---|---|
| | Linearly increasing | Constant |
| Rigid (free head) | $L \leq 2T$ | $L \leq 2R$ |
| Elastic (free head) | $L \geq 4T$ | $L \geq 3.5R$ |

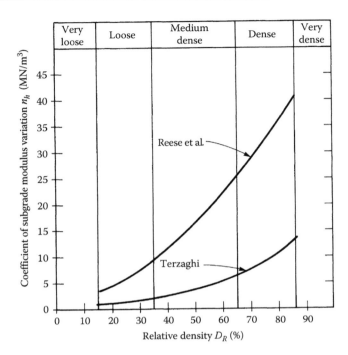

*Figure 6.20* Relationship between coefficient of modulus variation and relative density of sands. (After Garassino, A. et al., Soil modulus for laterally-loaded piles in sands and NC clays, *Proceedings of the Sixth European Conference*, ISSMFE, Vienna, Austria, Vol. I (2), pp. 429–434, 1976.)

The Brinch Hansen method[6.11] can be used to calculate the ultimate lateral resistance of short rigid piles. The method is a simple one which can be applied both to uniform and layered soils and is well suited to spreadsheet calculations. It can also be applied to longer semi-rigid piles to obtain a first approximation of the required stiffness and embedment length to meet the design requirements before undertaking the more rigorous methods of analysis for long slender piles described in Sections 6.3.4 and 6.3.5. The resistance of the rigid unit to rotation about point X in Figure 6.21a is given by the sum of the moments of the soil resistance above and below this point. The passive resistance diagram is divided into a convenient number $n$ of horizontal elements of depth $L/n$. The unit passive resistance of an element at a depth $z$ below the ground surface is then given by

$$p_z = p_{oz}K_{qz} + cK_{cz}$$   (6.14)

where
   $p_{oz}$ is the effective overburden pressure at depth $z$
   $c$ is the cohesion of the soil at depth $z$
   $K_{qz}$ and $K_{cz}$ are the passive pressure coefficients for the frictional and cohesive components respectively, at depth $z$

Brinch Hansen[6.11] established values of $K_q$ and $K_c$ in relation to the depth $z$ and the width of the pile $B$ in the direction of rotation, as shown in Figure 6.22.

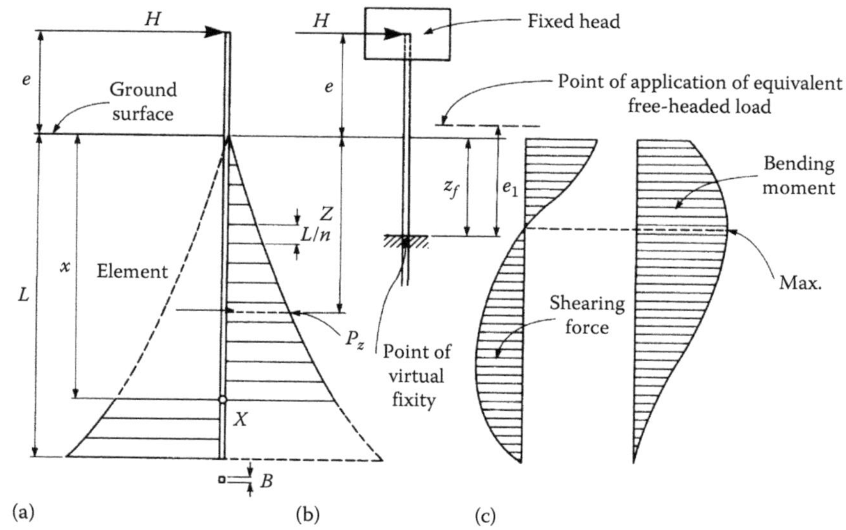

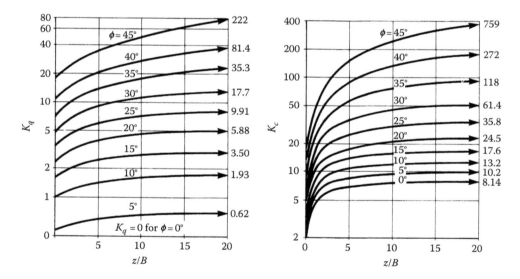

*Figure 6.21* The Brinch Hansen method for calculating ultimate lateral resistance of short piles: (a) soil reactions; (b) shearing-force diagram; (c) bending-moment diagram.

*Figure 6.22* The Brinch Hansen coefficients $K_q$ and $K_c$.

The total passive resistance on each horizontal element is $p_z \times L/n \times B$, and by taking moments about the point of application of the horizontal load,

$$\sum M = \sum_{z=0}^{z=x} p_z \frac{L}{n}(e+z)B - \sum_{z=x}^{z=L} p_z \frac{L}{n}(e+z)B \tag{6.15}$$

The point of rotation $X$ at depth $x$ in Figure 6.21a is correctly chosen when $\Sigma M = 0$, that is when the passive resistance of the soil above the point of rotation balances that below it.

Point $X$ is thus determined by a process of trial and adjustment. If the head of the pile carries a moment $M$ instead of a horizontal force, the moment can be replaced by a horizontal force $H$ at a distance $e$ above the ground surface where $M$ is equal to $H \times e$.

Where the head of the pile is fixed against rotation, the equivalent height $e_1$ above ground level of a force $H$ acting on a pile with a free head is given by

$$e_1 = \tfrac{1}{2}(e + z_f) \tag{6.16}$$

where

  $e$ is the height from the ground surface to the point of application of the load at the fixed head of the pile (Figure 6.21a)
  $z_f$ is the depth from the ground surface to the point of virtual fixity

The depth $z_f$ is not known at this stage, but for practical design purposes, it can be taken as 1.5 m for a compact coarse-grained soil or stiff clay (below the zone of soil shrinkage in the latter case) and 3 m for a soft clay or silt. The American Concrete Institute[6.12] recommends that $z_f$ should be taken as 1.4$R$ for stiff, over-consolidated clays and 1.8$T$ for normally consolidated clays, coarse-grained soils, and silt and peat (see Equations 6.11 and 6.12).

Having obtained the depth to the centre of rotation from Equation 6.15, the ultimate lateral resistance of the pile to the horizontal action $H_u$ can be obtained by taking moments about the point of rotation, when

$$H_u(e + x) = \sum_0^x p_z \frac{L}{n} B(x - z) + \sum_x^{x+L} p_z \frac{L}{n} + B(z - x) \tag{6.17}$$

The final steps in the Brinch Hansen method are to construct the shearing force and bending-moment diagrams (Figure 6.21b and c). The design bending moment, which occurs at the point of zero shear, should not exceed the design moment of resistance $M$ of the pile shaft. The appropriate partial factors are applied to the horizontal force $H_u$ to obtain the limiting permanent and variable actions.

When applying the method to layered soils, assumptions must be made concerning the depth $z$ to obtain $K_q$ and $K_c$ for the soft clay layer, but $z$ is measured from the top of the stiff clay stratum to obtain $K_c$ for this layer, as shown in Figure 6.23.

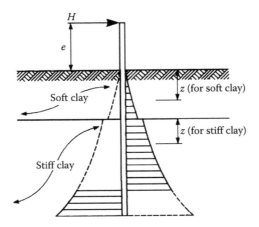

Figure 6.23 Reactions in layered soil on vertical pile under horizontal load.

The undrained shearing strength $c_u$ is used in Equation 6.14 for short-term loadings such as wave or ship-berthing forces on a jetty, but the drained effective shearing strength values ($c'$ and $\phi'$) are used for long-term sustained loadings such as those on retaining walls. A check should be made to ensure that undrained conditions in the early stages of loading are not critical. The step-by-step procedure using the Brinch Hansen method is illustrated in Worked Example 6.3. The Oasys ALP program (see Appendix C) applies the above-mentioned method and coefficents in an *elastic–plastic soil model*.

For short-term loading in uniform fine- and coarse-grained soils, the method of Broms (in Reese and van Impe[6.13]) may be used for the preliminary design of both short and long piles. The soil reaction is represented by the simplified diagrams in Figures 6.18 and 6.19 and the pile is designed using simple earth pressure principles.

### 6.3.2 Calculating the ultimate resistance of long piles

The lateral load and any applied bending moment which can be carried by a long pile are determined solely from the moment of resistance $M$ of the pile shaft. A simple method of calculating the ultimate load, which may be sufficiently accurate for cases of light loading on short or long piles of small to medium width, for which the cross-sectional area is governed by considerations of the relatively higher compressive loading, is to assume an arbitrary depth $z_f$ to the point of virtual fixity. Then from Figure 6.24,

$$\text{Lateral action on free-headed pile } H_u = \frac{M}{\left(e + z_f\right)} \tag{6.18}$$

$$\text{Lateral action on fixed-headed pile } H_u = \frac{2M}{\left(e + z_f\right)} \tag{6.19}$$

Arbitrary values for $z_f$ which are commonly used are given in the reference to the Brinch Hansen method.

It has already been stated that vertical piles offer poor resistance to lateral loads. However, in some circumstances, it may be justifiable to add the resistance provided by the passive resistance of the soil at one end of the pile cap and the friction or cohesion on the embedded sides of the cap. The pile cap resistance can be taken into account when the external loads

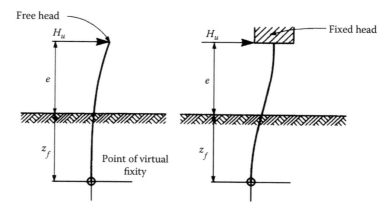

*Figure 6.24* Piles under horizontal load considered as simple cantilever.

are transient in character, such as wind gusts and traffic loads, but the resulting elastic deformation of the soil must not be so great as to cause excessive deflection and hence overstressing of the piles. The design of pile caps to resist lateral loading is discussed in Section 7.9.

### 6.3.3 Deflection of vertical piles carrying lateral loads

A simple method which can be used to check that the deflections due to small lateral loads are within tolerable limits and as an approximate check on the more rigorous methods described below is to assume that the pile is fixed at an arbitrary depth below the ground surface and then to calculate the deflection as for a simple cantilever either free at the head or fixed at the head but with freedom to translate.

Thus, from Figure 6.24,

$$\text{Deflection at head of free-headed pile} \, y = \frac{H(e + z_f)^3}{3EI} \tag{6.20}$$

and

$$\text{Deflection at head of fixed-headed pile} \, y = \frac{H(e + z_f)^3}{12EI} \tag{6.21}$$

where $E$ and $I$ are the elastic modulus and moment of inertia of the shaft, respectively, as before. Depths which may be arbitrarily assumed for $z_f$ are noted in Section 6.3.1.

### 6.3.4 Elastic analysis of laterally loaded vertical piles

The suggested procedure for using this Section and Section 6.3.2 is first to calculate the resistance of a pile of given cross section (or to determine the required cross sections for a given applied load) and then to apply the partial action factors to the $H_u$ values to obtain the limiting applied action $F_{trd}$ as Equation 6.10b. The alternative procedure is to calculate the deflection $y_0$ at the ground surface for a range of progressively increasing loads $H$ up to the value of $T_{trd}$. The limiting load is then taken as the load at which $y_0$ is within structural serviceability limits. As a first approximation, $H_u$ can be obtained by the Brinch Hansen method (Section 6.3.1) or from Equations 6.18 and 6.19. A preliminary indication of the likely order of pile head deflection under this load can be obtained from Equation 6.20 or 6.21 depending on the fixity conditions at the head.

It may be necessary to determine the bending moments, shearing forces and deformed shape of a pile over its full depth at a selected working load. These can be obtained for applied load conditions on the assumption that the pile behaves as an elastic beam on a soil behaving as a series of elastic springs. Calculations for the bending moments, shearing forces, deflections and slopes of laterally loaded piles are necessary when considering their behaviour as energy-absorbing members resisting the berthing impact of ships (see Section 8.1.1) or the wave forces in offshore platform structures (see Section 8.2).

Reese and Matlock[6.14] have established a series of curves for normally consolidated and cohesion-less soils for which the elastic modulus of the soil $E_s$ is assumed to increase from zero at the ground surface in direct proportion to the depth. The deformed shape of the pile and the corresponding bending moments, shearing forces and soil reactions are shown in Figure 6.25.

Coefficients for obtaining these values are shown for a lateral load $H$ on a free pile head in Figure 6.26a through e and for a moment applied to a pile head in Figure 6.27a through e.

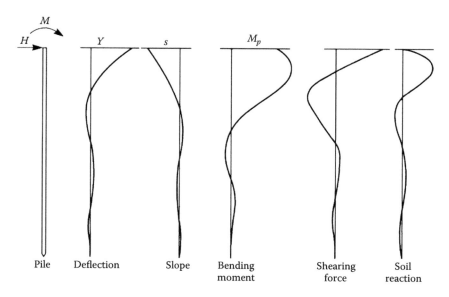

Figure 6.25 Deflections, slopes, bending moments, shearing forces and soil reactions for elastic conditions. (After Reese, L.C. and Matlock, H., Non-dimensional solutions for laterally-loaded piles with soil modulus assumed proportional to depth, *Proceedings of the Eighth Texas Conference on Soil Mechanics and Foundation Engineering*, Austin, TX, pp. 1–41, 1956.)

The coefficients for a fixed pile head are shown in Figure 6.28a through c. For combined lateral loads and applied moments, the basic equations for use in conjunction with Figures 6.26 and 6.27 are as follows:

$$\text{Deflection } y = y_A + y_B = \frac{A_y H T^3}{EI} + \frac{B_y M_t T^2}{EI} \tag{6.22}$$

$$\text{Slope} = s_A + s_B = \frac{A_s H T^2}{EI} + \frac{B_s M_t T}{EI} \tag{6.23}$$

$$\text{Bending moment} = M_A + M_B = A_m H T + B_m M_t \tag{6.24}$$

$$\text{Shearing force} = V_A + V_B = A_v H + \frac{B_v M_t}{T} \tag{6.25}$$

$$\text{Soil reaction} = P_A + P_B = \frac{A_p H}{T} + \frac{B_p M_t}{T^2} \tag{6.26}$$

For a fixed pile head, the basic equations are as follows:

$$\text{Deflection} = y_F = \frac{F_y H T^3}{EI} \tag{6.27}$$

$$\text{Bending moment} = M_F = F_m H T \tag{6.28}$$

$$\text{Soil reaction} = P_F = F_p \frac{H}{T} \tag{6.29}$$

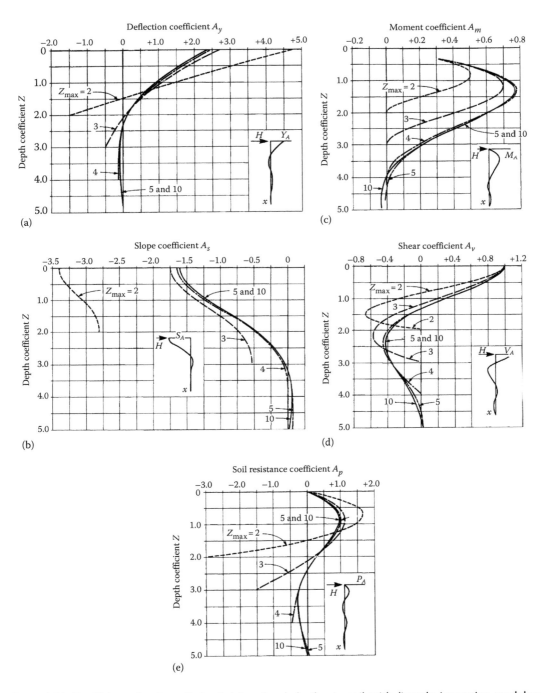

*Figure 6.26* Coefficients for laterally-loaded free-headed piles in soil with linearly increasing modulus: (a) coefficients for deflection, (b) coefficients for slope, (c) coefficients for bending moment, (d) coefficients for shearing force and (e) coefficients for soil resistance. (After Reese, L.C. and Matlock, H., Non-dimensional solutions for laterally-loaded piles with soil modulus assumed proportional to depth, *Proceedings of the Eighth Texas Conference on Soil Mechanics and Foundation Engineering*, Austin, TX, pp. 1–41, 1956.)

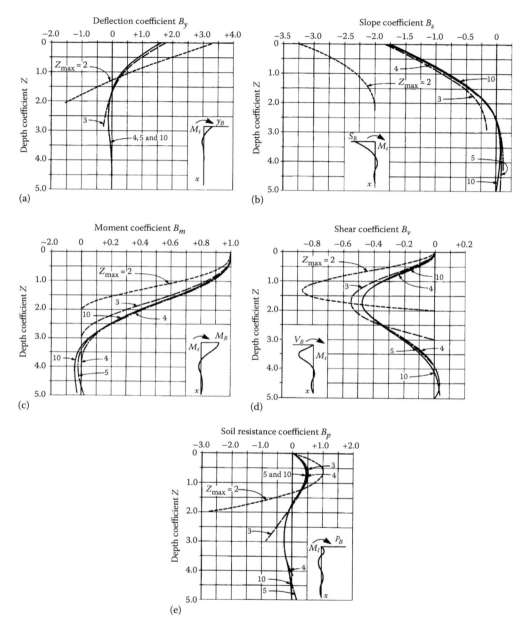

*Figure 6.27* Coefficients for piles with moment at free head in soil with linearly increasing modulus: (a) coefficients for deflection, (b) coefficients for slope, (c) coefficients for bending moment, (d) coefficients for shearing force and (e) coefficients for soil resistance. (After Reese, L.C. and Matlock, H., *Non-dimensional solutions for laterally-loaded piles with soil modulus assumed proportional to depth*, *Proceedings of the Eighth Texas Conference on Soil Mechanics and Foundation Engineering*, Austin, TX, pp. 1–41, 1956.)

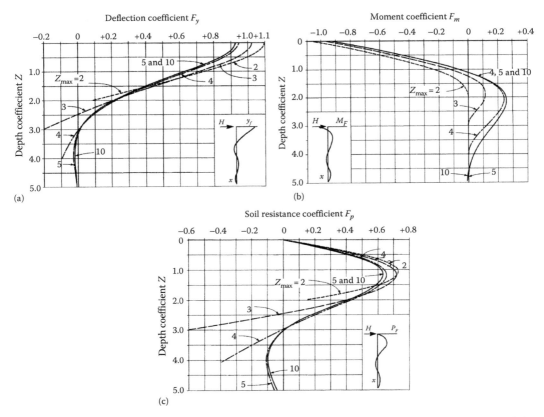

Figure 6.28 Coefficients for fixed-headed piles with lateral load in soil with linearly increasing modulus: (a) coefficients for deflection, (b) coefficients for bending moment, (c) and coefficients for soil resistance. (After Reese, L.C. and Matlock, H., Non-dimensional solutions for laterally-loaded piles with soil modulus assumed proportional to depth, *Proceedings of the Eighth Texas Conference on Soil Mechanics and Foundation Engineering*, Austin, TX, pp. 1–41, 1956.)

In Equations 6.22 through 6.29, $H$ is the horizontal load applied to the ground surface, $T$ (the stiffness factor) = $\sqrt[5]{EI/n_h}$ (as Equation 6.12), $M_t$ is the moment applied to the head of the pile, $A_y$ and $B_y$ are deflection coefficients (Figures 6.26a and 6.27a), $A_s$ and $B_s$ are slope coefficients (Figures 6.26b and 6.27b), $A_m$ and $B_m$ are bending-moment coefficients (Figures 6.26c and 6.27c), $A_v$ and $B_v$ are shearing-force coefficients (Figures 6.26d and 6.27d), $A_p$ and $B_p$ are soil-resistance coefficients (Figures 6.26e and 6.27e), $F_y$ is the deflection coefficient for a fixed pile head (Figure 6.28a), $F_m$ is the moment coefficient for a fixed pile head (Figure 6.28b) and $F_p$ is the soil-resistance coefficient for a fixed pile head (Figure 6.28c).

In Figures 6.26 through 6.28, the above coefficients are related to a depth coefficient $Z$ for various values of $Z_{max}$, where $Z$ is equal to the depth $x$ at any point divided by $T$ (i.e. $Z = x/T$) and $Z_{max}$ is equal to $L/T$. The use of curves in Figure 6.28 is illustrated in Worked Example 6.4. Further examples of the Reese and Matlock curves for lateral loads on piles at the mud line are provided in Reese and van Impe[6.13].

The case of a load $H$ applied at a distance $e$ above the ground surface can be simulated by assuming this to produce a bending moment $M_t$ equal to $H \times e$, this value of $M_t$ being used in Equations 6.22 through 6.29. The moments $M_a$ produced by load $H$ applied at the soil surface are added arithmetically to the moments $M_b$ produced by moment $M_t$ applied to the pile at the ground surface. This yields the relationship between the total moment and the depth below the soil surface over the embedded length of the pile. The deflection of a pile due to a lateral load $H$ at some distance above the soil surface is calculated in the same manner. The deflections of the pile and the corresponding slopes due to the load $H$ at the soil surface are calculated and added to the values calculated for moment $M_t$ applied to the pile at the surface. To obtain the deflection at the head of the pile, the deflection as for a free-standing cantilever fixed at the soil surface is calculated and added to the deflection produced at the soil surface by load $H$ and moment $M_t$, together with the deflection corresponding to the calculated slope of the pile at the soil surface. This procedure is illustrated in Example 8.2.

Davisson and Gill[6.15] have analysed the case of elastic piles in an elastic soil of constant modulus. The bending moments and deflections are related to the stiffness coefficient $R$ (Equation 6.11), but in this case, the value of $K$ is taken as Terzaghi's subgrade modulus $k_1$, using the values shown in Table 6.4. The dimensionless depth coefficient $Z$ in Figure 6.29 is equal to $x/R$. From these curves, deflection and bending-moment coefficients are obtained for free-headed piles carrying a moment at the pile head and zero lateral load (Figure 6.29a) and for free-headed piles with zero moment at the pile head and carrying a horizontal load (Figure 6.29b). These curves are valid for piles having an embedded length $L$ greater than $2R$ and different moment and deflection curves are shown for values of $Z_{max} = L/R$ of 2, 3, 4 and 5. Piles longer than $5R$ should be analysed for $Z_{max} = 5$. The equations to be used in conjunction with the curves in Figure 6.29 are as follows:

| Load on pile head | | For free-headed pile | | |
|---|---|---|---|---|
| Moment | $M$ | Bending moment | $= MM_m$ | (6.30) |
| Moment | $M$ | Deflection | $= My_m R^2/EI$ | (6.31) |
| Horizontal load | $H$ | Bending moment | $= HM_h R$ | (6.32) |
| Horizontal load | $H$ | Deflection | $= Hy_h R^3/EI$ | (6.33) |

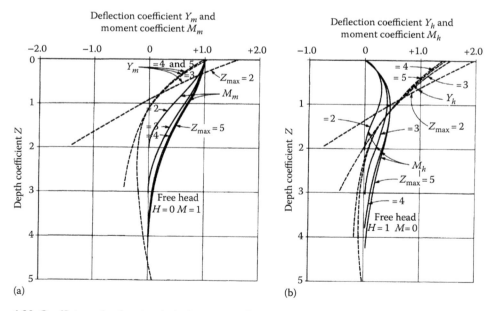

Figure 6.29 Coefficients for free-headed piles carrying lateral load or moment at pile head in soil of constant modulus: (a) coefficients for deflection and bending moment for piles carrying moment at head and zero lateral and (b) coefficients for deflection and bending moment for piles carrying horizontal load at head and zero moment. (After Davisson, M.T. and Gill, H.L., *J. Soil Mech. Div.*, 89(SM3), 63, 1963.)

The effect of fixity at the pile head can be allowed for by plotting the deflected shape of the pile from the algebraic sum of the deflections (Equations 6.31 and 6.33) and then applying a moment to the head which results in zero slope for complete fixity or the required angle of slope for a given degree of fixity. The deflection for this moment is then deducted from the calculated value for the free-headed pile. The use of the curves in Figure 6.29 is illustrated in Example 8.2. Conditions of partial fixity occur in *jacket-type* offshore platform structures where the tubular jacket member only offers partial restraint to the pile that extends through it to below seabed level.

Where marine structures are supported by long piles ($L \geq 4T$), Matlock and Reese[6.16] have simplified the process of calculating deflections by rearranging Equation 6.27 to incorporate a deflection coefficient $C_y$. Then

$$y = C_y \frac{HT^3}{EI} \tag{6.34}$$

where

$$C_y = A_y + \frac{M_t B_y}{HT} \tag{6.35}$$

Values of $C_y$ are plotted in terms of the dimensionless depth factor $Z$ ($= x/T$) for various values of $M_t/HT$ in Figure 6.30. Included in these curves are the fixed-headed case (i.e. $M_t/HT = -0.93$) and the free-headed case (i.e. $M_t = 0$).

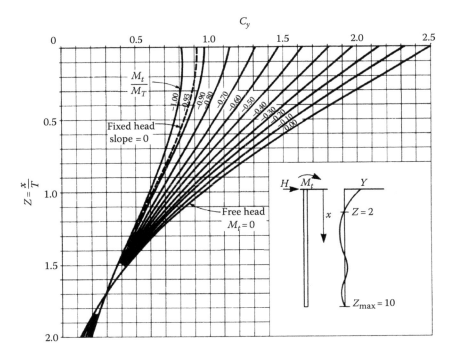

Figure 6.30 Coefficients for calculating deflection of pile carrying both moment and lateral load. (After Matlock, H. and Reese, L.C., Foundation analysis of offshore pile-supported structures, *Proceedings of the Fifth International Conference, ISSMFE, Paris, France, Vol. 2, pp. 91–97, 1961.*)

The elastic deflections of piles in layered soils, each soil layer having its individual constant modulus, have been analysed by Davisson and Gill[6.15] who have produced design charts for this condition.

### 6.3.5 Use of p–y curves

The analytical methods of Reese and Matlock[6.14] and Davisson and Gill[6.15] that are described in the previous section are applicable only to the deflections of piles which are within the range of the elastic compression of the soil caused by the lateral loading on the piles. However, these analytical methods can be extended beyond the elastic range to analyse movements where the soil yields plastically up to and beyond the stage of shear failure. This can be done by employing the artifice of p–y curves, which represent the deformation of the soil at any given depth below the soil surface for a range of horizontally applied pressures from zero to the stage of yielding of the soil in ultimate shear, when the deformation increases without any further increase of load. The p–y curves are independent of the shape and stiffness of the pile and represent the deformation of a discrete vertical area of soil that is unaffected by loading above and below it. This has led to the criticism that the method does not consider the soil as a *continuum*, but as Reese and van Impe[6.13] point out, a range of experiments with fully instrumented piles and case studies has shown good agreement between field results and the p–y computations.

The form of a p–y curve is shown in Figure 6.31a. The individual curves may be plotted on a common pair of axes to give a family of curves for the selected depths below the soil surface, as shown in Figure 6.31b. Thus, for the deformed shape of the pile (and also the induced bending moments and shearing forces) to be predicted correctly using the elastic analytical method described previously, the deflections resulting from these analyses must be compatible with those obtained by the p–y curves for the given soil conditions. The deflections obtained by the initial elastic analysis are based on an assumed modulus of subgrade variation $n_h$ and this must be compared with the modulus obtained from the pressures corresponding to these deflections, as obtained from the p–y curve for each particular depth analysed. If the soil moduli, expressed in terms of the stiffness factor $T$, do not correspond, the stiffness factor must be modified by making an appropriate adjustment to the soil

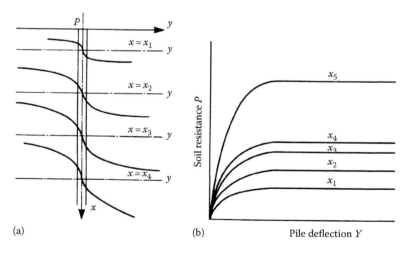

*Figure 6.31* p–y curves for laterally loaded piles: (a) shape of curves at various depths x below soil surface and (b) curves plotted on common axes.

modulus $E_s$ and from this to a new value of $n_h$ and hence to the new stiffness factor $T$. The deflections are then recalculated from the Reese and Matlock curves and the corresponding pressures again obtained from the $p$–$y$ curves. This procedure results in a new value of the soil modulus which is again compared with the second trial value and the process repeated until reasonable agreement is obtained.

Methods of drawing sets of $p$–$y$ curves have been established for soils which have a linearly increasing modulus, that is soft to firm normally consolidated clays and coarse soils. Empirical factors were obtained by applying lateral loads to steel tubular piles driven into soft to firm clays and sands. The piles were instrumented to obtain soil reactions and deflections over their full embedded depth.

The method of establishing $p$–$y$ curves for soft to firm clays is described by Matlock[6.17]. The first step is to calculate the ultimate resistance of the clay to lateral loading. Matlock's method is similar in concept to those described in Section 6.3.1, but the bearing capacity factor $N_c$ is obtained on a somewhat different basis.

Below a critical depth designated $x_r$, the coefficient $N_c$ is taken conventionally as 9. Above this depth, it is given by the equation

$$N_c = 3 + \frac{\gamma x}{c_u} + \frac{Jx}{B} \qquad (6.36)$$

where
  $\gamma$ is the density of the overburden soil
  $x$ is the depth below ground level
  $c_u$ is the undrained cohesion value of the clay
  $J$ is an empirical factor
  $B$ is the width of the pile

However, if $c_u$ varies with depth, then $N_c$ will become greater than 9 and Equation 6.36 should be used.

The experimental work of Matlock yielded values of $J$ of from 0.5 for a soft clay to 0.25 for a stiffer clay. The critical depth is given by the equation

$$x_r = \frac{6B}{(\gamma B/c_u) + J} \qquad (6.37)$$

The ultimate resistance above and below the critical depth is expressed in the $p$–$y$ curves as a force $p_u$ per unit length of pile, where $p_u$ is given by the pile width multiplied by the undrained shear strength $c_u$ and the above bearing capacity factor $N_c$, that is $p_u = N_c c_u B$.

Up to the point $a$ in Figure 6.32, the shape of the $p$–$y$ curve is derived from that of the stress/strain curve obtained by testing a soil specimen in undrained triaxial compression or from the load/settlement curve in a plate loading test (Figure 5.13). The shape of the curve is defined by the equation

$$\frac{p}{p_u} = 0.5 \sqrt[3]{\frac{y}{y_c}} \qquad (6.38)$$

where $y_c$ is the deflection corresponding to the strain $\varepsilon_c$ at a stress equal to the maximum stress resulting from the laboratory stress/strain curve.

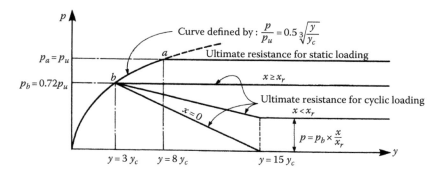

*Figure 6.32* Determining the shape of $p$–$y$ curve in soft to firm clay. (After Matlock, H., *Correlations for design of laterally loaded piles in soft clay*, Proceedings of the Offshore Technology Conference, Houston, TX, Paper OTC 1204, 1970.)

The strain $\varepsilon_c$ can also be obtained from the established relationship between $c_u$ and the undrained deformation modulus $E_u$ (see Section 5.2.2). Matlock[6.17] quotes values of $\varepsilon_c$ of 0.005 for 'brittle and sensitive clays' and 0.020 for 'disturbed or remoulded clays or unconsolidated sediments.' These values of $\varepsilon_c$ have been based on the established range of $E_u/c_u$ of 50–200 for most clays, and they can be applied to stiff over-consolidated clays, for example, the value of $E_u/c_u$ for stiff London Clay is 400. Matlock[6.17] recommends an average value of 0.010 for normally consolidated clays for use in the equation

$$y_c = 2.5\varepsilon_c B \qquad (6.39)$$

The effect of cyclic loading at depths equal to or greater than $x_r$ can be allowed for by cutting off the $p$–$y$ curve by a horizontal line representing the ultimate resistance $p_b$ of the clay under cyclically applied loads. From the experimental work of Matlock[6.17], the point of intersection of this line with the $p$–$y$ curve (shown in Figure 6.32 as point $b$) is given by

$$\frac{p_b}{p_u} = 0.72 \qquad (6.40)$$

The $p$–$y$ curves for cyclic loading with values of $y/y_c$ from 3 to 15 and for depths of less than $x_r$, and at greater than $x_r$ are shown in Figure 6.32.

There are little published data on values of the ultimate resistance, $p_b$, for various types of clay under cyclic conditions. The application of a static horizontal load after a period of cyclic loading, say, in a deep-sea structure where a berthing ship strikes a dolphin after a period of wave loading, produces a more complex shape in the $p$–$y$ curve and a method of establishing the curve for this loading condition has been described by Matlock[6.17].

The shape of a $p$–$y$ curve for a pile in sand as established by Reese et al.[6.9] is shown in Figure 6.33. It is in the form of a three-part curve up to the stage of the ultimate failure $p_u$. Calculations to determine the ultimate resistance per unit depth of the pile shaft at a given depth $x$ are obtained by using the angle of shearing resistance and density of the sand as determined by field or laboratory tests. The procedure for obtaining the shape of the curve and the trial-and-adjustment process using various assumed values of the coefficient of subgrade modulus variation $n_h$ to obtain the stiffness factor $T$ are more complex than those described previously for piles in normally consolidated clays.

It will be evident from the foregoing account of the construction and use of $p$–$y$ curves for laterally loaded piles in clays and sands that the procedure using longhand methods is

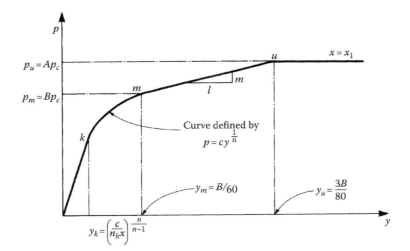

*Figure 6.33* Determining the shape of $p$–$y$ curve in sand. (After Reese, L.C. et al., Analysis of laterally loaded piles in sand, *Proceedings of the Offshore Technology Conference*, Houston, TX, Paper No. OTC 2080, 1974.)

extremely time consuming (see Worked Example 8.2). However, computer programs have been developed from which the required data on pile deflections, bending moments and soil resistances can be readily determined for varying pile diameters, depths and tubular wall thickness. The Oasys ALP program mentioned previously generates $p$–$y$ curves for soft clay based on the Matlock methods for both static and cyclic loading and can vary the ultimate resistance of the clay from $3c_u$ to $9c_u$ as given by Equation 6.36 (see Worked Example 6.7). For stiff clay, ALP applies the API[4.15] recommendations to produce the $p$–$y$ curves. ALP also follows the API recommendations for generating $p$–$y$ curves in sand. LPILE Plus (see Appendix C), developed from the work of Reese and others at the University of Texas at Austin, is a widely used program applying load and resistance factor design for $p$–$y$ curves in sand and stiff clay, under static and cyclic loading. This program (and ALP) incorporates later work by Reese[6.18] for rock sockets in weak rock with uniaxial compressive strength between 0.5 and 5 MN/m². LPILE can generate basic bilinear $p$–$y$ curves for strong rock (uniaxial compressive strength >6.9 MN/m²), but as pointed out in the extensive review by Turner[6.19], the input criteria do not account adequately for rock mass properties and the output is based on very limited correlation with field tests. It is therefore recommended that if the deflection using the 'strong rock' option is greater than 0.04% of the pile diameter, pile load testing is carried out. More complex 3D FEM programs, such as PLAXIS Foundation (see Appendix C), are needed to develop $p$–$y$ curves for large-diameter monopiles undergoing cyclic loading in marine clays, requiring careful selection of layered soil parameters.

The use of $p$–$y$ curves as described earlier is strictly applicable to piles in soils having a modulus which increases with depth (i.e. coarse soils and normally consolidated clays). In the case of stiff clays having a constant modulus of subgrade reaction $k_1$, Equation 6.36 can be used to obtain values of $N_c$ above the critical depth. The latter can be calculated from Equation 6.37 using a value of 0.25 for coefficient $J$. For soft clays where $c_u$ varies with depth, the critical depth in Equation 6.37 may be estimated from an average value of $J$, say, 0.33, and the average $c_u$. Also with increasing $c_u$, $N_c$ should be varied by substituting $\sigma'_v$, the vertical effective stress, for the term $\gamma x$ in Equation 6.36 to calculate values of $p_u$ above $x_r$. Values of $n_h$ are obtained by plotting the soil modulus $E_s$ against the depth, but the trial line is a vertical one passing through the plotted points, again with weight being given to depths of 0.5R or less. Cyclic loading can be a critical factor in stiff clays. The relationship in Equation 6.40 should preferably be established for the particular site by laboratory and field tests, but the factor of 0.72 may be used if results of such studies are not available.

Instead of relating the deflection $y_c$ to the strain $\varepsilon_c$ at a stress corresponding to the maximum stress obtained in the laboratory stress/strain curve for use in Equation 6.38, Reese and Welch[6.20] adopted the following relationship for stiff clays:

$$\frac{p}{p_u} = 0.5 \sqrt[4]{\frac{y}{y_{50}}} \tag{6.41}$$

where

$p$ and $p_u$ are as previously defined

$y_{50}$ is the deflection corresponding to the strain $\varepsilon_{50}$ at one-half of the maximum principal stress difference in the laboratory stress/strain curve, preferably obtained from isotropically consolidated undrained triaxial tests

If no value of $\varepsilon_{50}$ is available from laboratory tests, a figure of between 0.005 and 0.010 can be used in Equation 6.39 but substituting $y_{50}$ for $y_c$ and $\varepsilon_{50}$ for $\varepsilon_c$. The larger of these two values is the more conservative, but a value of 0.020 may be appropriate for over-consolidated clay, reflecting the higher $E_u/c_u$ value. Reese and Welch also describe a method for establishing $p$–$y$ curves for cyclic loading in stiff clay.

Zhang et al.[6.21] provide a non-linear approach to generating $p$–$y$ curves for rock sockets based on finite difference solutions, requiring the input of soil and rock parameters which will not be readily available in most design situations.

### 6.3.6 Effect of method of pile installation on behaviour under lateral loads and moments applied to pile head

The method of installing a pile, whether driven, driven and cast-in-place, or bored and cast-in-place, has not been considered in Sections 6.3.1 through 6.3.4. The effect of the installation method on the behaviour under lateral load can be allowed for by appropriate adjustments to the soil parameters. For example, if piles have to be driven through a soft sensitive clay to a bearing layer, then the resistance to lateral loads in the clay can be determined by using the remoulded shearing strength in conjunction with the Brinch Hansen method (Section 6.3.1). If the piles are not to be subjected to loading for a few months after driving, the full 'undisturbed' shearing strength can be used. There is unlikely to be much difference between the ultimate lateral resistance of short rigid piles driven into stiff over-consolidated clays and bored piles in the same type of soil. The softening effects for bored piles mentioned in Section 4.2.3 occur over a very short radial distance from the pile, and the principal resistance to lateral loads is provided by the undisturbed soil beyond the softened zone.

In the case of piles installed in coarse soils, the effect of loosening due to the installation of bored piles can be allowed for by assuming a low value of $\phi$ when determining $K_q$ from Figure 6.22. When considering the deflection of bored piles in coarse soils, the value of the soil modulus $n_h$ in Figure 6.20 should be appropriate to the degree of loosening which is judged to be caused by the method of installing the piles.

$p$–$y$ curves were developed primarily for their application to the design of long driven piles, mainly for offshore structures. Because such piles are required to have sufficient strength to cope with driving stresses, they have a corresponding resistance to bending stresses from lateral loading. On the other hand, bored and cast-in-place piles are required to have only nominal reinforcement, unless they are designed to act as columns above ground level or to carry uplift or lateral loading. Nip and Ng[6.22] investigated the behaviour of laterally loaded bored piles. They noted that while allowance can be made, arbitrarily, by assuming that the stiffness of a cracked reinforced pile section is 50% of that of an uncracked pile, this assumption can result in over-predicting the deflections and under-predicting the bending moments. By comparing the deflections measured in lateral load tests with predictions made by calculations using $p$–$y$ curves, they concluded that the latter can be used to predict deflections, bending moments and soil reactions of laterally loaded bored piles with varying $EI$ values corresponding to uncracked, partially cracked and fully cracked sections.

## 6.3.7 Use of the pressuremeter test to establish $p$–$y$ curves

The pressuremeter test (see Section 11.1.4) made in a borehole (or in a hole drilled by the pressuremeter device) is particularly suitable for use in establishing $p$–$y$ curves for laterally loaded piles. The test produces a curve of the type shown in Figure 6.34a. The initial portion represents a linear relationship between pressure and volume change, that is the radial expansion of the walls of the borehole. At the creep pressure $p_f$, the pressure/volume

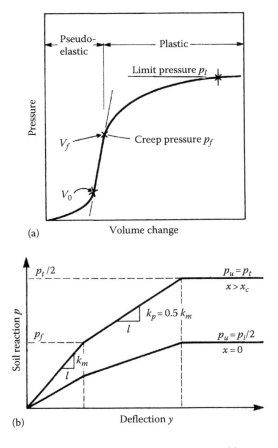

Figure 6.34 Obtaining soil reaction values from pressuremeter test: (a) pressure/volume change curve; (b) design reaction curve. (After Baguelin, F. et al., *The Pressuremeter and Foundation Engineering*, Trans Tech Publications, Clausthal, Germany, 1978.)

relationship becomes non-linear, indicating plastic yielding of the soil; at the limit pressure $p_l$, the volume increases rapidly without increase in pressure as represented by the horizontal portion of the $p$–$y$ curve.

Ménard used a Poisson's ratio of 0.33 to derive an expression for determining the pressuremeter modulus of the soil from the initial portion of the curve in Figure 6.34a. This equation as given by Baguelin et al.[6.23] is

$$E_m = 2.66V_m \frac{\Delta_p}{\Delta_v} \qquad (6.42)$$

where

$\Delta_p/\Delta_v$ is the slope of the curve between $V_0$ and $V_f$
$V_m$ is the midpoint volume

Baguelin et al.[6.23] give two sets of curves relating the response of the soil to lateral loading for the two stages in the pressuremeter tests as shown in Figure 6.34b. The upper curve is for depths below the ground surface equal to or greater than the critical depth, $x_c$, at which surface heave affects the validity of the calculation method. In fine-grained soils, $x_c$ is taken as twice the pile width, and in coarse soils, it is four times the width. Where there is a pile cap, there is no surface heave, $x_c$ is zero, and the lower curve in Figure 6.34b applies. The value of the coefficient of subgrade reaction, $k_m$ in Figure 6.34b, for pile widths greater than 600 mm is given by

$$\frac{1}{k_m} = \frac{2}{9E_m} B_0 \left( \frac{B}{B_0} \times 2.65 \right)^\alpha + \frac{\alpha B}{6E_m} \qquad (6.43a)$$

and for pile widths less than 600 mm

$$\frac{1}{k_m} = \frac{B}{E_m} \left( \frac{4(2.65)^\alpha + 3\alpha}{18} \right) \qquad (6.43b)$$

where

$E_m$ is the mean value of the pressuremeter modulus over the characteristic length of the pile
$B_0$ is a reference diameter (= 0.6 m)
$B$ is the pile diameter
$\alpha$ is the rheological factor varying from 1.0 to 0.5 for clays, 0.67 and 0.33 for silts, and 0.5 to 0.33 for sands

Clarke[6.24] quotes the Baguelin subgrade reaction equations for laterally loaded piles but comments that the method may over-predict the settlement near the surface, requiring a reduction of 0.5 in the Ménard ultimate resistance at the surface. The reduction only applies above a critical depth which for clays is $2B$ and for sands $4B$. He also provides data on other direct design methods using pre-bored and push-in pressuremeters.

Between the ground surface and the critical depth, $X_c$, the value of $k_m$ should be reduced by the coefficient $\lambda_z$, given by

$$\lambda_z = \frac{1 + (X/X_c)}{2} \qquad (6.44)$$

A simplified procedure in a homogeneous soil is to assume that there will be no lateral soil reaction between the ground surface and a depth equivalent to 0.5 to 0.75$B$ and then to use the full reaction given by the upper curve in Figure 6.34b.

Baguelin et al.[6.23] give the following equations for calculating deflections, bending moments and shears at any depth $z$ below the ground surface for conditions of a constant value of the pressuremeter modulus with depth

$$\text{Deflection } y(z) = \frac{2H}{Rk_mB} \cdot F_1 + \frac{2M_t}{R^2k_mB} \cdot F_4 \tag{6.45a}$$

$$\text{Moment } M(z) = \frac{H.R.F_3}{R} + M_tF_2 \tag{6.45b}$$

$$\text{Shear } T(z) = HF_4 - \frac{2M_t}{R}F_3 \tag{6.45c}$$

where
> $R$ is the stiffness coefficient given by Equation 6.11 (Baguelin refers to this as the transfer length, $l_0$)
> $H$ is the horizontal load applied to the pile head
> $M_t$ is the bending moment at the pile head
> $z$ is the dimensionless coefficient equal to $X/R$

Values of the coefficients $F_1$ to $F_4$ are given in Figure 6.35.

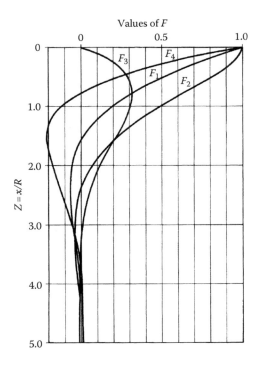

Figure 6.35 Values of the coefficients $F_1$ to $F_4$. (After Baguelin, F. et al., *The Pressuremeter and Foundation Engineering*, Trans Tech Publications, Clausthal, Germany, 1978.)

At the ground surface, the deflection becomes

$$y_0 = \frac{2H}{Rk_mB} + \frac{2M_t}{R^2k_mB} \tag{6.46}$$

and slope

$$y_0' = \frac{-2H}{R^2k_mB} - \frac{4M_t}{Rk_mB} \text{ rad} \tag{6.47}$$

If the head of the pile is fixed so that it does not rotate ($y_0 = 0$), Equations 6.45 through 6.47 become

$$y(z) = \frac{H}{Rk_mB} F_2 \tag{6.48a}$$

$$M(z) = \frac{-HR}{2} F_4 \tag{6.48b}$$

$$T(z) = H \times F_3 \tag{6.48c}$$

$$y_0 = \frac{H}{Rk_mB} \tag{6.48d}$$

$$M_t = \frac{-HR}{2} \tag{6.48e}$$

To draw the pile load–deflection curve, the deflections corresponding to soil reactions equal to the creep pressure, $p_f$, and the limit pressure, $p_l$, are calculated from the relationship $p = k_my$. The lateral pile loads then follow from Equations 6.45a, 6.46, 6.48a or 6.48d. For soil reactions between the limit pressure and creep pressure, the value of $k_m$ is halved as shown in Figure 6.34. The procedure is illustrated in Worked Example 6.6 where the pressuremeter tests show a linearly increasing soil modulus. The values of $n_h$ can be calculated from Equation 6.13 taking $K$ as $k_{mB}$. Deflections are calculated from the Reese and Matlock curves (Figures 6.26 through 6.28).

### 6.3.8 Calculation of lateral deflections and bending moments by elastic continuum methods

The method of preparing $p$–$y$ curves described in Section 6.3.5 was based on the assumption that the laterally loaded pile could be modelled as a beam supported by discrete springs. The springs would be considered as possessing linear or non-linear behaviour. In the latter case, the method could be used to model pile behaviour in strain conditions beyond the elastic range.

In many cases where lateral forces are relatively low and piles are stiff, the pile head movements are within the elastic range and it may be convenient to use the elastic continuum model to calculate deflections and bending moments.

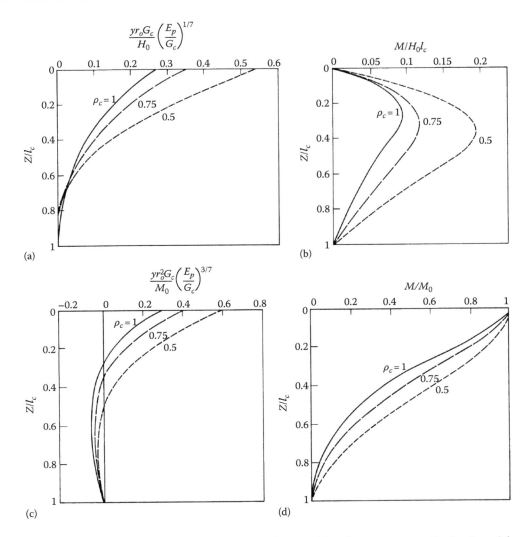

*Figure 6.36* Generalised curves giving deflected pile shape and bending-moment profile for lateral force and bending moment applied to pile head: (a) deflected pile shape for lateral force loading; (b) bending-moment profile for lateral force loading; (c) deflected pile shape for moment loading; (d) bending-moment profile for moment loading. (After Randolph, M.F., *Geotechnique*, 31(2), 247, 1981.)

Randolph[6.25] used finite element analyses to establish relationships between pile deflections and bending moments with depth for lateral force and moment loading as shown in Figure 6.36. The following notation applies to the parameters in this figure:

$y_0$ is the lateral displacement at ground surface
$z$ is the depth below ground level
$H_0$ is the lateral load applied at ground surface
$M$ is the bending moment in the pile
$M_0$ is the bending moment at ground surface
$r_0$ is the radius of the pile

$E'_p$ is the effective Young's modulus of a solid circular pile of radius $r_0$ (i.e. $4E_pI_p/\pi r_0^4$)
$G_c$ is the characteristic modulus of the soil, that is the average value of $G^*$ over depths less than $l_c$

$$G^* = G\left(1 + \frac{3}{4}v\right)$$

where
   $G$ is the shear modulus of the soil
   $v$ is Poisson's ratio
   $l_c$ is the critical length of the pile

$l_c = 2r_0(E'_p/G^*)^{2/7}$ for homogeneous soil

   $= 2r_0(E'_p/m^*r_0)^{2/9}$ for soil increasing linearly in stiffness with depth

$$m^* = m\left(1 + \frac{3v}{4}\right)$$

   $m = G/z$ where $G$ varies with depth as $G = mz$
   $\rho_c$ is a homogeneity factor

where

$$\rho_c = \frac{G^* \text{ at } l_c/4}{G^* \text{ at } l_c/2}$$

The use of the Randolph curves is illustrated in Worked Example 8.2.
   The Randolph method is useful where the shear modulus is obtained directly in the field using the pressuremeter. If Young's modulus values only are available, the shear modulus for an isotropic soil can be obtained from the equations

$$E_u = 2G(1 + v_u) \quad \text{and} \quad E' = 2G(1 + v') \tag{6.49}$$

where $v_u$ and $v'$ are the undrained and drained Poisson's ratios respectively.

## 6.3.9 Bending and buckling of partly embedded single vertical piles

A partly embedded vertical pile may be required to carry a vertical load in addition to a lateral load and a bending moment at its head. The stiffness factors $R$ and $T$ as calculated from Equations 6.11 and 6.12 have been used by Davisson and Robinson[6.26] to obtain the equivalent length of a free-standing pile with a fixed base, from which the factor of safety against failure due to buckling can be calculated using conventional structural design methods.
   A partly embedded pile carrying a vertical load $P$, a horizontal load $H$ and a moment $M$ at a height $e$ above the ground surface is shown in Figure 6.37a. The equivalent height $L_e$ of the fixed-base pile is shown in Figure 6.37b.

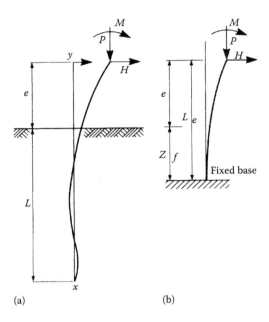

Figure 6.37 Bending of pile carrying vertical and horizontal loads at head: (a) partly embedded pile and (b) equivalent fixed-base pile or column.

For soils having a constant modulus:

$$\text{Depth to point of fixity } z_f = 1.4R \tag{6.50}$$

For soils having a linearly increasing modulus:

$$z_f = 1.8T \tag{6.51}$$

The relationships 6.50 and 6.51 are only approximate, but Davisson and Robinson[6.26] state that they are valid for structural design purposes provided that $l_{max}$, which is equal to $L/R$, is greater than 4 for soils having a constant modulus and provided that $z_{max}$, which is equal to $L/T$, is greater than 4 for soils having a linearly increasing modulus. From the earlier equations, the equivalent length $L_e$ of the fixed-base pile (or column) is equal to $e + z_f$ and the critical load for buckling is

$$P_{cr} = \frac{\pi^2 EI}{4(e + z_f)^2} \text{ for free-headed conditions} \tag{6.52}$$

and

$$P_{cr} = \frac{\pi^2 EI}{(e + z_f)^2} \text{ for fixed-(and translating-) headed conditions} \tag{6.53}$$

## 6.4 LATERAL LOADS ON RAKING PILES

The most effective way of arranging piles to resist lateral loads is to have pairs of piles raking in opposite directions as shown in Figure 6.5. The simple graphical method of determining the compressive and tensile forces in the piles by a triangle of forces assumes that the piles are hinged at their point of intersection and that the lateral loads are carried only in an axial direction by the piles. The tension pile will develop its maximum pull-out resistance with negligible movement, and the yielding of a properly designed compression pile of small to medium diameter is unlikely to exceed 10 mm at the working load. Thus, the horizontal deflections of the pile cap will be quite small.

For economy, the raking piles should be installed at the largest possible angle from the vertical. This depends on the type of pile used (see Section 3.4.11). Where raking piles are embedded in fill which is settling under its own weight (Figure 6.38a) or in a compressible clay subjected to a surcharge load or to superimposed fill (Figure 6.38b), the vertical loading on the upper surface of the rakers may induce high bending moments in the pile shaft. Because of this, raking piles may not be an appropriate form of construction in deep fill or compressible layers.

## 6.5 LATERAL LOADS ON GROUPS OF PILES

Loads on individual piles forming a group of vertical piles that is subject to horizontal loading or to combined vertical and horizontal loading can be determined quite simply (for cases where the resultant cuts the underside of the pile cap) by taking moments about the neutral axis of the pile group. Thus, in Figure 6.39, the vertical component $V$ of the load on any pile produced by an inclined thrust $R$, where $R$ is the resultant of a horizontal load $H$ and a vertical load $W$ given by

$$V = \frac{W}{n} + \frac{We\bar{x}}{\sum \bar{x}^2} \tag{6.54}$$

where
  $W$ is the total vertical load on the pile group
  $n$ is the number of piles in the group
  $e$ is the distance between the point of intersection of $R$ with the underside of the pile cap and the neutral axis of the pile group
  $\bar{x}$ is the distance between the pile and the neutral axis of the pile group ($\bar{x}$ is positive when measured in same direction as $e$ and negative when in the opposite direction)

This is a reasonable approximation provided that there is no interaction between the piles.

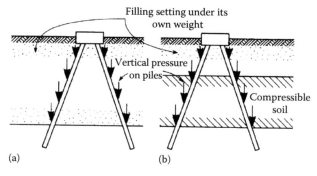

(a)  (b)

*Figure 6.38* Bending of slender raking piles due to loading from soil subsidence: (a) fill settling under own weight and (b) fill overlying compressible soil.

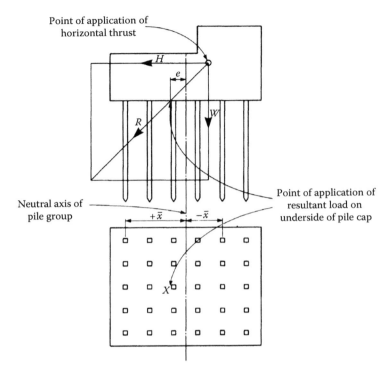

*Figure 6.39* Calculating load distribution on group of piles carrying vertical and horizontal loading.

Determination of the individual loads on groups of raking or combined raking and vertical piles is a complex matter if there are more than three rows of piles in the group. The latter case can be analysed by static methods if it is assumed that the piles are hinged at their upper ends, that horizontal loads are carried only by axial forces in the inclined piles, and that vertical piles do not carry any horizontal loading. Also there should be no interaction between piles. The forces in the piles are resolved graphically as shown in Figure 6.40. The same method can be used if pairs of piles or individual groups of three closely spaced piles are arranged in not more than three rows, as shown in Figure 6.41. To produce the polygon of forces, the line of action of the forces in the piles is taken as the centre line of each individual group.

The determination of the individual loads on piles installed in groups comprising multiple rows of raking or combined raking and vertical piles is a highly complex process which involves the analysis of movements in three dimensions, that is movements in vertical and horizontal translation and in rotational modes. The analysis of loadings on piles subjected to these movements requires the solution of six simultaneous equations, necessitating the use of a computer for practical design problems.

The reader is referred to the work of Poulos and Davis[4.31] and Poulos[6.27] for an account of their research into the behaviour of laterally loaded pile groups in an elastic medium. Randolph[6.25] gives expressions to determine the interaction factor between adjacent piles in groups carrying compression and lateral loading and compares them with values derived by Poulos and with results of model tests.

Poulos[6.28] describes the design of a piled raft foundation for a high-rise building in South Korea with a 5.5 m thick raft supported by 172, 2500 mm bored piles socketed into rockhead. The foundation is subject to lateral loading of 149 MN (in the $x$ direction) and 115

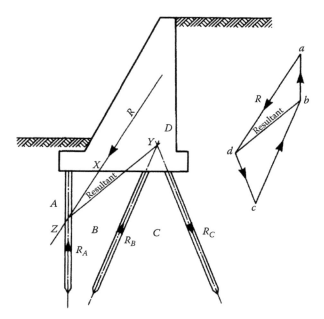

*Figure 6.40* Graphical method for determining forces on groups of vertical and raking piles under inclined loading.

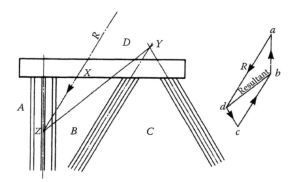

*Figure 6.41* Graphical method for determining forces on groups of closely spaced vertical and raking piles under inclined loading.

MN ($y$) which was analysed using the PLAXIS 3D computer program to assess the overall lateral stiffness of the foundation. Further programs were developed to analyse settlement and to assess the stiffness of the pile group assuming the raft is not in contact with the soil. The total lateral stiffnesses computed were 8958 MN/m ($x$) and 8435 MN/m ($y$), with lateral displacements of 17 and 14 mm, respectively.

The case of closely spaced groups of piles acting as a single unit when subject to lateral loads must also be considered. Prakash and Sharma[6.29] state that piles behave as individual units if they are spaced at more than 2.5 pile widths in a direction normal to the direction of loading and at more than 6–8 diameters parallel to this direction. Piles at a closer spacing can be considered to act as a single unit in order to calculate the ultimate resistance and deflections under lateral loads (Figure 6.42). In soft clays and sands, the effect of driving

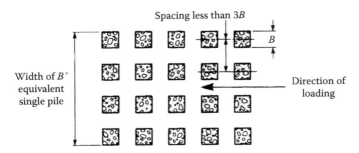

*Figure 6.42* Piles at close spacing considered as single unit. (After Prakash, S. and Sharma, H.D., *Pile Foundations in Engineering Practice*, John Wiley & Sons, Inc., New York, pp. 373, 1990.)

piles in groups at close spacing is to stiffen the soil enclosed by the group, thus increasing its capability as a single unit to resist movement when carrying horizontal loading. The grouping of piles in the centre of a raft to reduce settlement is considered in Section 4.9.4.

Calculations to determine the ultimate bearing capacity of pile groups carrying vertical and horizontal or inclined loading can be performed using the Brinch Hansen general Equation 5.1, assuming that the pile group takes the form of an equivalent block foundation. Alternatively, as noted in Section 5.4, the resistance of the group to compression loading can be calculated by assuming that the group acts as a single large-diameter pile. However, EC7 Clause 7.6.3.1 requires the resistance of a group subjected to tension loading to be provided by the frictional resistance of the soil enclosing a block foundation. No guidance is given in respect of pile groups carrying transverse loading. Clause 7.7.1(4)P merely requires group action 'to be considered'.

## WORKED EXAMPLES

### Example 6.1

The floor of a shipbuilding dock covers an area of $210 \times 60$ m. The 0.8 m floor is restrained against uplift by precast concrete shell piles having an overall diameter of 450 mm which are driven through 8 m of soft clay ($\bar{c}_u = 16$ kN/m$^2$) on to a strong shale ($\gamma = 2.3$ Mg/m$^3$). The piles are spaced on a 3 m square grid and each pile carries a permanent characteristic uplift load of 1100 kN. Design a suitable anchorage system for the dock floor using stressed cable anchors.

The application of current EC7 procedures to a stressed anchorage can be problematic; the allowable stress approach is generally preferred to determine loading and bond length:

From Figure 4.6, for $c_u/\sigma'_{vo} = 16/9.81 \times 0.8 \times 8 = 0.25$, $\alpha = 1.0$, and length factor $F$, for $L/B = 8/0.45 = 18$, of 1.0, Equation 4.11 omitting $\gamma_{Rd}$

$$Q_s = 1 \times 16 \times \pi \times 0.45 \times 8 = 181 \text{ kN}$$

For a safety factor of 2.5:

Allowable uplift resistance of the pile in soft clay = 181/2.5 = 72 kN
Thus, the load to be carried by anchorage in the shale = 1100 − 72 = 1028 kN

This load can be resisted by an anchor cable formed with seven Bridon Dyform 15.2 mm compacted strand, with a breaking load of 300 kN per strand. Therefore, working load = 1028/7 = 147 kN/m$^2$ which is 49% of the breaking load and satisfactory.

The approximate overall diameter of the cable is 45 mm. Therefore, for a bond stress between steel and grout of 1.0 N/mm² as BS 8081,

$$\text{Required bond length of cable} = \frac{1028 \times 1000}{\pi \times 45 \times 1.0 \times 1000} = 7.3 \text{ m}$$

Drill the cable hole to 9 m and provide an unwrapped and cleaned bond length of 7 m with compression fittings swaged on to the lower end. The cable can be fed down a 150 mm borehole for which

$$\text{Working bond stress between rock and grout} = \frac{1028 \times 1000}{\pi \times 150 \times 7.0 \times 1000} = 0.31 \text{ N/mm}^2$$

which is satisfactory for a strong shale (Table 6.3). The stress is not excessive if the anchors are stressed to 1.5 times the working load during installation.

From Figure 6.17a, the volume of a rock cone with a 30° half angle lifted by single anchor cable is $0.35 \times 9^3 = 255$ m³. The submerged weight of the rock cone = $1.3 \times 9.81 \times 255/1000 = 3.25$ MN.

Factor of safety against uplift = 3.25/1.028 = 3.1 which is satisfactory.

The anchorage of the whole dock floor requires 70 lines of anchors (at right angles to the centre line of the dock) and 20 ranks of anchors (parallel to the centre line of the dock) to form the 3 m square grid. Therefore, in Figure 6.17b, $N = 70$, $M = 20$ and $P = S = 3$ m. From Figure 6.17a, $m/L = n/L = 0.57$, and therefore, $m = n = 0.57 \times 9 = 5.1$ m. Then $P/n = S/m = 0.59$ so that, from Figure 6.17b, $\Delta V_n/V_c = \Delta V_m/V_c = 0.45$. Because $(P/n)^2 + (S/m)^2 = 2 \times 0.59^2 = 0.7$ is less than 4, there is composite overlapping of the rock cones, and the charts are not valid. The intersecting cones represent a rock volume roughly estimated to be $69 \times 3 \times 19 \times 3 \times 6 \times 70,794$ m³.

The total force resisting uplift is as follows:

Weight of dock floor = $(210 \times 60 \times 0.8 \times 2.4 \times 9.81)/1000 = 237.3$ MN
Submerged weight of soft clay = $(210 \times 60 \times 8.0 \times 0.8 \times 9.81)/1000 = 791.1$ MN
Submerged weight of anchored rock = $(70,794 \times 1.3 \times 9.81)/1000 = 902.8$ MN
Total = 1931.2 MN
Total uplift on underside of dock floor = $(70 \times 20 \times 1100)/1000 = 1540$ MN

Therefore,
Factor of safety against uplift = 1931.2/1540 = 1.25 which is satisfactory
(a more accurate assessment of the rock volume is not needed)
The UPL stability can be verified by the partial factors in EC7 as in Table 6.1, with $\gamma_{Gdst} = 1.1$ $\gamma_{Gstb} = 0.9$; hence,
Total destabilising uplift = $V_{dst\,d} = 1.1 \times 1540 = 1694$ MN
Permanent stabilising weight $G_{stb\,d} = 0.9 \times 1931.1 = 1738$ MN > 1694 MN and satisfactory

## Example 6.2

A piled dolphin carrying a horizontal pull of 1800 kN consists of a pair of compression piles and a pair of tension piles, raked at angles of 1 horizontal to 3 vertical. Design 'dead' anchors for the tension piles, which are driven through 3 m of weak weathered chalk to near refusal on strong rock chalk (having an average submerged density of 0.5 mg/m³).

From the triangle of forces (Figure 6.43), the uplift load on a pair of tension piles is 2800 kN. The load to be carried by a single pile is thus $0.5 \times 2800 = 1400$ kN and is treated as a variable action.

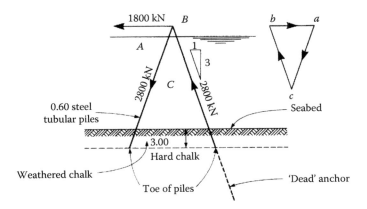

*Figure 6.43* Actions on piles for example 6.2.

From Table 4.12, the unit shaft friction in weathered chalk is 30 kN/m², so for a 600 mm diameter steel tubular pile penetrating 3 m,

Ultimate shaft resistance = $(\pi \times 0.6 \times \sqrt{(3^2 - 1^2)} \times 30 = 178$ kN

Assuming no pile tests for the driven pile, $\gamma_{st}$ = 2.0 from Table 4.3 and $\gamma_{Rd}$ = 1.4 giving

$$R_{sd\ pile} = \frac{178}{2.0 \times 1.4} = 64 \text{ kN}$$

The partial factor for the axial uplift load (variable action) on a single pile is $\gamma_{Qdst}$ = 1.5 as in Table 6.1:

Therefore, design value of anchorage = $P_d$ = 1.5 × 1400 = 2100 kN per pile
To satisfy the inequality $P_d \le R_{ad}$, the anchorage resistance must therefore be greater than 2100 – 64 = 2036 kN.

Use a steel tube in S355 grade having an outside diameter of 168.3 mm (6⅝ in.) and wall thickness 16 mm (⅝ in.) which has a cross-sectional area of 7600 mm²:

As $t \le 40$ mm $f_y$ = 355 N/mm² and $\gamma_{M0}$ = 1.0, then as in Section 7.10.2

$N_{Rd}$ = 7600 × 355/1.0 = 2698 kN > $P_d$ and satisfactory

The anchor will be installed in a 215 mm diameter drill hole. The grout to strong chalk unit bond stress of 0.8 N/mm² (Table 6.3) was based on pull-out tests, for which the standard practice of cycling the load would have been adopted. Table 6.2 gives an anchorage resistance partial factor of 1.4, but as required by the NA, this must be increased for unstressed anchors to conform to tension pile factors, giving $\gamma_a$ = 2.0. The anchor bond length to provide design resistance of 2036 kN is

$$L = \frac{2036 \times 1000}{(\pi \times 215 \times 0.8 \times 1000)/2} = 7.5 \text{ m}$$

Check bond between steel tube and cement grout:

$$= \frac{2036 \times 1000}{\pi \times 168.3 \times 7.5 \times 1000} = 0.52 \text{ N/mm}^2$$

Which is satisfactory for typical Class C25/30 grout with a design tensile strength of $f_{ctk}$ $_{0.05}/\gamma_C$ (Section 7.10.1), say, 1.2 N/mm².

Check the uplift resistance of the overlapping rock cones:

The bond length should be increased by approximately $L/2$ to comply with Figure 6.12b. Take a bond length over the cone of, say, 10 m below the surface of the weathered chalk and space the piles at 4 m centres. Then in Figure 6.17a, $V_c = 0.35 \times 10^3 = 350$ m³. Since $m/L = 0.61$, $m = 10 \times 0.61 = 6.1$ m. In Figure 6.17b, $S = 4$ m, so that $S/m = 4/7 = 0.66$, and thus $\Delta V_m/V_c = 0$. $M = 2$, $N = 1$ and $P = 0$, and therefore $\Delta V_n/V_c = 0$.

Rock volume anchored by pair of anchors = $350[(2 \times 1)-(1 \times 0.40)] = 560$ m³.
Weight of rock resisting uplift = $G_{stbk} = 560 \times 0.5 \times 9.81 = 2747$ kN.
With $\gamma_{G\ stab} = 0.9$ as in Table 6.1, then design value of weight of rock $G_{stbd} = 0.9 \times 2747 = 2472$ kN which is less than the 2800 kN uplift on the pair of piles.

Therefore, the frictional resistance on the sloping surfaces of the overlapping cones can be taken into account. As a rough approximation, assume that the two cones act as a rectangular block having a volume of 560 m³, say, $10 \times 8 \times 7.0$ m deep, and take the angle of shearing resistance of the chalk as 30° and take $K_0$ as 1.5:

Average unit frictional resistance on the vertical surfaces of the block

$$= 1.5 \times \tan 30° \times 9.81 \times 0.5 \times 3.5 = 14.9 \text{ kN/m}^2$$

Characteristic frictional resistance to uplift = $R_{sk} = (2 (10 + 8) \times 7.0 \times 14.9)/1.4 = 2682$ kN
With the UPL partial factor on shearing resistance $\gamma_{\phi'} = 1.25$ as in Table 6.2,

$$R_d = \frac{2682}{1.25} = 2146 \text{ kN}$$

With $\gamma_{Q\ dst} = 1.5$ for the variable action as in Table 6.1, design value $V_{dstd} = 1.5 \times 2 \times 1400 = 4200$ kN for the pair of piles. The vertical component of uplift

$$V_{dstd} = 4200 \times \sin 71.5 = 3983 \text{ kN}$$

Hence, for the inequality $V_{dstd} \leq G_{stbd} + R_d$ as Equation 6.3a

$$3983 < (2472 + 2146) = 4618 \text{ kN and satisfactory}$$

If shear connectors are to be provided, the BS EN ISO 19902 procedure can be used to calculate the required bond length. It is not strictly valid for the geometry of the connection but this example will illustrate the use of the equations. Assume an unconfined compression strength $f_{cu}$ of 25 N/mm² at 3 days and a modular ratio of 18. For a shear key upstand height of 10 mm and a spacing of 150 mm, the ratio $h/s = 0.067$.

From Equation 6.8a, stiffness factor

$$K = \frac{1}{18}\left(\frac{568}{200}\right)^{-1} + \left(\frac{168}{16} + \frac{600}{16}\right)^{-1}$$

$$= 0.04 \text{ which is greater than the limit of } 0.02$$

From Equation 6.8b, scale factor

$$C_p = \left(\frac{168}{1000}\right)^2 - \left(\frac{168}{500}\right) + 2$$

$$= 1.68 \text{ which is greater than the limit of } 1.5$$

From Equation 6.7a, $f_{g\,sliding}$ with the limiting values $C_p = 1.5$ and $K = 0.02$

$= 1.5[2 + 140\,(0.067)^{0.8}]0.02^{0.6} \times 25^{0.3} = 6.77$ N/mm²

From Equation 6.7b, $f_{g\,shear}$

$= [0.75 - 1.4(0.067)]5 = 3.28$ N/mm²

Therefore, for Equation 6.6, $f_g = 3.28$ N/mm² is the lower and $k = 1.0$, and the design interface transfer strength

$$f_d = \frac{3.28 \times 1.0}{2} = 1.64\,\text{N/mm}^2$$

$$\text{Required bond length over anchor} = \frac{1400 \times 1000}{\pi \times 168.3 \times 1.64}$$

$$= 1614 \text{ mm (using the characteristic uplift action)}$$

Therefore, provide 1614/150 = 10.8, say, 11 shear keys spaced at 150 mm centre over a distance of 1.6 m over anchor tube and pile.

As seen by the prescribed limits, the above equations are more applicable to large-diameter piles and jacket sleeves with a grouted annulus of 50–100 mm.

Checking the application of EC2-1 Table 3.1 concrete bond values for a C20/25 grout, $f_{ck} = 20$ N/mm² and $f_{ctk\,0.05} = 1.5$ (note $f_{ck\,cube}$ is used for BS EN ISO 19902, i.e. 25 N/mm² for a C20/25 grout), $f_{ctd} = 1.5/1.5 = 1.0$ N/mm² as Clause 3.1.6 and $f_{bd} = 2.25 \times 1.0 \times 1.0 = 2.25$ N/mm² as EC2-1-1 Clause 8.4.2, assuming that the shear connectors provide bond conditions as good as the referenced ribbed bars:

$$\text{Required bond length over anchor} = \frac{2036 \times 1000}{\pi \times 168.3 \times 2.25}$$

$$= 1710 \text{ mm (using the factored load and bond stress)}$$

The same 11 shear keys can be placed over 1.7 m to bond the anchor tube to the pile.

## Example 6.3

A vertical bored and cast-in-place pile 900 mm in diameter is installed to a depth of 6 m in a stiff over-consolidated clay ($\bar{c}_u = 120$ kN/m², $c' = 10$ kN/m², $\phi' = 25°$). Find the maximum permanent horizontal load which can be applied at a point 4 m above ground level. Also find the maximum applied load if the lateral deflection of the pile at ground level is limited to not more than 25 mm.

Consider first the ultimate horizontal load. For conditions of immediate application, that is using the undrained shearing strength, from Table 6.4 for $\bar{c}_u = 120$ kN/m², the soil modulus $k$ is 7.5 MN/m². If the elastic modulus of concrete is $26 \times 10^3$ MN/m² and the moment of inertia of the pile is $0.0491 \times (900)^4$ mm⁴, from Equation 6.11, the stiffness factor is

$$R = \sqrt[4]{\frac{26 \times 10^3 \times 0.0490 \times 0.9^4}{7.5 \times 0.9}} = 3.3 \text{ m}$$

$L$ is 6 m which is less than $2R$; therefore, the pile will behave as a short rigid unit, and the Brinch Hansen method can be used. The Brinch Hansen coefficients, as shown in Figure 6.22 with $c = c_u = 120$ kN/m$^2$ and $\phi = 0$, are tabulated as follows:

| z (m) | 0 | 1 | 2 | 3 | 4 | 5 | 6 |
|---|---|---|---|---|---|---|---|
| z/B | 0 | 1.1 | 2.2 | 3.3 | 4.4 | 5.5 | 6.6 |
| $K_c$ | 2.2 | 5.5 | 6.2 | 6.7 | 7.0 | 7.2 | 7.3 |
| $c_u K_c$ | 264 | 660 | 744 | 804 | 840 | 864 | 876 |

The soil resistance of each element 1 m wide by 1 m deep is plotted in Figure 6.44a. As a trial, assume the point of rotation $X$ is at 4.0 m below ground level. Then, taking moments about point of application of $H_u$,

$$\sum M = (462 \times 1 \times 4.5) + (702 \times 1 \times 5.5) + (774 \times 1 \times 6.5) + (822 \times 1 \times 7.5)$$

$$-[(852 \times 1 \times 8.5) + (870 \times 1 \times 9.5)] = +1629 \text{ kNm per metre width of pile}$$

If the point of rotation is raised to 3.9 m below ground level, $\sum M = +297\,\text{kNm}$, which is sufficiently close to zero for the purpose of this example.
Taking moments about the centre of rotation,

$$H_u \times 7.9 = (462 \times 3.4) + (702 \times 2.4) + (774 \times 1.4) + (820.2 \times 0.9 \times 0.45)$$

$$+ (838.2 \times 0.1 \times 0.05) + (852 \times 0.6) + (870 \times 1.6)$$

Thus, $H_u = 828$ kN per metre width. For a pile 0.9 m wide, $H_u = 0.9 \times 828 = 745$ kN.

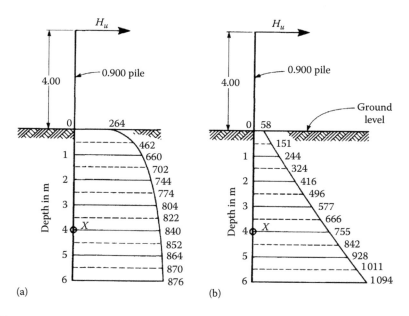

*Figure 6.44* Variation of Brinch Hansen coefficients with depth for Example 6.3 (a) undrained and (b) drained conditions.

Now consider the long-term stability under sustained loading, when the drained shearing strength parameters $c' = 10$ kN/m² and $\phi' = 25$ apply. From Figure 6.22, the Brinch Hansen coefficients for $K_c$ and $K_q$ are tabulated as follows:

| z (m) | 0 | 1 | 2 | 3 | 4 | 5 | 6 |
|---|---|---|---|---|---|---|---|
| z/B | 0 | 1.1 | 2.2 | 3.3 | 4.4 | 5.5 | 6.6 |
| $K_c$ | | 5.8 | 16 | 20 | 23 | 26 | 27 | 28 |
| $c'K_c$ | | 58 | 160 | 200 | 230 | 260 | 270 | 280 |
| $K_q$ | | 3.3 | 5.0 | 5.5 | 5.9 | 6.3 | 6.7 | 6.9 |
| $p_0$ (kN/m²) | 0 | 18.6 | 39.3 | 58.8 | 78.5 | 98.2 | 118 |
| $p_0K_q$ (kN/m²) | 0 | 93 | 216 | 347 | 495 | 658 | 814 |
| $c'K_c + p_0K_q$ (kN/m²) | 58 | 253 | 416 | 577 | 755 | 928 | 1094 |

The soil resistance of each element 1 m deep for a pile 1 m wide is plotted in Figure 6.44b. As a trial, consider the point of rotation X to be 4.0 m below ground level. Taking moments about the point of application of $H_u$,

$$\sum M = (155 \times 1 \times 4.5) + (335 \times 1 \times 5.5) + (496 \times 1 \times 6.5) + (666 \times 1 \times 7.5)$$

$$- [(842 \times 1 \times 8.5)] + (1011 \times 1 \times 9.5) = -6002 \text{ kNm}$$

If the centre of rotation is lowered to 4.5 m, then

$$\sum M = 10{,}759 + (798 \times 0.5 \times 8.25) - (885 \times 0.5 \times 8.75) + 9604)]$$

$$= 14{,}051 - 13{,}476 + 575 \text{ kN}$$

which is sufficiently close to zero for the purpose of this example. Then taking moments about the centre of rotation,

$$H_u \times 8.5 = (155 \times 4.0) + (335 \times 3.0) + (496 \times 2.0) + (660 \times 1.0)$$

$$+ (798 \times 0.5 \times 0.25) + (885 \times 0.3 \times 0.25) + (1011 \times 1 \times 1)$$

Thus, $H_u = 530$ kN per metre width. Therefore, the lowest value of the ultimate load results from drained shearing strength conditions. For a 900 mm pile, the ultimate horizontal load = $0.9 \times 530 = 477$ kN.

Calculating the allowable horizontal load which limits the lateral deflection at ground level to 25 mm, from Equation 6.50,

Depth to point of fixity = $1.4R = 1.4 \times 3.3 = 4.62$ m
From Equation 6.20 with $e = 0$, $H = 3 \times 0.025 \times 837.38 \times 10^3/4.62^3 = 637$ kN

Therefore, the allowable load is governed by the resistance of the pile to overturning. A factor of safety of 1.5 on the ultimate load of 477 kN will be appropriate giving an allowable load of 318 kN.

Checking against EC7 procedures:

A. Considering undrained shear strength and *DA1 combination 1*, the partial factor for a permanent action is $\gamma_G = 1.35$ and the M1 factor $\gamma_{cu} = 1.0$; hence, $c = c_u = 120$ kN/m$^2$. As before, from Figure 6.22 and applying the Brinch Hansen coefficients for $K_c$ and $K_q$ in Equation 6.14, the unit passive resistances are the same as Figure 6.44a. Also for a 0.9 m pile, $H_u = 745$ kN as before.

It is implied in EC7 that the model factor which is required for compressive loads on piles is not required for lateral loads. Therefore, for the inequality $F_{tr\,d} \le R_{tr\,d}$, and the maximum applied horizontal load $H_G = F_{tr\,d} = 745/(1.35) = 552$ kN

B. Considering drained shear strength and *DA1 combination 1*, the partial factor for a permanent action is $\gamma_G = 1.35$ and the M1 factor $\gamma_{c'} = 1.0$; hence, $c = c' = 10$ kN/m$^2$ and $\phi' = 25°$. As before, from Figure 6.22 and applying the Brinch Hansen coefficients for $K_c$ and $K_q$ in Equation 6.14, the unit passive resistances are the same as Figure 6.44b and $H_u = 477$ kN as before and the maximum applied horizontal load $H_G = F_{tr\,d} = 477/(1.35) = 353$ kN.

C. Considering undrained shear strength and *DA1 combination 2*, $\gamma_{cu} = 1.4$; hence, $c = c_{cu} = 120/1.4 = 86$ kN/m$^2$ and $\gamma_G = 1.0$.

The point of rotation is again approximately 3.9 m below ground level. Applying the Brinch Hansen factors and equation as mentioned earlier using these partial factors, the spread sheet for Figure 6.44a is slightly modified giving $H_u = 537$ kN per pile and the maximum applied horizontal load, $H_G = F_{tr\,d} = 537/(1.0) = 537$ kN.

D. Considering drained shear strength and *DA1 combination 2*, $\gamma_{c'} = 1.25$; hence, $c = c' = 10/1.25 = 8$ kN/m$^2$ and $\gamma_G = 1.0$.

The point of rotation is again approximately 4.5 m below ground level and the spread sheet for Figure 6.44b is modified so that $H_u = 435$ kN per pile and the maximum applied horizontal load $H_G = F_{tr\,d} = 435/(1.0) = 435$ kN.

Using EC7 factors, the lowest value of the applied load also results from drained shear strength conditions and the maximum load is governed by the resistance of the pile to overturning. For the inequality $F_{tr\,d} \le R_{tr\,d}$, DA1 combination 1 is the critical set and gives a value of 353 kN compared with the previous 318 kN (which had a factor of safety of 1.5).

## Example 6.4

A group of 36 steel box piles are spaced at 1.25 m centres in both directions to form six rows of six piles surmounted at ground level by a rigid cap. The piles are driven to a depth of 9 m into a medium-dense water-bearing sand and carry a permanent horizontal action of 240 kN on each pile. Calculate the bending moments, deflections and soil-resistance values at various points below the ground surface at the applied load. Calculate the horizontal deflection of the pile cap if the horizontal load is applied in the direction resisted by the maximum resistance moment of the piles. Moment of inertia of the pile in the direction of maximum resistance moment = 58,064 cm$^4$ and elastic modulus of steel = 21 MN/cm$^2$ as EC3.

From Figure 6.20, Terzaghi's value of $n_h$ for a medium-dense sand is 5 MN/m$^3$. Then from Equation 6.12, the stiffness factor is

$$T = \sqrt[5]{\frac{21 \times 58,064}{5 \times 10^{-6}}} = 189 \text{ cm}$$

Because the embedded length of 9 m is more than 4T, the pile behaves as a long elastic fixed-headed element.

Steel grade to BS EN 10025 is S275 with $f_{yk} = 275$ MN/m², modulus of section = 2950 cm³ and $\gamma_{M0} = 1.0$. Hence, design bending resistance $M_d = 2950 \times 0.0275/1.0 = 81$ MNcm = 810 kNm.

From Equation 6.51, depth to point of fixity = $1.8 \times 189 = 340$ cm.

From Equation 6.19, ultimate horizontal load = $H_u = 2 \times 81 \times 10^3/340 = 476$ kN.

Global factor of safety on applied load = 476/240 = 2.0, which is satisfactory if the pile head deflections and the pile group behaviour are within acceptable limits applying EC7 procedures.

Design action and resistances for lateral loads are determined using the partial factors from Tables 4.1 and 4.2 (EC7 Clause 2.4.7.3.1) with $\gamma_G = 1.35$ and M1 factors as unity for set A1. For SLS calculation, the partial factor is unity.

The deflections, bending moments and soil-resistance values for the single pile at the working load can be calculated from the curves in Figure 6.28.

From Equation 6.27:

$$\text{Deflection } y_F = \frac{240 \times 1.0 \times 189^3}{21 \times 10^3 \times 58{,}064} \times F_y = 1.329 F_y \text{ cm} = 13.29 F_y \text{ mm}$$

From Equation 6.28:

Bending moment $M_F = 240 \times 1.35 \times 189 \times F_m = 61{,}236 F_m$ kNcm = $612.4 F_m$ kNm

From Equation 6.29:

Soil reaction $P_F = 240 \times 1.35 F_p/189 = 1.71 F_p$ kN per cm depth = $171 F_p$ kN per m depth. $Z_{max} = L/T = 9.0/1.89 = 4.8$

Tabulated values of $y_F$, $M_F$ and $P_F$ using the above partial action factor are as follows:

| x (m) | 0 | 0.5 | 1.0 | 1.5 | 2.0 | 2.5 | 3.0 | 4.0 | 5.0 |
|---|---|---|---|---|---|---|---|---|---|
| z = x/T | 0 | 0.27 | 0.53 | 0.80 | 1.06 | 1.33 | 1.60 | 2.13 | 2.66 |
| $F_y$ | +0.92 | +0.90 | +0.82 | +0.71 | +0.61 | +0.50 | +0.37 | +0.18 | +0.04 |
| $y_F$ (mm) | +12.2 | +12.0 | +10.9 | +9.4 | +8.1 | +6.6 | +4.9 | +2.4 | — |
| $F_m$ | -0.91 | -0.65 | -0.40 | -0.18 | -0.03 | +0.10 | +0.19 | +0.25 | +0.21 |
| $M_F$ (kNm) | -557 | -398 | -245 | -110 | -18 | +61 | +116 | +153 | +129 |
| $F_p$ | 0 | +0.25 | +0.45 | +0.57 | +0.62 | +0.62 | +0.57 | +0.38 | +0.13 |
| $P_F$ (kN/m) | 0 | +42.7 | +76.9 | +97.5 | +106.0 | +106.0 | +97.5 | +65.0 | +22.2 |

From the above table, the pile head deflection is satisfactory and the inequality $M_d > M_F$ for bending of the pile is satisfied (design resistance of the pile 810 kNm > maximum bending moment of 557 kNm).

Because the piles are spaced at 125/46.7 = 2.67 diameters, the group will act as a single unit equivalent to a block foundation having a width of $5 \times 1.25$ m = 6.25 m and a depth below the ground surface of 9 m. The ultimate passive resistance to the horizontal thrust from a block foundation can be determined from the limit state Equation C.2 in Annex C of EC7 (parameters as given):

$$\sigma_p = K_p [\gamma z + q] + 2c \sqrt{K_p}$$

With the 9 m depth of block and $c$ and $q = 0$, passive resistance at the base of the block:

$$\sigma_p = 3.69 [1.3 \times 9.81 \times 9] = 423.5 \text{ kN/m}^2/\text{m}$$

Total resistance = $0.5 \times 9 \times 423.6 \times 6.25 = 11{,}912$ kN for the width of the block

A model factor is not applied to lateral load on piles, and with the partial action factor $\gamma_G = 1.35$ as before, the applied load on the pile group must be limited to $11,912/1.35 = 8824$ kN. The load on each pile must be limited to $8824/36 = 245$ kN and satisfactory.

It is also necessary to calculate the horizontal deflection of the pile group under the actual applied load of 240 kN per pile. The above-mentioned values of $P_F$ show that the horizontal load is effectively distributed over a depth of 4 m below the ground surface. Thus, the load on the group can be simulated by a block foundation having a width $B$ of 4 m, a length of 6.25 m and a depth of 6.25 m. The elastic modulus of a medium-dense sand can be taken as 20 MN/m$^2$. From Equation 5.22, with $H/B = 1000$, $L/B = 6.25/4 = 1.55$, $D/B = 6.25/4 = 1.55$, $\mu_1 = 0.85$ and $\mu_0 = 0.91$ as in Figure 5.18,

$$\text{Elastic settlement } \rho_i = \frac{0.85 \times 0.91 \times ((240 \times 36)/(4 \times 6.25)) \times 4 \times 1000}{20 \times 1000} = 52 \text{ mm}$$

This is within safe limits and, as would be expected, it is greater than the deflection of the single pile.

## Example 6.5

A tower is to be constructed on a site where 6 m of very soft clay overlie a very stiff glacial clay (undrained shearing strength = 190 kN/m$^2$). The tower and its base slab weigh 30,000 kN, and the tower is subject to a maximum horizontal wind force of 1500 kN with a centre of pressure 35 m above ground level. The base of the tower is 12 m in diameter. Design the foundations and estimate the settlements under the dead load and wind loading.

Because of the presence of the soft clay layer, piled foundations are required and the heavy vertical load favours the use of large bored and cast-in-place piles. A suitable arrangement of piles to withstand the eccentric loading caused by the wind force is 22 piles in the staggered pattern shown in Figure 6.45.

Allowable stress design will be used initially and then checked against EC7 recommendations.

The resultant of the vertical and horizontal forces has an eccentricity of $1500 \times 35/30,000 = 1.75$ m at ground level. From Equation 6.54, the vertical load on each of the outer four piles due to wind loading from an east–west direction is given by

$$V = \frac{30,000}{22} \pm \frac{30,000 \times 1.75 \times 6}{(4 \times 6^2) + (6 \times 4.5^2) + (4 \times 3^2) + (6 \times 1.5^2)}$$

$$= 1364 \pm 1000 \text{ kN}$$

Therefore, uplift does not occur on the windward side and the maximum pile load is 2364 kN. Checking the maximum pile load for wind in a north–south direction,

$$V = 1364 \pm \frac{30,000 \times 1.75 \times 5.20}{(8 \times 5.20^2) + (10 \times 2.60^2)} = 1364 \pm 962 \text{ kN}$$

Therefore, maximum pile load = 2326 kN.

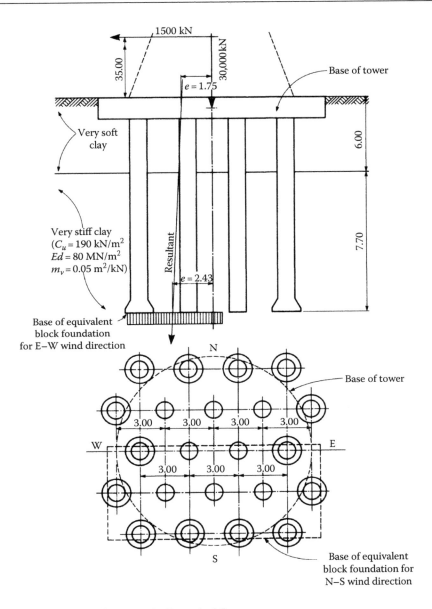

*Figure 6.45*  Piled foundation for tower in Example 6.5.

For piles with a shaft diameter of 1 m,

$$\text{Working stress on concrete} = \frac{2364 \times 1000}{\frac{1}{4}\pi \times 1000^2} = 3 \text{ N/mm}^2$$

which is within safe limits.

Adopting an under-reamed base to a diameter of 1.8 m, and applying Equation 4.7 (but without the model factor for allowable stress approach),

$$\text{Ultimate base resistance } Q_b = 9 \times 190 \times \tfrac{1}{4}\pi \times 1.8^2 = 4351 \text{ kN}$$

For a safety factor of 2 on the combined base resistance and skin friction, the required ultimate skin friction = (2 × 2364) – 4351 = 377 kN.

If the required depth of penetration into the glacial clay to mobilise the required ultimate resistance is $L$ m, ignoring the small skin friction in the very soft clay and adopting an adhesion factor of 0.3 (to allow for delays in under-reaming), then from Equation 4.10 (again without the model factor),

ultimate shaft resistance $Q_s = 0.3 \times 190 \times \pi \times 1 \times L$

and if $Q_s = 377$, $L = 2.10$ m. Thus, the allowable pile load for a factor of safety of 3 in base resistance and unity in skin friction is $\frac{1}{3}Q_b + Q_s = (\frac{1}{3} \times 4351) + 377 = 1827$ kN, which is insufficient. Taking $L$ as 4.9 m,

$$Q_s = 0.3 \times 190 \times \pi \times 1 \times 4.9 = 877 \text{ kN}$$

and the allowable pile load is $\frac{1}{3} \times 4351 + 877 = 2328$ kN which is satisfactory.

It is necessary to add two shaft diameters and the depth of the under-ream to arrive at the total penetration of the piles below ground level. Thus,

$$D = 6 \text{ m (soft clay)} + 4.9 + 2.0 + 0.8 = 13.7 \text{ m}$$

An adhesion factor of 0.5 is used for straight-shafted piles in a glacial clay (Figure 4.8). Therefore, the allowable load on a straight-shafted pile drilled to the same depth as the under-reamed piles and adopting a safety factor of 2 on combined end bearing and shaft friction is given by

$$Q_a = \frac{(9 \times 190 \times \frac{1}{4}\pi \times 1^2) + (0.5 \times 190 \times \pi \times 1 \times 7.7)}{2} = 1820 \text{ kN}$$

Therefore, straight-shafted piles can be used for the eight inner piles as shown in Figure 6.45. The maximum working load on these is one-half or less than one-half of the outer piles.

The overall depth to the base of the pile group of 13.7 m is only a little greater than the overall width of the group, that is 13 m to the outsides of the pile shafts. Therefore, it is necessary to check that block failure will not occur due to eccentric loading:

$$\text{Eccentricity of loading with respect to base of pile group} = \frac{500 \times (35 + 13.7)}{30,000}$$

$$= 2.43$$

From Equation 5.3a, the width of an equivalent block foundation for winds in a north–south direction = 10.40 – (2 × 2.43) = 5.54 m. The overall dimensions of this block foundation are thus 13 × 5.54 m. Tangent of the angle of inclination of the resultant force = tan $\alpha$ = 1500/30,000 = 0.05, and thus $\alpha$ = 2.87°.

From Figure 5.6, for $\phi = 0°$, $N_c = 5.2$; from Figure 5.7 for $B'/L' = 5.54/13.0 = 0.43$, $s_c = 1.1$; from Figure 5.9 for $D/B = 7.7/5.54 = 1.4$, $d_c = 1.2$. The horizontal force of 1500 kN in Figure 6.45 is less than $c_u B'L' = 190 \times 5.54 \times 13.0 = 13,684$ kN. Therefore, Equation 5.11 can be used to obtain the inclination factor $i_c = 1 - 1500/2 \times 190 \times 5.54 \times 13.0 = 0.95$. From Figure 5.6, $N_\gamma = 1.0$. From Figure 5.7, $s_\gamma = 0.95$; $d_\gamma = 1.0$. From Equation 5.12, $i_q = 1 - 1500/30,000 = 0.92$; therefore, from Equation 5.13, $i_\gamma = 0.92^2 = 0.85$.

The second term in Equation 5.1 is zero; therefore,

$q_{ub}$ = (190 × 5.2 × 1.1 × 1.2 × 0.95 × 1.0) + (0.5 × 9.81 × 1.8 × 5.54 × 1.0 × 1.0 × 0.95 × 0.85) = 1238 + 39 = 1277 kN/m²

$Q_{ub}$ = 1277 × 5.54 × 13.0 = 92,055 kN

Factor of safety against base failure = 92,005/30,000 = 3.1 which is satisfactory.

Checking for compliance with the EC7 procedures:

*For DA1 combination 1* with the maximum vertical action on a pile of 2364 kN as above, with 1364 kN being a permanent unfavourable action and 1000 kN a variable unfavourable action from the wind in the east–west direction. Table 4.1 gives $\gamma_G$ = 1.35 and $\gamma_Q$ = 1.5 hence the design action on the concrete

$F_d$ = 1.35 × 1364 + 1.5 × 1000 = 3341 kN

For piles with a shaft diameter of 1 m using C25/30 concrete, $f_{ck}$ = 25 N/mm², $\gamma_C$ = 1.5 × k and α = 0.84 (as Section 7.10.1) and applying the reduction factor 0.95 to the diameter as Table 4.6.

Design compressive strength of concrete = 0.85 × 25/(1.5 × 1.1) = 12.9 N/mm²

$$\text{Stress on concrete} = \frac{3341 \times 1000}{\frac{1}{4}\pi \times (950)^2} = 4.7 \text{ N/mm}^2 \text{ and satisfactory}$$

Checking dimensions and resistance of under-reamed pile:

*Applying DA1 combination 2* as likely to be critical, the action factors are $\gamma_G$ = 1.0 and $\gamma_Q$ = 1.3; hence,

$F_d$ = 1.0 × 1364 + 1.3 × 1000 = 2664 kN

Assuming the under-reamed base is 1.8 m as above for the outer piles, with the material factor $\gamma_{cu}$ = 1.0 and $\gamma_{Rd}$ = 1.4, then from Equation 4.7,
  Characteristic base resistance $R_{bk}$ = (9 × 190 × π/4 × 1.8²)/1.4 = 3108 kN
Take the length of shaft in the very stiff clay as 4.9 m as above but with an adhesion factor α = 0.5 and $\gamma_{Rd}$ = 1.4; then from Equation 4.10,
  Characteristic shaft resistance $R_{sk}$ = (0.5 × 190 × π × 1.0 × 4.9)/1.4 = 1045 kN
For design resistance of bored piles with 1% of working piles tested to 1.5 × applied load, $\gamma_b$ = 1.7, $\gamma_s$ = 1.4.
  $R_{cd}$ = (3108/1.7 + 1045/1.4) = 2575 kN ~ 2664 kN and will be acceptable with the length of pile increased to 5.1 m in the stiff clay. The overall depth is now

$D$ = 6 m (soft clay) + 5.1 + 2.0 + 0.8 = 13.9 m

*DA1 combination 2* will also be used to check the resistance of the 13.9 m long 1 m diameter, straight-shafted inner piles:

$R_{bk}$ = (9 × 190 × π/4 × 1.0²)/1.4 = 859 kN
$R_{sk}$ = (0.5 × 190 × π × 1.0 × (13.9 − 6))/1.4 = 1684 kN
and $R_{cd}$ = (859/1.7 + 1684/1.4) = 1707 kN

This can be considered satisfactory as the maximum load on these piles is less than half the load on the outer piles. *DA1 combination 1* is also satisfactory by inspection.

The vertical load on the pile group in respect of overturning is a permanent stabilising action and the horizontal wind loading is a variable unfavourable action.

*For DA1 combination 1* from Table 4.1, $\gamma_G = 1.0$ and $\gamma_Q = 1.5$ giving

Design actions $V_d' = 1.0 \times 30,000 = 30,000$ kN

and

$H_d' = 1.5 \times 1500 = 2250$ kN.

The group layout may be taken as that in Figure 6.45, that is, 13 m × 10.4 m, but the depth is now 13.9 m with 7.9 m into the clay.

Eccentricity of loading in respect of base of pile group = 2250(35 + 13.9)/30,000 = 3.66.

For winds in a north–south direction, width of equivalent block foundation = 10.4 − (2 × 3.66) = 3.08 m.

The overall dimensions of the transformed block foundation are 13.0 × 3.08 × 7.9 m deep. The material factor, $\gamma_{cu}$, for set M1 in Table 4.2 is 1.0, giving characteristic $c_u = 190$ kN/m². The resistance factors for spread foundations as NA Table A.NA.5 are $\gamma_{Rv} = \gamma_{Rh} = 1.0$ and no model factor is required.

Applying the Brinch Hansen bearing capacity factors for $\phi = 0°$ assuming an equivalent block foundation as above gives $N_c = 5.2$, $s_c = 1.05$, $d_c = 1.1$ and $i_c = 0.96$; $N_\gamma = 1.0$, $s_\gamma = 0.95$ and $d_\gamma = i_\gamma = 1.0$.

Characteristic unit base resistance = (190 × 5.2 × 1.05 × 1.1 × 0.96) + (0.5 × 9.81 × 1.8 × 3.08 × 0.95) = 1095 + 25 = 1120 kN/m².

Hence, $R_{bk} = 1120 \times 13.0 \times 3.08 = 44,845$ kN.

Ignoring the resistance of the perimeter of the block,
$R_{cd} = (44,845/1.0) = 44,845$ kN > $V_d' = 30,000$ kN and satisfactory.

*For DA1 combination 2* from Table 4.1, $\gamma_G = 1.0$ and $\gamma_Q = 1.3$ giving

Design actions $V_d' = 1.0 \times 30,000 = 30,000$ kN

and

$H_d' = 1.3 \times 1500 = 1950$ kN.

Eccentricity of loading = 1950(35.0 + 13.9)/30,000 = 3.18 m
Width of equivalent block foundation = 10.4 − 2 × 3.18 = 4.04 m
Dimensions of equivalent block foundation are 13.0 × 4.04 × 7.9 m deep

The material factor and block bearing factor are again unity. The Brinch Hansen factors are now modified to give

Characteristic unit base resistance = (190 × 5.2 × 0.9 × 1.3 × 0.86) + (0.5 × 9.81 × 1.8 × 4.04 × 0.94) = 994 + 27 = 1021 kN/m²

Hence, $R_{bk} = 1021 \times 13.0 \times 4.04 = 53,623$ kN
and $R_{cd} = 53,623/1.0) = 53,623$ kN > $V_d' = 30,000$ kN and satisfactory
Alternatively applying Equation D1 in EC7 NA D to the equivalent block,

$R/A = (\pi + 2)c_u \, b_c \, s_c \, i_c + q$   where $b_c = 1.0$, $i_c = 1.0$, and $s_c = 1 + 0.2(B/L)$

$R/A = 5.14 \times 190 \times (1 + 0.2 \times 3.08/13.0) + 1.8 \times 9.81 \times 13.9 = 1267$ kN/m²

$R_{cd} = 1267 \times 13 \times 3.08 = 50{,}730$ kN which, with the material factor and block bearing factor again being unity, is greater than $V'_d = 30{,}000$ kN

Checking compliance with EC7 with respect to sliding:

In the following calculations, the passive resistance of the soil to horizontal movement of the piles has been ignored.

*For DA1 combination 1*, the base area of the equivalent block using the factored values of $V'$ and $H'$ is $13.0 \times 3.08 = 40.0$ m$^2$. The horizontal resistance factor is $\gamma_{Rh} = 1.0$ and no model factor is required.

Therefore, design resistance to sliding $= 190 \times 1.0 \times 40.0 = 7600$ kN which is greater than $H'_d = 1.5 \times 1500 = 2250$ kN.

*For DA1 combination 2*, base area $= 13.0 \times 4.04 = 52.5$ m$^2$ and $R_{cd} = 1.0 \times 190 \times 52.5 = 9975$ kN which is greater than $H'_d = 1950$ kN.

It is also necessary to confirm that the total settlements and tilting of the structure are within safe limits. The following calculations are carried out using characteristic actions to verify the serviceability limit state.

Calculate first the immediate and consolidation settlements under permanent actions but exclude the wind load. Because the piles have under-reamed bases which carry the major proportion of the load, the base of the equivalent raft will be close to pile base level. Apply the results of the equivalent pile group with the 1.8 m under-ream, the approximate overall dimensions of the equivalent raft outside the toes of the pile bases are $13.8 \times 11.2$ m. Therefore,

$$\text{Overall base pressure beneath raft} = \frac{30{,}000}{13.8 \times 11.2} = 194 \text{ kN/m}^2$$

Assume a value of $E_u$ for the glacial clay of 80 MN/m$^2$ and a value of $m_v$ of 0.05 m$^2$/kN. From Figure 5.18 for $L/B = 13.8/11.2 = 1.2$, $H/B = \infty$ and $D/B = 7.9/11.2 = 0.7$ (ignoring the soft clay), $\mu_1 = 0.75$ and $\mu_0 = 0.92$. Therefore,

$$\text{Immediate settlement} = \frac{0.75 \times 0.92 \times 194 \times 11.2 \times 1000}{80 \times 1000} = 19 \text{ mm}$$

From Figure 5.11, the average vertical stress at the centre of a layer of thickness $2B$ is $0.3 \times 194 = 58$ kN/m$^2$.

The depth factor $\mu_d$ for $D/\sqrt{LB} = 0.63$ is 0.81 and the geological factor $\mu_g$ is 0.5. Therefore, from Equations 5.23 and 5.24,

$$\rho_c = \frac{0.5 \times 0.81 \times 0.05 \times 58 \times 2 \times 11.2 \times 1000}{1000} = 26 \text{ mm}$$

Part of the imposed loading will not be sustained and will not contribute to the long-term settlement. Thus, the total settlement under the vertical load of 30,000 kN will probably not exceed 30 mm.

It is necessary to estimate the amount of tilting which would occur under sustained wind pressure, that is the immediate settlement induced by the horizontal wind force of 1500 kN producing a pressure under the combined vertical and horizontal loading of $30{,}000/(13 \times 5.54) = 416$ kN/m$^2$ on the *equivalent raft* caused by the eccentric loading. For $L/B = 13/5.54 = 2.3$, $H/B = \infty$, and $D/B = 7.9/3.06 = 2.58$, $\mu_1 = 1.0$ and $\mu_0 = 0.9$ giving

$$\text{Immediate settlement} = \frac{1.0 \times 0.9 \times 416 \times 5.54 \times 1000}{80 \times 1000} = 26 \text{ mm}$$

Of this amount, 16 mm is due to vertical loading only, giving a tilt of 10 mm due to wind loading. A movement of this order would have a negligible effect on the stability of the tower.

The horizontal force on each pile if no wind load is carried by the pile cap is 1500/22 = 68 kN. A pile 1 m in diameter can carry this load without excessive deflection (see Example 6.3).

## Example 6.6

Pressuremeter tests made at intervals of depth in a highly weathered weak broken siltstone gave the following parameters:

Pressuremeter modulus $= E_m = 30$ MN/m$^2$

Limit pressure $= p_l = 1.8$ MN/m$^2$

Creep pressure $= p_f = 0.8$ MN/m$^2$

The above values were reasonably constant with depth. Draw the deflection curve for a horizontal load applied to the head of a 750 mm pile at the ground surface up to the ultimate load and obtain the deflection for a horizontal load of half the ultimate.

$$\text{Moment of inertia of uncracked pile} = \frac{\pi \times 0.75^4}{64} = 0.0155 \text{ m}^4$$

Modulus of elasticity of pile $= 26 \times 10^3$ MN/m$^2$

Take a rheological factor of 0.8; then from Equation 6.43a,

$$\frac{1}{k_m} = \frac{2 \times 0.6}{9 \times 30}\left(\frac{0.75}{0.6} \times 2.65\right)^{0.8} + \frac{0.8 \times 0.75}{6 \times 30}$$

$$k_m = 67 \text{ MN/m}^2$$

Over elastic range from $p = 0$ to $p = p_f$, then from Equation 6.11, stiffness factor is

$$R = \sqrt[4]{\frac{26 \times 10^3 \times 0.0155}{67 \times 0.75}} = 1.68 \text{ m}$$

To allow for surface heave, assume no soil reaction from ground surface to assumed surface at 0.5 × 0.75 = 0.375 m.

At creep pressure of 0.8 MN/m$^2$, corresponding deflection $= (0.8 \times 0.75)/67 = 0.0090$ m
From Equation 6.46, corresponding lateral load applied at assumed ground surface:

$$H = \frac{0.0090 \times 1.68 \times 67 \times 0.75}{2} = 0.380 \text{ MN}$$

From Equation 6.47, slope at assumed ground surface

$$= -\frac{2 \times 0.380}{1.68^2 \times 67 \times 0.75} = -0.0054 \text{ rad}$$

Deflection at real ground surface

$$= 0.009 - \frac{0.380 \times 0.375^2}{6 \times 26 \times 10^3 \times 0.0155} - (-0.0054 \times 0.375)$$

$$= 0.0110 \text{ m}$$

$$= 11 \text{ mm}$$

Between $p = p_f$ and $p = p_l$ the upper curve in Figure 6.34b gives $k_m = 67/2 = 33.5 \text{ MN/m}^2$ and

$$R = \sqrt[4]{\frac{26 \times 10^3 \times 0.0155}{33.5 \times 0.75}} = 2.00 \text{ m}$$

From upper curve in Figure 6.34b:

At limit pressure of 1.8 MN/m² corresponding deflection $= 1.8 \times 0.75/33.5 = 0.0403 \text{ m}$

Corresponding lateral load at assumed ground surface $= \dfrac{0.0403 \times 2.00 \times 33.5 \times 0.75}{2}$

$$= 1.012 \text{ MN}$$

Slope at assumed ground surface $= -\dfrac{2 \times 1.012}{2^2 \times 33.5 \times 0.75} = -0.0201 \text{ rad}$

Total deflection at real ground surface

$$= 0.0110 + 0.0403 - \frac{1.012 \times 0.375^2}{6 \times 26 \times 10^3 \times 0.0155} - (-0.0201 \times 0.375)$$

$$= 0.0588 \text{ m}$$

$$= 59 \text{ mm}$$

The load–deflection curve is shown in Figure 6.46. The deflection corresponding to an applied load of half the ultimate load of 1012 kN is 20 mm.

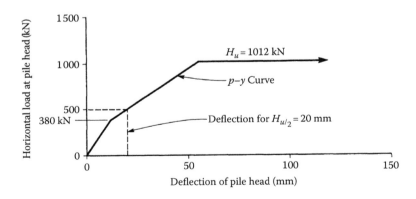

Figure 6.46 p–y curve for Example 6.6.

## Example 6.7

A lateral load of 100 kN is applied at ground surface to the free head of a 25 m long driven tubular steel pile, with an external diameter of 1300 mm and 30 mm wall thickness. The pile is fabricated from Grade S460 steel with an elastic modulus of $2.1 \times 10^5$ $MN/m^2$. The pile is driven into soft clay with an undrained strength profile of $c_u = 20 + 3z$, where $z$ is the depth below the top of the clay layer, and a unit bulk weight of 20 $kN/m^3$. Water table is taken at ground level. Using the $p-y$ method, calculate the deflection, shearing force and bending moment in the pile at a depth of 4.5 m below ground level.

*Solution 1. Using Oasys ALP program*
General data: soil model for soft clay; generated $p-y$ curves
  51 nodes selected at intervals of 0.5 m

$$EI = 2.1 \times 10^5 \times \pi(1.30^4 - 1.24^4)/64 = 50 \times 10^5 \text{ kNm}^2$$

Rate of change of undrained shear strength with depth, $dc_u/dz = 3$
100 kN horizontal load applied at first node, no restraining force, damping coefficient = 1
$E_{50} = 0.020$ being the strain at one-half the maximum stress for an undrained triaxial compression test, for a soft clay with no laboratory tests
  ALP calculates the ultimate soil resistance per unit length $(P_u)$ using $N_c$ from modified Equation 6.36 and using the nomenclature in ALP:

$$P_u = D\left\{3c_u + \sigma'_v + J\left(\frac{xc_u}{D}\right)\right\} \quad \text{for} \quad x \le x_r \quad \text{where } D \text{ is the pile diameter (1.3 m)}$$

$$P_u = 9c_u D \text{ for } x \ge x_r \text{ where } D \text{ is the pile diameter}$$

As $c_u$ varies with depth, these equations are solved at each depth until the second equation is less than the first to give $x_r$.
  The $p-y$ curve for the short-term *static* load cases is then generated at the following points:

| $P/P_u$ | $Y/Y_c$ |
|---|---|
| 0 | 0 |
| 0.29 | 0.2 |
| 0.50 | 1.0 |
| 0.72 | 3.0 |
| 1.0 | 8.0 |
| 1.0 | ∞ (2.5D) |

where $P$ is the soil resistance per unit length, $Y$ is the lateral deflection and $Y_c$ is $2.5E_{50}D$.

ALP calculates the deflection, bending moments and shear forces at each node along the pile and will present the results graphically. The key results for this example are summarised as follows:

| Node | Depth (m) | Deflection (mm) | Soil | Bending (kNm) | Shear (kN) |
|---|---|---|---|---|---|
| 10 | 4.5 | −2.7137 | Soft clay | 310.27 | −33.60 |
| Extreme values | | | | | |
| 1 | 0 | −5.8962 | | | |
| 15 | 7.0 | | | 346.50 | |
| 26 | 12.5 | | | | 35.739 |
| 34 | | 0.2269 | | | |

*Solution 2. Spreadsheet/hand calculation* (using the nomenclature from the text)
As $c_u$ varies with depth, Equation 6.37 will use a modified $J$ of 0.33 (between the Matlock values for soft and stiff clay) and an average value for $c_u$ over the length of the pile to give an initial estimate for the critical depth:

$$x_r = \frac{6 \times 1.3}{(20 \times 1.3)/57.5 + 0.33} = 10.0 \text{ m}$$

From Equation 6.39, the deflection at strain $\varepsilon_c$, $y_c = 2.5\varepsilon_c B$, and for $\varepsilon_c = 0.020$ as Matlock, $y_c = 0.065$ m.

Also as $c_u$ varies with depth, $Nc$ will be calculated using the ALP modification of Equation 6.36 for $p_u$ for $x \le x_r$ and $p_u = 9c_u B$ for $x \ge x_r$ where $B$ is the pile diameter.

The following table is a shortened form of the spreadsheet calculation for $p_u$. The values are as in the ALP calculation which used nodes at 0.5 m depth increments (the values of $p_u$ in italics used to calculate $p$).

| Depth x (m) | $c_u$ (kN/m²) | $\sigma'_v$ (kN/m²) | $N_c$ | $p_u$ x < $x_r$ (kN/m) | $p_u$ x > $x_r$ (kN/m) |
|---|---|---|---|---|---|
| 0 | 20 | 0 | 3.00 | 78.0 | 234 |
| 0.5 | 21.5 | 5.095 | 3.43 | 95.8 | 251.6 |
| 1.0 | 23.0 | 10.19 | 3.83 | *114.4* | 269.1 |
| 1.5 | 24.5 | 15.285 | 4.20 | *134.8* | 286.7 |
| 2.0 | 26.0 | 20.38 | 4.55 | *153.9* | 304.2 |
| 2.5 | 27.5 | 25.475 | 4.89 | *174.7* | 321.8 |
| 4.5 | 33.5 | 45.855 | 6.10 | *266.0* | 391.9 |
| 9.5 | 48.5 | 96.805 | 8.65 | *545.4* | 567.4 |
| 10.0 | 50.0 | 101.9 | 8.88 | 577.4 | 585.0 |
| 14.5 | 63.5 | 147.755 | 10.90 | 900.1 | 742.9 |
| 19.5 | 78.5 | 198.705 | 13.03 | 1329.8 | *918.5* |
| 24.5 | 93.5 | 249.655 | 15.09 | 1834.6 | *1093.9* |

A plot of $p_u$ using the separate equations from the above table is shown in Figure 6.47 confirming $x_r \sim 10$ m.

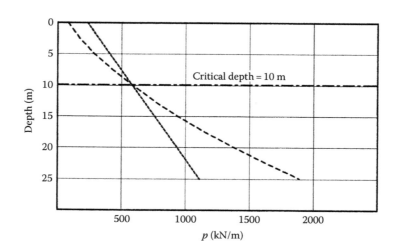

Figure 6.47    Depth v p to determine critical depth for Example 6.7.

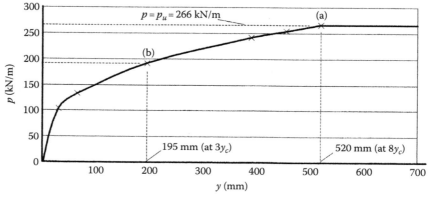

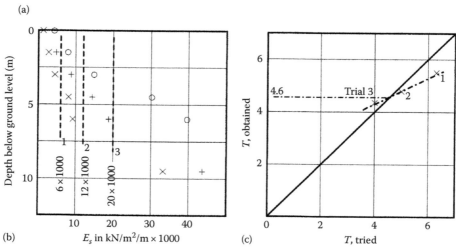

Figure 6.48    Determination of p–y curve for Example 6.7 (a) p–y curve at 4.5 m depth, (b) trial values of $E_s$ and (c) trial values of T.

For static loading, the $p-y$ curve shape in Figure 6.32 is defined by $p = 0.5p_u\sqrt[3]{y/y_c}$ as in Equation 6.38. Deflection at point (b) on Figure 6.32, $3y_c = 3 \times 65 = 195$ mm and at point (a) $8y_c = 520$ mm for $p = p_u$. In order to provide the $p-y$ curve at 4.5 m depth, Equation 6.38 is used in the following table to calculate $p$ for selected $y$ values at a depth of 4.5 m and shown in Figure 6.48a

| $y$ | $y/y_c$ | $p$ (kN/m) |
|---|---|---|
| 0 | 0 | 0 |
| 30 | 0.46 | 102.8 |
| 65 | 1.0 | 175.8 |
| 195 | 3.0 | 191.8 |
| 390 | 6.0 | 241.7 |
| 455 | 7.0 | 254.4 |
| 520 | 8.0 | 266.0 |

The first trial with stiffness factor $T$ for normally consolidated clays with linear increase in stiffness is defined in Equation 6.12:

$$T = \sqrt[5]{\frac{EI}{n_h}}$$

where
  $EI = 50 \times 10^5$ kNm$^2$
  $n_h = KB/x$, $K$ being the soil modulus

With a trial $n_h = 500$ kN/m$^3$ for soft clay, $T = 6.31$ and $L/T = 25/6.31 = 3.96$ (~4) indicating a long pile.

Then from Equation 6.34 for a 'long pile' and no applied moment, $y = (C_yHT^3)/EI$ and $C_y = A_y$ as given in Figure 6.26a. The relationship between $T$ and $y$ is recalculated as shown in the summary table of the spreadsheet:

| Depth $x$ (m) | 0 | 1.5 | 3.0 | 4.5 | 9.5 | 14.5 | 19.5 | 24.5 |
|---|---|---|---|---|---|---|---|---|
| $Z = x/T$ | 0 | 0.238 | 0.475 | 0.713 | 1.506 | 2.298 | 3.090 | 3.883 |
| $C_y = A_y$ | 2.4 | 1.85 | 1.5 | 1.1 | 0.4 | 0.05 | −0.1 | −0.1 |
| $y$ (m) | 0.0126 | 0.0093 | 0.0075 | 0.0055 | 0.0020 | 0.0003 | −0.0005 | −0.0005 |
| $p$ (kN/m) | 22.6 | 35.0 | 47.8 | 58.5 | 85.6 | 58.3 | −90.8 | −108.2 |
| $E_s$ (kN/m$^2$/m) | 1,381 | 2,899 | 4,877 | 8,139 | 32,758 | 178,505 | 139,013 | 165,574 |
| $n_h$ (kN/m$^3$) | 0 | 5,200 | 2,600 | 1,733 | 821 | 538 | 400 | 318 |
| New $T$ | 0 | 3.95 | 4.54 | 4.92 | 5.71 | 6.22 | 6.60 | 6.91 |

$E_s$ is plotted against depth and a new value with bias towards the top 9.5 m depth is selected as 6000 kN/m$^2$/m and $n_h$ recalculated as $n_h = E_s B/x$, to provide an 'obtained' $T$ which over the depth of pile averages 5.51.

Further iterations are tried with $T = 5.0$ and 4.0 which result in selecting $E_s = 12,000$ kN/m$^2$/m and 20,000 kN/m$^2$/m respectively (Figure 6.48b). These trials are plotted on the 'tried v obtained' graph which intersects the equality line at approximately $T = 4.6$ (Figure 6.48c).

If further iterations are carried out for successive $E_s$ values, $T$ settles at 4.54.

Using Equations 6.22, 6.24 and 6.25 at 4.5 m depth, for $Z = x/T = 0.99$, hence $A_y = 1.0$, $A_m = 0.7$ and $A_v = 0.3$:

Deflection $= (1.0 \times 100 \times 4.54^3)/50 \times 10^5 = 0.0019$ m (cf ALP 0.00271 m)
Bending moment $= 0.7 \times 100 \times 4.54 = 317$ kNm (cf ALP 310.5 kNm)
Shear force $= 0.3 \times 100 = 30$ kN (cf ALP 33.6 kN)

The Matlock and Reese charts provide a reasonable agreement with ALP, but the method is subject to interpolations. (See also Worked Example 8.2.)

## REFERENCES

6.1 Radhakrishna, H.S. and Adams, J.I. Long-term uplift capacity of augered footings in fissured clay, *Canadian Geotechnical Journal*, 10 (4), November 1973, 647–652.
6.2 Meyerhof, G.G. and Adams, J.I. The ultimate uplift capacity of foundations, *Canadian Geotechnical Journal*, 5 (4), November 1968, 225–244.
6.3 (a) The strength of grouted pile-sleeve connections for offshore structures: Static tests relating to sleeve buckling. Wimpey Offshore Engineers and Constructors Ltd. HM Stationery Office, Offshore Technology Series, OTH.85.223, 1986. (b) The strength of grouted pile-sleeve connections. A composite report for DoE. HMSO Offshore Technology Services, OTH.86.210, 1986. (c) A study of length, longitudinal stiffening and size effects on grouted pile-sleeve connections. Wimpey Offshore Engineers and Constructors Ltd. HMSO Offshore Technology Series, OTH.86.230, 1987.
6.4 Harwood, R.G., Billington, C.J., Buitrago, J., Sele, A.B. and Sharp, J.V. Grouted pile to sleeve connections: Design provisions for the new ISO Standard for Offshore Structures, *Offshore Mechanics and Arctic Engineering (OMAE) Conference*, Florence, Italy, June 1996.
6.5 Littlejohn, G.S. Soil anchors, *Proceedings of the Conference on Ground Engineering*, Institution of Civil Engineers, London, UK, 1970, pp. 41–44.
6.6 Hutchinson, J.N. Discussion on ref. 6.5, *Conference Proceedings*, pp. 85.
6.7 Wimpey Laboratories Ltd., unpublished report.
6.8 Freeman, C.F., Klajnerman, D. and Prasad, G.D. Design of deep socketed caissons into shale bedrock, *Canadian Geotechnical Journal*, 9 (1), February 1972, 105–114.
6.9 Reese, L.C., Cox, W.R. and Koop, F.B. Analysis of laterally loaded piles in sand, *Proceedings of the Offshore Technology Conference*, Houston, TX, 1974, Paper No. OTC 2080.
6.10 Garassino, A., Jamiolkowski, M. and Pasqualini, E. Soil modulus for laterally-loaded piles in sands and NC clays, *Proceedings of the Sixth European Conference*, ISSMFE, Vienna, Austria, 1976, Vol. I (2), pp. 429–434.
6.11 Hansen, J.B. The ultimate resistance of rigid piles against transversal forces, *Danish Geotechnical Institute*, Bulletin No. 12, 1961, 5–9.
6.12 Guide to the design, manufacture and installation of concrete piles, American Concrete Institute, Report ACI 543R-12, 2012.
6.13 Reese, L.C. and van Impe, W.F. *Single Piles and Pile Groups under Lateral Loading*, 2nd ed. Taylor & Francis Group, Abingdon, England, 2010.
6.14 Reese, L.C. and Matlock, H. Non-dimensional solutions for laterally-loaded piles with soil modulus assumed proportional to depth, *Proceedings of the Eighth Texas Conference on Soil Mechanics and Foundation Engineering*, Austin, TX, 1956, pp. 1–41.
6.15 Davisson, M.T. and Gill, H.L. Laterally-loaded piles in a layered soil system, *Journal of the Soil Mechanics Division*, 89 (SM3), May 1963, 63–94.
6.16 Matlock, H. and Reese, L.C. Foundation analysis of offshore pile-supported structures, *Proceedings of the Fifth International Conference*, ISSMFE, Paris, France, Vol. 2, 1961, pp. 91–97.

6.17 Matlock, H. Correlations for design of laterally loaded piles in soft clay, *Proceedings of the Offshore Technology Conference*, Houston, TX, 1970, Paper OTC 1204.

6.18 Reese, L.C. Analysis of laterally loaded piles in weak rock, *Journal of the Geotechnical and Geoenvironmental Engineering*, 123 (11), 1997, 1010–1017.

6.19 Turner, J. Rock-socketed shafts for highway structure foundations, in *NCHRP Synthesis 360*. Transport Research Board, Washington, DC, 2006, 54–99.

6.20 Reese, L.C. and Welch, R.C. Lateral loading of deep foundations in stiff clay, *Journal of the Geotechnical Engineering Division*, 101 (GT7), 1975, 633–640.

6.21 Zhang, L., Ernst, H. and Einstein, H.H. Non-linear analysis of laterally loaded rock-socketed shafts, *Journal of the Geotechnical and Geoenvironmental Engineering*, 126 (11), 2000, 955–968.

6.22 Nip, D.C.N. and Ng, C.W.W. Back-analysis of laterally-loaded bored piles, *Geotechnical Engineering*, 158 (GE2), 2005, 63–73.

6.23 Baguelin, F., Jezequel, J.F. and Shields, D.G. *The Pressuremeter and Foundation Engineering*. Trans Tech Publications, Clausthal, Germany, 1978.

6.24 Clarke, B.G. *Pressuremeters in Geotechnical Design*. Blackie Academic, Glasgow, UK, 1995.

6.25 Randolph, M.F. The response of flexible piles to lateral loading, *Geotechnique*, 31 (2), June 1981, 247–259.

6.26 Davisson, M.T. and Robinson, K.E. Bending and buckling of partially embedded piles, *Proceedings of the Sixth International Conference, ISSMFE*, Montreal, Quebec, Canada, Vol. 2, 1965, pp. 243–246.

6.27 Poulos, H.G. Behavior of laterally loaded piles: I – Single piles, *Proceedings of the American Society of Civil Engineers*, Vol. SM5, May 1971, pp. 711–731.

6.28 Poulos, H.G. The design of foundations for high-rise buildings, *Proceedings of the Institution of Civil Engineers*, Vol. 163, November 2010, pp. 27–32.

6.29 Prakash, S. and Sharma, H.D. *Pile Foundations in Engineering Practice*. John Wiley & Sons, Inc., New York, 1990, pp. 373.

# Chapter 7

# Some aspects of the structural design of piles and pile groups

## 7.1 GENERAL DESIGN REQUIREMENTS

Piles must be designed to withstand stresses caused during their installation and subsequently when they function as supporting members in a foundation structure. Stresses due to installation occur only in the case of piles driven as preformed elements. Such piles must be capable of withstanding bending stresses when they are lifted from their fabrication bed and pitched in the piling rig. They are then subjected to compressive, and sometimes to tensile, stresses as they are being driven into the ground and may also suffer bending stresses if they deviate from their true alignment. Piles of all types may be subjected to bending stresses caused by eccentric loading, either as a designed loading condition or as a result of the pile heads deviating from their intended positions. Differential settlement between adjacent piles or pile groups can induce bending moments near the pile heads as a result of distortion of the pile caps or connecting beams.

The designer may need to consider the effects of unseen pile breakage caused during driving in selecting the design stresses when difficult driving conditions are expected; in other conditions, the possible imperfections in concrete cast in situ and the long-term effects of corrosion or biological decay may have to be accounted for.

Pile caps, capping beams and ground beams are designed to transfer loading from the superstructure to the heads of the piles and to withstand pressures from the soil beneath and on the sides of the capping members. These soil pressures can be caused by settlement of the piles, by swelling of the soil and by the passive resistances resulting from lateral loads transmitted to the pile caps from the superstructure.

In addition to guidance on structural design and detailing, matters of relevance to the design of piled foundations in EC2-1-1 include the following:

1. Dimensional tolerances of cast-in-place piles (see Table 4.6)
2. Partial factors for the ultimate limit state (ULS) of materials
3. The influence of soil–structure interaction caused by differential settlement
4. Strength classes of concrete and reinforcement cover for various exposure conditions
5. Slenderness and effective lengths of isolated members
6. Punching shear and reinforcement in pile caps
7. Limits for crack widths
8. Minimum reinforcement for bored piles

Many of these items have been dealt with in the previous chapters. Structural analysis, design and detailing of reinforced concrete and prestressed concrete members will not, in general, be covered in this chapter, but a particular point to be noted is that EC2 does not permit a reduction in design stresses for temporary works. BS EN 12699 allows for an increase in compressive stress generated during driving.

## 7.2 DESIGNING REINFORCED CONCRETE PILES FOR LIFTING AFTER FABRICATION

The reinforcement of piles to withstand bending stresses caused by lifting has to be considered only in the case of precast reinforced (including prestressed) concrete piles. Bending takes place when the piles are lifted from their horizontal position on the casting bed for transportation to the stacking area. Overstressing will occur when the concrete is immature. Timber piles in commercially available lengths which have a cross-sectional area sufficiently large to withstand driving stresses will not be overstressed if they are lifted at the normal pick-up points. Splitting could occur if attempts were to be made to lift very long piles fabricated by splicing together lengths of timber, but there is no difficulty in designing spliced joints so that the units can be assembled and bolted together while the pile is standing vertically in the leaders of the piling frame. Steel piles with a cross-sectional area capable of withstanding driving stresses will not be overstressed when lifted in long lengths from the horizontal position in the fabrication yard.

Reinforced concrete and prestressed concrete piles have a comparatively low resistance to bending, and the stresses caused during lifting may govern the amount of longitudinal reinforcing steel needed. These considerations are principally concerned with piles cast on the project site using the techniques described in Section 2.2.2. In the United Kingdom, driven precast concrete piles are usually factory made in either single lengths or as the proprietary jointed types described in Section 2.2.3, where specially designed facilities for handling and transport are available. The safe lifting points should be marked on the pile.

The length and cross section of the pile are first obtained from consideration of the resistance of the soil or rock as described in Chapters 4 and 6. Then for a given length and cross section, the pick-up point is selected, having regard to the type of piling plant and cranage to be employed and the economies which may be achieved by lifting the piles from points other than at the ends or the centre, as shown in Figure 7.1. The bending moment due to the factored self-weight of the pile is calculated corresponding to the selected pick-up point. The design bending resistance $M$ of the pile as a beam is then determined using EC2 rules and the partial factors for concrete and reinforcement as given in Table 7.3. The applied design bending moment $M_{Ed}$ is then compared with $M$ so that $M_{Ed} \leq M$.

Table 7.1 gives the bending moments due to self-weight when square piles are lifted at the various pick-up points shown in Figure 7.1a through h. Table 7.2 shows the maximum lengths of square-section piles for given reinforcement for a selection of pick-up points. The table is based on C500 reinforcing bars with a characteristic steel strength $f_{yk}$ of 500 kN/mm$^2$ and a C40/50 grade concrete with a characteristic strength $f_{ck}$ of 40 kN/mm$^2$. The table relies only on bottom steel in tension and no account is taken of top steel in compression. If longer piles are required, then doubly reinforced beams (i.e. taking account of top and bottom steel) or increased concrete size and grade may be considered. Transverse steel (links) should follow the code requirements given in Section 2.2.2 at the head and toe of the pile, but a check should be made for shear resistance at the pick-up point—which may require an increase in link diameter or decrease in spacing. Figure 7.2 shows the bending moments for

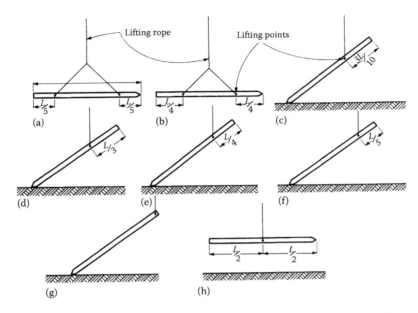

*Figure 7.1* Methods of lifting reinforced concrete piles. See Table 7.1 for (a) through (h) descriptions.

*Table 7.1*  Bending moments induced by lifting and pitching piles

| Condition | Maximum static bending moment |
|---|---|
| Lifting by two points at L/5 from each end | WL/40 (Figure 7.1a) |
| Lifting by two points at L/4 from each end | WL/32 (Figure 7.1b) |
| Pitching by one point 3L/10 from head | WL/22 (Figure 7.1c) |
| Pitching by one point L/3 from head | WL/18 (Figure 7.1d) |
| Pitching by one point L/4 from head | WL/18 (Figure 7.1e) |
| Pitching by one point L/5 from head | WL/14 (Figure 7.1f) |
| Pitching from head | WL/8 (Figure 7.1g) |
| Lifting from centre | WL/8 (Figure 7.1h) |

300 and 450 mm square piles at lengths from 5 to 40 m and the ultimate bending moments for the designated steel reinforcement as listed in Table 7.2.

Although bars are placed in the corners, cracks may appear during lifting, but by using the ULS resistance factors, these should close up once the piles are pitched. In addition to handling concerns, the concrete strength may have to be decided by the need to resist driving stresses (Section 2.2.2) or to give durability in aggressive conditions (Chapter 10 and Clauses 4 and 7 of EC2). A situation may arise in which a moment is induced in a pile due to the pile being driven just within the specified vertical tolerance (usually 1 in 75 as noted in Section 3.4.13). BS EN 12699 at Clause 7.4.3 requires that in such cases the pile performance should be reassessed. This can be done by checking a square pile as a column with an axial load and moment using the design charts in Narayanan and Beeby[7.1] (similar to BS 8110 Part 3 charts but with C500 reinforcing steel). See also Section 6.3.9 for partly embedded piles.

*Table 7.2* Maximum lengths of square section precast concrete piles for given reinforcement

| Pile size (mm) | Main reinforcement (mm) | Head and toe A | 0.33 × length from head B | 0.2 × Length from head and toe C |
|---|---|---|---|---|
| | | *Maximum length in metres for pick-up at* | | |
| 300 × 300 | 4 × 20 | 12.3 | 20.4 | 27.4 |
| | 4 × 25 | 14.9 | 24.7 | 33.3 |
| 350 × 350 | 4 × 20 | 11.7 | 19.4 | 26.2 |
| | 4 × 25 | 14.3 | 23.7 | 32.1 |
| | 4 × 32 | 17.5 | 29.0 | 39.1 |
| 400 × 400 | 4 × 25 | 13.8 | 22.8 | 30.8 |
| | 4 × 32 | 17.0 | 28.2 | 38.0 |
| | 4 × 40 | 20.3 | 33.6 | 45.4 |
| 450 × 450 | 4 × 25 | 13.2 | 21.9 | 29,5 |
| | 4 × 32 | 16.4 | 27.2 | 36.7 |
| | 4 × 40 | 19.8 | 32.9 | 44.4 |

Notes: Concrete grade C40/50; steel grade $f_{sk}$ 500 kN/mm²; cover 40 mm to link steel.

Transverse steel depends on lifting point, but generally 6 and 8 mm bars are suitable.

The above lengths could be shortened by a 'dynamic' factor of 1.1 in difficult handling conditions.

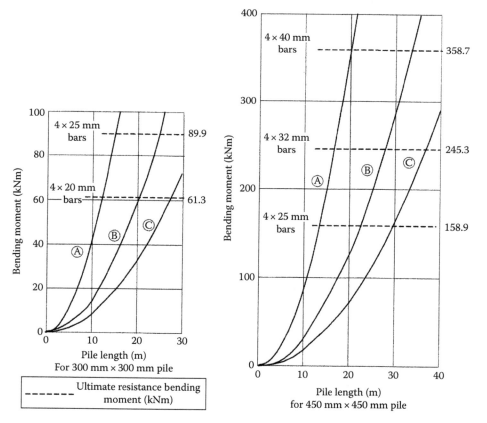

*Figure 7.2* Diagrams showing required lifting points for reinforced concrete piles of various cross sections. Pick-up points A, B and C as in Table 7.2.

## 7.3 DESIGNING PILES TO RESIST DRIVING STRESSES

It is necessary to check the adequacy of the designed strength of a pile to resist the stresses caused by the impact of the piling hammer. Much useful data to aid the estimation of driving stresses came from the initial research of Glanville et al.[7.2] and the development of pile driving analysers (PDAs) has greatly assisted the on-going research. Stress recorders are embedded in or attached to piles to measure the magnitude and velocity of the stress wave induced in the pile by blows from the hammer. The stress wave travels from the head to the toe of the pile and is partly reflected from there to return to the head. If the pile is driven onto a hard rock, the sharp reflection of the wave at the toe can cause a compressive stress at the toe which is twice that at the head, but when long piles are driven into soil of low resistance, the tensile stress wave is reflected, causing tension to develop in the pile. It can be shown from simple impact theory that the *magnitude* of the stress wave depends mainly on the height of the drop. This is true for a perfectly elastic pile rebounding from an elastic material at the toe. In practice, there is plastic yielding of the soil beneath the toe, and the pile penetrates the soil by the amount described as the 'permanent set.' The mass of the hammer is then important in governing the length of the stress wave and hence the efficiency of the blow in maintaining the downward movement of the pile.

The simplest approach to ensuring that driving stresses are within safe limits is to adopt design stresses under static loading such that heavy driving is not required to achieve the depth of penetration required for the calculated ultimate bearing capacity. The usual practice is to assume that the dynamic resistance of a pile to its penetration into the soil is equal to its ultimate static load-bearing capacity and then to calculate the permanent set in terms of blows per unit penetration distance to develop this resistance, using a hammer of given rated energy or mass and height of drop. The driving stress is assumed to be the ultimate driving resistance divided by the cross-sectional area of the pile, and this must not exceed the design stress on the pile material. As already stated in Section 1.4, the dynamic resistance is not necessarily equal to the static load-bearing capacity. However, if soil mechanics calculations as described in Chapter 4 have been made to determine the required size and penetration depth necessary to develop the ultimate bearing capacity, then either a simple dynamic pile driving formula or, preferably, stress wave theories can be used to check that a hammer of a given mass and drop (or rated energy) will not overstress the pile in driving it to the required penetration depth. If at any stage of penetration the stresses are excessive, a heavier hammer must be used, but if greater hammer mass and lesser drops still cause overstressing, then other measures, such as pre-boring, drilling below the pile toe or using an insert pile having a smaller diameter, must be adopted.

It is important to note that in many instances the soil resistance to driving will be higher than the value of ultimate bearing capacity as calculated for the purpose of determining the pile design capacity. This is because calculations for ultimate bearing capacity are normally based on average soil parameters or, where data are limited, more conservative parameters are assumed. Hence, when determining resistance to driving, the possible presence of soil layers stronger than the average must be considered in a separate calculation of ultimate bearing capacity. Also, in cases where negative skin friction is added to the applied load to give the unfavourable action on the pile, the soil strata within which the downdrag is developed will provide resistance to driving at the installation stage.

A widely used method of calculating driving stresses is based on the stress wave theory developed by Smith[7.3]. The pile is divided into a number of elements in the form of rigid masses. Each mass is represented by a weight joined to the adjacent element by a spring as shown in the case of modelling a pile carrying an axial compression load in Figure 4.29. The hammer, helmet and packing are also represented by separate masses joined to each other and to the pile by springs. Shaft friction is represented by springs and dashpots attached to the sides of the masses which can exert upward or downward forces on them. The end-bearing spring can act only in compression. The resistance of the ground at toe is assumed to act as a resisting force to the downward motion of the pile when struck by the hammer. Friction on the pile shaft acts as a damping force to the stress wave as determined from the side springs and dashpots. For each blow of the hammer and each element in the hammer–pile system, calculations are made to determine the displacement of the element, the spring compression of the element, the force exerted by the spring, the accelerating force and the velocity of the element in a given interval of time. This time interval is selected in relation to the velocity of the stress wave and a computer is used to calculate the successive action of the weights and springs as the stress wave progresses from the head to the toe of the pile. The output of the computer is the compressive or tensile force in the pile at any required point between the head and toe.

The objectives of applying computer software to the wave equation analysis are to aid equipment selection, provide a 'driveability' analysis, develop driving criteria (i.e. set per blow) and determine bearing capacity of the pile. The input to the computer comprises factual data such as the length and weight of the pile and the weight and fall of the hammer; other data rely on estimates such as the hammer efficiency, the quake (elastic compression) and damping properties of the soil, and the elastic modulus of the soil.

The efficiency and energy versus blow rate of the hammer are obtained from the manufacturer's rating charts, but these can change as the working parts become worn. The elastic modulus and coefficient of restitution of the packing may also change from the commencement to the end of driving. The elastic compression of the ground is usually taken as the elastic modulus under static loading, and this again will change as the soil is compacted or is displaced by the pile. Thus, the wave equation can never give exact values throughout all stages of driving, but as a result of the large amount of data now available and well-researched correlations between calculated stress values and observations of driving stresses in instrumented piles, the principle is widely accepted.

The basic Smith idealisation represents a pile being driven by a drop hammer or a single-acting hammer. Diesel hammers have to be considered in a different manner because the energy transmitted to the pile varies with the resistance of the pile as it is being driven down. At low resistances, there are low energies per blow at a high rate of striking. As the pile resistance increases, the energy per blow increases and the striking rate decreases. When predictions are being made of the ability of a particular diesel hammer to drive a pile to a given resistance, consideration should be given to the range of energy over which the hammer may operate. Hydraulic hammers are now generally preferred over diesel hammers as they provide a more constant energy per blow for use in the analysis. Goble et al.[7.4] have published details of the GRLWEAP computer program (see Appendix C) which models diesel and other hammer behaviour realistically. The program proceeds by iterations until compatibility is obtained between the pile–soil system and the energy/blows per minute performance of the hammer. Smith[7.3] states that the commonly accepted values for quake and the damping constants for the toe and sides of the pile are not particularly 'sensitive' in the calculations and, in certain analyses, may be omitted (see Section 11.4.1).

Pile driving resistance can be computed from field measurements of acceleration and strain at the time of driving by using the dynamic PDA in conjunction with the CAPWAP® program[7.5] (Appendix C). Pairs of accelerometers and strain transducers are mounted near the pile head and the output of these instruments is processed to give plots of force and velocity versus time for selected hammer blows as shown in Figure 7.3a. The second stage of the method is to run a wave equation analysis with the pile only modelled from the instrument

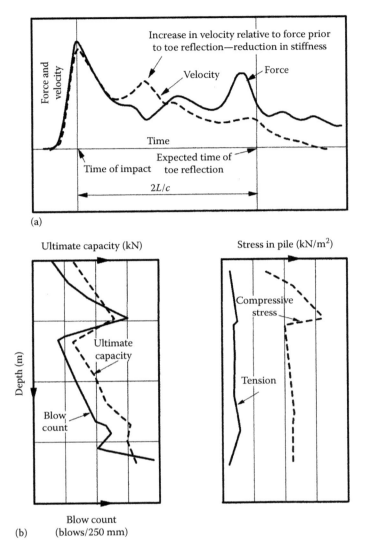

*Figure 7.3* (a) Typical output from CAPWAP indicating pile damage during driving (*L*, pile length; *c*, wave velocity). (b) Typical drivability output from GRLWEAP analysis.

location downwards. Values of soil resistance, quake and damping are assigned and the measured time-varying velocity is applied as the boundary condition at the top of the pile model. The analysis generates a force versus time plot for the instrument location and this is compared with the measured force versus time plot. Adjustments are made to the values of resistance, quake and damping until an acceptable match is reached between computed and measured values. At this stage, the total soil resistance assigned in the analysis is taken as the resistance at the time of driving. The latter is a reliable assessment of the static resistance in coarse-grained soils and rocks where time effects are negligible.

The instrumentation and field processing equipment described above provide a regular method of on-the-spot control of pile driving producing blow count and transferred energy data versus depth and are used in routine load testing applications. The GRLWEAP program will provide estimates of the tension and compression stresses in the pile during driving and

CAPWAP® will indicate pile integrity from examination of the peaks and troughs in the force and velocity printout (Figure 7.3a and b). The drivability charts for a specified hammer–pile combination can be produced to predict blow count and set, provided that the soils data input is appropriate for the application.

When assessing the results of wave equation analyses made at the project planning stage for the purpose of predicting the capability of a particular hammer to achieve the required penetration depth, due account should be taken of the effects of time on pile resistance as discussed in Section 4.3.8. Sufficient reserve of hammer energy should be provided to overcome the effects of set-up (increase in driving resistance) when re-driving a partly driven pile after a delay period of a few hours or days. If pile driving tests are made at the planning stage, it is helpful to make re-strike tests in conjunction with wave equation analyses at various time intervals after the initial drive.

The benefits from the output of the field processing system and the associated computer programs are maximised by rigorous analysis of the data by experienced engineers. Wheeler[7.6] described experiences of a field trial competition in the Netherlands when a number of firms specialising in dynamic pile testing were invited to predict the ultimate bearing capacity of four instrumented precast concrete piles driven through sands and silts to penetrations between 11.5 and 19 m. A wide range of predicted capacities was obtained. More recent comparative research was carried out by Butcher et al.[7.7] on a series of specially installed 'identical' 450 mm CFA piles 9.5 m deep in London Clay. The piles were tested using dynamic (drop weight) and rapid load (Statnamic) methods and compared with maintained load (ML) and constant rate of penetration (CRP) static tests (see Section 11.4). The analysis program, together with engineering interpretations, indicated that dynamic and rapid load testing predicted ultimate bearing capacity between 18% and 5% of the static load. While these results were clearly an improvement on the earlier predictions, it must be concluded that further work is needed to understand the mechanisms involved in rapid load testing on piles in clay and to provide improved analytical models. It is essential that the hammer blow imparts sufficient energy to the pile to overcome the resistance mobilised by the soil; a 'rule of thumb' is for the hammer mass to be 1/50 of the ultimate pile-bearing capacity depending on hammer efficiency. Paikowsky[11.39] in his extensive report on pile testing methods found that dynamic testing compared well with static loading and that Statnamic testing in rock and sand (allowing for rate effect factors of 0.96 and 0.91, respectively) gave good comparisons. His rate factors for stiff clay are not reliable[11.40].

## 7.4 EFFECTS ON BENDING OF PILES BELOW GROUND LEVEL

Slender steel tubular piles and H-section piles may deviate appreciably off line during driving. As noted in Section 2.2.4, the ill effects of bending or buckling of tubular piles below ground level could be overcome by inserting a reinforcing cage and filling the pile with concrete, but such a procedure could not be adopted with H-piles. Therefore, where long H-piles are to be driven in ground conditions giving rise to bending or buckling, a limiting value must be placed on their curvature.

It is not usual to take any special precautions against the deviation of reinforced concrete piles other than to ensure that the joints between elements of jointed precast pile systems (see Section 2.2.3) are capable of developing the same bending strength as the adjacent unjointed sections. Reinforced concrete piles without joints cannot in any case be driven to very long lengths in soil conditions which give rise to excessive curvature. It is possible to inspect hollow prestressed concrete piles internally and to adopt the necessary strengthening by placing in situ concrete if they are buckled.

It is impossible to drive a pile with a sufficient control of the alignment such that the pile is truly vertical (or at the intended rake) and that the head finishes exactly at the designed position. Tolerances specified in various codes of practice are given in Section 3.4.13. If the specified deviations are exceeded, to an extent detrimental to the performance of the piles under working conditions, the misaligned piles must be pulled out for re-driving or additional piles driven. Calculations may show that minor deviations from the specified tolerances do not cause excessive bending stresses as a result of the eccentric loading. In the case of driven and cast-in-place or bored and cast-in-place piles, it may be possible to provide extra reinforcement in the upper part of the pile to withstand these bending stresses. For this reason, Fleming and Lane[3.24] recommend that checks on the positional accuracy of in situ forms of piling should be made before the concrete is placed. The methods described in Section 6.3.9 can be used to calculate the bending stresses caused by eccentric loading. The effect of the deviation is expressed as a bending moment $Pe$, where the load $P$ deviates by a distance $e$ from the vertical axis of the pile.

## 7.5 DESIGN OF AXIALLY LOADED PILES AS COLUMNS

Normally, a buckling check of axially loaded piles terminating at ground level in a pile cap or ground beam is not required; EC7 in Clause 7.8 states thats this is the case where the $c_u$ of the soil exceeds 10 kN/m². Thus, such piles need not be considered as long columns for the purpose of structural design. However, it is necessary to consider the column strength of piles projecting above the soil line, as in jetties or piled trestles.

EC2-1-1 Clause 5.8.3 defines the parameters for considering concrete piles as long columns and provides equations for calculation of the effective length to determine the buckling load. Figure 5.7 of EC2 gives examples of the effective lengths as:

Restrained at both ends in position and direction:                    $0.5L$
Restrained at both ends in position and one end in direction:         $0.7L$
Restrained at both ends in position but not in direction:             $1.0L$
Restrained at one end in position and direction and at the other end
in direction but not in position:                                     $1.0L$
Restrained at one end in position and direction and free at the other end: $2.0L$

It is then necessary to calculate the slenderness ratio $\lambda = l_o/i$ (where $l_o$ is the effective length and $i$ is the radius of gyration), and if this is lower than $\lambda_{min}$ as given in EC2 Equation 5.13N, then buckling need not be considered. An example of the calculations for the buckling load and buckling moment for a slender column is given in Narayanan and Beeby[7.1].

EC3-5 Clause 5.3.3(5) gives the critical buckling length of steel piles acting as long columns as $kH$ where $H$ is the pile length in water and soft soil and $k$ is defined in Figure 5.8 of EC2 as follows:

Connection at pile head to concrete or steel, translation fixed and rotation free:  1.0
Connection at pile head to concrete or steel, translation fixed and rotation fixed: 0.7
Connection at pile head to concrete or steel, translation free and rotation fixed:  2.0

The effects of local buckling on fully concreted cased tubular steel piles with steel grades of S235–S460 may be neglected subject to a maximum diameter to wall thickness ratio of 90 for S235 steel (EC4 for composite structures).

The 'relative slenderness' for timber piles considered as columns is defined in EC5 at Clause 6.3.2. Typical effective lengths are given by McKenzie and Zhang[2.7].

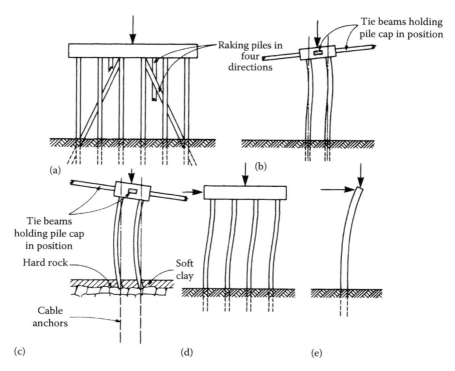

*Figure 7.4* Conditions of restraint for vertical piles: (a) restrained at top and bottom in position and direction; (b) restrained at bottom in position and direction, restrained at top in position but not in direction; (c) restrained at top and bottom in position but not in direction; (d) restrained at bottom in position and direction, restrained at top in direction but not in position; (e) restrained at bottom in position and direction, unrestrained at top in position or direction.

A pile embedded in the soil can be regarded as properly restrained in position and direction at the point of virtual fixity in the soil. The restraint at the upper end depends on the design of the pile cap and the extent to which the pile cap is restrained against movement by its connection with adjacent pile caps or structures. Some typical cases of the restraint of piles are shown in Figure 7.4a through e which correspond to the previous EC2 examples.

## 7.6 LENGTHENING PILES

Precast (including prestressed) concrete piles can be lengthened by cutting away the concrete to expose the main reinforcement or by splicing bars for a distance of 40 bar diameters. The reinforcement of the new length is then spliced to the projecting steel, the formwork is set up and the extension is concreted. It is usual to lengthen a prestressed concrete pile by this technique in ordinary reinforced concrete. The disadvantage of using the method is the time required for the new length to gain sufficient strength to allow further driving.

A rapid method of lengthening which can be used where the piles carry compressive loads or only small bending moments is to place a mild steel sleeve with a length of four times the pile width over the head of the pile to be extended. The sleeve is made from 10 mm plates and incorporates a central diaphragm which is bedded down on a 10–15 mm layer of dry sand–cement mortar trowelled onto the pile head. After setting the sleeve, a similar layer of mortar is placed on the upper surface of the diaphragm and rammed down by a

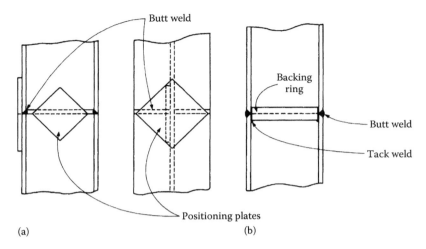

*Figure 7.5* Splicing steel piles: (a) positioning plates for H-pile and (b) backing ring for tubular piles.

square timber. The extension pile with a square end is then dropped down into the sleeve and driving commences without waiting for the mortar to set. An epoxy resin–sand mortar can be used instead of sand–cement mortar. An epoxy resin joint can take considerable tensile or bending forces, but the length of time over which the adhesion of the resin to the concrete is effective is indeterminate. The bond may be of rather short duration in warm damp conditions.

Another method of lengthening piles is to drill holes into the pile head. Then bars projecting from the extension piece are grouted into these holes using a cement grout or an epoxy resin mortar.

Timber piles are lengthened by splicing as shown in Figures 2.3 and 2.4, and steel piles are butt welded to lengthen them (Figure 7.5a and b). Backing plates or rings are provided to position the two parts of the pile while the butt weld is made, but the backing plates for the H-piles (Figure 7.5a) may not be needed if both sides of the pile are accessible to the welder. The backing ring for the tubular pile shown in Figure 7.5b is deliberately made thin so that it can be 'sprung' against the inside face of the pile. When lengthening piles in marine structures, the position of the weld should be predetermined so that, if possible, it will be situated below the seabed level and thus be less susceptible to corrosion than it would if located at a higher elevation.

The specification adopted for making welded splices in steel piles should take into account the conditions of loading and driving. For example, piles carrying only compressive loading and driven in easy to moderate conditions would not require a stringent specification with non-destructive testing for welding below the soil line. However, piles carrying substantial bending moments in marine structures would require a specification similar to that used for welding boilers or pressure vessels. Advice on specifications suitable for given conditions of loading and driving should be sought from the manufacturers of the piles[2.14].

## 7.7 BONDING PILES WITH CAPS AND GROUND BEAMS

Where simple compressive loads without bending or without alternate compressive and uplift loading are carried by precast or cast-in-place concrete piles, it is satisfactory to trim off the pile square so that the head without any projecting reinforcement is set some

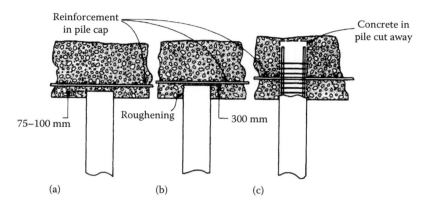

*Figure 7.6* Bonding reinforced concrete piles into pile caps: (a) compressive loading only on piles, (b) compressive loading alternating with light to moderate uplift loading on piles and (c) bending moments or heavy uplift loads on piles.

75–100 mm into the cap (Figure 7.6a). Some uplift (but not bending) can be carried if the sides of the pile are roughened over a distance of about 300 mm and cast into the cap (Figure 7.6b). Where bending moments are to be transferred from the cap to the piles (or vice versa), the concrete must be cut away to expose the reinforcing steel or prestressing tendons, which are then bonded into the cap (Figure 7.6c). It is sometimes the practice to provide steel splicing bars in the heads of prestressed concrete piles, which are exposed by cutting away the concrete after driving is complete. Alternatively, couplers can be set flush with the pile head to which further tendons or bars are attached for bonding into the cap. Splicing bars or couplers are satisfactory if the depth of penetration of the pile can be predicted accurately. If the upper part of the pile has to be cut away, they no longer have any useful function, but they can serve as a means of lengthening a pile should this be necessary.

Hydraulic pile croppers and breakers (Section 3.4.6) can either break off the excess length of a concrete pile at the required level or nibble the concrete leaving the reinforcement exposed.

Steel box, tubular or H-section piles carrying only compressive loads can be terminated at about 100–150 mm into the pile cap without requiring any special modifications to the pile to provide for bonding (Figure 7.7a). There must, however, be a sufficient thickness of concrete in the pile cap over the head of the pile to prevent failure in punching shear. Provided that the concrete in the pile cap is of adequate thickness and if the reinforcement is correctly disposed to withstand shearing and bending forces, there is no need to provide a bearing plate or other devices for transferring load at the head of an H-pile. However, where steel piles are carrying the design load permitted by the material in cross section, the thickness of concrete in the pile cap to resist punching shear may be uneconomically large. In such cases, the head of the pile should be enlarged by welding on a capping plate (Figure 7.7b) or by threading steel bars through close-fitting holes drilled in the pile (Figure 7.7c). The capping arrangements shown in the latter two figures can be used to bond the pile to the cap when uplift loads or bending moments are carried by the pile, or alternatively bonding bars can be welded to the pile. Load transfer from large-diameter tubular piles to pile caps can be achieved by welding rectangular plates around the periphery of the pile at its head.

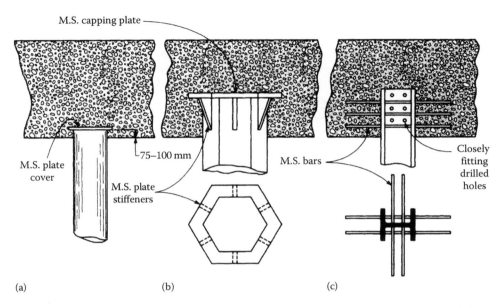

Figure 7.7  Bonding steel piles into pile caps: (a) compressive loads only on steel tubular piles, (b) hexagonal box pile carrying heavy compressive loads or uplift loads and (c) H-pile carrying uplift loading or bending moments.

## 7.8 DESIGN OF PILE CAPS

A pile cap has the function of spreading the load from a compression or tension member onto a group of piles so that, as far as possible, the load is shared equally between the piles. The pile cap also accommodates deviations from the intended positions of piles, and by rigidly connecting all the piles in one group by a massive block of concrete, the ill effects of one or more defective piles are overcome by redistributing the loads. The minimum number of small-diameter piles under an *isolated* pile cap is three. Caps for single piles should be interconnected by ground beams in two directions and for twin piles by ground beams in a line transverse to the common axis of the pair. Recommendations for the spacing of piles are given in Section 5.2.1.

A single large-diameter pile carrying a column does not necessarily require a cap. Any weak concrete or laitance at the pile head can be cut away and the projecting reinforcing bars bonded to the starter bars of the column reinforcement. Where a steel column is carried by a single large-diameter pile, the concrete is cut down and roughened to key to the pedestal beneath the column base. The heads of large-diameter piles are cast into the ground floor or basement floor concrete in order to distribute the horizontal wind forces on the superstructure to all the supporting piles.

Design of the pile cap can be considered in three ways in EC2: as a beam (Clause 9.7), as a solid slab (Clause 9.3) and as a truss (the strut and tie method in Clauses 5.6.4 and 6.5). Clause 9.8.1 deals specifically with pile cap design.

Deep pile caps are desirable for providing the stiffness necessary to distribute heavy concentrated column loads onto a pile cluster as shown in Figure 7.8. By adopting this arrangement, the column load is transferred directly into the pile heads in compression. The bending

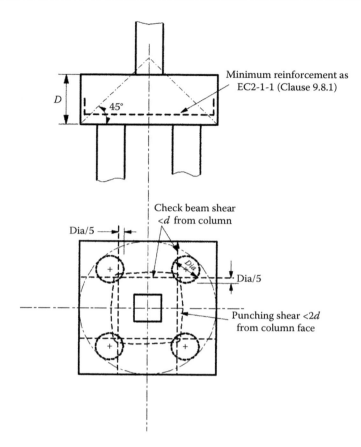

*Figure 7.8* Load transfer from column to deep four-pile cap and critical shear perimeters.

and shearing forces are negligible, requiring only the minimum proportion of steel in two directions at the bottom of the cap. The distance from the outer edge of the pile to the edge of the pile cap should be sufficient to allow the tie forces in the cap to be properly anchored; otherwise, large-radius bends may have to be provided in the reinforcement. The extent of the compressive zone can be allowed for when determining the anchorage length of the main reinforcement. This is most efficiently concentrated in the 45° stressed zone between the tops of the piles as in Figure 7.9. The minimum diameter of reinforcement is 8 mm. Pile caps constructed over large groups of piles as in Figures 7.10 and 7.11 can be designed as solid slabs in accordance with EC2.

The bending moments on rectangular pile caps will be greater than for the deep beam in Figure 7.8 and are assumed to act from the centre of the pile to the face of the nearest column or column stem (Figure 7.12a). When calculating bending moments, an allowance should be made for deviations in the positions of the pile heads, up to the specified maximum tolerance (see Section 3.4.13). Where columns carry a compressive load combined with a unidirectional bending moment, the line of action of the column load should be made to coincide with the centroid of the pile group in order to obtain a uniform distribution of load on the piles. Where an eccentric column load is applied to a non-symmetrical rectangular cap as in Figure 7.12b, then the loads in the individual piles should be calculated and the

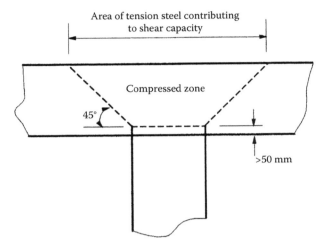

*Figure 7.9* Distribution of compressive stress from pile head to pile cap.

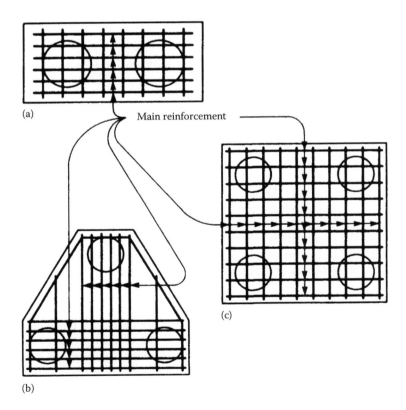

*Figure 7.10* Arrangement of reinforcement in pile caps.

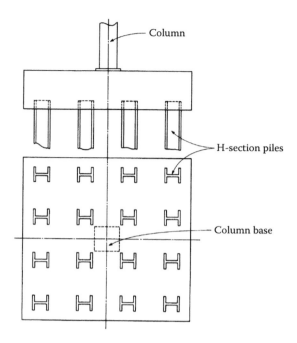

*Figure 7.11* Solid slab cap for 16-pile group.

bending moments checked as a continuous beam with top and bottom steel. As given by Mosley et al.[7.8], the distribution may be calculated from

$$F = \frac{N}{n} \pm \frac{Ne_{xx}}{I_{xx}} y_n \pm \frac{Ne_{yy}}{I_{yy}} x_n \tag{7.1}$$

where
$F$ is the load on an individual pile
$N$ is vertical load on the pile group
$n$ is the number of piles
$e_{xx}$ and $e_{yy}$ are the eccentricities about the respective centroid axes
$I_{xx}$ and $I_{yy}$ are the second moments of area about the centroid axes
$x_n$ and $y_n$ are the distances of an individual pile from the centroid axes

The checks required for both beam shear and punching shear for foundations are different from those for slabs. In EC2, the punching shear perimeter has rounded corners as shown in Figure 7.8 and the critical perimeter has to be determined iteratively, within the limit of twice the effective depth ($d$) from the column face. Beam shear is checked within the effective depth of the cap from the column face. Only the tension steel placed within the compressed zone should be considered as contributing to the shear capacity.

As an alternative design method, Figure 7.13 shows a simple strut and tie arrangement for a vertically loaded pile cap supported by two piles as given by Mosley et al. If the load from the column is $F$ and the load in each of the piles is $F/2$, then the tension $T$ in the bottom

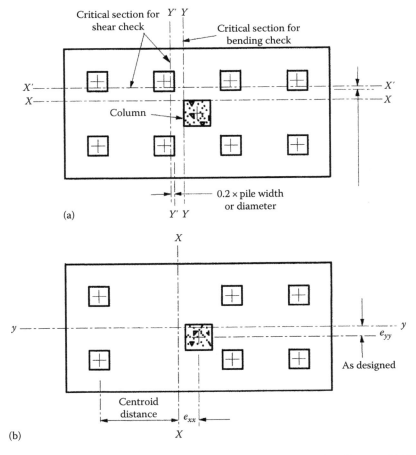

*Figure 7.12* Calculation of bending moments and shearing forces on rectangular pile caps. (a) Central vertical load on a rectangular pile cap, (b) eccentric vertical load on a pile cap for a non-symmetrical pile group. (Reproduced from Mosley, W. et al., *Reinforced Concrete Design to Eurocode 2*, 7th ed., Palgrave MacMillan, Basingstoke, UK, 2012. With permission.)

reinforcement is $Fl/2d$ where $2l$ is the distance between the centres of the pile and $d$ is the effective depth of the reinforcement. The area of reinforcement is

$$A_s = \frac{Fl}{2d} 0.87 f_{yk} \tag{7.2}$$

where $f_{yk}$ is the characteristic yield strength of the steel reinforcement.
    For a four-pile cap,

$$A_s = \frac{Fl}{4d} 0.87 f_{yk} \tag{7.3}$$

and this reinforcement should be provided in both directions at the bottom of the pile cap as in Figure 7.10. In a simple cap, steel in the top and sides of the cap will generally be nominal to control thermal cracking (Clause 7.3 of EC2) and the whole formed

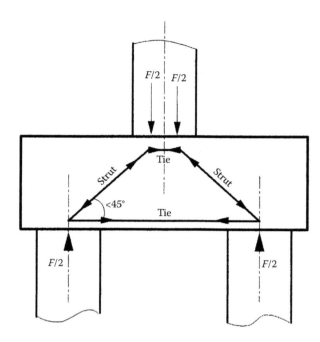

*Figure 7.13* Strut and tie model for pile cap with two piles. (Reproduced from Mosley, W. et al., *Reinforced Concrete Design to Eurocode 2*, 7th ed., Palgrave MacMillan, Basingstoke, UK, 2012. With permission.)

into a cage with horizontal links. Anti-crack steel at the faces is especially important in aggressive ground conditions.

The method is useful for deep pile caps where the strut angle can be around 45°; angles less than 30° are unrealistic as they could involve high compatibility strains. The depth of the pile cap to satisfy shear resistance will depend on the load, the diameter of the piles and the distance between them. A guide for the depth of a cap with up to six piles is 2.2–2.4 times the pile diameter, and Viggiani et al.[4.14] recommend that strut and tie design is applied to caps where the depth of the cap is greater than half the centre-to-centre distance of the piles.

Thompson et al.[7.9] describe the design of large pier pile caps (19 m × 11 m × 4 m) supported by six bored piles of 2–2.5 m diameter for the Stonecutters Bridge in Hong Kong using this truss analogy to determine the area of reinforcement. The main tower pile caps were larger and analysed with both rigid and flexible elements, using computer programs PIGLET for the rigid case and SAFE for the flexible case (Appendix C).

Where pile caps are needed, they should preferably be constructed prior to in situ ground beams which then can pass over the cap rather than frame into the cap. The connection between the cap and the ground beam is provided by starter bars and by the friction and bond between cap and beam. The concrete forming the caps may then be placed in one operation and without the inconvenience and potential weakness that result from the formation of pockets to receive the ground beams. If the beams must frame into the cap sides, an alternative to providing pockets is to place the concrete in the caps in two operations, a horizontal construction joint being formed in each cap at the level of the underside of the ground beams.

Provision often has to be made for services to pass through a foundation. If the ground beams are all situated on top of the pile caps, the routes of the services are not obstructed by any pile caps, since the services may pass over the cap through holes or sleeves left in the ground beams. The apparent economy in materials and excavation gained by framing ground beams into the sides of pile caps can easily be lost by the inconvenience it causes to other operations.

The cover to all reinforcement depends on the exposure condition and the grade of concrete being used in the pile cap, and reference should be made to Clause 4 of EC2. In particular, where concrete is cast directly against the earth, the cover should not be less than 75 mm. In all cases, the aim should be to pre-assemble reinforcement cages for pile caps and ground beams in order to avoid difficulties for steel fixers working in confined conditions in pits and trenches.

Deep pile caps can sometimes cause construction difficulties in unstable soils where the groundwater level is at a shallow depth below the ground surface. It is desirable, on the grounds of cost, to avoid construction expedients such as a well point groundwater lowering system to enable the pile cap to be constructed in dry conditions. Consideration should therefore be given to raising the level of the pile cap to bring it above groundwater level or to such a level that sump pumping from an open excavation will not cause instability by upward seepage.

The dimensions of a number of standardised types of cap for use in design using the Whittle and Beattie[7.10] methods and the RC Pile Cap software (Appendix C) are shown in Figure 7.14. The design and construction of pile caps at over-water locations is discussed in Section 9.6.3.

## 7.9 DESIGN OF PILE CAPPING BEAMS AND CONNECTING GROUND BEAMS

Pile capping beams have the function of distributing the load from walls or closely spaced columns onto rows of piles. For heavy wall loading in conjunction with transverse bending moments, the piles are placed in transverse rows surmounted by a wide capping beam (Figure 7.15a). The piles may be placed in a staggered row for walls carrying a compressive loading with little or no transverse bending moments (Figure 7.15b). A lightly loaded wall can be supported by a single row of piles beneath the centre line, provided that the beam capping the piles is restrained by tying it to transverse capping beams carrying cross walls in the structure. Attention should be given to providing adequate restraint to transverse movement and bending where ground beams are supported by micropiles. The structural designer of load-bearing brick walls will determine if the wall acts compositely with the ground beam, which, provided that the floor slab is carried by the soil, will allow some reduction in beam bending moments and load transferred to the pile cap. Any later structural alterations to the wall-beam arrangement will compromise the composite action.

When designing pile capping beams by limit state principles, it is seldom necessary to consider the serviceability limit state. However, an examination of the limit state of cracking is necessary if the beam is to be exposed to soil or groundwater which can be expected to be corrosive. The limit state of deflection should be checked if the beam is to support a wall faced with a material such as mosaic tiles, which are particularly susceptible to cracking due to small movements.

Uplift pressures due to soil swelling against the underside of floor slabs and pile capping beams cast directly onto susceptible soil must be considered. In clay soils where mature trees or hedges have been removed, the clay may swell up to 100 mm over a long period of years and soils with a plasticity index of 40–60 can swell up to 150 mm. Swelling of pyritic mudstones and shales can occur due to the growth of gypsum crystals within the laminations of these rocks. Gypsum growth can be caused by chemical and

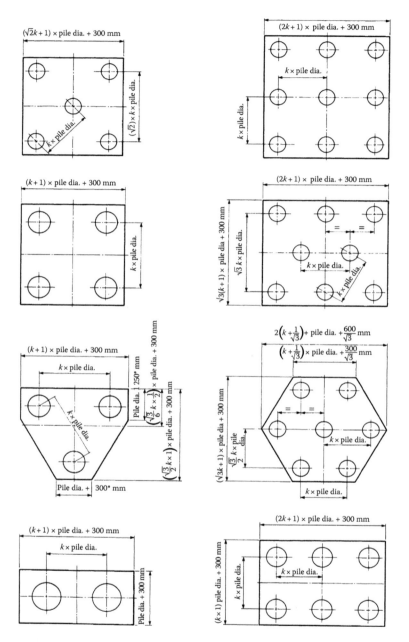

*Figure 7.14* Standard pile caps. (After Whittle, R.T. and Beattie, D., *Concrete*, 6, 34; 6, 29, 1972.)

microbiological changes consequent on changed environmental conditions[7.11]. Swelling pressures, if the upward movement of the soils is resisted by a reinforced concrete capping beam, can be of a magnitude which will cause the piles to fail as tension members or which will lift the piles out of the soil. Cracking and failure of piles, ground beams and superstructures to low-rise buildings have occurred on swelling clays in recent years,

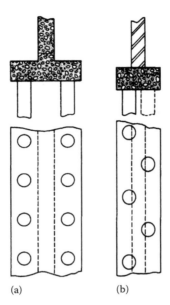

(a)                              (b)

*Figure 7.15* Arrangement of piles in capping beams: (a) heavy wall loading with transverse bending moments and (b) light wall loading with little or no transverse bending.

caused mainly by deficiencies in design such as inadequate tension reinforcement and lack of proper provision for uplift on ground beams.

Where piling and pile caps are considered for new foundations on sites previously occupied by foundries or furnaces, consideration must be given to the potential for baked and desiccated clayey soils to swell. In these conditions, the soil will not have been subjected to the seasonal swelling and drying sequence one sees with expansive clays where the effects extend to a depth of around 0.8 m below ground in the United Kingdom (ignoring the presence of roots; see Section 6.1). The desiccation cracks can be deeper and will form preferential paths for water ingress leading to differential swelling and changes in the soil shear strength. It may be feasible to remove the affected soil to a level where the pile cap can be reliably founded, that is where the $c_u$ strength is similar to that remote from the affected area. If this is not possible, then measures are needed to isolate the structure from the expansion and anchor it to stable strata[7.12].

In swelling conditions, it is essential to insert a layer of compressible material such as *Clayboard* or special low-density polystyrene to provide a void between the soil and underside of the capping beam to reduce the uplift forces transferred to the piles (Figure 7.16). *Cellcore HX* moulded void formers will compress to accommodate swelling movements up to 150 mm and support the self-weight of concrete ground beams up to 900 mm deep. Two potential modes of failure must be examined – lifting of the beam off the pile cap and bending and shear failure.

Horizontal swelling forces can also impose loads on pile capping beams due to the restraint provided by the beam to the expansion of the mass of the soil. To avoid excessive swelling forces on the inner sides of beams, they should not be left in contact with the clay (Figure 7.16). *Cellform* protection around the sides of the ground beam in contact with the swelling ground is a standard means of accommodating horizontal movement

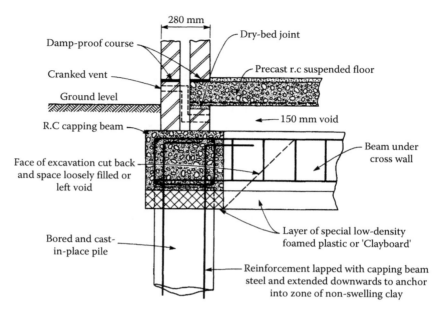

*Figure 7.16* Design of pile capping beam for swelling clay soils.

and may include the Cellcore former on the underside of the beam. Reference should be made to the latest edition of NHBC Standards[7.13] covering foundation design for newly built housing.

Ground beams are provided to act as ties or compression members between adjacent pile caps, so providing the required restraint against sidesway or buckling of the piles under lateral or eccentric loading. Ground beams and pile capping beams may have to withstand horizontal loading from the soil due to the tendency to movement of vertical piles under lateral loading. They may also be subjected to bending in a vertical direction due to differential settlement between adjacent groups of piles.

It may be permissible to allow the passive resistance of the soil against the sides of pile caps and ground beams to supplement the resistance of the piles to lateral loading. However, in clay soils, the ground will shrink away from the sides of shallow members in dry weather conditions. Trenching for building services alongside pile caps must also be considered a possibility. Although appreciable yielding of the soil must take place before its passive resistance is fully mobilised, the movement may be sufficient to cause bending failure of vertical piles.

The superimposed loading on the ground beams or pile capping beams is transferred to the piles by bonding the longitudinal reinforcing steel in the beams to the pile caps. This is straightforward for concrete piles with starter bars as previously mentioned, but driven steel piles may have deviated so that the beam reinforcement does not line up adequately with the pile. Arrangements showing the main steel in ground beams or ground floor slabs extending across steel piles are shown in Figure 7.17a and b. Subject to vertical and transverse loading conditions, the ground beams may be precast concrete units fabricated off-site to predetermined lengths to span between the pile caps as shown

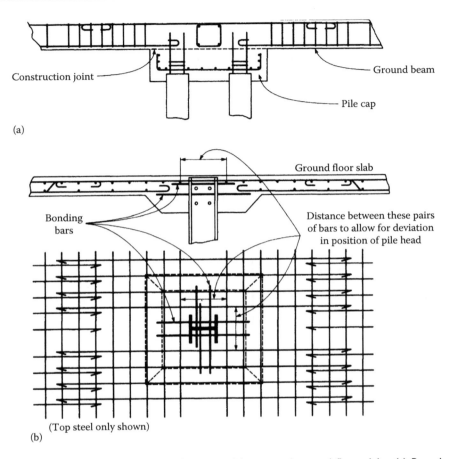

(a)

(b)

(Top steel only shown)

*Figure 7.17* Arrangement of reinforcing steel in ground beams and ground floor slabs. (a) Bored piles and (b) H-pile.

in Figure 7.18. In situ concrete is required to stitch the ends of the beam over the pile cap and to the dowel in the pile cap.

## 7.10 VERIFICATION OF PILE MATERIALS

Eurocodes provide factors for determining the design values of compressive resistance and strength in materials as summarised in Table 7.3.

The general requirement for verifying all materials is

$$X_d = \frac{X_k}{\gamma_M}$$

where
   $X_d$ is the material design value
   $X_k$ is the characteristic material property
   $\gamma_M$ is the material partial factor

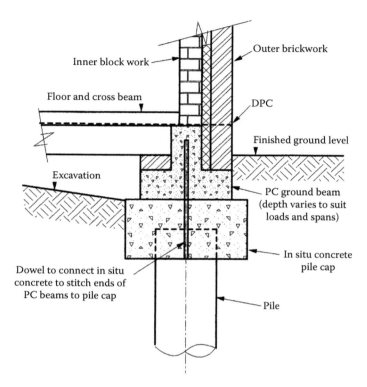

*Figure 7.18* Precast concrete ground beams connecting pile caps for low-rise building. (Courtesy of Roger Bullivant Ltd., Burton-upon-Trent, UK.)

*Table 7.3* Partial factors for reinforced concrete in compression for ULS verification as EC2-1-1 Table 2.1N

| Design situations | $\gamma_c$ for concrete | $\gamma_s$ for reinforcing steel | $\gamma_s$ for prestressing steel |
|---|---|---|---|
| Persistent and transient | 1.5 | 1.15 | 1.0 |
| Accidental | 1.2 | 1.0 | 1.0 |

The design resistance of structural elements must follow

$$\frac{N_{Ed}}{N_{Rd}} \leq 1.0$$

where
$N_{Ed}$ is the design value of the compressive force applied to the section
$N_{Rd}$ is the design resistance of the section in uniform compression

Similarly, the design moment of resistance of the section, $M_{Rd}$, must be greater than the applied moment, $M_a$.

## 7.10.1 Reinforced concrete

The design compressive strength of concrete is calculated from

$$f_{cd} = \frac{\alpha_{cc} f_{ck}}{\gamma_C}$$

where, as EC2-1-1 Clause 3.1.6, $\alpha_{cc}$ is a coefficient to take account of long-term effect, given as 0.85 in the UK National Annex (NA) for axial and flexure loading, $f_{ck}$ is the characteristic compressive cylinder strength at 28 days and $\gamma_C$ is the material partial factor as Table 7.3. For cast-in-place piles without permanent casing, $\gamma_C$ should be multiplied by a factor $k_f = 1.1$ as EC2-1-1 Clause 2.4.2.5(2).

The applied compressive stress $\sigma_{cd}$ is then compared so that

$$\sigma_{cd} = \frac{F_{cd}}{A} \leq f_{cd}$$

where
   $F_{cd}$ is the design action
   $A$ is the section area under load

The cross section should be verified by

$$N_{Rd} = A f_{cd}$$

The design compressive strength for the reinforcing steel is calculated from

$$f_{cd} = \frac{f_{yk}}{\gamma_S}$$

where $f_{yk}$ is the characteristic yield strength of reinforcement in BS EN 10080 (valid for yield strength range from 400 to 600 N/mm²). $\gamma_S$ is the material partial factor as Table 7.3.

The design tensile strength of concrete is calculated from

$$f_{ctd} = \frac{\alpha_{ct} f_{ctk0.05}}{\gamma_C}$$

where
   $\alpha_{ct}$ is a coefficient to take account of long-term effects, given as 1.0 in the NA
   $f_{ctk\,0.05}$ is the characteristic 5% proof stress of concrete (Table 3.1 of EC2-1-1)
   $\gamma_C$ is the material partial factor as Table 7.3

The design value for ultimate bond stress $f_{bd}$ for ribbed and other reinforcement is calculated from $f_{ctd}$ as given in EC2-1-1 Clause 8.4.2, depending on bond conditions.

The bond strength of pretensioned tendons $f_{bdp}$ is calculated from $f_{ctd}$ as given in EC2-1-1 Clause 8.10.2.3, depending on tendon type and bond conditions.

### 7.10.2 Steel

The cross section should be verified by

$$N_{Rd} = \frac{A\,f_y}{\gamma_{M0}}$$

where $f_y$ is the nominal value of the yield strength of the reinforcing steel as stated in EC3-1-1 Tables 3.1 and 3.2 and $\gamma_{M0} = 1.0$ as EC3-1-1 Clause 6.2.4.

### 7.10.3 Infilled steel tubes

Applicable to steel grades S235–S460 and concrete classes C20/25 to C50/60:
The cross section should be verified by adding the design resistances of the components:

$$N_{Rd} = \frac{A_a f_y}{\gamma_{M0}} + \frac{A_c \alpha_{cc} f_{ck}}{\gamma_C} + \frac{A_s f_{yk}}{\gamma_S}$$

where
  $A_a$ is the area of the steel tube
  $f_y$ is the nominal value of the yield strength of the steel $\gamma_{M0} = 1.0$
  $\alpha_{cc}$ may be taken as 1.0
  $f_{ck}$ is the characteristic compressive cylinder strength at 28 days
  $f_{yk}$ is the characteristic yield strength of reinforcement
  $\gamma_C$ and $\gamma_S$ are the material partial factors
  $A_c$ and $A_s$ are the areas of concrete and reinforcing steel respectively

### 7.10.4 Timber

The design compressive strength parallel to the grain is calculated from

$$f_{cd} = \frac{f_{c0k} \times k_{\mathrm{mod}} \times k_{sys}}{\gamma_M}$$

where
  $f_{c0k}$ is the characteristic compressive strength parallel to the grain (see examples of strength class in Table 2.1)
  $k_{\mathrm{mod}}$ is a factor depending on 'service class', usually 0.6 for piles ($k_{sys}$ may be ignored as it is a factor for laminated timber)
  $\gamma_M$ is the material partial factor for solid treated timber given as 1.3 in NA to EC5-1

The applied compressive stress $\sigma_{cd}$ is then compared so that

$$\sigma_{cd} = \frac{F_{cd}}{A} \leq f_{cd}$$

The cross section should be verified by

$$N_{Rd} = A\,f_{cd}$$

# REFERENCES

7.1  Narayanan, R.S. and Beeby, A. *Designers' Guide to EN1992-1-1 and EN1992-1-2, Eurocode 2: Design of Concrete Structures: General Rules and Rules for Buildings and Structural Fire Design.* Thomas Telford, London, UK, 2005.

7.2  Glanville, W.G., Grime, G. and Davies, W.W. The behaviour of reinforced concrete piles during driving, *Journal of the Institution of Civil Engineers*, 1, 1935, 150.

7.3  Smith, E.A.L. Pile driving analysis by the wave equation, *Journal of the Soil Mechanics Division, American Society of Civil Engineers*, 86 (SM4), 1960, 35–61.

7.4  Goble, G.G., Rausche, F. and Likins, G.E. Jr. The analysis of pile driving – A state of art, *Proceedings of the International Seminar on the Application of Stress Wave Theory on Piles*, A. A. Balkema, Stockholm, Sweden, 1980, Rotterdam, the Netherlands, 1981.

7.5  Goble, G.G. and Rausche, F. Pile driveability calculations by CAPWAP, *Proceedings of the Conference on Numerical Methods in Offshore Piling*, Institution of Civil Engineers, London, UK, 1979, pp. 29–36.

7.6  Wheeler, P. Stress wave competition, and making waves, *Ground Engineering*, 25 (9), 1992, 25–28.

7.7  Butcher, A.P., Powell, J.J.M., Kightley, M. and Troughton, V. Comparison of behaviour of CFA piles in London clay as determined by static, dynamic and rapid testing methods, in *Deep Foundations on Bored and Auger Piles*, W. F. Van Impe and P. Van Impe (ed.), Taylor & Francis Group, London, UK, 2009, pp. 205–212.

7.8  Mosley, W., Bungay, J. and Hulse, R. *Reinforced Concrete Design to Eurocode 2*, 7th edn. Palgrave MacMillan, Basingstoke, UK, 2012.

7.9  Thompson, P.A., Lee, D., Kite, S. and Colwill, R.D. Stonecutters bridge: Design of foundations, *Proceedings of the Institution of Civil Engineers*, 165 (BE1), 2012, 42–51.

7.10 Whittle, R.T. and Beattie, D. Standard pile caps, *Concrete*, 6 (1), 1972, 34–36 and 6 (2), 1972, 29–31.

7.11 Hawkins, A.B. and Pinches, G.M. Cause and significance of heave at Llandough Hospital, Cardiff – A case history of ground floor heave due to gypsum growth, *Quarterly Journal of Engineering Geology*, 20, 1987, 41–57.

7.12 Nelson, J.D. and Miller, D.J. *Expansive Soils: Problems and Practice in Foundation and Pavement Engineering*. Wiley, New York, 1992.

7.13 NHBC Standards 2013. *National House-Building Council*. Milton Keynes, UK, 2013

# Chapter 8

# Piling for marine structures

## 8.1 BERTHING STRUCTURES AND JETTIES

Cargo jetties consist of a berthing head at which the ships are moored to receive or discharge their cargo and an approach structure connecting the berthing head to the shore and carrying the road or rail vehicles used to transport the cargo. Where minerals are handled in bulk, the approach structure may carry a belt conveyor or an aerial ropeway. In addition to its function in providing a secure mooring for ships, the berthing head carries cargo-handling cranes or special equipment for loading and unloading dry bulk cargo and containers.

Berthing structures or jetties used exclusively for handling crude petroleum and its products are different in layout and equipment from cargo jetties. The tankers using the berths can be very much larger than the cargo vessels. However, the hose-handling equipment and its associated pipework are likely to be much lighter than the cranage or dry bulk-loading equipment installed on cargo jetties serving large vessels. The approach from the shore to a petroleum loading jetty consists only of a trestle for pipework and an access roadway. Where the deep water required by large tankers commences at a considerable distance from the shoreline, it is the usual practice to provide an island berthing structure connected to the shore by pipelines laid on the seabed.

In spite of the considerable differences between the two types of structure, piling is an economical form of construction for cargo jetties as well as for berthing structures and pipe trestles for oil tankers. The berthing head of a cargo jetty is likely to consist of a heavy deck slab designed to carry fixed or travelling cranes and the imposed loading from vehicles and stored cargo. The berthing forces from the ships using the berths can be absorbed by fenders sited in front of and unconnected to the deck structure (Figure 8.1a), but it is more usual for the fenders to transfer the berthing impact force to the deck and in turn to the rows of supporting piles. The impact forces may be large, and because the resistance of a vertical pile to lateral loading is small, the deck is supported by a combination of vertical and raking piles (Figure 8.1b). These combinations can also be used in structures of the open trestle type such as a jetty head carrying a conveyor (Figure 8.2).

The piles in the berthing head of a cargo jetty are required to carry the following loadings:

1. Lateral loads from berthing forces transmitted through fendering
2. Lateral loads from the pull of mooring ropes
3. Lateral loads from wave forces on the piles
4. Current drag on the piles and moored ships
5. Lateral loads from wind forces on the berthing head, moored ships, stacked cargo and cargo-handling facilities

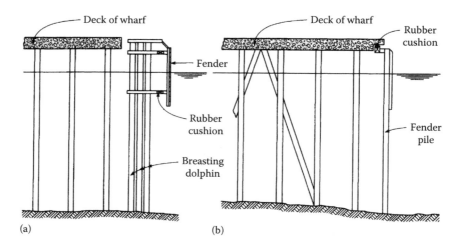

Figure 8.1 Fender piles for cargo jetties: (a) in independent breasting dolphin; (b) attached to main deck structure.

Figure 8.2 Raking and vertical piles used to restrain berthing forces in bulk-handling jetty.

6. Compressive loads from the dead weight of the structure, cargo-handling equipment and imposed loading on the deck slab
7. Compressive and uplift forces induced by overturning movements due to loads 1–5 above
8. In some parts of the world, piles may also have to carry vertical and lateral loads from floating ice and loading from earthquakes

These forces are not necessarily cumulative. Whereas wind, wave and current forces can occur simultaneously and in the same direction, the forces due to berthing impact and mooring rope pull occur in opposite directions. Berthing would not take place at times of maximum wave height, nor would the thrust from ice sheets coincide with the most severe wave action. Where containers are stored on the deck slab, the possibility of stacking them in tiers above a nominal permitted height must be considered.

This section briefly describes the various forces acting on near-shore structures. For off-shore structures in deep water and exposed conditions, the principles in BS EN ISO 19901-1 should be applied; Randolph and Gourvenec[8.1] provide information on a wide range of offshore structures.

## 8.1.1 Loading on piles from berthing impact forces

The basic equation used in calculating the force on a jetty or independent berthing structure due to the impact of a ship as it is brought to rest by the structure is

$$\text{Kinetic energy } E_k = \frac{m_s V^2}{2g} \tag{8.1}$$

where
   $m_s$ is the displacement of the ship and the mass of water moving with the ship
   $V$ is the velocity of approach to the structure

The whole of the energy as represented by Equation 8.1 is not imparted directly to the jetty piles. Kinetic energy is also absorbed by the deformation of the hull of the ship and by the compression of the fenders and of the cushioning between the fenders and their support-ing structure. Ships normally approach the jetty at a narrow angle to the berthing line, and the kinetic energy in the direction parallel to this line is generally retained in kinetic form, but a part may be lost in overcoming the resistance of the water ahead of the ship's bows, in friction against the fenders and in the pull on the mooring ropes if these are used to restrain longitudinal movement. A full consideration of the complexities involved in calculating the magnitude and direction of berthing forces cannot be dealt with adequately in this book, and the reader is referred to BS 6349-1 for guidance on design of 'maritime' structures (as opposed to 'offshore' structures). As noted in Appendix B, the full suite of codes in this standard is to be extensively revised by 2016. BS 6349-2 for the design of jetties and dol-phins provides, in Table A1, partial factors for permanent and variable actions compliant with the Eurocode limit state design approaches. Persistent variable actions include wind loads, berthing and mooring loads, and wave and current loads; reference should be made to Eurocode BS EN 1990 Clause 6.4.3 for combining actions which are considered to occur simultaneously.

On the assumption that the kinetic energy of the ship transverse and parallel to the berth-ing line has been correctly calculated, the problem is then to assess the manner in which the energy is absorbed by the fenders and their supporting piles. Taking the case of a verti-cal pile acting as a simple cantilever from the point of virtual fixity below the seabed, and receiving a blow from the ship with a force $H$ applied at a point A (Figure 8.3a), the distance moved by the point A can then be calculated by the simple method shown in Equation 6.20 and repeated here for convenience, namely,

$$\text{Distance moved } y = \frac{H(e+z_f)^3}{3EI} \tag{8.2}$$

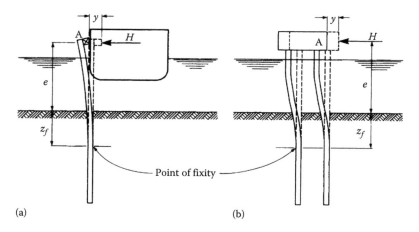

*Figure 8.3* Lateral movement of fender piles due to impact force from berthing ship: (a) single free-headed pile; (b) group of fixed-headed piles.

If the ship is brought to rest by the vertical pile as it moves the pile head over the distance $y$, then the work done by the force $H$ over this distance is given by

$$\text{Work done} = \frac{1}{2}Hy \quad = \frac{H^2\left(e+z_f\right)^3}{6EI} \tag{8.3}$$

The bending moment $M$ on the pile is equal to $H(e + z_f)$; therefore,

$$\text{Work done} = \frac{M^2\left(e+z_f\right)}{6EI} \tag{8.4}$$

If required, the more rigorous methods described in Sections 6.3.3 and 6.3.4 can be used to calculate the deflection of the pile head and hence the work done in bringing the ship to rest.

The bending moment which can be applied to a pile is limited by the design stress on the material forming the pile for normal berthing impacts or by the yield stress with abnormal berthing velocities. Thus, if the design resistance moment $M$ is used in Equation 8.4, the capacity of the pile to absorb kinetic energy can be calculated and compared to the kinetic energy of the moving ship which must be brought to rest. If the capacity of the pile is inadequate, the blow from the ship must be absorbed by more than a single pile. In practice, vertical piles are grouped together and linked at the head and at some intermediate point (Figure 8.1a) to form a single berthing dolphin or are spaced in rows or *bents* in the berthing head of a jetty structure. In the latter case, the kinetic energy of the ship may be absorbed by a large number of piles. In the case of a pile fixed against rotation by the deck slab of a structure (Figure 8.3b), it was shown in Equation 6.21 that

$$\text{Distance } y \text{ moved at point A} = \frac{H\left(e+z_f\right)^3}{12EI} \tag{8.5}$$

The bending moment caused by a load at the fixed head of a pile is equal to ½ $H(e + z_f)$, and thus the work done is the same as shown in Equation 8.4.

BS 6349-1 points out that in the case of a piled wharf erected parallel to a sloping shore line, the piles supporting the rear of the deck, being more deeply embedded than those at the front, will resist a much higher proportion of the horizontal forces imposed on the fendering. It may be necessary to consider sleeving the rearward piles to equalise the flexural resistance. If the rear of the deck is abutting a retaining wall such as a sheet pile wall, virtually the whole of the horizontal forces on the deck will be transmitted to the wall.

Where medium to large vessels are accommodated, the berthing impact is not absorbed directly by a pile or by a deck structure supported by piles. Means are provided to cushion the blow, thus reducing the risk of damaging the ship and limiting the horizontal movement of the jetty. It is also more economical during design to provide cushioning devices than to absorb forces directly on the structure. It must be noted that whereas independent berthing dolphins can be allowed to deflect over a considerable distance (and large deflections are the most efficient means of absorbing kinetic energy), the deck slab of a cargo jetty cannot be permitted to move to an extent which would cause instability in travelling cranes, stacked containers or mechanical elevators. This limitation restricts the allowable movement of such cargo jetties to a very small distance.

Where energy-absorbing fenders are provided, work Equation 8.4 is modified. Taking the simplified case shown in Figure 8.4 of a fender pile backed by a cushion block transmitting the impact to a bent of piles transverse to the berthing line, the work equation becomes the kinetic energy of moving ship absorbed by the system as shown in Figure 8.4:

$$= \tfrac{1}{2} \times H \times \Delta = \tfrac{1}{2} \times H \left( \Delta_1 + \Delta_2 \right) \tag{8.6}$$

where
    $H$ is the impact force of the first blow on the fender
    $\Delta$ is the distance moved in bringing the ship to rest after the first impact
    $\Delta_1$ is the distance moved by the compression of the cushion block
    $\Delta_2$ is the distance moved by the pile bent

In a practical design case, a limit is placed on $\Delta_2$ by the operating conditions on the jetty. Then if the cushion block is to be fully compressed by the ship moving at the maximum design approach velocity, $\Delta_1$ is known and $\Delta$ is the sum of $\Delta_1$ and $\Delta_2$. Hence, knowing the kinetic energy of the moving ship, the impact force $H$ can be calculated. This force is the

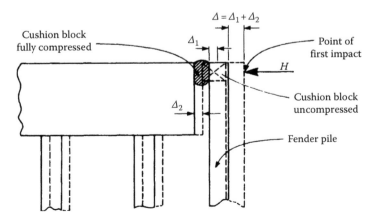

Figure 8.4 Energy absorption of fender pile cushioned at head.

sum of the force in the cushion block and the shearing force at the head of the pile. The bending moment induced in the fender pile by the action of force $H$ over distance $\Delta$ is compared with the moment of resistance of the selected pile, and the energy-absorbing capacity of the cushion block is checked to ensure that the force required for full compression is not exceeded by the force $H$. The condition shown in Figure 8.4, of a single fender pile transmitting the full force of a moving ship to a single pile bent, does not occur in practice. In a cargo jetty, the fender piles are spaced at equal distances along the berthing face and the impact is absorbed by a number of piles, depending on the closeness of their spacing and the extent to which they are tied together by intercostal beams or by a longitudinal berthing beam. An approximate rule is to assume that the blow is absorbed over a length of berthing face equal to twice the width of the jetty. BS 6349-2 recommends a minimum distance between ships moored along a jetty of 15 m.

The design process is one of trial and adjustment to determine the most economical combination of vertical fender piles with rubber or spring cushion blocks that will limit the movement of the protected jetty structure to the desired value. If the impact is delivered at a point below the head (Figure 8.5), some of the energy is absorbed by the soil, some by the deflection of the pile considered as a beam fixed at the lower end and with a yielding prop at the upper end, and some by the yielding at the prop position (i.e. the yielding of the cushion block).

As alternatives to the system of fender piles, each backed by a cushion block as shown in Figure 8.4, a group of piles can carry a rubber fender (Figure 8.6a) or a link-suspended clump fender (Figure 8.6b). For these designs, the energy transmitted to the supporting piles is equal to the kinetic energy of the moving ship, less the energy expended in compressing, displacing and raising the fender from its neutral position.

Forces act in a direction parallel to as well as normal to the berthing line. Assuming that there are no objects projecting beyond the side of the ship, the force acting parallel to the berthing line is equal to the coefficient of friction between ship and fender times the reaction normal to the berthing line. The longitudinal force tends to cause the twisting of fender piles and of pile bents set transversely to the berthing line. The rotational force on the pile bents is a maximum when the ship makes contact near the end of the jetty, and it is desirable to provide piles raking in a longitudinal direction at the two ends of the structure. The end piles in a jetty head are vulnerable to impact below the waterline from the bulbous bows of vessels provided with bow-thrust propellers.

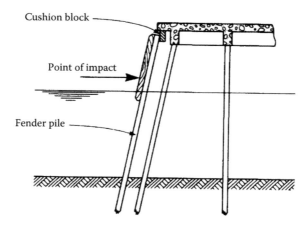

*Figure 8.5* Impact force below head of raking fender pile.

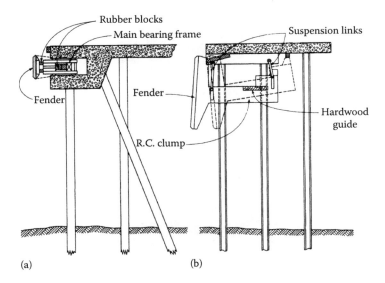

*Figure 8.6* Pile-supported fendering systems: (a) rubber-cushion fender; (b) link-suspended clump fender.

Damage to fender piles or their connections to the main structure by longitudinal forces can be avoided by spiking timber rubbing strips onto the faces of the fenders. These will be torn off by a severe impact but the pile will remain relatively undamaged.

Rubber fenders are designed to deflect in a longitudinal as well as a transverse direction and are thus capable of absorbing impact energy from both directions. Suspended fenders are given a degree of freedom to swing in a longitudinal direction and they fall clear as the ship sheers off after the first impact. Fenders can also be provided with rollers mounted on vertical axles to reduce the longitudinal frictional force on the structure.

As already noted, the facilities provided at the berthing head of an oil jetty or island berthing structure are limited to hose-handling gear and pipework. A relatively small deck area is required and the berthing structure can take the form of two main fenders spaced at a distance equal to about 0.3 times the length of the largest tanker using the berth, with two or more secondary fenders having a lower energy-absorbing capacity sited between them to accommodate smaller vessels (Figure 8.7). Frequently, the main and secondary fenders are

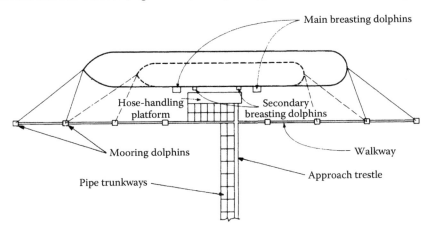

*Figure 8.7* Layout at berthing head of oil jetty.

*Figure 8.8* Steel tubular breasting dolphin.

sited in front of the hose-handling platform and pile trestles to allow them to take the full impact of the tanker without transmitting any thrust to these structures. The independent breasting dolphins, as shown in Figure 8.7, are designed so that their collapse load is not exceeded by the thrust due to the maximum berthing velocity expected.

The type of piling required for independent breasting dolphins depends on the soil conditions. Where rock, stiff clay or granular soils offering a good resistance to lateral loads are present at or at a short distance below the seabed, the dolphin can consist of a group of large-diameter circular or box-section vertical steel piles, linked together by horizontal diaphragms (Figure 8.8) and carrying a timber fender with rubber cushion blocks on the front face of the group. The face area of the fender should be large enough to prevent concentrated loading from damaging the hull of the ship. The horizontal bracing members are not rigidly connected to the pile group. This is to allow the piles to deflect freely to the maximum possible extent (BS 6349-2 suggests as much as 1.5 m) while performing their function of bringing the ship to rest.

The layout shown in Figure 8.7 can sometimes restrict the size and numbers of vessels using the berth. It can be more economical to adopt a berthing structure of the type used for cargo handling (Figure 8.1b). The berthing forces are transmitted directly to the deck so permitting vessels to berth in any position along the face. Pairs of rakers resisting the ship impact are spaced at intervals along the deck or are grouped to form 'strong points' with the deck slab acting as a horizontal beam.

Breasting dolphins for the oil loading terminal of Abu Dhabi Marine Areas Ltd., at Das Island, were designed by BP to consist of groups of vertical steel tubular piles. The main outer dolphins were formed from a group of seven piles, and the inner secondary dolphins were in three-pile groups. The conditions at seabed level, which consisted of a layer of shelly limestone cap rock underlain by a stiff calcareous marl and then a dense detrital limestone, favoured the adoption of vertical piles to absorb the berthing forces. The 36.6 m piles varied in outside diameter from 800 to 1300 mm and were drilled and socketed into rock followed by grouting of the annulus.

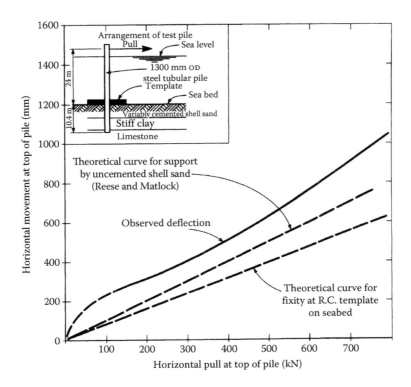

**Figure 8.9** Load–deflection curve for 1300 mm OD steel tubular pile due to horizontal load at head of pile. (After Broadhead, A., *Quart. J. Eng. Geol.*, 3, 73, 1970.)

Broadhead[8.2] described a pulling test made on a mooring dolphin pile to confirm that the lateral resistance of the weak rocks below the seabed would not be exceeded at the working load. The test pile had a bottom diameter of 1300 mm and the pull was applied at a point 24 m above the seabed. The load/deflection curve obtained at a measuring point 22.86 m above the seabed is shown in Figure 8.9 and is compared with the theoretical deflection curve assuming fixity at seabed level or support from an uncemented shell sand below seabed, using the elastic analysis of Reese and Matlock (see Section 6.3.4).

### 8.1.2 Mooring forces on piles

Mooring structures are not required to carry any pull from ropes during the operation of berthing ships other than a restraining longitudinal movement at the final stages of the berthing operation.

When the ship is fully moored, four ropes are attached to bollards or bitts fixed to the jetty structure or mounted on independent mooring dolphins in positions such as those shown in Figure 8.7. Using this type of layout, the ship is restrained from excessive ranging against the fenders and also from moving away from the berth under the influence of offshore waves or currents. The load on any individual rope due to winds or currents acting on the ship or to checking the way of a ship during berthing cannot be calculated with any accuracy. It depends on the tensioning of the rope and its angle to the berthing line.

The wind and current forces on the ship can be calculated using the equations given below for calculating the current force on a pile (Equation 8.9) or the wind force on a pile (Equation 8.13).

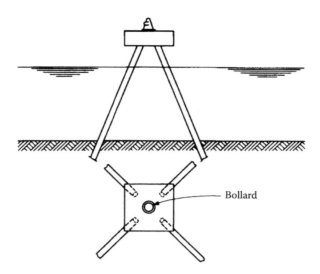

*Figure 8.10* Mooring dolphin with piles raked in two directions.

Mooring dolphins should be designed to be as rigid as possible. This is to restrict the ranging of ships which is exaggerated by the lifting and sagging of the mooring ropes. Independent mooring dolphins can take the form of pile groups set back from the berthing line as shown in Figure 8.7 or placed beyond the ends of the berthing head. Piles in mooring dolphins can be raked in two directions to resist longitudinal, transverse and torsional pulls (Figure 8.10). Where rock is present at or at a short distance below the seabed, the installation of raking piles can be difficult and anchorages are required to withstand the uplift on tension piles as described in Section 6.2.4. A vertical *jacket-type* structure comprising large-diameter vertical tubular piles, drilled and socketed into rockhead with inserted vertical dead anchors similar in construction to Figure 6.11, is to be preferred. The group of three or four piles is connected at deck level with the jacket, either steel or concrete, to provide a composite structure. For the breasting and mooring dolphins at the BP tanker terminal in the Firth of Forth, a full-face drilling bit drilled out soil and rock by reverse circulation to install four 2000 mm diameter vertical piles for each jacket. This was followed by drilling in a 560 mm steel tubular dead anchor to a depth of 15 m into the rock to provide an uplift resistance of 7.4 MN.

Guidance on the design of mooring structures and fendering is given in BS 6349-4 (undergoing comprehensive revisions in 2014).

## 8.1.3 Wave forces on piles

Jetties are normally sited in sheltered waters or in locations selected as not being subject to severe storm waves or swell. Consequently, the forces on piles due to wave action are considerably less severe than those caused by the impact from berthing or the pull from mooring ropes. Also, berthing operations are not expected to take place when heavy wave action is occurring. Therefore, it is the usual practice to disregard wave forces on piles forming the berthing head of a jetty and any associated independent dolphin structures where these are sited in sheltered waters. However, in the case of island berthing structures for large vessels, which are sited in deep and relatively unsheltered waters, the wave forces may represent a significant proportion of the total force required to be calculated. Also, piles supporting the

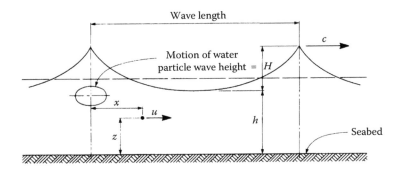

*Figure 8.11* Shape of breaking wave.

approach trestle to a jetty are not required to withstand berthing impact forces. Thus, wave forces, even in fairly sheltered waters, when combined with wind pressures on the super-structure and current drag on the piles, may produce substantial loading transverse to the axis of the trestle.

A simple approach to the calculation of wave forces on fixed structures is to assume that the maximum wave force can be expressed as the equivalent static force caused by a solitary wave of the shape shown in Figure 8.11. This shape is representative of a breaking wave. An oscillatory wave has a different shape but the factors given in Figure 8.12 and Table 8.1 for use with Equations 8.7 and 8.8 are applicable only to breaking wave conditions. Drag and inertial forces are exerted on the structure by the water particles which move in an elliptical path as shown. From the work of researchers in America in the 1950s, it is possible to calculate the water particle velocity $u$ at any point having coordinates $x$ horizontally from the wave crest and $z$ vertically above the seabed. The water particle velocity can be related to the velocity of advance of the wave crest (the wave celerity $c$) and expressed in terms of $(u/c)^2$ and $1/g \times du/dt$ for various ratios of $x$ and $z$ to the height $h$ of the trough of the wave above the seabed.

The solitary-wave theory is limited in its application to a range of conditions defined by the ratio of the wave period to the water depth. Because the equations given below are applicable only to breaking wave conditions, they represent the maximum force which can be applied to a structure. Breaking wave conditions are unlikely to occur in deep-water berths for large tankers, and these conditions are likely to be found only in fairly shallow water on exposed jetty sites, for example along the line of the approach structure from the shore to a deep-water berth. However, as noted by Newmark[8.3], the solitary-wave theory is often applied to situations beyond its strict range of validity for want of a better theory. For deep-water structures, the solitary-wave theory gives over-conservative values of wave force. However, Equations 8.7 and 8.8 based on this theory together with the dimensionless graphs are simple and easy to use. It is suggested that the equations are used for all parts of a deep-water berthing-head structure and for the shallow-water approach whenever it is necessary to calculate wave forces. If these forces together with current drag, wind forces and berthing impact forces do not produce excessive bending stresses on the piles, then the calculations need not be further refined. It must be kept in mind that the cross-sectional area of a pile may be governed by considerations of corrosion and driving stress rather than the stress resulting from environmental forces. Where the wave forces calculated by the solitary-wave theory are a significant factor in the design of the piles, more detailed calculations should be made taking into account the relationship between wave height, water depth and wave period.

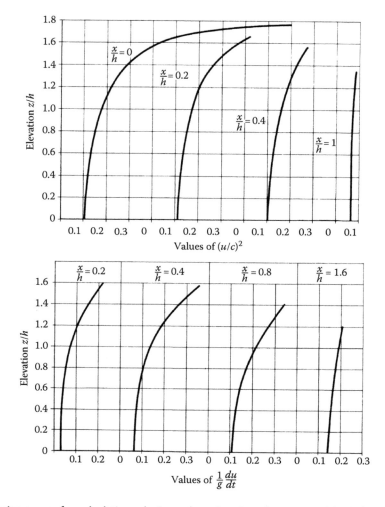

Figure 8.12 Design curves for calculating velocity and acceleration of water particles in breaking wave.

In general wave theories, the wave force on a fixed structure is taken as the sum of the drag and inertial forces exerted by the wave. These are expressed by the commonly used Morison equation[8.4]:

$$f = f_D + f_I = C_D \frac{wu^2}{2g} + C_M \frac{w\pi D}{g4} \cdot \frac{du}{dt} \tag{8.7}$$

where

$f$, $f_D$ and $f_I$ are the wave force, drag force and inertial force, respectively, per unit area of object in the path of the wave
$C_D$ is a drag coefficient
$w$ is the density of water
$g$ is the gravitational acceleration
$u$ is the horizontal particle velocity of water
$C_M$ is a coefficient of inertia force
$D$ is the diameter of the cylindrical object
$du/dt$ is the horizontal acceleration of a water particle

Table 8.1 Surface elevations, velocities and accelerations for solitary breaking wave

| Distance from crest, $x/h$ | Surface elevation, $z_s/h$ | Values of $(u/c)^2$ | | | | | Values of $\dfrac{1}{g} \times \dfrac{du}{dt}$ | | | | |
|---|---|---|---|---|---|---|---|---|---|---|---|
| | | At surface | At $z = h$ | At bottom | Average value | Height to centroid | At surface | At $z = h$ | At bottom | Average value | Height to centroid |
| 0 | 1.78 | 1.000 | 0.176 | 0.109 | 0.226 | 1.19 | 0 | 0 | 0 | 0 | |
| 0.2 | 1.67 | 0.430 | 0.170 | 0.106 | 0.181 | 1.03 | 0.242 | 0.073 | 0.031 | 0.081 | 1.14 |
| 0.4 | 1.57 | 0.276 | 0.156 | 0.099 | 0.150 | 0.92 | 0.347 | 0.137 | 0.060 | 0.133 | 1.02 |
| 0.6 | 1.48 | 0.201 | 0.133 | 0.092 | 0.123 | 0.83 | 0.380 | 0.184 | 0.087 | 0.164 | 0.93 |
| 0.8 | 1.41 | 0.138 | 0.106 | 0.078 | 0.097 | 0.80 | 0.357 | 0.214 | 0.110 | 0.180 | 0.88 |
| 1.0 | 1.35 | 0.092 | 0.082 | 0.070 | 0.077 | 0.70 | 0.321 | 0.225 | 0.127 | 0.186 | 0.78 |
| 1.2 | 1.29 | 0.062 | 0.063 | 0.058 | 0.061 | 0.65 | 0.280 | 0.225 | 0.140 | 0.187 | 0.73 |
| 1.4 | 1.25 | 0.041 | 0.046 | 0.048 | 0.047 | 0.61 | 0.243 | 0.209 | 0.146 | 0.182 | 0.68 |
| 1.6 | 1.21 | 0.029 | 0.032 | 0.038 | 0.035 | 0.59 | 0.209 | 0.192 | 0.148 | 0.173 | 0.65 |
| 1.8 | 1.18 | 0.020 | 0.023 | 0.029 | 0.027 | 0.56 | 0.174 | 0.171 | 0.145 | 0.159 | 0.62 |
| 2.2 | 1.13 | 0.009 | 0.011 | 0.018 | 0.014 | 0.50 | 0.122 | 0.128 | 0.130 | 0.130 | 0.57 |
| 2.6 | 1.08 | 0.004 | 0.005 | 0.009 | 0.007 | 0.50 | 0.088 | 0.091 | 0.109 | 0.102 | 0.53 |
| 3.0 | 1.05 | 0.002 | 0.002 | 0.004 | 0.003 | 0.50 | 0.065 | 0.067 | 0.084 | 0.078 | 0.51 |
| 3.4 | 1.03 | 0.001 | 0.001 | 0.002 | 0.002 | 0.50 | 0.049 | 0.049 | 0.062 | 0.058 | 0.50 |
| 5.0 | 1.01 | 0.000 | 0.000 | 0.000 | 0.000 | 0.50 | 0.012 | 0.012 | 0.017 | 0.016 | 0.50 |

Table 8.2 Drag force and inertia coefficients for square section piles

| Flow direction | Figure no. | $C_D$ | $C_M$ |
|---|---|---|---|
| Perpendicular to face | 8.13a | 2.0 | 2.5 |
| Against corner, in direction of diagonal | 8.13b | 1.6 | 2.2 |
| Perpendicular to face, rounded corner, $r/y_s = 0.17$ | 8.13c | 0.6 | 2.5 |
| Perpendicular to face, rounded corner, $r/y_s = 0.33$ | 8.13c | 0.5 | 2.5 |

The data compiled by Wiegel et al.[8.5] for the drag and inertia coefficients show considerable scatter. However, as pointed out by Sarpkaya[8.6], while the Morison equation has given rise to discussion as to what values should be used for the two coefficients, they work well for engineering purposes when either drag or inertia is the sole dominant force. The values for $C_D$ and $C_M$ shown in Table 8.2 can be used in the version of the Morison equation given in Equations 8.7 and 8.8.

Newmark[8.3] reduced Equation 8.7 to a simple expression which in SI units is

$$f = \left[ 7.8 C_D h \left( \frac{u}{c} \right)^2 + 8 C_M D \cdot \frac{1}{g} \cdot \frac{du}{dt} \right] \text{ kN/m}^2 \tag{8.8}$$

Values of $(u/c)^2$ and $1/g \cdot (du/dt)$ for different positions relative to the location of the wave crest are shown in Figure 8.12 and Table 8.1. This table also lists the average values of $(u/c)^2$ and $1/g \ (du/dt)$ together with the heights to the centroid of the two components. The wave forces and moments applied to each increment of height of pile projecting above the scoured seabed up to wave crest level, and on any underwater bracing or jacket members, are integrated to obtain the total horizontal force on the pile or group of piles and also the overturning moment about the point of fixity below the seabed.

For use with Equations 8.8, Newmark[8.3] recommends a value for $C_D$ of 0.5–0.6 for *cylindrical* members and 1.5–2.0 for the inertia coefficient $C_M$. For *rectangular*, H-, and I- sections, $C_D$ can be taken as up to 2.0. Theoretically, $C_D$ is related to the Reynolds number ($R_e$) as discussed in the following section. Newmark also recommends that shielding effects produced by closely spaced piles or bracing members should be disregarded when calculating wave forces.

Clause 9.5.2 in BS EN ISO 19902 expresses the Morison equation in a somewhat different form for cylindrical piles to offshore platforms where the ratio of the wave length to pile diameter is >5. The 'hydrodynamic' coefficients depend on whether the member is *smooth* when the recommendations are $C_D = 0.65$ and $C_M = 1.6$ or *rough* when the coefficients are 1.05 and 1.2, respectively. Annex A.9.5.2 in this Standard provides further information on the assessment of the coefficients in a variety of conditions.

The rough coefficient may be supplemented by allowing for an increase in the pile diameter. It has been reported[8.7] that marine growths more than 200 mm in thickness have occurred around steel piles of the southern North Sea gas production platforms after about 8 years of exposure. The growths extend down to seabed where the water depths were about 25 m. If drag forces due to marine growths are excessive, provision can be made for the members to be cleaned periodically by divers.

## 8.1.4 Current forces on piles

The velocities and directions of currents (or tidal streams) affecting the structure are obtained by on-site measurements which should include the determination of the variation in current velocity between the water surface and the seabed. Current meters and float tracking are suitable. A curve is plotted relating the velocity to the depth and the current drag force is calculated for each increment of height of the pile above the seabed. Potential scour below the seabed should be provided for.

Current forces are calculated from the equation in BS 6349-1 (to be included in a new BS 6349-1-2 for the assessment of actions):

$$F_D = 0.5 C_D \rho V^2 A_n \qquad (8.9)$$

The components are defined as follows:
$F_D$ is the steady drag force (kN)
$C_D$ is the dimensionless time-averaged drag force coefficient
$\rho$ is the water density (tonne/m$^3$)
$V$ is the incident current velocity (m/s)
$A_n$ is the area normal to flow (m$^2$)

$C_D$ is related to the Reynolds number, which for cylindrical members and normal water temperatures with a kinetic viscosity of $1.075 \times 10^{-6}$ m$^2$/s is given by the equation

$$R_e = 9.3 \ VD \times 10^5 \qquad (8.10)$$

Section 5 of BS 6349-1 includes graphs relating $C_D$ for cylindrical members to their surface roughness and Reynolds number. They show that $C_D$ for rough members is in the range of 0.4–0.6 for Reynolds numbers between $10^5$ and $10^6$. This code gives values for $C_D$ and $C_M$ for square section piles as shown in Figure 8.13 and Table 8.2.

If piles or other submerged members are placed in closely spaced groups, shielding of current forces in the lee of the leading member will occur. Shielding can be allowed for by modifying the drag coefficient. Values of the shielding coefficient have been established by Chappelaar[8.8].

Where currents are associated with waves, it may be necessary to add the current velocity vectorially to the water particle velocity $u$ to arrive at the total force on a member. Also, the possibility of an increase in the effective diameter and roughness of a submerged member due to barnacle growth must be considered.

Having calculated the current force on a pile, it is necessary to check that oscillation will not take place as a result of vortex shedding induced by the current flow. This oscillation

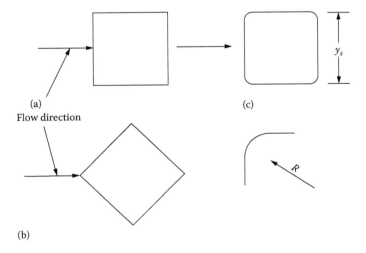

Figure 8.13 Flow conditions for determining drag conditions. (See Table 8.2).

occurs transversely to the direction of current flow when the frequency of shedding pairs of vortices coincides with the natural frequency of the pile.

Determination of the critical velocity for the various forms of flow-induced oscillation of cylindrical members is given in BS 6349-1 by the equation

$$V_{crit} = Kf_N W_s \tag{8.11}$$

where $K$ is a constant equal to the following:
  1.2 for onset of in-line motion
  2.0 for maximum amplitude of in-line motion
  3.5 for onset of crossflow motion
  5.5 for maximum amplitude of crossflow motion
  $f_N$ is the natural frequency of the cylinder
  $W_s$ is the diameter of the cylinder

The natural frequency of the member is given by the equation

$$f_N = \frac{K'}{L^2}\sqrt{\frac{EI}{M}} \tag{8.12}$$

where
  $K'$ is a constant
  $L$ is the pile length
  $E$ is the elastic modulus
  $I$ is the moment of inertia
  $M$ is the effective mass per unit length of pile
  $W_s$ should take into account the possibility of barnacle growth

$K'$ is equal to 0.56, 2.45 and 3.56 respectively, for cantilevered, propped and fully fixed piles. The elastic modulus is expressed in units of force. In the case of a cylindrical pile, the effective mass $M$ is equal to the mass of the pile material plus the mass of water displaced by the pile. Where hollow tubular piles are filled with water, the mass of the enclosed water must be added to the mass of the material. In the case of a tubular steel pile with a relatively thin wall, the effective mass is approximately equal to the mass of the steel plus twice the mass of the displaced water.

BS 6349-1 provides graphs relating $V_{crit}$ in Equation 8.11 to $L'/W_s$ where $L'$ is the overall pile length from deck level, where the pile is assumed to be pin jointed, to the level of apparent fixity below seabed.

Very severe oscillations were experienced during the construction of the Immingham Oil Terminal. At this site in the Humber Estuary, piles were driven through water with a mean depth of 23 m and where ebb currents reach a mean velocity of 2.6 m/s (5 knots). The piles were helically welded steel tubes with outside diameters of 610 and 762 mm and a wall thickness of 12.7 mm. Before the piles could be braced together, they developed a crossflow motion which at times had an amplitude of ±1.2 m. Many of the piles broke off at or above the seabed. A completed dolphin, consisting of a cap block with a mass of 700 tonnes supported by 17 piles, swayed with a frequency of 90 cycles/min and an amplitude of ±6 mm.

Moored ships can transmit forces due to current drag onto the piles supporting the mooring bollards. The current drag on the ship is calculated from Equation 8.9.

## 8.1.5 Wind forces on piles

Wind forces exerted directly on piles in a jetty structure are likely to be small in relation to the quite substantial wind forces transmitted to the piles from deck beams, cranes, conveyors, stacked, containers, sheds and pipe trunkways. In a jetty approach, the combined wind and wave forces which usually act perpendicularly to the axis of the approach can cause large overturning moments on the pile bents, particularly when the wind forces are acting on pipe trunkways or conveyor structures placed at a high elevation, say, at a location with a high tidal range. Wind forces on moored ships also require consideration, and allowance should be made where necessary for the accretion of ice on structures.

Wind forces on structures generally are defined in EC1-1-4, Clause 5, and are calculated using the force coefficients or from surface pressures as given in the National Annex. The basic relationship between wind velocity and wind force (action) on cylindrical piles is given in Clause 9.7.2 of BS EN ISO19902 as

$$F = 0.5U^2C_sA \tag{8.13}$$

where

$F$ is the wind force

$U$ is the sustained wind velocity at the elevation of the portion of the structure under consideration

$C_s$ is a shape (or drag coefficient)

$A$ is the projected area of the object (including an allowance for ice accretion or barnacle growth)

This expression can be applied to inshore structures with the shape coefficient varied depending on the Reynolds number and pile roughness, that is for cylindrical piles, $C_s$ ranges from 0.65 to 1.2. Shielding coefficients[8.8] can be applied for closely spaced members.

Wind velocities can be corrected for height by means of the equation

$$U_2 = U_1\left(\frac{H_2}{H_1}\right)^{1/7} \tag{8.14}$$

where $H_2$ and $H_1$ are the two elevations concerned. It should be noted that wind velocities based on short-duration gusts may be over-conservative when considering wind forces on large ships.

## 8.1.6 Forces on piles from floating ice

Forces on piles caused by floating ice have characteristics somewhat similar to those from berthing ships, the principal difference being the length of time over which the ice forces are sustained. Ice floes are driven by currents and wind drag on the surface of the floe. Typically, a floe consists of a consolidated layer, which may be up to 3 m thick in subarctic waters, underlain by a mass of 'rubble' in the form of loose blocks, and wholly or partly covered by loose debris and snow. When designing a structure to resist ice forces, it is necessary to determine the dominant action, that is whether it is the pressure of the wind and current driven floe against the structure or the resistance offered by the structure in splitting the advancing consolidated layer. In an extensive review of the subject, Croasdale[8.9] stated that only on relatively small bodies of water will the wind-induced forces govern the design load.

Wind forces can be calculated from Equation 8.9. Croasdale advises omitting the factor 0.5 when using this equation and gives values for $C_D$ as 0.0022 for rough ice cover, $0.00335 < C_D < 0.00439$ for unridged ice and 0.005 for ridged Arctic sea ice. In Equation 8.9, the values for $C_D$ are appropriate to m/s units of the wind velocity at the 10 m level. Croasdale gives a typical force on a 4 m diameter cylindrical pier as 10 MN caused by an ice sheet 4.15 × 4.15 km in area, driven by a wind velocity of 15 m/s.

On striking a vertical pile which is restrained from significant yielding, the consolidated ice layer is crushed at the point of impact. With further movement of the floe, radial cracks are propagated in the ice sheet followed by buckling. The buckling dissipates the energy of the moving mass which is brought to rest locally against the pile. The surrounding cracked ice sheet and the underlying loose rubble are diverted to flow past the pile, and in doing so, they generate frictional forces on the contact surfaces. The force is likely to be at a maximum at the time of initial cracking of the ice sheet followed by lesser peaks due to jamming of the packed ice and adfreezing of the ice onto the structure (Section 9.4).

The American Petroleum Institute specification API RP2N[8.10] and its commentary give comprehensive procedures for calculating ice pack loads on offshore concrete, steel and hybrid structures for level ice and ice ridges in Arctic conditions. The basic equation for the ice crushing force on a narrow rigid structure is given as

$$F = p_e \, D \, t \qquad (8.15)$$

where

$p_e$ is the effective crushing pressure (as given in the API design chart)
$t$ is the ice thickness
$D$ is the width of structure

Floe splitting is considered in detail in API RP2N with the effective pressure conditional on the ice fracture toughness and length of floe, as defined in the design charts. This code assumes that the load is limited by ice failure and applies load factors as stated to give the design load depending on load conditions, for example the likely frequency of a crushing event.

For wedges splitting at an angle of 45° to the edge of the ice sheet, the equation for calculating the effective ice stress from Croasdale is

$$p = \sigma_c \left( 1 + 0.304 \frac{t}{D} \right) \qquad (8.16)$$

A contact factor of 0.5 should be applied in Equation 8.16 for continuously moving ice and 1.0 or more for ice frozen around a structure. The compression strength is difficult to determine by laboratory testing. It depends on the crystal structure, strain rate, temperature and sample size.

The forces on the pile from the rubble have been mentioned earlier. Frictional forces from loose blocks can be assumed to act as a granular material. Where the blocks are frozen together, the stresses on the pile will be lower than that of the consolidated ice sheet because the bonds between the blocks will fracture at low strain levels.

It is evident that a single large pile or cylinder will be more effective in resisting ice forces than a cluster of smaller piles. A more efficient structure has a conical shape as shown in Figure 8.14. The impact force from the ice sheet is distributed in directions normal and tangential to the sloping face. Energy is dissipated as the ice sheet is levered up and cracked

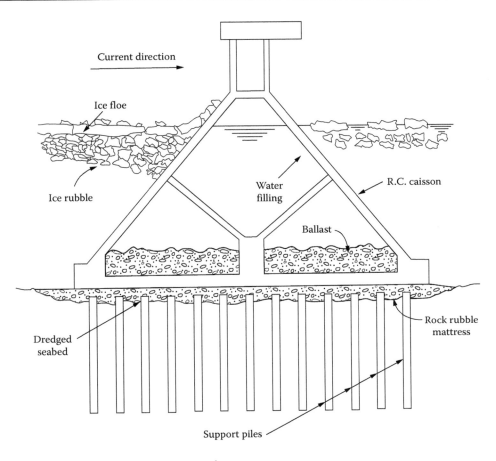

*Figure 8.14* Conical structure for resisting ice forces.

circumferentially. Further energy is dissipated as the broken blocks are pushed up the slope. Methods of calculating ice forces on conical structures are discussed in a paper by Croasdale and Cammaert[8.11], which include 3D analysis, and more recently by Brown[8.12].

The structure shown in Figure 8.14 is designed for weak ground conditions needing support by a piled raft to resist horizontal and vertical forces. The shape is unsuitable for berthing large ships, but it is suitable as a single-point mooring or as a foundation for a wind generator.

### 8.1.7 Materials for piles in jetties and dolphins

For jetties serving vessels of light to moderate displacement tonnage and of shallow draught, timber is the ideal material for fender piles. It is light and resilient and easy to replace. As already noted, the face of a timber fender pile can be protected by a renewable timber rubbing strip. The type of timber used for fender piles is governed by considerations of the attack by organisms present in the seawater. Suitable species of timber are described in Chapter 10.

For jetties and berthing structures in deep water serving large vessels, either steel or pre-stressed concrete tubular piles can be used. Steel piles have the advantage that they can withstand rough handling while being loaded onto barges and lifted into the leaders of the floating piling frame or jack-up platform. They can withstand hard driving to attain the penetration depths necessary to achieve the required uplift and lateral resistance. However, they require expensive cleaning and coating treatments above the soil line, supplemented by cathodic protection to enable them to resist corrosion in seawater. Losses in thickness of the pile section over the design life of the structure caused by corrosion need to be considered in relation to the design stresses under operating conditions. The types of steel suitable for piling in marine structure are discussed in Section 2.2.6.

Prestressed concrete piles also possess considerable resilience, but repair is a difficult problem if they are subjected to accidental heavy-impact damage. Prestressed concrete piles are suitable for approach structures and for jetty heads protected by independent berthing structures. Problems of seawater attack on steel and concrete structures are discussed in Chapter 10.

Bored and cast-in-place piles are not suitable for marine structures unless used in a composite form to extend the penetration of a driven tubular pile. BS 6349-1-4 provides general information on materials for maritime structures in line with current Eurocodes.

## 8.2 FIXED OFFSHORE PLATFORMS

Because of their location, frequently in deep water exposed to severe wave action, the forces acting on fixed platform structures are different in character from those on jetties in relatively shallow and sheltered waters. Whereas in berthing structures the dominant forces are those caused by the berthing of ships, the offshore platform is served only by small vessels and the environmental forces resulting from waves, winds and currents have a dominating influence on design. In very deep water, the environmental forces can account for three-quarters of the total load on a main supporting member.

The economics in the design and construction of offshore platforms for petroleum and gas production and wind farms are viewed from a standpoint very different from that applied to jetty design. In the case of jetties, the main requirements are low capital cost, ease of maintenance and a long life; the construction time is not usually a critical factor in design. The design of offshore platforms must not only ensure stability in the most severe exposure conditions in deep water but also take account of the need for rapid installation. While the large floating cranes and other construction plant which are now deployed can operate in more severe conditions than in the early days of oil and gas field developments, the weather and sea state will still control the time available to 'pin' the basic structure to the seabed. Gerwick[3.9] provides comprehensive information on design and construction methods in these conditions for large structures. He also highlights the effects of vortex shedding and scour around piles and cyclic dynamic conditions.

The initial platforms constructed in the United Kingdom and western Europe were mainly in the relatively shallow waters (25–150 m) of the North Sea requiring the multi-pile foundations of the type shown in Figure 8.15. These jacket-type steel structures continue to be installed worldwide as being economical and proven. The need to develop oil fields in more exposed and deeper waters (up to 1300 m depth) led to the design and construction of the *tension leg* floating platform (TLP)[8.13], tethered to piled templates or gravity bases on the seabed by vertical cables and held in tension by the buoyancy of the platform hull. Semi-submersible rigs have been in use for drilling for many years in water depths up to 3000 m and can also be used as production platforms; buoyancy is provided by submerged pontoons and the deck is supported above wave action on four large-diameter legs attached to the pontoons. Temporary position keeping is by sophisticated thrusters, with anchored cables for a more permanent installation.

The demand for wind-produced energy in the United Kingdom and Europe has led to major developments in the use of offshore turbines. The wind farms are required, from consideration of visual intrusion to be located at least 5 km from the shore line, but it has

*Figure 8.15* Float-out of platform for the Ninian Field (North Sea), showing pile guides and sleeves for clusters of piles around each leg.

been possible to find sea areas having water depths sufficiently shallow to permit the use of jack-up platforms to construct piled foundations for the turbines. The design and construction of wind farms present severe problems for the engineer which have been reviewed by Bonnett[8.14] and by ffrench et al.[8.15] The examples they give of wind turbines with rotor diameters up to 90 m, weighing with the associated machinery, some 250 tonne mounted at a height of 70 m above sea level, are now being exceeded in order to provide generating capacity of 5 MW. At peak wind force conditions, the dynamic forces generated by the turbines can act concurrently with peak wave action on the supporting structure to cause cyclic overturning moments on the foundations. A dominant design problem is in providing sufficient stiffness in the combined machinery and foundation system so that its natural frequency exceeds that of the excitation forces.

The majority of the present generation of wind turbines are erected on a single large-diameter pile (monopile) foundation. Penetration depths of piles are determined from considerations of resistance of the soil to dynamically applied horizontal and vertical forces taking into account the possibility of seabed scour and soil degradation increasing the overturning moments. Tubular steel piles 5.4 m in diameter have been driven in water depths up to 20 m using equipment of the type shown in Figure 3.7. The connection of the turbine mast to the monopile is usually by grouting the annulus between the pile and the mast sleeve; internal shear connectors are essential.

The review of piles for wind turbines by Gavin et al.[8.16] indicates that the CPT design methods for open-ended steel piles (see Section 4.3.7) provide a more reliable design approach than the earth pressure method for large-diameter piles. However, they note that the tension loads applied to the foundations were much higher than those considered in the calibration of these offshore design methods. They suggest that the semi-empirical factors, which depend on the pile geometry and are currently applied to the CPT tests, overestimate the radial stress distribution of a pile in loose sand and underestimate the profile in dense sand.

*Figure 8.16* GeoSea's *Goliath* jack-up barge with a tripod seabed template slung ready for deployment at the Borkum wind farm in the North Sea. (Courtesy of GeoSea and DEME, Zwijndrecht, Belgium.)

This indicates that if improved measurements of radial effective stress under cyclic loading can be made, there is potential for savings in piled foundations.

As a result of the new offshore wind farm developments inevitably requiring larger turbines in locations where water depths are greater than 30 m, the technical limitations of the dynamically sensitive monopile foundations will render them uneconomic. This will lead to the deployment of alternative *multipod* foundations, jacket-type platforms or shallow gravity foundations which avoid piling. The Borkum wind farm utilised tripod templates for piles seen in Figure 8.16, designed using the ICP method and EC7 procedures[4.48].

Certifying authorities for oil and gas production usually demand a specific safety factor for a 100-year wave combined with the corresponding wind force and maximum current velocity, referred to as the 'design' environmental conditions. The maximum forces due to operations on the platform such as drilling are combined with specified wind and sea conditions and are known as the 'operating' environmental conditions. The API RP2A working stress design specification[4.15] requires the safety factors on the ultimate bearing capacity of piled foundations not to be less than the guidance given in Table 8.3. This specification is being phased out and the load and resistance factor design version of API RP2A has been withdrawn; the factors in ISO 19902 are now applied.

*Table 8.3* Minimum working stress safety factors for various loading conditions

| Loading condition | Minimum safety factor |
|---|---|
| 1. Design environmental conditions with appropriate drilling loads | 1.5 |
| 2. Operating environmental conditions during drilling operations | 2.0 |
| 3. Design environmental conditions with appropriate producing loads | 1.5 |
| 4. Operating environmental conditions during producing operations | 2.0 |
| 5. Design environmental conditions with minimum loads | 1.5 |

## 8.3 PILE INSTALLATIONS FOR MARINE STRUCTURES

Where marine structures are connected to the shore, as in the case of a jetty head with a trestle approach, the piles may be driven either as an *end-on* operation, with the piling equipment mounted on girders cantilevering from the completed pile bents, or as an operation from a floating or jack-up barge (Figure 8.17). In tidal waters, there is usually sufficient water depth to float a barge with a draft of 1–1.5 m to a location close inshore. However, this can be inconvenient where tidal flats or saltings cover a long depth of the approach or where it is unsafe to ground the barge on the seabed at low water.

Where the end-on method is used, the spacing between pile bents is limited by the ability of the girders to cantilever when carrying the weight of the piling frame, hammer and suspended pile. Loading can be minimised by utilising the buoyancy of tubular piles with permanently or temporarily closed ends or by using trestle guides of the types shown in Figures 3.6 and 3.8 in conjunction with a pile-mounted hammer and a crane barge for lifting and pitching the piles.

The range of piling barges and crane vessels for deep-water locations has expanded significantly. Cargo barges capable of carrying up to 20,000 tonnes are typically 120 m long, 30 m wide and 7 m deep and may be fitted out for foundation works. Semi-submersible crane vessels, such as the Heerema *Balder* and *Hermod*, are multifunctional, dynamically positioned vessels which can install foundations and moorings; *Hermod's* two cranes are capable of placing topside structures with a tandem lift of 8100 tonnes. The draft of this vessel is 11 m but when working is ballasted down to 25 m for stability.

*Figure 8.17* Shallow draft barge end-on to jetty driving 813 mm tubular piles in Mombasa with a CG240 hydraulic hammer. (Courtesy BSP International Foundations Ltd., Ipswich, England.)

Modular jack-up barges are built up of units in the size of an ISO freight container and are easy to transport by rail and road. When assembled, they form a working platform approximately 30 m × 17 m with a maximum payload of 400 tonnes for working in water depth up to 25 m. Large monohull jack-up barges with self-elevating platforms up to 70 m long × 40 m wide and payload capability of 3000 tonnes can operate in water depths up to 50 m. They operate most efficiently when provided with mechanically adjustable pile guides installed either by cantilevering from the side of the barge or spanning a *moon pool* inset in the barge hull.

If possible, piles should be driven to their full design penetration without the need to weld on additional pile lengths, to drive insert piles, or to clean out the soil plug or drill below the initial refusal level of an open-ended tubular pile. Gerwick[3.9] gives an example of the times required for welding add-on lengths of 1.37 m OD tubular piles; they varied from 3¼ h for 25 mm wall thickness to 10½ h for 64 mm thickness. Bhattacharya et al.[8.17] examined data from 1980 on the short-term set-up of 1.4 m diameter open-ended steel piles driven in the North Sea, subject to delays in driving of between 1 and 100 h. From a back analysis of the driving records of 53 cases, the increase in total soil resistance for hard clay ($c_u$ = 500–800 kN/m$^2$) was 20%–60% for piles with toes at 35–50 m below mud line and of the order of 60% for piles 17–27 m below mud line, both for delays of 24 h. This study was based on steam hammers and does not claim to predict long-term changes in soil resistance, but will be useful for selecting the appropriate, more efficient hydraulic piling hammer.

Removal of the soil plug in such stiff clays is not particularly effective in improving driving where only a small proportion of the resistance is obtained from end bearing, and it does not reduce the external friction of the surrounding clay. There are situations where the required penetration cannot be achieved without the use of the 'drill-and-drive' technique, but the successive operations of driving the pile to refusal, removing the hammer, assembling the drilling gear then drilling, and removing the equipment can be very protracted; the aim should be to restrict the drilling phase to only one operation. Clean-out is effective in reducing end-driving resistance to obtain deep penetration in coarse-grained soils in order to develop uplift resistance, to avoid excessive settlement due to vibration effects or to reach rockhead. Suitable equipment for these operations is described in Section 3.3. Where rotary methods are used, centralisers are required to keep the drilling pipes in line with the pile axis.

Insert piles can be used where piles driven to their full design penetration fail to attain a satisfactory resistance or where drilling-and-driving techniques are unable to achieve the required penetration. The transfer of load from the insert pile to the main (primary) pile, and from the main pile to the pile sleeve on the leg of a jacket platform, is made by grouting the annular space between the members (Figure 8.18) or, in special cases, by welded joints at the pile heads. The grout bond between the pile and sleeve or between primary and insert pile is described in Section 6.2.5 and in BS EN ISO 19902. The grout is prevented from flowing out from the pile-sleeve annulus by means of 'active' inflatable packers or, more usually, by 'passive' Crux *wiper* seals built into the bottom of the pile sleeve. The insert pile is usually drilled to a specified depth, requiring grouting of the soil–pile annulus, rather than relying on driving to a set using a *slimline* underwater hydraulic hammer. To ensure that the annulus is grouted uniformly, transverse ducts may be installed at intervals up the insert pile. An alternative to the grouting shoe shown at the base of the insert pile in Figure 8.18 is a diaphragm plate at the base of the pile fitted with a non-return valve to which the stinger is connected.

The research work[6.3,6.4] on large-scale grouted pile–sleeve connections described in the reports in Section 6.2.5 demonstrated that mechanical shear keys would allow reductions in the bonded sleeve length and still provide the necessary API working stress safety factors on

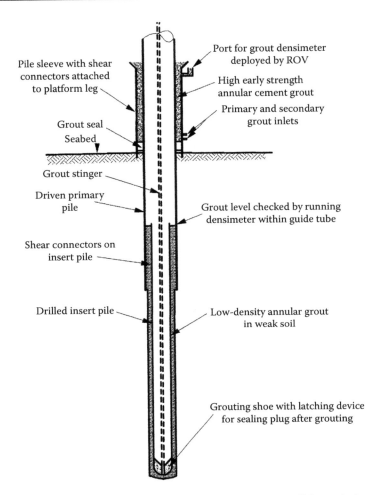

*Figure 8.18* Schematic of driven primary pile and drilled insert pile for fixing offshore platform to weak seabed.

the bond stress. In order to transfer the loads from the platform to the piles effectively, the annulus between the pile and sleeve should be kept to the minimum, say 75–100 mm. It may be difficult to determine where to provide the shear keys of the driven primary pile to coincide with the sleeve. Where shear keys are provided only on the inner surface of the sleeve, it is essential that, in a raking sleeve, they are designed so that they do not cause obstructions when the primary pile is lowered through the sleeve. Shear keys may be required on the outside of the insert pile.

The use of a rotary under-reaming tool operating below the toe of an open-ended steel tubular pile to produce an enlarged base provides both increased resistance to compressive loads and a positive anchorage against uplift. However, the method is losing favour in offshore conditions due to difficulties in retracting the large expanding cutter (not experienced to the same extent with the casing under-reamer).

When open-end piles are driven into deep granular soil deposits, the driving resistance may be very low for the reasons described in Section 4.3.3. As a result, calculations of resistance to axial compression loads based on dynamic testing are correspondingly low, indicating that a very deep penetration of the pile is needed to achieve the required resistance. These penetrations are often much greater than those required for fixity against lateral loading. Although base resistance to axial loading can be achieved by grouting beneath the pile toe as described in Section 3.3.9, the operations of cleaning out the pile and grouting are slow and relatively costly. An alternative method of developing base resistance of open-end piles which has been used on a number of marine projects is to weld a steel plate diaphragm across the interior of the pile. The minimum depth above the pile toe for locating the diaphragm is the penetration below seabed required for fixity against lateral loading. However, a further penetration is necessary to compact the soil within the plug and to develop the necessary base resistance. It is not possible to achieve a resistance equivalent to a solid-end pile but the penetration depths are much shorter than those required for an open-end pile.

The diaphragm method was used for the piling at the Hadera coal unloading terminal near Haifa[8.18]. Open-end piles 1424 and 1524 mm OD were proposed but initial trial driving showed that very deep penetrations, as much as 70 m below seabed in calcareous sands, would be needed to develop the required axial resistance. The blow count diagram in Figure 8.19 showed quite low resistance at 36 m below seabed. Another trial pile was

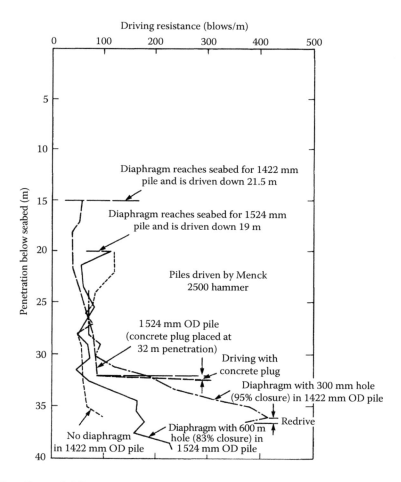

Figure 8.19 The effects of different methods of plugging steel tubular piles driven with open ends, Hadera coal unloading terminal.

driven to 32 m, cleaned out and plugged at the toe with concrete. An acceptable driving resistance of about 300 blows per metre was obtained by driving the plugged pile, but it was appreciated that the plugging operations would be costly and would seriously delay completion of the project. Trials were then made of the diaphragm method. A diaphragm with a 600 mm hole giving 83% closure of the cross section was inserted 20 mm above the toe. This increased the driving resistance at 39 m below seabed and another trial with a 300 mm hole (95% closure) gave a higher resistance at 37 m.

The diaphragm method is ineffective if a very deep penetration is required because the long plug cannot compress sufficiently to mobilise the end-bearing resistance of the diaphragm and settlements at the working load would be excessive. It is also ineffective in clays or where clays are overlying the coarse soil bearing stratum. A hole is necessary in the diaphragm for release of water pressure in the soil plug and to allow expulsion of silt. Stresses on the underside of the diaphragm are high during driving and radial stiffeners are needed (Figure 8.20). The pile wall below the diaphragm must be sufficiently thick to prevent bursting by circumferential stresses induced by compression of the soil in the plug.

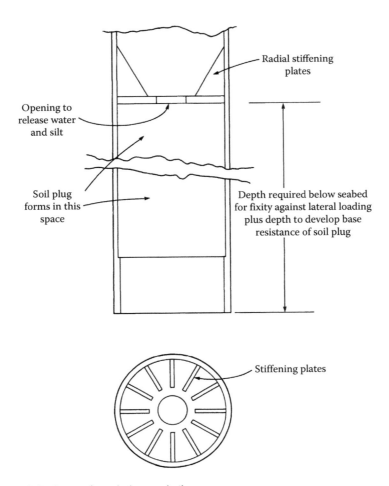

*Figure 8.20* Internal diaphragm for tubular steel pile.

## WORKED EXAMPLES

### Example 8.1

A breasting dolphin is constructed by linking at the head four 350 × 350 mm reinforced concrete piles which are driven through 2.5 m of soft clay into a stiff clay to a total penetration below seabed of 9.0 m. Find the kinetic energy which can be absorbed by the pile group for an impact at a point 8 m above the seabed. The maximum energy absorption value is to be taken as the figure which stresses the piles to their yield point.

The piles can be considered as fixed at the surface of the stiff clay stratum, and the ultimate resistance moment of each pile at the yield point is 125 kNm. Therefore, from Equation 8.4, work done in deflecting piles to yield point

$$= \frac{4 \times 125^2 (8 + 2.5)}{6 \times 26 \times 10^6 \times 0.0833 \times 0.35^4} = 3.37 \text{ kJ}$$

### Example 8.2

A steel tubular pile having an outside diameter of 1300 mm and a wall thickness of 30 mm forms part of a pile group in a breasting dolphin. The pile is fabricated from high-tensile alloy steel to BS 10210-1 Grade S460. The piles are driven into a stiff over-consolidated clay ($c_u$ = 150 kN/m²). Calculate the maximum cyclic force which can be applied to the pile at a point 26 m above the seabed at the stage when the failure in the soil occurs at seabed level, the deflection of the pile head at this point, and the corresponding energy absorption value of the pile.

Steel to Grade S460 has a minimum yield strength of 460 N/mm² (for steel less than 40 mm thick) and an elastic modulus of 2.1 × 10⁵ MN/m².

$$\text{Moment of inertia of pile} = \pi \frac{\left(1.30^4 - 1.24^4\right)}{64} = 0.024 \text{ m}^4$$

$$\text{Moment of resistance of pile at yield point} = \frac{460 \times 0.024}{0.65} = 17 \text{ MNm}$$

The first step is to establish the $p$–$y$ curves. In this example, spreadsheet calculations are used to demonstrate the principles of the method, but because the data provided are limited, solution by a basic computer program may not be feasible. In Equations 6.36 and 6.37, the submerged density of the soil is 1.2 Mg/m³, and a value of 0.25 can be taken for the factor $J$.

*At seabed level*

$$\text{Critical depth } x_r = \frac{6 \times 1.3}{1.2 \times 9.81 \times 1.3 / 150 + 0.25} = 22.1 \text{ m}$$

$N_c = 3 + 0 + 0 = 3$ as Equation 6.36
$p_u = 3 \times 150 \times 1.3 = 585$ kN/m depth
(note that $c_u$ is unfactored for lateral load)
For cyclically applied loading as Equation 6.40, take
$p_b = 0.72 p_u = 0.72 \times 585 = 421$ kN/m depth

In the absence of laboratory compression tests, the appropriate value of $\varepsilon_c$ in Equation 6.39 can be taken as 0.01, and the $p-y$ curves will be derived in the same manner as for a normally consolidated clay.

Therefore,

$$y_c = 2.5 \times 0.01 \times 1.3 = 0.0325 \text{ m} = 32.5 \text{ mm}$$

The deflection corresponding to $p_b$ is $3y_c = 3 \times 32.5 = 97$ mm.

Other points of the $p-y$ curve are calculated from Equation 6.38. Thus, for $y = 15$ mm,

$$p = 0.5 \times 585 \times \sqrt[3]{\frac{15}{32.5}} = 226 \text{ kN/m depth}$$

Similarly for

$$y = 25 \text{ mm}, \quad p = 268 \text{ kN/m depth}$$

$$y = 50 \text{ mm}, \quad p = 338 \text{ kN/m depth}$$

$$y = 75 \text{ mm}, \quad p = 386 \text{ kN/m depth}$$

Beyond the critical point at $3y_c$, the $p-y$ curve decreases linearly from $p_b = 0.72p_u$ to zero at $y = 15y_c = 15 \times 32.5 = 487$ mm for $x/x_r = 0$.

The $p-y$ curve at seabed level for the six points established earlier is shown in Figure 8.21a.

*At 0.5 m below seabed*

$$N_c = 3 + \frac{1.2 \times 9.81 \times 0.5}{150} + \frac{0.25 \times 0.5}{1.3} = 3.13$$

$$p_u = 3.13 \times 150 \times 1.3 = 610 \text{ kN/m depth}$$

$$p_b = 0.72 \times 610 = 439 \text{ kN/m at } y = 97 \text{ mm}$$

For $y = 15$ mm, $p = 610 \times 0.5 \sqrt[3]{15/32.5} = 236$ kN/m depth. Similarly,

$$y = 25 \text{ mm}, \quad p = 280 \text{ kN/m depth}$$

$$y = 50 \text{ mm}, \quad p = 352 \text{ kN/m depth}$$

$$y = 75 \text{ mm}, \quad p = 406 \text{ kN/m depth}$$

The $p-y$ curve falls linearly at $15y_c = 487$ mm to a value of $p = 0.72 \times 610 \times 0.5/22.1 = 10$ kN/m, as in Figure 6.32 for cyclic loading.

The $p-y$ curve for $x = 0.5$ m is also plotted in Figure 8.21a and the curves for values of $x$ of 1.0, 1.5, 2.0 and 2.5 m below seabed, established in a similar manner, are also shown on this figure.

The value of $p_b = 421$ kN/m represents the pressure at which yielding of the soil at the seabed occurs. Therefore, the unfactored bending moment at seabed level is

$$M_t = 26 \times 0.421 = 10.9 \text{ MNm}$$

The deflections at various points below the seabed are obtained from Figure 6.29a and b, taking as a first trial $R = 3.78$, corresponding to a $k$ value from Equation 6.11 of about

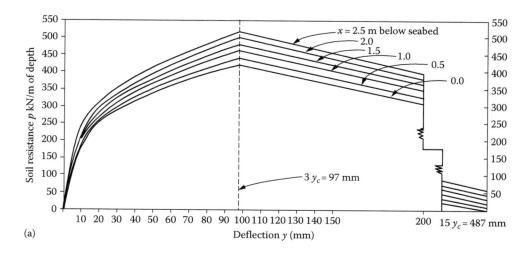

(a)

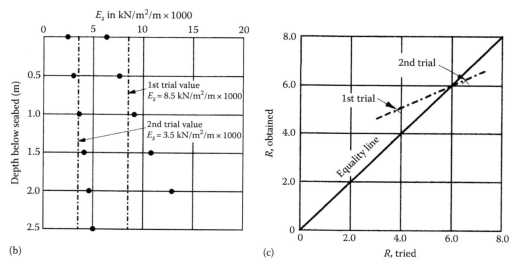

(b)                                                          (c)

*Figure 8.21* Determination of p–y curves for example 8.2 (a) p–y curves at depth, (b) trial values of $E_s$ and (c) trial values of T.

20 MN/m². Required penetration depth, as in the table in Section 6.3.1 for constant soil modulus and assuming an elastic pile, is $3.5 \times 3.78 = 13.2$, say, 14 m. Then

$$Z_{max} = \frac{L}{R} = \frac{14}{3.78} = 3.7$$

From Equation 6.30, $M_A = 10.9\, M_m$ MNm

From Equation 6.32, $M_B = 0.421 \times 3.78 \times M_h = 1.6 M_h$ MNm (where $H = p_b$)

From Equation 6.31, $y_A = \dfrac{10.9 \times 3.78^2 \times 1000}{2.1 \times 10^5 \times 0.024} y_m = 30.8 y_m$ mm

From Equation 6.33, $y_B = \dfrac{0.421 \times 3.78^3 \times 1000}{2.1 \times 10^5 \times 0.024} y_h = 4.5 y_h$ mm

| x (m) | Z = x/R | $y_m$ | $y_A = 30.8y_m$ (mm) | $y_h$ | $y_B = 4.5y_h$ (mm) | $y_A + y_B = y$ (mm) |
|---|---|---|---|---|---|---|
| 0 | 0 | +1.0 | +30.8 | +1.40 | +6.3 | +37.1 |
| 0.5 | 0.13 | +0.78 | +24.0 | +1.32 | +5.9 | +30.0 |
| 1.0 | 0.26 | +0.63 | +19.4 | +1.15 | +5.2 | +24.6 |
| 1.5 | 0.40 | +0.50 | +15.4 | +1.00 | +4.5 | +19.9 |
| 2.0 | 0.53 | +0.40 | +12.3 | +0.90 | +4.1 | +16.4 |
| 2.5 | 0.66 | +0.32 | +9.9 | +0.80 | +3.6 | +13.5 |

The previous values of $y$ are referred to the $p$–$y$ curves to obtain the corresponding values of $p$ and hence to obtain the soil modulus $E_s$ from the linear relationship $E_s = p'/y$, as tabulated in the following:

| x (m) | y (mm) | p (kN/m) | $p' = p/1.3$ (kN/m²) | $E_s = p'/y$ (kN/m²/m) |
|---|---|---|---|---|
| 0 | 37.1 | 303 | 233 | 6.3 |
| 0.5 | 30.0 | 295 | 227 | 7.6 |
| 1.0 | 24.6 | 290 | 223 | 9.1 |
| 1.5 | 19.9 | 280 | 215 | 10.8 |
| 2.0 | 16.4 | 274 | 211 | 12.9 |
| 2.5 | 13.5 | 269 | 207 | 15.4 |

The values of $E_s$ are plotted against depth in Figure 8.21b, from which an average constant value of $E_s$ of $8.5 \times 10^3$ kN/m²/m is obtained (with weight given to depths $\leq 0.5R$ as Section 6.3.5). From Equation 6.11,

$$R(\text{obtained}) = \sqrt[4]{\frac{2.1 \times 10^5 \times 0.024}{8.5}} = 4.9$$

This value of $R(\text{obtained})$ is plotted against $R(\text{tried})$ in Figure 8.21c, from which a second trial value of $R$ of 6.5 is taken. This higher value requires a deeper penetration of the pile, that is, $L > 3.5 \times 6.5 = 22.75$, say, 23 m. Thus, $Z_{max} = 23/6.5 = 3.5$, and from Equation 6.31,

$$y_A = \frac{10.9 \times 6.5^2 \times 1000}{2.1 \times 10^5 \times 0.024} y_m = 91.3 y_m \text{ mm}$$

From Equation 6.33,

$$y_B = +\frac{0.421 \times 6.5^3 \times 1000}{2 \times 10^5 \times 0.024} y_h = 23.0 y_h \text{ mm}$$

From Figure 6.29a and b, the computed deflections are tabulated in the following:

| x (m) | Z = x/R | $y_m$ (mm) | $y_A = 91.3y_m$ (mm) | $y_h$ | $y_B = 23y_h$ m (mm) | $y_A + y_B = y$ (mm) |
|---|---|---|---|---|---|---|
| 0 | 0 | +1.00 | +91.3 | +1.45 | +33.4 | 124.7 |
| 0.5 | 0.08 | +0.85 | +77.6 | +1.37 | +31.5 | 109.1 |
| 1.0 | 1.15 | +0.75 | +68.5 | +1.30 | +29.9 | 98.4 |
| 1.5 | 0.23 | +0.65 | +59.3 | +1.20 | +27.6 | 86.9 |
| 2.0 | 0.31 | +0.57 | +52.0 | +1.11 | +25.3 | 77.3 |
| 2.5 | 0.38 | +0.52 | +47.5 | +1.05 | +24.2 | 71.6 |

From the *p–y* curve

| x (m) | y (mm) | p (kN/m) | p′ = p/1.3 (kN/m²) | $E_s = -p'/y$ (kN/m²/m) |
|---|---|---|---|---|
| 0 | 124.7 | 391 | 301 | 2.4 |
| 0.5 | 109.1 | 426 | 327 | 3.0 |
| 1.0 | 98.4 | 458 | 352 | 3.6 |
| 1.5 | 86.9 | 461 | 355 | 4.1 |
| 2.0 | 77.3 | 461 | 355 | 4.6 |
| 2.5 | 71.6 | 466 | 358 | 5.0 |

From Figure 8.21b, the second trial value of $E_s = 3.5 \times 10^3$ kN/m², and

$$R(\text{obtained}) = \sqrt[4]{\frac{2.1 \times 10^5 \times 0.024}{3.5}} = 6.2$$

This is sufficiently close to the equality line for 6.5 to be accepted as the final value of $R$ (see Figure 8.21c). Where feasible, iterations in computer programs will provide more precise correlation.

For serviceability limit states (SLS) calculations, the actions are not factored. The *deflection* of the pile head at the loading for the critical value of $H = 421$ kN for soil rupture is the sum of the following deflections (a)–(c):

(a) Deflection of pile considered as cantilever fixed at seabed

$$= \frac{0.421 \times 26^3 \times 1000}{3 \times 2.1 \times 10^5 \times 0.024} = 489 \text{ mm}$$

(b) Deflection of pile at seabed due to soil compression (from table earlier) = 124.7 mm
(c) Deflection of pile head due to slope of pile below seabed

This can be obtained from the difference of the deflections at the seabed and 1.0 m below the seabed. From the previous table, the deflection at 1 m below seabed = 98.4 mm. Therefore, slope below seabed = 124.7 − 98.4 = 26.3 mm in 1 m. Thus, deflection at pile head = 26 × 26.3 = 684 mm.

Total deflection at pile head = 489 + 125 + 684 = 1298 mm.

It is necessary to check the *bending moments* at and below the seabed to ensure that the resistance moment of the pile section is not exceeded. From Figure 6.29a and b, for $Z_{max} = 23/6.5 = 3.5$ and applied in the table below.

For ULS, take the lateral load as being the result of current, wave, berthing and mooring loads with $\gamma_Q = 1.4$ as in Table A1 of BS 6349 for variable persistent actions. The favourable permanent action factor is zero. Then as per Equation 6.10 of BS EN 1990, the combination action factor to apply is 1.4.

| x (m) | Z = x/R | $M_m$ | $M_A = 10.9 \times 1.4 M_m$ (MNm) | $M_h$ | $M_B = 0.421 \times 1.4 \times 6.5 M_h$ (MNm) | $M = M_A + M_B$ (MNm) |
|-------|---------|-------|-----------------------------------|-------|-----------------------------------------------|------------------------|
| 0 | 0 | +1.00 | +15.3 | 0 | 0 | +15.3 |
| 0.5 | 0.08 | +0.98 | +15.0 | +0.10 | +0.4 | +15.4 |
| 1.0 | 1.15 | +0.97 | +14.8 | +0.15 | +0.6 | +15.4 |
| 1.5 | 0.23 | +0.95 | +14.5 | +0.20 | +0.8 | +15.3 |
| 2.0 | 0.31 | +0.94 | +14.3 | +0.27 | +1.0 | +15.3 |
| 4.0 | 0.62 | +0.85 | +13.0 | +0.40 | +1.5 | +14.5 |
| 8.0 | 1.23 | +0.55 | +8.4 | +0.45 | +1.7 | +10.1 |

From the above table the maximum bending moment of 15.4 MNm <17 MNm, the design resistance of the steel pile and satisfactory.

From Equation 8.3, the *kinetic energy* absorption value of the pile for horizontal movement at the stage of soil rupture at seabed level

$$= \tfrac{1}{2}p_b y = \frac{0.5 \times 421 \times 1298}{1000} = 273 \text{ kJ}$$

In a similar manner to that set out earlier, it is possible to obtain pile head deflections and bending moments for various stages of horizontal loading up to the stage of yielding of the steel and hence to draw curves of deflection and energy absorption against horizontal load.

The *deflection* of the pile at seabed level caused by a lateral force of 421 kN applied at the seabed can be calculated using Randolph's curves (Section 6.3.8).

Effective Young's modulus of equivalent solid section pile:

$$= E'_p = \frac{4 \times 2.1 \times 10^5 \times 0.024}{\pi \times 0.65^4} = 36 \times 10^3 \text{ MN/m}^2$$

An average constant soil modulus of 3.5 MN/m² from Figure 8.21b was used to calculate pile deflections and bending moments. For undrained loading, take Poisson's ratio $v_u = 0.5$.

$$\text{Shear modulus} = G_c = \frac{3.5}{2(1+0.5)} = 1.17 \text{ MN/m}^2$$

$$G^* = 1.17(1 + 0.75 \times 0.5) = 1.6$$

$$\text{Critical length} = l_c = 2 \times 0.65 \left( \frac{36 \times 10^3}{1.6} \right)^{2/7} = 22.8 \text{ m}$$

Homogeneity factor $\rho_c = 1$
In Figure 6.36a,

$$\frac{y r_o G_c}{H_o} \left( \frac{E'_p}{G_c} \right)^{1/7} = \frac{y \times 0.65 \times 1.5}{0.421} \left( \frac{36 \times 10^3}{1.6} \right)^{1/7} = 10.4y$$

At 0.5 m below seabed,

$$z/l_c = 0.5/22.8 = 0.022 \text{ m}$$

giving $\dfrac{yr_oG_c}{H_o}\left(\dfrac{E'_p}{G_c}\right)^{1/7} = 10.4y = 0.26$ from Figure 6.36a;

hence, $y = \dfrac{0.26\times10^3}{10.4} = 25$ mm

## Example 8.3

A cross section of an approach trestle giving roadway access to a cargo jetty is shown in Figure 8.22. The trestle is sited at right angles to the direction of maximum current velocity and travel of storm waves. The distribution of current velocity with depth is shown on the cross section. The deck slab and other components of the superstructure impose a total horizontal wind force of 25 kN on each pile bent. Storm waves have a maximum height from crest to trough of 3 m. Determine the distribution of current and wave forces on the pile bent and calculate the bending moments on the piles produced by these forces.

The maximum horizontal force on the piles will be due to the combined current and wave action at HWST (+6.0 m). At this stage of the tide, the storm wave crest will be at +7.5 m. The underside of the transom beam is at 8.0 m and therefore the wind force on the exposed

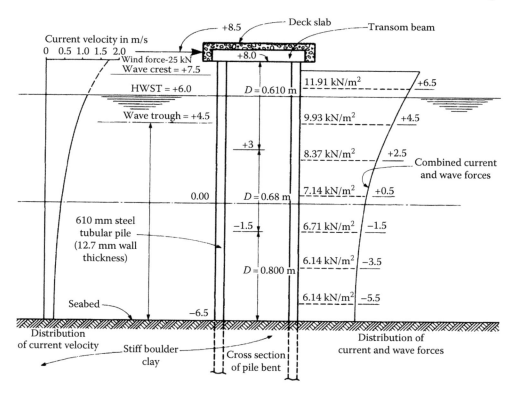

Figure 8.22 Layout of piled trestle and plots of current velocity and current and wave forces v water depth for Example 8.3.

Table 8.4 Calculations for current and wave forces in Example 8.3

| Elevation (m) (CD) | Average height above seabed (m) | Pile diameter (m) | Current | | $\dfrac{z}{h}$ | Wave drag | | Wave inertia | | Total force $(kN/m^2)$ | Horizontal load on element (kN) | Bending moment (kNm) |
|---|---|---|---|---|---|---|---|---|---|---|---|---|
| | | | Velocity (m/s) | Force $(kN/m^2)$ | | $(u/c)^2$ | Force $(kN/m^2)$ | $\dfrac{1}{g}\cdot\dfrac{du}{dt}$ | Force $(kN/m^2)$ | | | |
| +7.5 to +5.5 | 13 | 0.610 | 1.2 | 0.36 | 1.18 | 0.21 | 9.01 | 0.26 | 2.54 | 11.91 | 14.53 | ×14.5 = 210.69 |
| +5.5 to +3.5 | 11 | 0.610 | 0.8 | 0.16 | 1.00 | 0.18 | 7.72 | 0.21 | 2.05 | 9.93 | 12.15 | ×12.5 = 151.87 |
| +3.5 to +1.5 | 9 | 0.680 | 0.6 | 0.09 | 0.82 | 0.15 | 6.43 | 0.17 | 1.85 | 8.37 | 11.38 | ×10.5 = 119.49 |
| +1.5 to −0.5 | 7 | 0.680 | 0.4 | 0.04 | 0.64 | 0.13 | 5.58 | 0.14 | 1.52 | 7.14 | 9.71 | ×8.5 = 82.53 |
| −0.5 to −2.5 | 5 | 0.680 | 0.3 | 0.02 | 0.45 | 0.12 | 5.15 | 0.12 | 1.31 | 6.48 | 8.81 | ×6.5 = 57.25 |
| −2.5 to −4.5 | 3 | 0.800 | 0.25 | 0.01 | 0.27 | 0.11 | 4.72 | 0.11 | 1.41 | 6.14 | 9.82 | ×4.5 = 44.19 |
| −4.5 to −6.5 | 1 | 0.800 | 0.25 | 0.01 | 0.09 | 0.11 | 4.72 | 0.11 | 1.41 | 6.14 | 9.82 | ×2.5 = 24.55 |

Maximum bending moment (at −6.5 m) = 690.57 kN m

length of pile from +7.5 to +8.0 m will be relatively small and can be neglected. It is convenient to divide the length of the pile into 2 m elements. Allowance is made for barnacle growth on the piles. Thus,

From +7.5 to +3.0 m: no increase in diameter (i.e. $D$ = 0.61 m)
From +3.0 to –1.5 m: increase of 70 mm ($D$ = 0.68 m)
From –1.5 to seabed: increase of 190 mm ($D$ = 0.80 m)

Taking Newmark's values, a drag force coefficient of 0.5 is used to calculate the current and wave drag forces, and an inertia coefficient of 2.0 is used to calculate the wave inertia forces. Thus, in Equation 8.9 for $\rho$ = 1 tonne/m$^3$,

$$F_D = 0.5 \times 0.5 \times \rho \times V^2 \times A_n = 0.25 \ V^2 A_n \ \text{kN}$$

and in Equation 8.8

$$f = 7.8 \times 0.5 \times 11 \left(\frac{u}{c}\right)^2 + 8 \times 2 \times D \left(\frac{1}{g} \cdot \frac{du}{dt}\right) = 42.9 \left(\frac{u}{c}\right)^2 + 16D \left(\frac{1}{g} \cdot \frac{du}{dt}\right)$$

The calculated wave and current forces are shown in Table 8.4 and Figure 8.20. The bending moments shown in Table 8.4 have been calculated on the assumption of virtual fixity of the pile at a point 1.5 m below the seabed in the stiff boulder clay. Scour would not be expected around the piles in this type of soil. From Table 8.4, the combined wave and current forces produce a maximum bending moment at the point of fixity of 690.57 kNm.

Bending moment due to wind force on deck slab:

$$0.5 \times 25 \times (15.0 + 1.5) = 206.25 \ \text{kNm}$$

Total bending moment = 896.82 kNm/pile
Moment of inertia of pile section = $\pi (0.6100^4 - 0.5846^4)/64 = 1.063 \times 10^{-3} \ \text{m}^4$

$$\text{Extreme fibre stress of pile} = \frac{896.82 \times 0.305}{1.063 \times 10^{-3} \times 10^3} = 257 \ \text{MN/m}^2$$

The direct stress resulting from the dead load of the deck slab and self-weight of the pile is added to the bending stress calculated previously. It is also necessary to calculate the susceptibility of the pile to current-induced oscillations.

Assuming the pile to be filled with fresh water, the effective mass is approximately equal to the mass of metal plus twice the mass of the displaced water. Therefore,

$$M = 187 + (2 \times \pi/4 \times 0.61^2 \times 1000) = 771.5 \ \text{kg/m}$$

When the pile is in an unsupported condition cantilevering from the seabed, from Equation 8.12 the frequency

$$f_N = \frac{0.56}{14^2} \sqrt{\frac{200 \times 10^9 \times 1.063 \times 10^{-3}}{771.5}} = 1.50 \ \text{Hz}$$

From Equation 8.11, critical velocity for onset of crossflow oscillation = 5.5 × 1.5 × 0.61 = 5 m/s.

Therefore, crossflow or in-line oscillations should not take place for the flow velocities shown in Figure 8.22.

# REFERENCES

8.1   Randolph, M. and Gourvenec, S. *Offshore Geotechnical Engineering*. Spon Press, Abingdon, UK, 2011.

8.2   Broadhead, A. A maritime foundation problem in the Arabian Gulf, *Quarterly Journal of Engineering Geology*, 3 (2), 1970, 73–84.

8.3   Newmark, N. The effect of dynamic loads on offshore structures, *Proceedings of the Eighth Texas Conference on Offshore Technology*, Houston, TX, Paper No. 6, September 1956.

8.4   Morison, J.R., O'Brien, M.P., Johnson, J.W. and Schaaf, S.A. The force exerted by surface waves on piles, *Petroleum Transactions, American Institute of Mining and Metallurgical Engineers*, 189 (2846), 1950, 149–154.

8.5   Wiegel, L., Beebe, K.E. and Moon, J. Ocean wave forces on circular cylindrical piles, *Proceedings of the American Society of Civil Engineers*, 83 (HY2) 1957, 1199/1–36.

8.6   Sarpkaya, T. *Wave Forces on Offshore Structures*. Cambridge University Press, Cambridge, MA, 2010.

8.7   Steady pounding of North Sea gas platforms, *Ocean Industry*, 10 (8), 1975, 64–71.

8.8   Chappelaar, J.G. Wave forces on groups of vertical piles, *Journal of Geophysical Research, American Geophysical Union*, 64, 1959.

8.9   Croasdale, K.R. Ice forces on fixed rigid structures, US Army Cold Regions Research and Engineering Laboratory, Special Report No. 80-26, 1980.

8.10  American Petroleum Institute. *Planning Designing and Constructing Structures and Pipelines for Arctic Conditions*, 2nd edn. API RP2N, Washington, DC, 1997, reaffirmed 2007.

8.11  Croasdale, K.R. and Cammaert, A.B. An improved method for the calculation of ice loads on sloping structures in first year ice, *Hydrotechnical Construction*, 28 (3), 1994, 174–180.

8.12  Brown, T.G. Ice loads on the piers of the Confederation Bridge, Canada, *The Structural Engineer*, 78 (5), 2000, 18–23.

8.13  American Petroleum Institute. *Recommended Practice for Planning, Designing and Constructing Tension Leg Platforms*, 3rd ed. Publication API RP2T, Washington, DC, 2010.

8.14  Bonnett, D. Wind turbine foundations – Loading, dynamics and design, *The Structural Engineer*, 83 (3), 2005, 41–45.

8.15  ffrench, R., Bonnett, D. and Sandon, J. Wind power – A major opportunity for the UK, *Proceedings of the Institution of Civil Engineers*, 158 (Special Issue 2), 2005, 20–27.

8.16  Gavin, K., Igoe, D. and Doherty, P. Piles for offshore wind turbines: A state-of-the-art review, *Proceedings of the Institution of Civil Engineers, Geotechnical Engineering*, 164 (4), 2011, 245–256.

8.17  Bhattacharya, S., Carrington, T. and Aldridge, T. Observed increases in offshore pile driving resistance, *Proceedings of the Institution of Civil Engineers, Geotechnical Engineering*, 162 (1), 2009, 71–80.

8.18  Yaron, S.L. and Shimoni, J. The Hadera offshore coal unloading terminal, *Proceedings of the Offshore Technology Conference*, Houston, TX, Paper OTC 4396, 1982.

# Chapter 9

# Miscellaneous piling problems

## 9.1 PILING FOR MACHINERY FOUNDATIONS

### 9.1.1 General principles

The foundations of machinery installations have the combined function of transmitting the dead loading from the machinery to the supporting soil and of absorbing or transmitting to the soil in an attenuated form the vibrations caused by impacting, reciprocating or rotating machinery. In the case of impacting machinery or equipment such as forging hammers or presses, and reciprocating machines, piston compressors and diesel engines, the dynamic loads transmitted to the soil take the form of thrusts in a vertical, horizontal or inclined direction. Rotating machinery, such as gas and steam turbines, creates a torque on the shaft, resulting in lateral loads or moments applied to the foundation block. Rock crushers and metal shredders produce random dynamic loads as a result of rotating imbalances depending on the particular operation. Dynamic loading from hammers or presses or from low-speed reciprocating engines has a comparatively low frequency of application, but the vibrations resulting from out-of-balance components in high-speed rotating machinery can have a high frequency.

The higher the frequency of dynamic loading, the less is the amplitude which can be permitted before damage to the machinery occurs or before damage to nearby structures, and noise and discomfort to people in the vicinity becomes intolerable. When the frequency of vibration of a machine and its foundations approaches the natural frequency of the supporting soil, resonance occurs and the resulting increased amplitude may result in damage to the plant and excessive settlement of the soil. The latter is particularly liable to occur when the vibrations are transmitted to loose or medium-dense coarse-grained soils and the combined frequency hits resonance. Repeated pounding of such soils by the dynamic loading of a drop forge can also cause foundation failure.

When the mass of the machine and its foundations and the vibration characteristics of the soils are known, it is possible to calculate the resonant frequency of the combined machine–foundation–soil system. In order to avoid resonance, the frequency of the applied dynamic loading should ideally not exceed 50% of the resonant frequency for most impact hammers or reciprocating machinery. In the case of high-speed rotating machinery, it is probable that the applied frequency will be higher than the resonant frequency of the machine–foundation–soil system. Dynamic loading will also cause degradation of the foundation stiffness, which will move the natural frequency closer to the impact frequency leading to resonance. For this condition, the aim should be to ensure that the applied frequency is at least 1.5 times the resonant frequency. The need for the wide divergence is to allow for the starting-up and shutting-down periods when the frequency of the machine passes through the resonant stage. If the applied frequency is too close to the resonant frequency, the stage of resonance at the acceleration or slowing down of the machine might be too protracted. Operating frequencies of large compressors vary up to 190 Hz and gas turbines up to 250 Hz.

When designing shallow foundations for machinery, vibrations that might cause damage or nuisance to the surroundings can be absorbed or attenuated by increasing the mass of the foundation block. There are old 'rules of thumb' that require the ratio of the mass of the foundation to the mass of the machine to be in the range of 1.5:1 for rotating machines to 4:1 for reciprocating machines (with the same ratios applied to pile caps firmly anchored to the piles with shear connectors). Anyaegbunam[9.1] provides a basic model for determining the minimum foundation mass required to limit vertical machine vibration amplitude, and BS 2012-1 is the Code of practice for machinery foundations. However, the resulting required mass of the foundation may be excessive for loose or weak soils leading to excessive settlement, even under static loading conditions, and necessitating the provision of a piled foundation. Also, it may be necessary to employ piles on sites where the water table is at a depth of less than one-half of the width of the block below the underside of the base or even within a depth of twice the width of the block. This is because water transmits amplitudes of vibration almost undamped over long distances, which might result in damaging effects over a wide area surrounding the installation. Similarly, piles may be desirable if a rigid stratum of rock or strongly cemented soil exists within a depth of 1.5 times the block width. Such a stratum reflects energy waves and magnifies their amplitude of vibration.

Generally, the effect of providing a piled foundation to a reciprocating or rotating machine is to increase the natural frequency of the installation in the vertical, rocking, pitching and also possibly longitudinal modes. This is because of the behaviour of the mass of soil enclosed by the pile group acting with the pile cap and the piles themselves. The soil mass may be relatively small where the piles act in end bearing or large in the case of friction piles. The natural frequency may be decreased in the lateral and yawing modes of vibration because of the low resistance of piles to lateral loads at shallow depths. As noted in the following discussion for static foundations, it is important to ensure that the centre of gravity of the dynamic loading coincides with the centre of gravity of the support system; otherwise, in this case, the foundation design will have to deal with the torsional and rocking modes introduced.

To ensure that the ratio of the frequency of the disturbing moment or disturbing force applied by the machinery to the natural frequency of the machine–foundation–soil system is either greater or less than the required value, it is necessary to calculate the natural frequency of the system. This is a complex matter, particularly for piled foundations, and is beyond the scope of this book. The reader is referred to the publications of Hsieh[9.2] and Prakash and Sharma[9.3] for general guidance on empirical analysis. The American Concrete Institute[9.4] presents various design criteria and methods of analysis, design and construction as currently applied to dynamic equipment foundations. BS 2012-1 is applicable to the design and construction of block foundations for reciprocating machinery in the low-to-medium frequency ranges (<25 Hz); piled foundations are also referenced.

## 9.1.2 Pile design for static machinery loading

Piles and pile groups carrying static loads from machinery should be designed by the methods described in Chapters 4 and 5. Particular attention should be paid to the avoidance of excessive differential settlement of the pile cap; the differential movement should not exceed 8 mm. The centre of gravity of the machine combined with the pile cap and supporting piles should be located as nearly as possible on a vertical line through the centroid of the pile group, and the eccentricity of the combined masses should not be greater than 5% of the length of the side of the pile group. If possible, the centre of gravity of the machine and soil mass should be below the top of the pile cap.

### 9.1.3 Pile design for dynamic loading from machinery

Generally, it can be stated that the effect of applying dynamic loads to piles in fine-grained soils is to reduce their shaft friction and end-bearing value, that is to reduce their ultimate carrying capacity, and the effect in coarse-grained soils is to reduce their shaft friction but to increase their end-bearing resistance at the expense of increased settlement under working load.

The reduction in the shaft friction and end-bearing resistance of piles in fine-grained soils is the result of a reduction in the shearing strength of these soils under cyclic loading. The amount of reduction for an infinite number of load repetitions depends on the ratio of the applied stress to the ultimate stress of the soil. It is the usual practice to double the global safety factor on the combined shaft friction and end bearing to allow for the dynamic application of load. For design to Eurocode rules, the 'combinations of actions for persistent or transient situations' as given in BS EN 1990 Clause 6.4.3 will have to be considered using the partial action factors in the National Annex, together with the EC7-1 partial factors to determine the necessary increase in pile resistance.

The torque of rotating machinery can cause lateral loading on the supporting piles. The deflection under lateral loading can be calculated by the methods described in Chapter 6. To allow for dynamic loading, the deflections calculated for the equivalent static load should be doubled. However, as pointed out by Bhatia[9.5], the evaluation of the dynamic characteristics of a pile group remains a complex problem that calls for many assumptions to be made leading to associated uncertainties. He suggests a design approach based on 'equivalent pile springs' and gives an example of its application.

The type of pile, whether driven, driven and cast-in-place, or bored and cast-in-place, is unlikely to have any significant effect on the behaviour of piles installed wholly in fine-grained soils. It is possible that the lateral movements of driven precast concrete or steel H-piles will be greater than those of cast-in-place piles, because of the formation of an enlarged hole around the upper part of the shaft (see Figure 4.5).

The frictional resistance of a pile to static compressive loading in a coarse-grained soil is relatively low. This resistance is reduced still further when the pile is subjected to vibratory loading, and it is advisable to ignore all frictional resistance on piles carrying high-frequency vibrating loads. If such piles are terminated in loose to medium-dense soils, there will be continuing settlement to a degree that is unacceptable for most machinery installations. It is, therefore, necessary to drive piles to a dense or very dense coarse soil stratum, and even then the settlements may be significant, particularly when high end-bearing pressures are adopted. This is due to the progressive attrition of the soil grains at their points of contact. The continuing degradation of the soil particles results in the slow but continuous settlement of the piles. If possible, piles carrying vibrating machinery should be driven completely through a coarse soil stratum for termination on bedrock or within a stiff clay.

Spacing of piles should be as large as possible, at least five times the diameter. Applied pile stress should be kept well below the design stress. The ACI report[9.4] considers the complex interaction of piles in a group under dynamic loading when piles are closer than 20 diameters and recommends suitable computer programs to consider group dynamic stiffness and damping effects in such cases. For example, Ensoft Inc. has developed a program (DynaPile) for the analysis of pile foundations under dynamic loading for single piles and pile groups (see Appendix C).

At a site in Glasgow, where gear cutting machinery had to operate to an accuracy of 0.009 mm with each machine enclosed in separate units under conditions of constant temperature and humidity, it was essential to avoid settlement caused by vibrations of these machines and at new adjacent workshops. The possibility of compaction of the 18.6 m deep sand layer over glacial till due to driving a group of closely spaced piles led to the choice of small displacement piles formed from Larssen BP2 box section and driven with a double-acting hammer into the till-bearing layer. The surrounding plant was supported on driven and cast-in-place piles terminated 4.5 m into the sand stratum.

## 9.2 PILING FOR UNDERPINNING

### 9.2.1 Requirements for underpinning

Underpinning of existing foundations may be required for the following purposes:

1. As a remedial measure to arrest the settlement of a structure
2. As a precautionary measure carried out in advance to prevent the excessive settlement of a structure when deep excavations are to be undertaken close to its foundations
3. As a strengthening measure to enable existing foundations to carry increased loading or to replace the deteriorating fabric of a foundation

Before underpinning by piling or any other method is considered, it is essential to determine the cause of structure–foundation instability and confirm the ground conditions at depth. For piling solutions in difficult ground, either preliminary test piles should be considered or means of checking pile capacity and integrity once installed should be available. The potential for causing distress to the structure due to the method of construction and mobilisation of the load transfer should be examined. If party walls are to be underpinned, all affected owners should be advised, and if work is to encroach on space below adjacent property, then it is essential that a specialist is consulted on liabilities and insurance prior to commencing work.

An example of the use of piling as a *remedial* measure is shown in Figure 9.1a. The column has settled exclusively due to the consolidation of the soft clay beneath its base. Piles are installed on each side of the base and the load transferred to the pile heads by needle beams inserted below the base.

A typical use of piles as a *precautionary* underpinning measure is shown in Figure 9.1b, where a deep basement is to be constructed close to an existing building on shallow strip foundations. Underpinning of the foundation adjacent to the basement is required since yielding of the ground surface as a result of the relief of lateral pressure due to the excavations would cause excessive settlement.

Piling as a *strengthening* measure is shown in Figure 9.1c. Here, pits are excavated beneath the existing foundation, and piles are jacked down to a bearing on a hard incompressible stratum. Underpinning of the foundations may be required where the existing piles have deteriorated due to attack by aggressive substances in the soil or groundwater. New piles can be installed in holes drilled through the cap or raft (Figure 9.1d). The new pile heads are bonded to the reinforcement of the existing substructure.

The application of piling methods directly under an existing foundation as in Figure 9.1c will be limited because it is usually necessary to excavate pits by hand below the existing substructure to place supporting beams or pads. If the excavation depth is in excess of 1.2 m, the pit will need support and the Confined Spaces Regulations may apply. However, in a high proportion of the cases where remedial or strengthening works are required, a suitable bearing stratum exists at no great depth. In such cases, it is cheaper to take the pits down to this stratum and to backfill the void with mass concrete rather than installing piles in restricted working conditions. Also a considerable force may be required to jack down an underpinning pile, and there may be insufficient mass in the existing structure to provide the required reaction to this jacking force. When using open-ended tubular steel piles in pits, it will be difficult to remove the soil plug to ease jacking and, where steel sections have to be welded on to reach the required stratum or resistance, the alignment and welding quality can be difficult to control. In low-rise

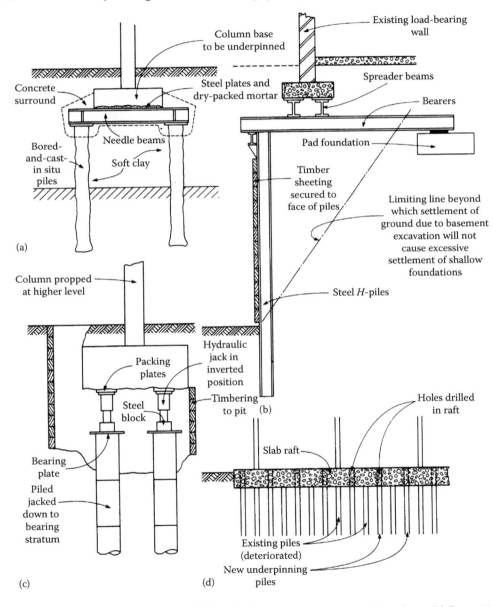

Figure 9.1  Use of piles in underpinning: (a) Remedial measures to support column base. (b) Precautionary measures in underpinning strip foundation adjacent to deep excavation. (c) Jacked piles to strengthen column base. (d) Drilled piles to replace existing piles beneath raft slab.

(continued)

(e)

*Figure 9.1* (continued) Use of piles in underpinning: (e) Braced lateral support underpinning shallow foundation. (Courtesy of Macro Enterprises Ltd. Massapequa, NY.)

buildings, the mass concrete strip foundations and brick footing walls are unlikely to have sufficient bending strength to withstand the jacking load, even though a spreader beam is used
between the jack and the foundation. Another consideration for jacked piles in clay soils is the
potential loss of capacity due to medium- to long-term pore pressure dissipation.

## 9.2.2 Piling methods in underpinning work

Bored piles, using the methods for mini- and micropiles as described in Section 2.6, are suitable for underpinning a variety of structures and can be installed outside the periphery of the
existing foundations as shown in Figure 9.1a. Where piles have to be installed inside buildings as in Figure 9.2, spoil disposal will be an issue. Precast reinforced concrete sections or

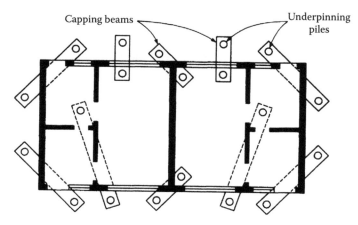

*Figure 9.2* Layout of piles for light structures.

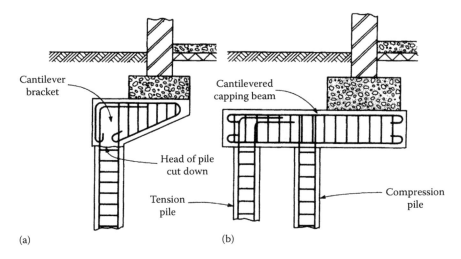

*Figure 9.3* Cantilevered brackets for supporting light structures: (a) From single piles. (b) From pairs of piles.

steel H-piles can be concreted or grouted into the pile boreholes in cases where it is desired to transfer the loading to underpinning piles as quickly as possible after installing them.

Light structures can be underpinned from a single row of bored piles located outside the building. After concreting the piles, cantilever brackets are cast onto their heads as shown in Figure 9.3a. The bending resistance of a small-diameter pile is relatively low, and therefore, the form of construction is limited to strip foundations of light buildings or to lightly loaded columns. Heavier structures can be underpinned by pairs of piles located outside the building but carrying a cantilevered bracket as shown in Figure 9.3b. This system can cause difficulties in pile design. The compression pile is required to carry heavy loading, and there may be problems in achieving the required resistance to uplift in shaft friction on the tension pile.

The *Pali Radice* system[9.6], operated in the United Kingdom by the Fondedile piling division of Keller Geotechnique, and equivalent propriety systems are well-proven techniques of internally reinforcing and stitching existing structures and foundations as a retrofitting operation. The method is also used to construct new foundations using *reticulated* piles where space is limited and to strengthen existing foundations in order to carry heavier loads. Compact rotary drilling machines are available to operate with relatively little vibration in a working space of only 2 m × 1.5 m and headroom as low as 1.8 m. Pile diameters range from 100 to 300 mm with temporary or permanent lining to accommodate either a single reinforcing bar or cage. The bore may be filled with tremie concrete or pressure injected with cement grout to produce frictional and end-bearing resistance.

Figures 9.4 and 9.5 show the installation of 56 × 280/220 mm diameter vertical and raking Pali Radice piles used to stabilise the pier and abutment of a rail underbridge that had suffered significant settlement adjacent to a canal. The 10.5–18.5 m long piles were founded in the underlying dense sand and had axial capacities ranging from 823 kN compression to 447 kN tension.

Heavily loaded foundations can be underpinned by jacking piles down to the bearing stratum using the dead load of the existing foundations and superstructure as the reaction to the jacking operation. The Abbey Pynford *Presscore* precast jacked-in pile is described in Section 2.2.3 and Figure 2.14. The piles require a pit excavation beneath the foundation and a hole in the floor of the pit to receive the bottom-pointed unit of the pile. A careful sequence of jacking, strutting, and packing is used to press the pile to the bearing stratum or until the desired

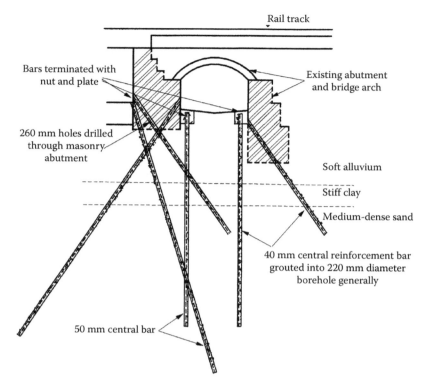

*Figure 9.4* Schematic of Pali Radice underpinning as shown in Figure 9.5. (Courtesy of Keller Geotechnique, Coventry, UK Tata Steel Projects. York, UK.)

*Figure 9.5* Keller piling rig installing Pali Radice piles to underpin the foundations to a rail underbridge. (Courtesy of Keller Geotechnique, Coventry, UK Tata Steel Projects. York, UK.)

preload has been attained. The precast elements are next bonded together by inserting short steel bars into the longitudinal central hole and grouting them with cement. On the completion, short lengths of steel beam are driven hard into the space between the pile head and the foundation or between the pile head and the spreader beams. Grout bags, which are inflated and pressurised with cement grout are a convenient alternative to the steel packing.

An alternative method that *pretests* the jacked-in pile once it reaches the bearing stratum, or the desired value of preload has been attained, requires a pair of hydraulic jacks to be inserted between the head of the pile and a bearing plate packed up to the underside of the existing foundation. The thrust on the rams of these jacks is adjusted to apply a load of 1.5 times the working load onto the pile. When downward movement of the pile has ceased, a short length of steel H-section with end-bearing plates is wedged tightly into the space between the jacks (Figure 9.6). The latter can then be removed and used for the same procedure on the adjoining piles. Where piles are installed in rows or closely spaced groups by preloading or pretesting methods, the operation of jacking an individual pile relieves some of the load on the adjacent piles that have already been installed and wedged up. It then becomes necessary to replace the jacks and reload these piles, after which the inserted struts are re-wedged. Alternatively, all the pretesting jacks can remain in position until the last pile in the group or row is jacked down. Then, all the loads on the jacks are balanced, the struts installed and the jacks removed. The final operation is to encase the struts and pile heads in concrete well rammed up to the underside of the existing foundation.

Whichever system of jacked piles is used, it is essential to maintain the load on the jack until the packing is completed to avoid any rebound of the pile head and subsequent settlement when the load from the structure is transferred to the piles. Safeguards are needed to avoid a sudden drop in the ram due to the loss of oil pressure. Also care must be taken to restrain the existing foundation, or the rows of jacks and struts, from moving horizontally due to lateral or eccentric thrusts. Raking shores to the superstructure, strutting of the existing foundation to the walls of the underpinning pit, or bracings between jacks and pile heads can be used to restrain lateral movement.

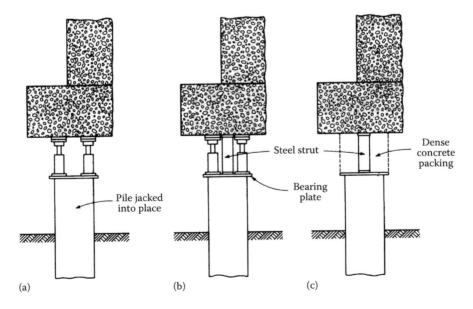

*Figure 9.6* Underpinning with pretest load: (a) Jacking down underpinning pile. (b) Insertion of steel strut. (c) Steel strut wedged into place before encasement in concrete.

Where H-section piles are used to provide underpinning combined with lateral support to a deep excavation, as shown in Figure 9.1b, they can be installed by placing them in stable holes previously drilled by mechanical auger to minimise vibration. In potentially unstable soils, cased bores will be necessary and precautions taken to avoid jamming of the auger and loss of ground. The building should be shored temporarily with supports bearing on the ground outside the zone of potential subsidence. Below the level planned for the base of the excavation, the space between the pile and the borehole is filled with concrete to provide passive resistance to lateral loads on the piles from the retained structure; weak mortar may be used if the H-sections are to be removed when permitted by the planned sequence of underpinning and construction of the permanent work. The H-piles may need tiebacks or bracing to support the lateral loads as shown in Figure 9.1e.

An alternative to the excavation support in Figure 9.1b is shown in the grouting solution in Figure 9.7 for a shallow basement excavation. The grouted block is formed by the injection of cement and chemical grouts as appropriate through tubes à manchette, and once it has reached the required strength, the excavation for the mass concrete underpinning blocks is carried out in a hit-and-miss operation.

Closely spaced bored piles are regularly used to form retaining walls for deep basement excavations and can minimise the settlement of adjacent existing buildings by acting as underpinning support to foundations, similar to the scheme in Figure 9.1b. The single row of piles may be constructed so that they virtually touch each other, known as *contiguous* piles (Figure 9.8a), or as a single row of interlocking piles—*secant* piles (Figure 9.8b). Contiguous piles are cheaper to install, but there are usually gaps present between adjacent piles, which can allow sand and silt below water table to bleed through the gaps causing a considerable loss of ground. While jet grouting between contiguous piles can deal with such seepage, contiguous piles are best suited to underpinning and excavation support in firm to stiff clays or damp silts and sands above the water table.

In water-bearing, coarse soils, secant piles are preferred to avoid loss of ground. Here, alternate piles are first installed by conventional drilling and casting relatively weak concrete in situ.

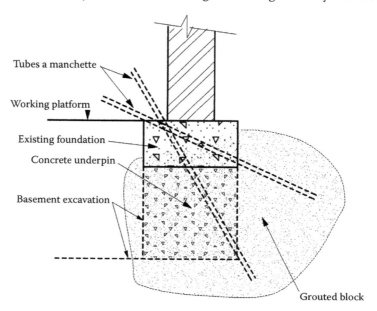

*Figure 9.7* Concrete underpinning with grouted support.

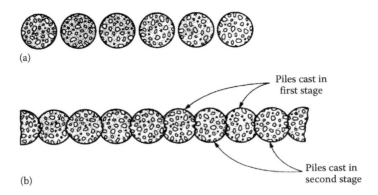

*Figure 9.8* Bored piles used for combined underpinning and lateral support: (a) Contiguous piles. (b) Secant piles.

The soil in the space between the pile shafts is then drilled out with a *secant* being cut into the wall of the 'soft' pile on each side, using appropriate drilling tools, including heavy-duty continuous flight auger (CFA) techniques. Structural concrete is placed to fill the drill hole, thus forming the interlocked and virtually watertight wall. Longitudinal reinforcement is provided in the 'hard' piles to the extent necessary to carry vertical loading, eccentric loads from the underpinning bearers and lateral loading from earth and hydrostatic pressure. In confined spaces, minipiles can be designed as retaining structures, subject to inherent stiffness limitations.

Helical plate screw piles as described in Section 2.3.6 are suitable for underpinning light buildings in confined conditions. They are screwed into the ground adjacent to the foundation using rig-mounted hydraulic rotary drives or handheld drives where access is restricted. The foundation is supported either on a steel bracket attached to the pile shaft or as shown in Figure 9.3a for the larger loads; hence, lateral loading and bending have to be taken into account in design. The International Code Council of the United States in its standard AC358 'Acceptance Criteria for Helical Foundations and Devices' proposes that the capacities of the top bracket and helix should be determined separately in addition to the ultimate load from static tests. The components are usually galvanised, but their use is not advised where corrosion from high organic soil and landfill may be expected and where softening of clays may occur due to perched groundwater passing along the helix path.

## 9.3 PILING IN MINING SUBSIDENCE AREAS

The form in which subsidence takes place after extracting minerals by underground mining depends on the particular technique used in the mining operations. In Great Britain, the problems of subsidence mainly occur in coal-mining areas where the practice in the remaining working collieries is to extract the coal by *longwall* methods. Using this technique, the entire coal seam is removed from a continuously advancing face, with the roof of the workings supported by multiple rows of hydraulically operated props. As the face moves forwards, the props in the rear are systematically lowered to allow the roof to sink down to the floor. The overlying rock strata and overburden soil follow the downward movement of the roof, and the consequent subsidence of the ground surface is in the form of a wave that advances parallel to and at approximately the same rate as the advancing coal face. This results in substantial horizontal strains of the ground surface, with tensile strains at the crest

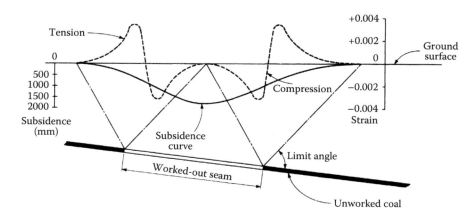

Figure 9.9 Profile of ground subsidence over longwall mine workings.

of the wave and compressive at the trough as shown in Figure 9.9. The magnitude of the strain can range from as much as 0.8% of the overburden thickness above shallow workings to 0.2% over deep seams.

The horizontal ground movements make it virtually impossible to use piled foundations in areas where longwall mining is proposed or is currently being practised. The horizontal shearing forces accompanying the strains are so high that it is quite uneconomical to attempt to resist them by heavily strengthening the pile shaft. Predicting the time between the completion of mining and the occurrence of subsidence is difficult. In areas where there is little faulting, the trough subsidence following longwall mining can cease within a few months, and piled foundations may be acceptable. Delayed fault reactivation[9.7] can result in a long period of residual subsidence, and such movements may be substantial near the boundary of the worked-out seam. In cases where limited residual movement may occur, say 10% of the total, piles could terminate in a soil layer overlying rockhead, as shown on the left-hand side of Figure 9.10. The soil acts as a cushion, preventing any concentration of load on the broken rock strata. If the workings are shallow, piles should be taken down through the collapsed overburden to intact rock layers below the coal seam as shown on the right-hand side of Figure 9.10. Bored and cast-in-place piles are used for this purpose, but it is essential to isolate the shaft of the pile from the overburden above the coal seam in order to avoid high compressive loading caused by downdrag from the collapsing strata. This isolation is achieved by placing the concrete within a shell formed from light-gauge steel tubing terminating at the base of the coal seam. Below this level, the concrete can be cast against the surface of the stable strata to form a rock socket, as shown in Figure 9.11. Most Coal Measure rocks will carry a load in end bearing equal to the design strength of the concrete pile. The space between the shell and the wall of the drill hole through the overburden should be filled with bentonite slurry to prevent emission of mine gases from the coal seam. A minimum clearance of 150 mm should be provided to accommodate minor lateral movements as the rock strata adjust themselves to their equilibrium position. In areas of recent longwall extraction, consideration should also be given to potential rotation of the pile cap being transmitted to the superstructure.

The grid of galleries and coal pillars that were formed from the old *pillar and stall* methods of coal and mineral extraction methods can cause considerable surface subsidence. Pillars may have been left intact or were wholly or partially removed as the coal extraction operations retreated towards the shaft. Where the pillars were wholly removed, the pattern of subsidence followed that of longwall mining (Figure 9.9). The unpredictable stability of pillars that were left in place continues to cause complex settlement problems in buildings

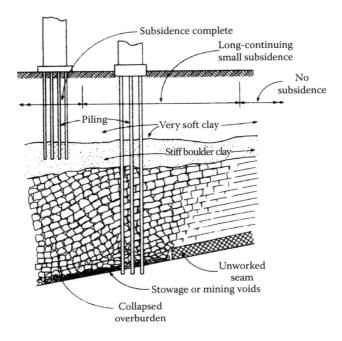

*Figure 9.10* Piling through collapsed ground over longwall mine workings.

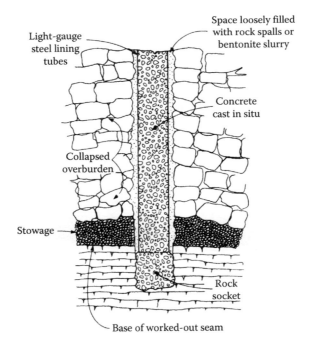

*Figure 9.11* Isolating shafts of bored piles from surrounding collapsed ground.

over abandoned mine workings, and *crownholes* are frequent hazards in urban areas of Britain[9.8]. The greatest problems are likely to occur where old pillared workings exist with less than 40 m of overburden cover. Chalk was mined in southeast England for flints and agricultural purposes from prehistoric times until the 1950s. The mining was usually in the form of a rather haphazard pillar and stall method, and numerous shallow cavities still exist.

The instability of coal pillars may be due to the slow decay of the coal, to changes in the groundwater regime in flooded workings, to increased loading on the ground surface, to an increase in the load transferred to pillars due to the collapse of neighbouring areas, or to longwall mining in deeper coal seams. If massive rock strata such as the thick sandstones of the Coal Measures are overlying the partly worked seam, they may form a bridge over the cavities such that the collapse of the weak strata forming the roof of the working will not extend above the base of the massive rock stratum (Figure 9.12a). Provided that the coal pillars themselves do not decay, the workings may remain in a stable condition for centuries, and it will be quite satisfactory to construct piled foundations overlying them.

Where massive rock strata are not present and the overburden consists only of weak and thinly bedded shales, mudstones and sandstone bands overlain by soil, a collapse of the roof will eventually work its way up to the ground surface to form a chimney-like cavity known as a *crownhole* (Figure 9.12b). Piling should be avoided above these unstable, or potentially unstable, areas, but if the workings lie at a fairly shallow depth, it is possible to install bored and cast-in-place piles completely through the overburden, terminating them in a stable stratum below the coal seam as shown in Figure 9.12b. The pile shaft must be isolated from the soils and rocks of the overburden in the manner illustrated in Figure 9.11. Any collapse of the strata over pillar and stall workings usually takes place in a vertical direction with little lateral movement, but nevertheless a generous space (a minimum of 150 mm) should be allowed between the pile shaft and the walls of the lined drill hole. Large-diameter piles are

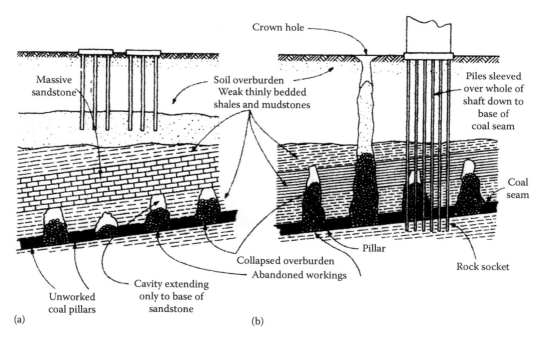

*Figure 9.12* Piling in areas of abandoned *pillar and stall* mine workings: (a) Where massive rock forms stable roof over workings. (b) Where roof over workings is weak and unstable.

preferable to small sections because of their higher resistance to lateral loading that may be due to local distortions of the rock strata. Driven piles may precipitate ground movements and should be avoided in most conditions.

Where the consolidation of open or partially stowed shallow workings, say down to 30 m deep, using drilling and grouting methods is appropriate to assist piling or prevent void migration, the pattern of grout holes, the choice of drill-flushing medium and grout materials should ensure maximum penetration of the grout but minimal flow away from the area being treated. In situations where design criteria impose strict settlement and distortion criteria, say for high-speed rail tracks, remedial and mitigation works may have to include both consolidation of the workings and a reinforced concrete slab supported on bored micropiles.

The revised CIRIA publication 'Abandoned Mineworkings Manual'[9.9] describes the essential steps to check for old workings prior to undertaking site development. This includes determination of the geology and physical conditions of the soils and rocks, the location of shafts, depth of workings and the state of pillars, stowage and backfill. Particularly important is a review of the old mineral mining plans, many of which are available from the British Geological Survey or through the UK Coal Authority. The Coal Authority will undertake searches and may provide indemnity against collapse of workings. It is also necessary to check for the presence of mine water and mine gases, taking due care to observe stringent safety measures. Modern exploration techniques can provide data for 3D modelling of the rock strata and coal seams, and geophysical surveys can highlight changes in subsurface profiles and locate old mine shafts. The magnitude and extent of subsidence can then be calculated reasonably accurately, but predicting the timescale for movement at the surface remains difficult.

## 9.4 PILING IN FROZEN GROUND

### 9.4.1 General effects

In most parts of the United Kingdom, the depth of penetration of frost into the ground does not exceed 0.6 m, and consequently frozen soil conditions are not detrimental to piled foundations. However, in countries lying in the northern latitudes with continental-type climates, the penetration of frost below the surface gives rise to considerable problems in piling work. In the southern regions of Canada and in Norway, the frost penetrates to depths of 1.2–2.1 m. In far-northern latitudes, the ground is underlain by great depths of permanently frozen soil known as *permafrost*. About 49% of the land mass of the former USSR is a permafrost region, which generally lies north of latitude 50°. The depth of permafrost extends to 1.5 km in some areas. Permafrost regions are also widespread in northern Canada, Alaska and Greenland.

In areas where frost penetration is limited to a deep surface layer overlying non-frozen soil, the effect on pile foundations is to cause uplift forces on the pile shaft and on the pile caps and ground beams. These effects occur in frost-susceptible soils, that is soils which exhibit marked swelling when they become frozen, such as silts, clays and sand–silt–clay mixtures. The formation of ice lenses below foundations and frozen soil adhering to pile shafts causes uplift forces, referred to as *adfreezing*, which must be counteracted to avoid structural instability.

The foundation problems presented by permafrost are much more severe, because of the extreme conditions of instability of this material within the depths affected by piling work. The permanently frozen ground is overlain by an *active layer* that is subject to seasonal freezing and thawing. In winter, adfreezing occurs on foundations sited within frost-susceptible soils in the active layer. In summer, there is rapid and massive collapse of thawing ice lenses in the active zone. Severe freeze–thaw conditions in highly frost-susceptible soils can result in the formation of dome-shaped ice caverns as much as 6 m high above the permafrost. The thickness of the active layer is not constant but varies with cyclic changes in the climate of the region, with changes in the cover of vegetation such as mosses and lichens, and with the effects of buildings and roads constructed over the permafrost. The laws governing the physical, chemical and mechanical properties of frozen soil have been reviewed by Andersland and Ladanyi[9.10], and they provide extensive soil mechanics data for frozen ground conditions with worked examples of a variety of foundation support systems. They also offer a comprehensive theoretical solution to the heave rate and mobilised adfreeze stress taking account of climate data, soil thermal properties, frost penetration and creep.

Tsytovich[9.11] has described three modes of formation of permafrost: these are when water-bearing soils are frozen through, when ice and snow are buried, and when ice is formed in layers in the soil, 'recurrent vein ice'. This ice can contain layers of unfrozen water within the permanently frozen soil, and foundation pressure applied to such ground can result in substantial settlement. Because of the variation in thickness of the active layer, the upper zone of the permafrost can undergo considerable changes such as major heaving, the collapse of ice caverns, and the migration of unfrozen water.

### 9.4.2 Effects of adfreezing on piled foundations

Penner and Irwin[9.12] reported results of uplift forces caused by adfreezing on steel pipes anchored into unfrozen soil in the Leda clay of Ontario where deep penetration of frost occurs below the ground surface. The formation of ice lenses in the soil caused a surface heave of 75–100 mm where the frost penetrated to a depth of 1.2 m. The adfreezing force on the steel pipe was 96 kN/m$^2$. Other tests showed adfreezing forces ranging from 86 kN/m$^2$ on timber columns to 600 kN/m$^2$ on epoxy resin–coated concrete pipe. Clays and coarse soils with low moisture content exhibit lower adfreeze strength than sandy soils.

Andersland and Ladanyi[9.10] quote Dalmatov's adfreezing equation:

$$F = Lh_a(c - 0.5b\ T_m) \tag{9.1}$$

where
   $F$ is the total upward force due to frost heave (kgf)
   $L$ is the perimeter of the foundation in contact with the soil in centimetres
   $h_a$ is the thickness of the frozen zone in centimetres
   $T_m$ is the surface temperature (°C)
   $b$ and $c$ are constants determined experimentally, indicated as 0.1 and 0.4, respectively

Design for frost heave must ensure that uplift forces are not sufficient to cause movement of the structure and that the adfreeze bond is not ruptured causing creep and an increased rate of uplift in the permafrost zone. Piles in ice-rich frozen soil can be expected to creep at a steady rate at stresses below the adfreeze strength.

### 9.4.3 Piling in permafrost regions

Piled foundations are generally employed where structures in permafrost regions are sited in areas of frost-susceptible soils. Shallow foundations cannot normally be used because of the massive volume changes that take place in the active layer under the influence of seasonal freezing and thawing.

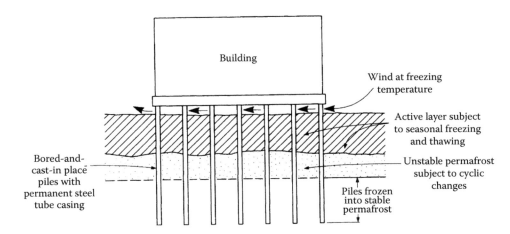

*Figure 9.13* Piling into permafrost.

The general principle to be adopted when designing piled foundations is to anchor the piles securely into a zone of stable permafrost (which can be difficult to locate) or into non-susceptible material such as well-drained sandy gravel or relatively intact bedrock. Where the piles are anchored into the permafrost layers, their stability must be maintained by conserving as far as possible the natural regime that existed before construction was commenced in the area. Thus, buildings must be supported well clear of the ground (Figure 9.13) to allow winds at sub-zero temperatures to remove the heat from beneath the buildings and so prevent thawing of the active layer in winter season.

The depth to which piles should be taken into the permafrost depends on the state of stability of this zone. Consideration must be given to the recurrence of cyclic changes in the upper layers, to the presence of layers of unfrozen water and to the pretreatment that can be given to the permafrost by thawing, compaction of the soil and refreezing.

Compressive loads on the piles are carried almost entirely by adfreezing forces on the pile shaft in the permanently frozen zone. Little end-bearing resistance is offered by the frozen ground due to the repacking and recrystallisation of ice under pressure and the migration of unfrozen water. Uplift forces on the piles that occur as a result of adfreezing in the active layer in winter season must be allowed for. Where pile caps are not placed above ground, the effect of uplift forces can be reduced by interposing a layer of compressible material, such as low-density expanded polystyrene, between the cap and the soil as shown in Figure 7.16.

Generally, it is not recommended to drive piles into permafrost at temperatures less than −5°C since this will cause splitting of the frozen ground, allowing thawing waters to penetrate deeply into the cracks, and so upsetting the stable regime. Adfreeze occurs earlier in driven piles, but driving resistance should not be used to calculate long-term capacity of piles in permafrost.

Drilled and cast-in-place piles are feasible but the concrete must not be placed in direct contact with the frozen ground. North American practice is to use powered rotary augers to drill into the permafrost to the required depth, but wear on bits will be high in silts and sands. A permanent steel casing is then placed in the drill hole and filled with concrete. The heat of hydration thaws the surrounding ground, and as the concrete cools, the freezing of the melt water bonds the pile permanently to the permafrost.

Timber piles installed in predrilled holes or driven in conjunction with steam jetting have been used for many years in northern Canada. Timber piles will generally remain well preserved in permafrost but must be protected against deterioration in the active zone.

*Thermal piles* or *refrigerated piles* are piles utilising natural convection (*passive* systems) or forced circulation cooling systems (*active*) where difficulties arise in maintaining adfreeze in thawing or unstable permafrost. Zarling[9.13] describes three systems as air cooled, liquid cooled using a refrigerant and *thermopiles* cooled in two phases. The latter is the most stable and uses exposed fins or the pile shell to remove heat when the air is colder than the ground in contact with the pile, together with internal refrigerant pipes charged with carbon dioxide.

## 9.5 PILED FOUNDATIONS FOR BRIDGES ON LAND

### 9.5.1 Selection of pile type

Bridge construction in built-up, urban conditions is subject to many constraints concerned with access to sites and environmental conditions. These have an important influence on the selection of a suitable pile type and equipment for installation. In more open country and remote locations, the constraints are fewer and selection of suitable pile types is influenced mainly by the ground conditions.

When constructing new main highways, it is desirable to complete under- and overbridges at an early stage in the overall construction programme in order to facilitate the operation of earthmoving and paving equipment along the length of the highway without the need for detours or the use of existing public highways by construction equipment. Hence, access to bridges will be difficult at this early stage, and it may be impossible to route the piling equipment and material deliveries along the cleared highway alignment without interfering with the early earthmoving operations.

In the case of small bridges, such as those carrying minor roads over or beneath the main highway, it is desirable to use light and easily transportable equipment to install a number of small- or medium-diameter piles rather than a few large-diameter piles requiring heavy equipment. Suitable types are precast concrete or steel sections, which have the advantage over bored piles of the facility to drive them on the rake, thus providing efficient resistance to lateral forces, which are an important consideration in most bridge structures. Only small angles of rake are feasible with bored piles (see Section 3.4.11), and it is usually preferable to provide only vertical bored piles suitably reinforced to resist horizontal loads and bending moments.

Some of the most difficult access problems are involved with bridges in deep cuttings where the bridge is constructed in an isolated excavation in advance of the main earthmoving operations. It is possible to install the piling for piers of bridges with spill-through abutments at the toe of the cutting and for piers in the median strip from plant operating from ground level before bulk excavation is commenced for the bridge. Initial excavation of the cutting to a temporary steep slope to enable piles to be driven at the toe using trestle guides should be undertaken with care. If the piles are pitched by a crane standing at the crest of the cutting, there is a risk of instability of the slope due to surcharge load and, in the case of clay slopes, to excess pore pressures caused by soil displacement.

Bridge construction, or reconstruction, in urban areas involves piling in severely restricted sites with the likely imposition of noise abatement regulations. Driven piles have the advantage of speed and simplicity. Compliance with noise regulations may be possible by adopting a bottom-driven type (see Sections 2.3.2 and 3.2) in conjunction with sound-absorbent screens surrounding the piling equipment. If possible, pile caps should be located above groundwater level in order to avoid sump pumping from excavations which could cause loss of ground or settlement of adjacent buildings due to general drawdown of the water table; when dewatering is necessary, the use of controlled well points is preferred.

Piling over or beneath railways involves special difficulties. The presence of overhead electrification cables will probably rule out any form of bored or driven pile requiring the use of equipment with a tall mast or leaders. The railway authority will insist on piling operations being limited to restricted periods of track possession by the contractor if there is any risk of equipment or materials falling onto the track. Soil disturbance by large-displacement driven piles may cause heave or misalignment of the rails. If it is at all possible, the design of the bridge should avoid the need for piling the foundations.

While many of the constraints described in the preceding paragraphs do not apply to bridges in remote locations, access to such sites should always be investigated. Equipment should be capable of being transported over poor roads and across weak bridges of limited width.

## 9.5.2 Imposed loads on bridge piling

The various types of loading imposed on bridge foundations have been reviewed by Hambly[9.14] in a wide-ranging report that is published by the Building Research Establishment and provides a useful checklist for current design:

Dead and live loads on superstructure
Dead load of superstructure
Earth pressure (including surcharge pressure) on abutments
Creep and shrinkage of superstructure
Temperature variations in superstructure
Traffic impact and braking forces on bridge deck (longitudinal and transverse)
Wind and earthquake forces on superstructure
Impact from vehicle collisions, locomotives and rail wagons
Construction loads including falsework

Bridge design is governed by the structural Eurocodes: BS EN 1990, giving the basis of design, EC1 for relevant actions, EC2 for concrete bridges, EC3 for steel bridges and EC4 for composite structures. In addition, geotechnical design and seismic design for bridge substructures must conform to EC7 and EC8, respectively. The Highways Agency's (HA) 'Design Manual for Roads and Bridges' (DMRB) has not been fully updated to conform to the Eurocodes (due in 2014), but 'Interim Advice Note' IAN/124/11[9.15] provides guidance on implementation of Eurocodes for the design of highway structures. The previous BS for bridges have been withdrawn (Code of practice BS 5400 – all parts), but a series of 'Published Documents' (PDs) has been produced by British Standards Institute (BSI) giving complementary and additional guidance on the application of Eurocodes to structural design. For example, PD 6694-1 provides comprehensive 'recommendations on the design of structures subject to traffic loading', which is 'noncontradictory' with EC7-1. DMRB Section BA 42/96[9.16], which covers the design of integral bridges, has been partly superseded by PD 6694 in respect of EC7 requirements, and will be phased out.

The permanent, variable and accidental actions on bridges have to be considered separately and jointly in relation to allowable differential settlements between piers or between piers and abutments in longitudinal and transverse directions. Allowable settlements are often poorly defined or not defined at all by bridge designers. Hambly[9.14] states that foundations for simply supported deck bridges are frequently designed for differential settlements of up to 1 in 800 relative rotations (25 mm in a 20 m span). In reasonably homogeneous soils, differential settlements between adjacent foundations are often assumed to be half the total settlement; thus, a total settlement of 50 mm would be permissible under this criterion.

Differential settlements of the order of 1 in 800 in a continuous deck bridge are required to be treated as a load producing bending moments in the superstructure. This can add to the cost of the bridge, but it should also be noted that the limitation of total settlement to 5–10 mm is difficult to achieve with spread foundations on soils of moderate-to-low compressibility. Some designers expect the rotation to be limited to 1 in 4000, which is equivalent to a differential settlement of only 5 mm in a 20 m span bridge. This would be difficult to ensure for bridges with longer spans even when supported by piles taken down to a competent bearing stratum. Larger rotations have to be anticipated in special conditions such as bridges in mining subsidence areas.

Although Eurocodes provide guidance, the distribution of variable actions when assessing total and differential settlement can be a matter of judgement. Full live load on the whole or part of the spans should be allowed for when calculating immediate settlements, but the contribution of live load to consolidation settlement may be small in relation to that from the dead loading. The AASHTO specification[4.96] requires consideration to be given to a wide variety of transient loads depending on the limit state being analysed and site-specific details. Figure 9.14 shows the loading on a typical pier foundation for the 4 km long elevated section of the Jeddah–Mecca Expressway designed by Dar al-Handasah, consulting engineers. The piers support the 36 m continuous spans of the three-lane carriageway. It will be noted that the predominant horizontal force on the piers was in a longitudinal direction, the resulting bending moments increasing the loads on the outer piles of the eight-pile group by about 25% above the combined vertical dead and live loads. It was possible to carry the horizontal forces and bending moments by 770 mm diameter bored and cast-in-place base-grouted piles of the type described in Section 3.3.9 using the *flat-jack* process.

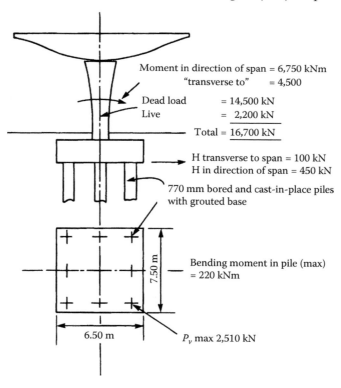

Figure 9.14 Vertical and horizontal loads on viaduct piers of Jeddah–Mecca Expressway.

Horizontal earth and surcharge pressures on *free-head* (*flexible*) bridge abutments are resisted more efficiently by raking piles than vertical piles but have several drawbacks (as noted in PD 6694 below). Rakers provide a high degree of rigidity to the foundations in a horizontal direction, which may require designing for at rest earth pressures ($K_0$) rather than the lower active pressures ($K_a$), which depend on yielding of the retained structure. Hence, rakers are most effective when used to restrict forward rotation of high retaining walls subjected to heavy compaction of the backfill. Where used, the angle of rake should be varied as shown in Figure 9.15 to spread the load on the bearing stratum. *Buildability* is another factor when considering the use of rakers. If the vertical piles in a group are to be bored, then forming the adjacent rakers as bored piles will present difficulties with placing casing and reinforcement. A better combination would be for all piles to be driven, even allowing for the reduced efficiency when driving on the rake. Section 6.5 demonstrates the basic methods of determining individual pile loads in groups of vertical and raking piles. For bridge foundations, the three types of actions on the piles (permanent, variable and accidental) have to be considered both as coexisting and as separate variables in order to obtain the maximum resultant for the axial load and bending moment in the rakers.

In the case of bridges with *spill-through* abutments and embanked approaches, the piles supporting the flexible *bank seats* are best installed from the surface of the completed embankment (Figure 9.16a). In this way, the downdrag forces from the settling embankment and any underlying compressible soils are carried preferably by vertical piles. The downdrag force can be minimised by using slender steel sections. If the piles are constructed at ground level with the bank seat supported on columns erected on a pile cap, the latter will act as a 'hard-spot' attracting load from the embankment fill (Figure 9.16b). Unless precautions are taken, the higher loading on the piles supporting the low-level pile cap will result in greater tendency for them to settle relatively to the piles supporting the adjacent bridge pier with consequent differential movement in the bridge deck.

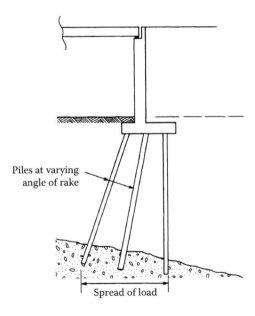

Piles at varying angle of rake

Spread of load

*Figure 9.15* Bridge abutment supported by raking piles.

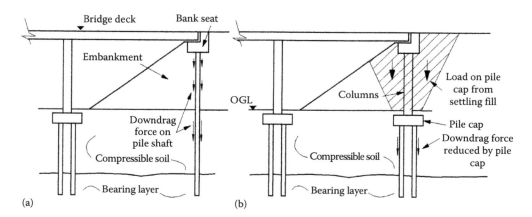

*Figure 9.16* Piling for bridges with spill-through abutments: (a) Bank seat carried by piles driven from completed embankment. (b) Bank seat carried by columns with pile cap at original ground level.

Vertical piles are preferable to rakers for supporting free-head abutments constructed on the ground underlain by a soft deformable layer, whether or not the abutments are of the spill-through type or in the form of vertical full-height retaining walls and inclined wing walls. The flexible abutment is only partially restrained from moving forwards under the influence of the retained soil. A small degree of restraint is provided at the top of the wall by friction or rotation in the bearings supporting the deck structure. At pile cap level, higher restraint is provided by the stiffness of the supporting piles, but the amount of forward movement should, theoretically, result in earth pressure on the back of the abutment corresponding to the *active state*. Heavy compaction of the embankment filling is required to prevent settlement of the road surface, such that the earth pressure, particularly near the top of the wall, can be higher than the $K_0$ condition.

Bending moments and deflections in rows of vertical piles caused by earth pressure on the abutment can be calculated by the methods described from Sections 6.3 through 6.5. Where the abutment is underlain by a weak deformable layer such as soft clay, horizontal and vertical movements take place in the soft clay layer under the loading of the embankment. The vertical movements are restrained if there is a stiff underlying layer, but the only restraint to horizontal movement is shear resistance between the soft clay and the underside of the pile cap and at the interface between the soft clay and the stiff layer. As the embankment loading increases, plastic deformation occurs in the soft clay which flows horizontally away from the abutment. In effect, the clay layer is extruded between the piles accompanied by horizontal pressure on the *upstream* face of the piles and an upward pressure on the underside of the pile cap. The horizontal pressure is low at pile cap level because the pile and soil are moving together. It is also low at the interface with the stiff layer because the pile movement at this level is relatively small and the stiff layer is also moving forwards as a result of shear stress on it from the soft clay.

Springman and Bolton[9.17] undertook research on behalf of the Department of Transport firstly into the behaviour of a single vertical free-head model pile subjected to one-sided surcharge pressure caused by placing fill on a weak deformable layer, underlain by a stiffer but yielding stratum. Later, Springman et al.[9.18] dealt with the case of a full-height bridge abutment supported by two rows of three vertical piles in each driven through a soft clay layer into a dense sand stratum (Figure 9.17). Centrifuge modelling of two load cases was generally confirmed by finite element analysis to give the pressure and bending moment distributions shown. The data in Figure 9.17 are for the surfaces of the central pile furthest from the embankment. They show a marked difference in the magnitude of deflection and pressure between the

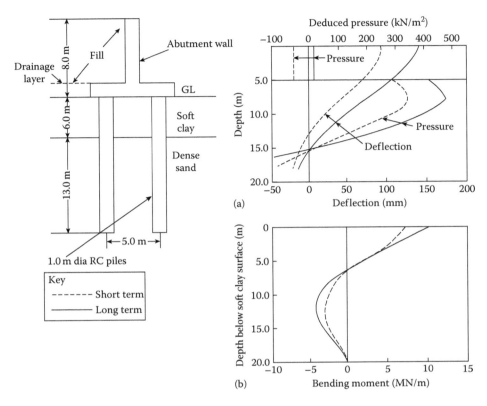

*Figure 9.17* Lateral pressure distribution on full-height bridge abutment supported by two rows of piles driven through soft clay into dense sand: (a) Deflection. (b) Bending moments. (After Springman, S.M. et al., Centrifuge and analytical studies of full height bridge abutment on piled foundation subject to lateral loading, Project Report TRL98, Transport Research Laboratory, Wokingham, UK, 1995; Crown copyright 1995. Reproduced by permission of HM Stationery Office.)

short-term (end of construction) and long-term (125 weeks after end of construction) simulated loading periods. The pressure on the front pile surface within the soft clay was negative at the end of construction as a result of the large deflections causing the pile to pull away from the soil. With time, the clay closed up against the pile causing a small positive pressure to develop. Generally, the measurements on the model piles showed increases in maximum bending moments over the 125-week (prototype) loading period of 30% for the rear (furthest from the embankment) and 15% for the front row of piles. The Springman and Bolton equations and design charts relating to Figure 9.17 are given in Tomlinson and Boorman[9.19].

The above-mentioned research was essential for dealing with the types of flexible support abutments in Figures 9.15 and 9.16. The HA's DMRB Section BD 57/01[9.20] now recommends that all bridges with lengths not exceeding 60 m and skews not exceeding 30° should be designed as *integral bridges*, with abutments connected directly to the bridge deck. The reason is the improved durability as bearings and expansion joints are eliminated; also there is improved seismic performance. However, the resulting expansion and contraction of the monolithic structure causes progressive long-term increase in the soil pressure on the abutment ('strain ratcheting'). Springman et al.[9.21] and other researchers such as England et al.[9.22] carried out further investigation into the cyclic loading on integral abutments, and the results of this work have been incorporated into the BSI document PD 6694. Analysis by limit equilibrium methods, as provided for in PD 6694, requires assessment of the earth pressure coefficient ($K^*$) produced during thermal movements resulting in rotation and/or flexure of a full-height integral abutment. $K^*$ is calculated as a function of at rest earth pressure ($K_0$) and passive pressure ($K_p$), but must not exceed the 'maximum unfavourable' value of $K_{pt}$ (as PD 6694 Table 8), to produce the design value, ($K_d^*$). For example, for translational thermal movements (with rotation),

$$K_d^* = K_0 + \left(\frac{Cd_d'}{H}\right)^{0.6} K_{pt} \qquad (9.2)$$

where

H is the vertical height from the ground level to the level at which the abutment is assumed to rotate

$d_d'$ is the design horizontal displacement at $H/2$ when the end of the deck expands $d_d$

C is a coefficient depending on the soil under the pile cap, varying from 20 where the modulus E is $\leq 100$ kN/m² to 66 where the cap is on rock or soil with $E \geq 1000$ kN/m²

A similar expression is given for an abutment undergoing thermal movements without rotation and allows for an assumed triangular pressure diagram at depth $z$ of $\gamma z K_d^* \gamma_G$. As noted by Lehane[9.23], care has to be taken when selecting the $K_p$ parameter (EC7 Annex C) to ensure it is derived using the correct $\phi_{peak}$ triaxial value in order to avoid underestimating $K^*$ pressures at the upper third of the abutment.

Under the PD 6694 procedures, limit equilibrium analysis may be applied to integral abutments founded on spread footings and those seated on pile caps supported by more than one row of piles. Provided the sway at pile cap level is small and $K_0$ can be considered as acting at pile cap level, then the pressure diagram in Figure 9.18 can be applied (with minor modification of $K^*$ as given in BA 42/96). The additional stiffness from the pile group will generally reduce lateral movements so that soil–structure interaction (SSI) effects are small.

PD 6694 requires the design of full-height frame abutments founded on a single row of vertical piles and embedded wall abutments to be based on SSI analysis. This also applies to all abutments where there are cohesive and layered soils and over-consolidated backfill.

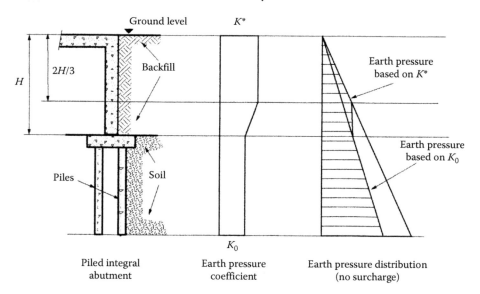

Figure 9.18 Earth pressure distribution for full-height, piled integral bridge abutment. (After Highways Agency, *The Design of Integral Bridges, Design Manual for Roads and Bridges*, Vol. I, Section 3, BA 42/96, Department for Transport, London, UK, 2003.)

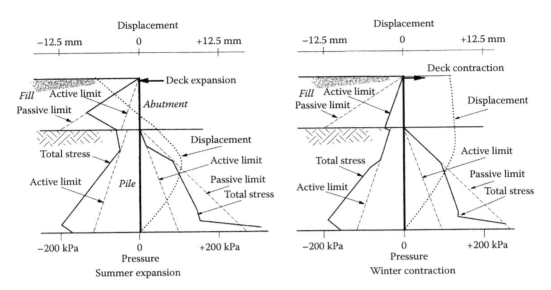

*Figure 9.19* Representations of the SSI analysis from FREW showing the effects of thermal expansion and contraction of a bridge deck on an embedded integral abutment. (Courtesy of Oasys Ltd., Newcastle-upon-Tyne, UK. – An Arup subsidiary.)

The analysis must account for all potential variables including soil parameters that reflect the changes in soil stiffness that occur after 120 cycles of winter contraction to summer expansion of the deck and the consequent effects of the different earth pressures on the retained wall and the foundations. SSI analysis therefore relies on computer programs (such as FREW, see Appendix C) which calculate the necessary iterations of $d'_d$ and $H$ to give combinations of maximum and minimum earth pressures with maximum expansion and contraction of the bridge deck. It is also necessary to find the zone of soil affected by the deck movement in order to determine $d'_d$. Figure 9.19 represents a typical numerical model output from FREW based on the SSI analysis in Annex A of PD 6694 and described by Denton et al.[9.24] Numerical models should be calibrated against comparable experience, laboratory testing or relevant historical data.

Integral bank seat abutments supported on a single row of piles should be designed using the SSI analysis as given above to determine earth pressure, with due account taken of the earth slope in front of the piles. Initially, it would be appropriate to apply $K_0$ to both faces of the pile which would allow the program to determine the soil forces by means of the soil stiffness, checking the output for the active and passive limits and applying these if the program is not sufficiently sophisticated to recognise active and passive limits. The integral bank seat may also be supported by piers fixed to a pile cap at ground level. In both cases, the pile should be embedded in the bank seat for a minimum of two pile widths to achieve fixity. Where possible, pile dimensions should be selected so that bending stresses are reduced for a given displacement, making it easier to achieve fixed-head behaviour in the pile.

Cyclic loading of integral abutments and bank seats will affect the soil around the fixed-head piles as they flex; in granular backfill the soil will be loosened and cohesive soil softened. This will determine the degree of downdrag on the piles, which for long bank seat piles could be large (see Section 4.8). The distribution of downdrag and lateral forces on the piles will vary depending on their distance from the embankment crest and location beneath the pile cap. The cyclic loading may also induce rocking on a pile cap supporting a pile group founded in stiff, but compressible strata. Where piles are not founded on hard rock, it would be advisable to make a structural joint near the base of the pier above the cap to satisfy the rotation principle in Equation 9.2. Flexible support abutments, which have a pile cap integral with the deck but with the piles in sleeves to allow the piles to flex, can be analysed by limit equilibrium methods in Chapter 4. A group of vertical/raked piles to support full-height integral abutments and prevent rocking provides a high degree of rigidity at the pile cap level and could result in earth pressures exceeding the $K_{pt}$ maximum permitted. If the piles bear on rock, the degree of fixity is increased. PD 6694 therefore allows for the use of rakers only when the 'pile/pile cap configuration does not form a mechanism if the piles are considered to be pinned top and bottom'. Previous advice in BA 42/96 stated that 'raking piles should not be used for foundations that move horizontally', and while this is still appropriate for simple structures, rakers could be used if analysed in an SSI program. For piled abutments in cohesive backfills, the effects of strain ratcheting may be ignored.

Springman and Bolton[9.17] recommended that the embankment–pile–soil system should be designed to ensure that the ratio of the mean horizontal soil pressure ($p_m$) to the undrained shear strength ($c_u$) should lie within the pseudo-elastic zone shown in the interaction diagram (Figure 9.20). In this diagram, the ratio $p_m/c_u$ is plotted as the ordinate with an upper limit of 10.5. This is similar to the earlier Randolph and Houlsby[9.25] proposal that the maximum horizontal pressure which could be applied to piles within a soft clay is $9.14c_u$ for a perfectly smooth pile and $11.94c_u$ for a perfectly rough pile. At this stage, the clay flows plastically around the pile and cannot exert any higher pressure. As noted in Section 9.10, there is a critical spacing to diameter ratio ($s/d$) above which soil flow can be expected. Elastic behaviour of the system is defined by the limits of the height/pile diameter ratio, $h/d$, being between 4 and 10. Plastic yielding of the soil beneath the embankment is reached when the ratio $q/c_u = (2 + \pi)$, where $q$ is the embankment surcharge pressure. Hence, to avoid excessive deformation of the embankment causing soil to flow between the piles supporting the abutment, there should be adequate resistance against base failure. Provided that the pile section is designed with adequate resistance to the vertical and horizontal forces from the abutment, then consideration of the additional forces on the pile caused by soil movements may not be necessary.

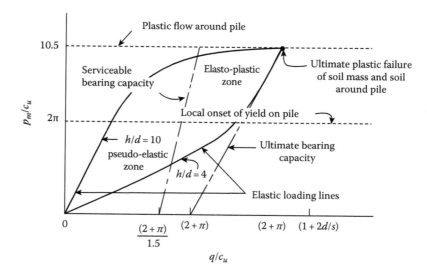

Figure 9.20 Interaction diagram for horizontal soil pressure on vertical pile driven through soft clay into an underlying stiff stratum. (After Springman, S.M. and Bolton, M.D., The effect of surcharge loading adjacent to piles, Contractor Report, Transport and Road Research Laboratory, Wokingham, UK, 1990; Crown copyright 1995. Reproduced by permission of HM Stationery Office.)

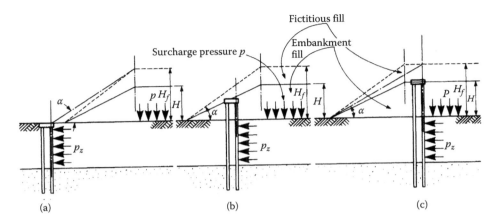

*Figure 9.21* Calculation of lateral pressure on vertical piles due to unsymmetrical surcharge loading. (a) Piles at ground level, (b) Piles within embankment fill, and (c) Piles at top of embankment fill. (After De Beer, E. and Wallays, M., Forces induced by unsymmetrical surcharges on the soil around the pile, *Proceedings of the fifth European Conference on Soil Mechanics and Foundation Engineering,* Madrid, Spain, Vol. I, 1972, pp. 325-332.)

When calculating lateral forces on the piles for a range of values of $c_u$, the higher values should be used to obtain the bending moments and pile deflections and the lower values for assessing the stability of the embankment. It is also important to ensure that the side slopes of the embankment have an adequate resistance against rotational shear failure.

De Beer and Wallays[9.26] established a method of calculating the lateral pressure on vertical piles due to unsymmetrical surcharge loading. The surcharge is represented by a fictitious fill of height $H_f$ with a sloping front face, as shown for three arrangements of piles and embankment loading in Figure 9.21a through c. The height $H_f$ is given by

$$H_f = H\frac{\gamma}{1.8} \qquad (9.3)$$

where $\gamma$ is the density of the fill in tonne/m³.

The fictitious fill is assumed to slope at an angle $\alpha$, which is drawn by one of the methods shown in Figure 9.21a through c, depending on the location of the surcharge loading in relation to the piles.

The lateral pressure on the piles is then given by

$$p_z = fp \qquad (9.4)$$

where $f$ is a reduction factor given by

$$f = \frac{\alpha - 0.5\phi'}{90° - 0.5\phi'} \qquad (9.5)$$

where
  $p$ is the surcharge pressure
  $\phi'$ is the effective angle of shearing resistance of the soil applying pressure to the pile

It should be noted that when $\alpha \le 0.5\phi$, the lateral pressure becomes negligible. De Beer and Wallays point out that the method is very approximate. It is a useful guide to the maximum bending moments and, as it is based on undrained conditions, the moments experienced during construction. The method should not be used to obtain the variation in moments along the pile shaft. They also make the important point that the calculation method cannot be used if the global safety factor for conditions of overall stability of the surcharge load is less than 1.6. The method of Brinch Hansen in Section 6.3.1 may be used to obtain the ultimate lateral resistance of the piles and hence their contribution to the restraint of the surcharge fill against slipping. However, it is preferable to apply the SSI procedures summarised above and consider the factors outlined in Section 9.10 for more precise determinations.

Driving piles within or close to the toe of clay slopes can result in the development of excess pore pressure, which may cause the slope to slip. Massarsch and Broms[9.27] developed a method of predicting the excess pore pressures induced by the soil displacement.

It is very difficult to avoid relative settlement between a piled bridge abutment and the fill material forming an embanked approach behind the abutment. Settlement of the fill often occurs even when well-compacted granular material is used. Relative settlement can be large where the embankment is placed on a compressible clay. Means of limiting settlement and ground movement and ensuring that piles are not oversized due to soil-induced lateral load include the use of lightweight backfill, reinforcement of the fill and ground improvement below the fill, such as preloading, excavation and replacement, vertical drains, stone columns and other appropriate construction methods as discussed by Seaman[9.28]. The concept of allowing piles to yield under load was adopted by Reid and Buchanan[9.29] for the purpose of reducing the relative settlement of a piled bridge abutment and the approach embankment that was founded on soft compressible clay. The arrangement of piles is shown in Figure 9.22, with closely spaced piles beneath the embankment near to the abutment,

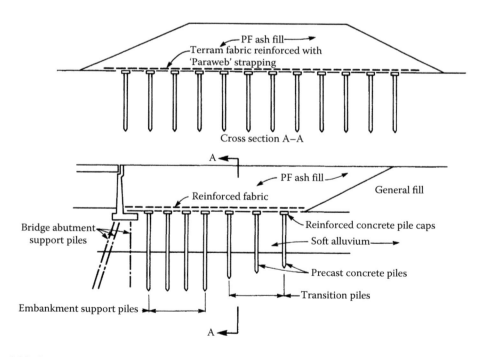

Figure 9.22 Arrangement of settlement-reducing piles beneath bridge approach embankment. (After Reid, W.M. and Buchanan, N.W., Bridge approach support piling, in *Proceedings of the Conference on Piling and Ground Treatment*, Institution of Civil Engineers, London, UK, 1983, pp. 267–274.)

designed to carry the whole of the embankment load. After the first four rows, the spacing was increased to a 3–4 m grid, and the piles were made successively shorter so that they would yield under a progressively increasing proportion of the embankment load. Loading from the embankment was distributed to the pile heads by a flexible membrane consisting of two layers of Terram plastics fabric reinforced with Paraweb strapping. If piles are used to support a bridge approach slab, the embankment design and construction and the subsoil conditions will affect the downdrag on the piles.

## 9.6 PILED FOUNDATIONS FOR OVER-WATER BRIDGES

### 9.6.1 Selection of pile type

Because of the desirability of avoiding different types of piling on the same bridge project, the piling used for piers constructed in over-water locations will usually dictate the type to be used for the abutments. Driven piles are the favoured type for over-water piers. The installation of bored piles is limited to work carried out either in a pumped-out cofferdam or in a permanent casing driven below riverbed. In fast-flowing rivers, the casing will have to be taken down to a sufficient depth below the riverbed to obtain fixity against overturning particularly in conditions of bed scour. Tubular steel piles or precast concrete piles of cylindrical section are preferred to H-sections in order to minimise current drag and eddies causing bed scour. The need for raked piles for efficient resistance of lateral forces again favours a driven type of pile. Where precast prestressed cylindrical piles are used in deep-water locations or for deep penetrations below bed level, there can be problems with handling long heavy piles. Also, forming joints to extend partly driven piles can cause difficulties and delays.

Attrition by soil particles of the exterior surface of piles at the sea- or riverbed can be a factor influencing the material of the pile and its wall thickness. This is more likely to be a problem where the bed level is constant or changing over a limited range rather than rivers where seasonal floods cause wide variations in bed contours.

A notable example of precast concrete piling for bridge works is the over-water sections of the 25 km causeway between Saudi Arabia and Bahrain Island[9.30]. The bridge sections of the causeway form a total length of 12.5 km and were constructed in water depths ranging from 5 to 12 m. A single 3.50 m OD × 0.35 m wall thickness precast concrete cylinder supports the 50 m span box girder carrying the two-lane carriageway of the dual carriageway bridge (Figure 9.23). The cylinders were cast vertically in short sections at the shore-based casting yard. The sections were then formed into complete piles by longitudinal prestressing and transported to the bridge locations by a 1000 tonne crane barge, for installation by a reverse-circulation pile-top rig operated from a jack-up platform.

The foundations for the cable-stayed Sutong Bridge[9.31] over the lower Yangtze River had to deal with water depths of 30 m with maximum flow rates of 3 m/s and layers of silty sands and silty clays extending up to 270 m below river level to bedrock. 131 drilled shafts, 2.8/2.5 m in diameter, with ultimate capacity of 92 MN, support the two main pylon piers constructed on a 13 m deep pile cap. Construction of the shafts was carried out from a steel platform fixed over the pier 3 m above high water and the 2.8 m casings driven by vibratory hammers at the north pier and diesel hammer at the south pier to depths of around 60 m. Eight rotary drills, using a variety of soft formation drill tools 2.5 m diameter, were used on each platform to extend the shafts to depths of 114–117 m using bentonite slurry to maintain hole stability. Reinforcement cages were inserted and a batching plant rated at 100 m³/h

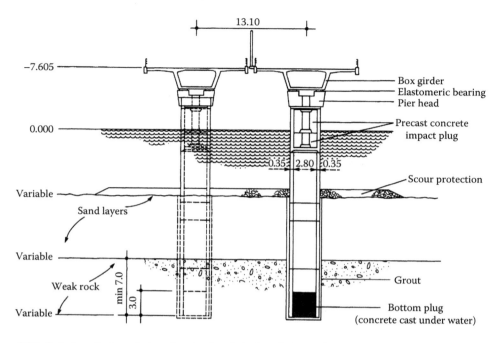

*Figure 9.23* Cylinder pile foundations for the Saudi Arabia–Bahrain Causeway. (After Beetstra, G.W. et al., Bridges and viaducts, in *The Netherlands Commemorative Volume*, E.H. de Leew, ed., *Proceedings of the 11th International Conference, ISSMFE*, San Francisco, CA, 1985; Courtesy of Ballast Nedam, Nieuwegein, the Netherlands.)

moored downstream of the platform supplied concrete. Post-grouting of the pile tip was carried out using methods similar to those shown in Figure 3.39, increasing pile capacity by 20% as indicated by before and after tests.

## 9.6.2 Imposed loads on piers of over-water bridges

In addition to the loadings listed in Section 9.5.2, the piles of over-water bridges are required to withstand lateral forces from current drag and wave action, pressure from floating flood debris or ice, and impact from vessels straying from the designated navigation channels.

*Wave forces and current drag* can be calculated using the methods described in Sections 8.1.3 and 8.1.4. The profile of the current velocity with depth varying from a maximum at the water surface to a minimum at bed level must be considered in relation to the bending moments on piles in deep fast-flowing rivers. Current-induced oscillation can also be a problem in these conditions. It is also necessary to calculate the lateral deflections in the direction of the river flow at pile head level because these can induce bending of the bridge superstructure in the horizontal plane.

The depth of *scour* below riverbed around piles at times of peak flood must be estimated for the purpose of calculating bending moments due to current drag forces and wave action on piles. The scour consists of three components: (1) general scour from changes in bed levels across the width of the channel, (2) formation of troughs in *sand waves* that move downstream with the passage of the flood and (3) local scour around the piles. *Riprap*, armouring, cable-tied concrete block mats and grout bag mats are used to protect piers and abutment foundations. Care has to be taken to prevent failure due to *winnowing* of sediments between the mats and blocks, causing uplift and rolling up of the leading edge of the mat if not anchored. May et al.[9.32] reviewed the causes and effects of, and remedies for, scour around bridge piers.

An extreme example of the influence of bed scour on bridge foundations is given by the design of the foundations of the multipurpose bridge over the Jamuna River near Sirajganj in Bangladesh[4.42,4.43]. The bridge provides a dual two-lane roadway, a metre gauge railway, pylons carrying a power connector and a high-pressure gas pipeline. At the bridge location, the river was 15 km wide. The waterway had a braided configuration with numerous deep scour channels and shifting sandbanks. In order to limit the overall length of the bridge, the waterway was narrowed by constructing massive armoured training bunds on each bank which reduced the width to 4.8 km. It was calculated that the result of constriction of flow would cause the riverbed to scour to a depth of 40–45 m below bank level at the time of a 1 in 100-year flood discharging 63,000 m³/s. An additional 10 m of scour was estimated to occur around the foundation piles.

The bridge structure consists of 52 segmental box girder spans carried on piers, each pier being supported by a pair of raking piles (Figure 9.24). The 40/60 mm wall thickness piles were driven with open ends and have outside diameters of 2.50 and 3.15 m depending on the location relative to the training bunds. The piles were driven to a depth of about 70 m below bank level into a loose becoming medium-dense to dense silty medium to fine sand containing up to 5% of micaceous particles. Support to the piles is provided partly by shaft friction and partly by base resistance. The maximum load in compression on a 3.15 m pile

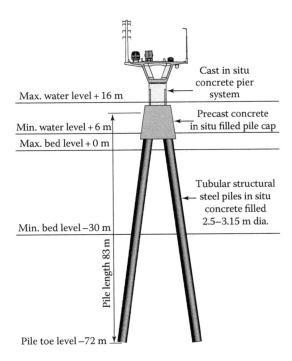

*Figure 9.24* Two-pile bent supporting intermediate piers of Jamuna River Bridge, Bangladesh. (After Tappin, R.G.R., van Duivandijk, and Haque, M. The design and construction of Jamuna Bridge, Bangladesh, *Proc. Inst. Civil Eng.*, 126, 1998, 162.)

was estimated to be 57.1 MN resulting from the bridge loading combined with current drag forces caused by the 1 in 100-year flood and by earthquake forces. The maximum lateral load on each pile was calculated to be 1.5 MN.

At the time of a major flood, more than half the shaft friction available from the soil below riverbed level under dry conditions could be lost due to scour. Furthermore, the frictional resistance in the upper part of the piles could be reduced as a result of relief of overburden pressure (see Section 4.3.6). These conditions could not be produced at the site of the pre-construction trial piling, nor could conventional loading tests to failure be contemplated on piles with such large diameters. Accordingly, tests were made on 762 mm tubular piles instrumented to measure the distribution of shaft resistance during driving and test loading. The driving test measurements were analysed by the CAPWAP® method (see Section 7.3) to confirm that the hammer selected to drive the piles was adequate for the purpose. This was a MENCK 1700T hydraulic hammer with a 102 tonne ram delivering 1700 kJ of energy per blow. The damping constants and other characteristics obtained from the driving tests were used to correlate the dynamic measurements made at the time of driving the permanent piles. The results of the measurements of shaft friction resistance on the trial piles are discussed in Section 4.3.7.

On completion of driving the permanent piles, the sand within the shafts was cleaned out by reverse-circulation drilling to within 3 m of the toe. A grid of tubes à manchette was placed on the levelled sand surface, and the pile was filled with concrete followed by grouting with cement through the tubes at a pressure of 50 bar.

Scour protection at the main piers is a major feature of the Sutong Bridge[9.31] where the steel casings for the piles are exposed above the riverbed level. The initial inner protection zone, extending 20 m around the piles, comprises sand-filled geotextile bags (1.6 m × 1.6 m × 0.6 m) dumped on the riverbed, through which the pile casings were driven. On completion of piling, protection was provided by layers of quarry-run filter and 1 m of rock armour with a density of 2.65 tonne/$m^3$. The outer zone, 20 m around the inner, consists of a layer of sandbags topped by a filter layer and 1 m rock armour. A *falling apron*, in which the material in the apron is intended to fall down a scoured slope to form a stable profile, forms the next variable width zone, set at 1.5 times the expected scour depth and comprises quarry-run stone overlain by armour with a $D_{50}$ of 0.4–0.6 m (Figure 9.25). Dumping of the materials was monitored by echo sounders.

Grout-filled mattresses, formed from woven nylon fabric sown into a series of pillow-shaped interconnected compartments injected with cement grout, produce flexible, articulated bedding that can provide effective scour protection at bridge piers and abutments. The benefit of this *fabric formwork* is that it can be quickly made and deployed and injected on the riverbed without the need for major construction equipment.

*Impact by ships* can be a severe problem in the design of bridge support piles in situations where impact cannot be absorbed by massive structures such as caissons or piers constructed inside cofferdams. It is difficult to achieve an economical solution to the problem particularly at deep-water locations. The incidence of random collisions between ships straying from the navigable channel and bridge piers has not decreased since the introduction of shipborne radar. In fact, it may have increased because of the false sense of security given by such equipment.

Three possible methods of protecting piled foundations may be considered. In shallow water not subject to major bed changes and with a small range between high and low water, the pile group can be surrounded by an *artificial island* protected against erosion by rockfill. Figure 9.26 shows a cross section of one of four islands protecting the piers of the Penang Island Bridge[9.33]. The Muroran Bridge Bay Bridge in Hokkaido features a 67 m diameter man-made island formed by placing self-setting fly ash slurry underwater on the soft seabed within a cofferdam. These forms of protection have the added advantage of preventing local

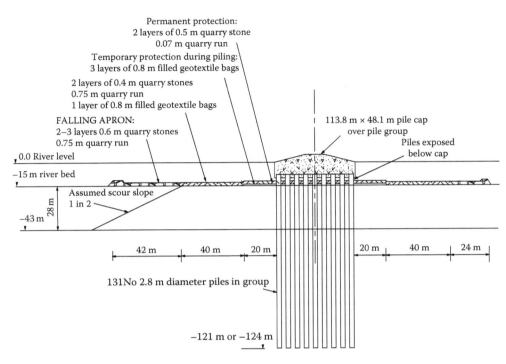

*Figure 9.25* Sutong Bridge, scour protection at main pylons. (After Bittner, R.B. et al., Design and construction of the Sutong bridge foundations, *Deep Foundations Institute, Marine Foundations Speciality Seminar*, New York, pp. 19–27. Copyright Deep Foundations Institute, 2005.)

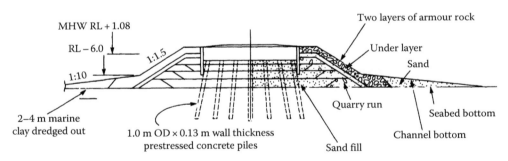

*Figure 9.26* Artificial islands protecting the piers of the Penang Island Bridge, Malaysia. (After Chin Fung, K. and McCabe, R., *Proc. Inst. Civil Eng.*, 88, 531, 1990.)

scour around the foundations. The island must be large enough to prevent impact between the overhanging bows of a ship and the bridge pier or pile if the vessel should ride up the slope of the island when drifting out of control in a fast-flowing river.

Piles can be strengthened against buckling under direct impact by increasing the wall thickness, and a group of piles can be given lateral restraint by a diaphragm connecting them at some point between the cap and bed levels. The cylinder piles of the Bahrain Causeway Bridge were strengthened by the insertion of precast concrete elements to increase the thickness over the zone of possible impact (Figure 9.23).

*Fender piles* constructed independently of the piers can be installed in deep-water locations. Piles are required to protect the sides of the piers as well as the ends in the case of

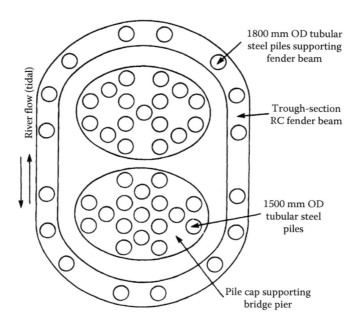

River flow (tidal)

1800 mm OD tubular
steel piles supporting
fender beam

Trough-section
RC fender beam

1500 mm OD
tubular steel
piles

Pile cap supporting
bridge pier

*Figure 9.27* Fender beam and piles protecting the river piers of the Sungai Perak Bridge, Malaysia. (After Stanley, R.G., *Proc. Inst. Civil Eng.*, 88, 571, 1990.)

impact at an angle to the axis of the pier. The arrangement of fender piles capped by a massive reinforced concrete ring beam to protect the piers of the Sungai Perak Bridge[9.34] in Malaysia is shown in Figure 9.27. The ring beam was constructed by placing precast concrete trough sections on the piles, sealing the joints between the sections and placing the reinforcement and concrete infill in dry conditions. The loading on fender piles is calculated in the same way as fender piles for berthing structures (see Section 8.1.1).

### 9.6.3 Pile caps for over-water bridges

It can be advantageous to locate pile caps at or below low river or low tide level. It avoids floating debris build-up between piles and ensures that if collision by vessel does occur, the impact will be on a massive part of the substructure instead of directly on a pile. Also a vessel is likely to sheer off at the first impact with a pile cap, whereas it might become trapped when colliding with a group of piles. Aesthetically, pile cap at or below water level is preferable to one exposed at low water. However, high-level pile caps are economical for a bridge requiring a high navigation clearance, but such an arrangement would have to be restricted to approach spans in water too shallow to be navigable by vessels, which could demolish piles supporting a high-level deck bridging the navigation channel.

Pile caps partly submerged or wholly below water level can be constructed within sheet pile cofferdams (Figure 9.28a). The sheet piles can be cut off at low water to give protection against scour. Alternatively, if a heavy lifting barge is available, a precast concrete cap in the form of an open-topped box can be lowered onto collars welded to the heads of the piles and prevented from floating by clamps. The annulus between the pile wall and the opening in the box can be sealed by quick-setting concrete or by rubber rings. The box is then pumped out and reinforcement and concrete is placed in dry conditions. The concrete seal is used in tidal conditions where a sufficient period of time is available for the concrete to set before the

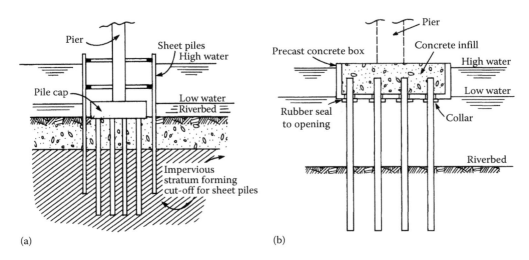

*Figure 9.28* Construction of submerged pile caps: (a) In cofferdam. (b) In open-topped box.

bottom of the box is submerged. Arrangements should be made to flood the box to equalise pressures above and below the seal until the concrete has hardened (Figure 9.28b).

Where piers are located in deep water and there is a risk of ship collision, it is desirable to construct the pile cap at bed level in order to eliminate any unsupported length of piling. This arrangement is also desirable if lateral forces from earthquakes are transmitted from the bridge superstructure and piers onto the piles. Several methods have been successfully used:

- The pier and pile cap can be constructed on shore as a single buoyant unit lowered onto the seabed followed by driving piles through peripheral skirts in a manner similar to the piled foundations of offshore drilling platforms.
- The piles can be driven in the form of a raft with their heads projecting above a rock blanket or geotextile mattress. A prefabricated pier unit is then lowered over the pile group and the connection between the two formed by underwater concrete as for the construction of the 15 piers for the bridges between Sjaelland and Falster in Denmark[9.35]. The availability of heavy-lift cranes on barges or jack-up platforms favours this type of design.
- The concrete pile cap is constructed at the site over the predriven piles and lowered to the seabed using the *lift-slab* technique. The method is described and illustrated by Elazouni and El-Razek[9.36] and was used for the construction of the Dapdapia Bridge in Bangladesh in the fast-flowing, 13 m deep Kirtonkhola River. Figure 9.29 shows the basic principles of forming the base of the pile cap above high water level, supported initially on RSJs sitting on the tops of extensions of the predriven piles. The cruciform lifting beam is concreted into the top of each casing and the pile cap box cast in stages on the soffit formwork, allowing openings for the box to slide over the piles. Lifting rods are set into the box base and connected to the hydraulic jacks on the lifting beam. Steel caissons are erected on the box to form working chambers. The box is lifted off the formwork, the platform removed, and, using the jacks and connecting rods, the box lowered down the piles to its final level. The caissons are sealed and pumped out to allow the cap and piers to be cast in dry conditions.

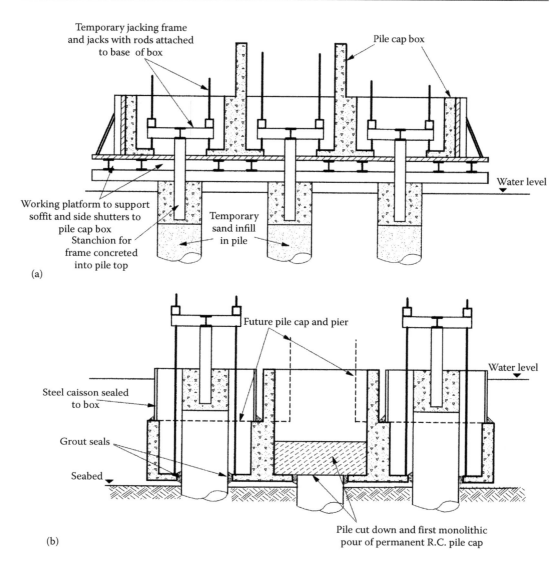

*Figure 9.29* Schematic of lift-slab method for pile cap: (a) Erecting temporary platform, jacking frames and concreting pile box. (b) Working platform removed, caissons fitted and pile cap box lowered on rods to final level and commencing monolithic infill concrete for cap and pier in pumped-out caissons. (After Elazouni, A.M. and El-Razek, M.E., *J. Constr. Eng. Manage.*, 126, 149, 2000.)

## 9.7 PILED FOUNDATIONS IN KARST

The design and construction of piles for structures on land underlain by limestone formations, which exhibit karst conditions such as wide fissures and solution cavities, present several unique challenges. Because variations in rockhead and cavitation can occur over short distances, it is difficult to produce an overall geological model of the site to determine if shallow foundations can be used or whether piles can be founded on 'competent rock'. The first requirements are, therefore, to assess the depth and strength of the overburden, the extent of cavities and the degree of infilling under each foundation by drilling a series of closely spaced probe holes using a combination of rotary-percussive rigs capable of installing casing and rotary coring drills. Waltham and Fookes[9.37] give an engineering classification of karst as a means of identifying foundation difficulties, but they point out that there is no simple answer to the number of probes which may be required to assess the hazards. The probes are usually taken to a depth of at least 3 m below rockhead and any void encountered or to a similar depth below the anticipated depth of rock socket of each pile. Because of the possibility of vertical faces in the rockhead and cavities, it is advisable to include a percentage of raked probe holes in the investigation. Where it is necessary to investigate a large area of potential karst features, the use of geophysical methods, such as microgravity to locate caverns or seismic tomography to reveal fissures, can reduce the number of probe holes needed.

The selection of the pile installation method is critical, as it may be necessary to overcome random boulders in the overburden, remove and replace weak material in cavities through which the pile has to pass and finally found on competent rock, or form a socket in rock, ensuring that sound rock also exists within the bearing zone. Large driven piles are not usually feasible and the most effective method is the drilled and cast-in-place pile, with permanent steel casing sealed into the rock at the top of the socket. For pile diameters up to 1200 mm, rotary-percussive rigs which can simultaneously install permanent casing (duplex drilling) are generally considered the most cost-effective installation method. For larger diameter piles, the use of a powerful casing oscillator and a drilling method to clean out the pile and form the rock socket is recommended (see Section 3.3.2); above this diameter, shaft sinking or caisson construction techniques may be necessary. Whichever method is used, it is essential to probe below the base of the pile to check for cavities.

The removal of cavity and fissure infill debris and replacement with cement grout to allow uncased holes to be drilled for piles is expensive and rarely achieves the desired results. Flushing/grout holes are required at less than 1 m centres under and around the pile group, and flushing water is necessary in quantities greater than 150 L/min and pressure greater than 10 bar – potentially causing pollution of surrounding water courses. If sufficient grout can then be injected, it may be possible to place concrete in the open pile hole, or as temporary casing is withdrawn, without the loss of fluid concrete. Jet grouting could be used to consolidate any cavity infill within the bearing zone below the sound rock socket – again high grout pressure and volume (450 bar and 350 litres/min) will be required with adequate venting to the surface and pollution control.

Drilling *slim* holes, with or without simultaneous casing, or driving long H-piles in karstic conditions can cause significant problems due to deviations compromising the axial capacity of the piles. Concreting or grouting open holes or while withdrawing a temporary casing runs a risk of loss of material into weak cavity infill or undetected voids requiring pre-grouting using a low slump mix injected in several stages and re-drilling.

Micropiles can be effective in karst conditions if precautions are taken to avoid contamination of the bond zone by the drilling method. For example, Uranowski et al.[9.38] describe the use of micropiles with capacities of 890 and 1160 kN for bridge piers by inserting 245 mm diameter thick-walled steel tubes into grout-filled holes drilled by *Tubex* casing (see Section 2.3.5) in karstic dolomite. 160 micropiles up to 59 m deep replaced the original proposal for forty 1371 mm diameter steel *caisson* piles at each of three piers. At another location where the karstic conditions were less variable, a down-the-hole rotary-percussive drill was used to drill 305 mm diameter holes up to 23 m deep without casing to insert the specified 245 mm steel tube – with the assistance of a D5 pile hammer (Figure 9.30). The pile holes in each case were grouted using a tremie pipe, ensuring that the grout level was stable at the top of the hole prior to inserting the permanent tube.

Natural overburden and decomposed debris overlying the karst formation can be treated by various ground improvement techniques prior to piling—such as vibroflotation,

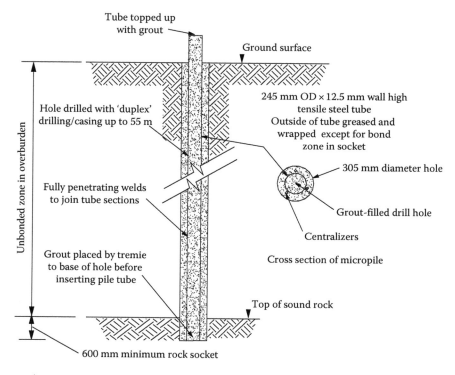

*Figure 9.30* Micropile in karst. (After Uranowski, D. et al., Micropiles in karstic dolomite; similarities and differences of two case histories, in *Geo-Support 2004, Drilled Shafts, Deep Mixing, Remedial Methods and Speciality Foundation Systems*, Turner, J.P. and Mayne, P.W., ed., American Society of Civil Engineers, Geotechnical Special Publication No 124, pp. 674–681, 2004.)

compaction grouting, and jet grouting. Fischer[9.39] describes the foundations for a nuclear power plant on karst terrain, which comprised tubular steel piles driven into relatively flat limestone bedrock, with a 5 m deep probe at each pile tip to locate cavities. The pile hole was extended by under-reaming where the probes located cavities and the tube re-driven as necessary to sound rock and filled with concrete. The overburden sand, up to 20 m deep, was treated by vibroflotation to improve the relative density to 85%–90% in order to reduce liquefaction potential.

## 9.8 PILED FOUNDATIONS IN SEISMIC REGIONS

In seismically active regions, the pile designer will need to assess the maximum seismically induced bending and shear forces on the pile in addition to providing resistance to the axial loads. The pile behaviour is significantly affected if the soil liquefies during an earthquake resulting in large lateral displacements, possible failure due to buckling and significant downdrag.

A seismic risk assessment, initially based on a probabilistic method, should determine the detrimental outcomes from the magnitude (*severity*) of the possible adverse consequences and the likelihood (*probability*) of the occurrence of each consequence. The site can then be classified as to risk level and the appropriate ground investigation strategy developed to assess liquefaction potential. Eurocode EC8-1 at Table 3.1 provides the descriptions of ground types (A to $S_2$) categorised by the average shear wave velocity, the standard penetration test

(SPT) $N$-values and $c_u$. The seismic hazard is related to a single parameter in Type A ground ('rock'), namely, the peak ground acceleration, $a_{gR}$, which for the United Kingdom is given in the zoned maps in BSI document PD 6698. In cases of low or very low seismicity as defined in EC8-1, Clause 3.2.1(4) allows for simplified structural design methods using the $a_g$ values in the National Annex.

Cyclic or softening liquefaction can occur during earthquakes when cyclic loading in an undrained situation causes a reduction in effective stress to near zero. It can occur in almost all saturated sands if the cyclic loading and shaking is long enough and in clays when the applied cyclic shear stress is close to the undrained shear strength.

In the methodology initially proposed by Seed and Idriss in the 1970s, if the *cyclic stress ratio* (CSR), incorporating the site-specific or design horizontal acceleration at the ground surface, is greater than the *cyclic resistance ratio* (CRR), then cyclic liquefaction is likely. Eurocode EC8-5, which supplements EC7 in respect of foundation design in seismic conditions, provides a simplified chart in Annex B relating the CRS parameter $\tau_e/\sigma'_{vo}$ to SPT $N$-values. The cone penetration test (CPT) and seismic cone penetrometer test (SCPT) are considered more reliable indicators of the CRR and the potential for cyclic liquefaction as described by Robertson[9.40]. EC8-5 at Clause 4.1.4(8) suggests that liquefaction may be neglected in certain soil conditions based on grading and plasticity index.

EC8-5 Clause 5.4.2(1)P requires piles to be designed to resist the inertial forces from the superstructure and the kinetic forces from the deformation of the surrounding soil due to the seismic wave; in all cases, the inequality $R_d \geq E_d$ should be satisfied. Partial factors for soil strengths, $\gamma_M$ set M2, are as EC7 with an added factor, $\gamma_{t\,cy\,u} = 1.25$, for cyclic undrained shear strength, $\tau_{cy\,u}$. Clause 4.1.4(14) cautions against the use of pile foundations alone where the loss of lateral support occurs due to liquefaction, and Clause 5.4.2(4)P requires that side resistance in soils subject to liquefaction or degradation should be ignored. Piles should be designed to be elastic but may have a plastic hinge at the pile head as stated in EC8-1 for structural design.

The seismic loading will dominate the calculations for bending resistance of the pile, and depending on the degree of liquefaction in the susceptible soil, 'lateral spreading' or displacement of soil will result in the loss of support for the pile making it susceptible to buckling. The soil–pile interaction in these conditions is complex with many uncertainties that are difficult to deal with in analytical models. Hence, two basic simplified empirical design methods based on extensive observations of earthquakes in recent years are frequently adopted. Puri and Prakash[9.41] comment on limit equilibrium analysis based on the lateral pressure assumptions in Figure 9.31 and $p$–$y$ analysis (as Section 6.3.5).

From Figure 9.31a, the liquefied layer is assumed to apply a pressure that is about 30% of the total overburden pressure, and the maximum pseudo-elastic bending moment is assumed to occur at the interface between the liquefied and non-liquefied layers. The ALP soil–pile interaction program (Appendix C) may be used for the $p$–$y$ analysis, but Puri and Prakash[9.41] note that in American practice, the $p$–$y$ curves are modified by a 'p-multiplier' ranging from 0.3 to 0.1 depending on the increase in pore pressure due to the seismic accelerations, with 0.1 applying when excess pore pressure is 100%. The factor ranges from 0.1 to 0.2 for sand with a relative density of about 35% and from 0.25 to 0.35 for a relative density of 55%.

Tabash and Poulos[9.42] provide a simple means of making preliminary estimates of maximum bending moments and shear in a single pile embedded in a 'linearly elastic clay layer' subject to seismic actions. The design charts are based on bedrock acceleration of 0.1 g, but as the analysis is elastic, they comment that values for higher accelerations may be prorated.

Bhattacharya and Bolton[9.43] consider the development of pile buckling before and after the soil becomes fully liquefied. Before lateral spreading starts, the bending moments and

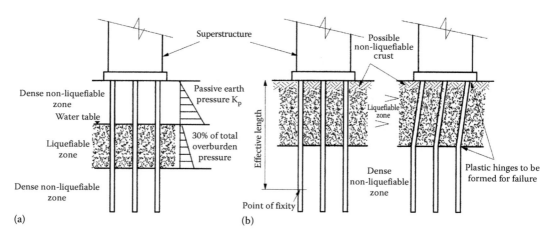

*Figure 9.31* Simplified pseudo-elastic analyses due to seismic-induced lateral flow in liquefiable soils. (a) Pressure distribution due to lateral spread. (After Puri, V.K. and Prakash, S., Pile design in liquefying soil, in *Proceedings of 14th World Conference on Earthquake Engineering*, Paper 301, Beijing, China, 2008.) (b) Buckling and collapse mechanism of pilled foundation due to lateral spread. (After Bhattacharya, S. and Bolton, M., Errors in design leading to pile failures during seismic liquefaction, *Proceedings of fifth International Conference on Case Histories in Geotechnical Engineering*, Prakash, S and Puri, V.K., ed., Paper 12A-12, New York, 2004.)

shear forces are due to the inertial effects of the earthquake, and the pile will start losing its shaft resistance, shedding axial loads onto the base; if base resistance is exceeded, settlement will occur. As lateral spreading starts, slender piles will be prone to axial instability, and buckling may occur under the high transient forces. They propose that the slenderness ratio of piles be kept below 50 and the ratio of axial load to the elastic critical load be below 0.35 for steel piles and 0.15 for concrete. The structural design of the pile is critical to ensure that if structural plastic hinges form (Figure 9.31b), the axial load is still fully supported.

For the foundations at the North Morecambe gas terminal in the United Kingdom, Raison[2.2] describes the measures undertaken to deal with design seismic accelerations of 0.05 and 0.2 g at bedrock level in variable glacial soils overlain by 7–8 m of fly ash from settlement lagoons, which previously covered the site. The fly ash was partially removed and the remaining fly ash and loose granular soils which were susceptible to liquefaction were treated using vibro-densification comprising 20 m long stone columns. Precast piles of 320 mm and 380/480 mm reinforced cast-in-place piles were installed up to 15 m into the treated zone to support the structural loads and lateral loads with the estimated settlement of 50 mm and lateral displacement of 100 mm for the 0.2 g earthquake.

The foundations for the Rion–Antirion cable-stayed bridge across 2500 m of the Gulf of Corinth in Greece had to withstand significant seismic and tectonic disturbances. Teyssandier et al.[9.44] describe the innovative solutions developed to cope with the deep-sea location, weak foundation strata consisting of soft alluvial deposits in excess of 500 m thick, seismic accelerations of 0.48 g at seabed, and tectonic (fault) movements of up to 2 m in any direction between adjacent piers. Liquefaction was not considered to be a problem except on the north shore where material was excavated. The three main pier foundations consist of 90 m diameter caissons resting on the seabed which required strengthening to accommodate the large seismic inertial forces. This was achieved by reinforcing the top 20 m with steel 'inclusions' comprising 2 m diameter steel tubes driven 25–30 m into the weak strata. Two hundred and fifty inclusions were used for each pier and topped with a 3 m thick gravel bed on which the caisson rests. The inclusions do not, therefore, act as piles, and the transfer of inertial shear forces on the superstructure during an earthquake to the foundation is limited. The caissons are free to slide providing additional isolation of seismic forces.

## 9.9 GEOTHERMAL PILES

Ground temperatures in much of Europe are reasonably constant at 10°C–15°C (and in the tropics as high as 20°C–25°C) below a depth of 10 m. This near-surface geothermal energy potential is being exploited to provide a consistent low-level, but cost-effective and environmentally friendly, source of heating for buildings, using the thermal properties of the building foundations. Concrete has a high thermal storage capacity and good thermal conductivity, and heat from the ground taken up by the pile, diaphragm wall or other foundation can be transferred from the concrete to a heat exchanger coil buried within the concrete and moved by a simple heat pump to heat the building. Conversely, in suitable soils, the heat from the building can be transferred to the concrete and ground for cooling during summer. Brandl[9.45] describes the heat transfer mechanisms in the ground and between the absorber fluid in the exchanger pipework and the structural concrete and provides recommendations for the design and operation of geothermal piles and other 'earth-contact' concrete elements. The geothermal properties of the ground (thermal conductivity and capacity) and groundwater flow and direction have to be determined as described by Clarke et al.[9.46] in order to carry out the complex heat exchange calculations using 3D FEM analyses. Loveridge et al. [9.47] provide a review of the design and construction of geothermal piles and describe field tests used to determine thermal conductivity together with typical values.

The main purpose of piles must be to resist the applied structural loads; unless a sound economic case can be made, the designed pile diameter and length should not be increased to suit the geothermal requirements. The NHBC Design Guide[2.21] points out that this technology could provide a significant proportion of the heating demands of low-rise housing and achieve high levels of saving as required in the UK government's 'Code for Sustainable Homes' in respect of energy used and $CO_2$ emissions.

The primary heat exchange circuit within the pile comprises absorber pipes of high-density polyethylene plastic, 25–30 mm diameter and 2–3 mm wall thickness, formed into several closed-end coils or loops and fixed evenly around the inside of a rigid, welded reinforcement cage for the full depth. Typically, loops of eight vertical runs would be provided in a 600 mm diameter pile. The geothermal effectiveness of piles less than 300 mm diameter is much reduced due to lower surface area and the limited number of loops which can be fitted; the economically minimum depth of a geothermal pile is about 6 m – suitable for house foundations. Each loop is filled with the heat transfer fluid, such as water with antifreeze or saline solution and fitted with a locking valve and manometer at the top of the pile cage. This may necessitate off-site fabrication. The piling method must produce a stable hole for the careful insertion of the cage and absorber pipework. Bored piles, with or without drilling fluid support, or a cased or withdrawable tube method is acceptable for most schemes. Before concreting, the absorber pipes are pressurised to around 8 bar for an integrity test to prevent collapse due to the head of fluid concrete. The pressure has to be maintained until the concrete has hardened and then re-applied before the primary circuit is finally enclosed. Concreting should be by tremie pipe placed to the base of the pile to avoid damaging the pipework.

The primary circuits in each pile are connected via header pipes to manifold blocks, which in turn are connected, usually through a heat pump, to the secondary circuit embedded in the floors and walls of the building. Using a heat pump with a coefficient of performance of 4 (the ratio of the energy downstream of the heat pump to the energy input of the pump), the ground temperature of 10°C–15°C can be raised to between 25°C and 35°C at the building. Depending on soil properties and installation depth of the absorbers, Brandl notes that 1 kW heating needs between 20 m² of saturated soil and 50 m² of dry sand in contact with the pile surface; clay soils will require a larger contact area. The ground temperature around the pile in a heat extraction system using brine will be lowered by around 5°C. If excessive heat is extracted using a lower-temperature refrigerant as the absorber, temperature around the foundation can drop to near freezing.

Laloui and Di Donna[9.48] carried out full-scale tests into the thermo-mechanical behaviour of a 10 m deep, end-bearing geothermal pile at Lausanne. They found that additional stresses and strains induced by the temperature changes appear within the pile and that the temperature changes affect the soil–pile interface. A maintained load-cyclic thermal test carried out on a 1200 kN test pile at Lambeth College, London[9.49], confirms the Lausanne result that the response to thermal loading was elastic in that changes in the pile response during thermal cycles occur as a result of the pile expanding and contracting, but the effects appear to be reversible. The ratio of the stress in the concrete to the characteristic strength had increased from 0.14 before the heating cycle to 0.24 after. If the factor of safety on the shaft resistance is low, steps need to be taken to ensure that thermal changes in the mobilised shaft resistance do not lead to adverse movement of the pile. However, any temperature-induced settlement/heave at the usual operating cyclic temperatures is likely to be less than the displacements due to the applied loads on the foundations. The design should consider the overall serviceability and structural forces within the pile. Numerical analyses that apply the conventional soil parameters to the thermo-mechanical behaviour of a pile are now available (e.g. THERMO-PILE and the 2012 version of PILE by Oasys Ltd; see Appendix C).

Cementation Skanska have installed 130, 52 m deep, 1500 mm diameter rotary-bored foundation 'energy piles' as part of a heating/cooling system for an apartment block in London. The absorber loops and reinforcement cage were placed into the bentonite support fluid prior to concreting. The system was designed to provide 760 kW of heating and 650 kW of cooling. CFA methods were used by BAM Construction to install 10 m deep, 450 mm diameter geothermal piles, with a single U-shaped absorber pipe plunged into the fluid concrete. Attaching multiple absorber loops to the reinforcement cage to plunge into the CFA concrete is also feasible but more risky. Care is needed to ensure that the pipe is pressure tested before and after insertion and allowance made in the scheme for failures.

Geothermal piles are considered 'closed systems'. 'Open systems' supplying geothermal energy to buildings are based on deep wells that utilise the heat in groundwater pumped to heat exchangers on the surface. The benefits of a well system are that the depth can be greater than that required for a structural–geothermal pile; fewer boreholes are needed and may be retrofitted. The disadvantages are that planning consent and an extraction licence are required and there is a continuing power demand for pumping. In addition, there can be problems with biofouling in the well.

## 9.10 USE OF PILES TO SUPPORT SLOPES

The technique of using 'spaced piles' to steepen the sides of existing slopes and to repair and realign slopes is increasingly being used by the HA for motorway and highway maintenance and widening. The principle is to place discrete piles at spacings of 2–5 times the pile diameter in a row along the slope so that potentially unstable soil arches between the piles; the piles should extend below a potential or existing slip plane. Ito and Matsui[9.50] produced charts showing the effects of $\phi$, $c$ and the pile spacing and diameter on the force acting on the pile from their theoretical stability analysis of a slope containing piles. This method, based on plastic deformation, is further described in the HA research report summarised by Carder[9.51]. Ellis et al.[9.52] also refer to the Ito and Matsui work but provide a more simple approach to pile spacing in order to avoid the risk of soil flow through the gaps in a row of piles. They use the basic Barton equation quoted by Fleming et al.[4.23] for the limiting load per unit length of a pile, $P_{p,ult} = K_p^2 \sigma'_{vo} d$, on a single isolated pile and the modified expression $P_{p,ult} = (K_p - K_a)\sigma'_{vo} s/d$, where $K_p$ and $K_a$ are the Rankine passive and active earth pressure coefficients in front and behind a row of piles spaced at $s$, respectively, $\sigma'_{vo}$ the vertical effective stress and $d$ the pile diameter. Where these two equations intersect, a 'critical spacing' exists to ensure that the soil 'arches' between adjacent piles, such that

$$\left(\frac{s}{d}\right)_{crit} = \frac{K_p^2}{K_p - K_a} \tag{9.6}$$

For $\phi$ values of 20° and 35°, $(s/d)_{crit}$ is 2.7 and 4.0 respectively, which ties in with current practice. The stability of a generic slope containing a row of piles which intercept a potential slip plane at various locations in the slope was examined by Ellis et al. using FLAC 2D and 3D analyses (see Appendix C). A model of the stabilising force combined with a model of limiting pile row interaction demonstrated that the increase in slope stability factors of safety, $\Delta F$, ranged between 0.05 and 0.27 within the $s/d_{crit}$ ratios given above.

The preferred location for the row of piles is at or above midslope, but this may depend on the construction method available. Consideration must be given to the possibility of slip planes not intercepted by the piles and to the lateral pressure on piles below the potential rupture zone, particularly if the pile is short and founded on a stiff stratum. Viggiani[9.53] analysed six different failure modes of piles used to stabilise slopes in fine-grained soil to provide equations for maximum bending moment and shear force. While these solutions are simple to apply, there are indications that they underestimate the restraining parameters[9.54]. The WALLAP program (using the single pile option) will calculate bending moments and shear at ULS in a two dimensional analysis. A 3D program such as PLAXIS will more accurately simulate the arching of the soil between piles. Pore pressure changes due to pile driving through a clay slope also need to be examined[9.27].

## 9.11 REUSE OF EXISTING PILED FOUNDATIONS

As the redevelopment of city sites continues, it is inevitable that many will be underlain with deep and complex foundations from the previous buildings. A foundation system that has already been tested and 'proved' by supporting the existing load could provide considerable economic advantage for a new structure on the same site.

A desk study of the design drawings, calculations and specification for the existing structure and foundations, together with as-built records, is essential before embarking upon intrusive investigation and testing. Unfortunately, such records are likely to be incomplete for buildings finished before 1994 when the *Construction (Design and Management) Regulations*[9.55] (CDM Regulations, updated 2007, with revisions due in 2015) legally required building owners to retain a set of construction records. EC7 also gives explicit requirements for retention of foundation documentation. The introduction of *Building Information Modelling* (BIM)[9.56], designed to provide 'whole life asset management' of structures, will make significant changes to record keeping from the earliest conception through as-built documentation to final demolition. Shared CAD programs, such as Autodesk BIM (see Appendix C), are just one part of the process to create and manage sustainable buildings and infrastructure projects faster, more economically and efficiently with minimal environmental impact.

Prior to demolition, the existing building should be surveyed to determine if any structural damage was due to inadequate original foundation capacity. The procedures for a forensic investigation, during and after demolition, to examine the existing foundations and the condition of the concrete and reinforcement are given in CIRIA Report C653[9.57]. The Building Research Establishment Handbook on the *RuFUS* research project[9.58] also gives guidance on technical risk, investigations and design of new foundations alongside old.

Depending on the information revealed by the detailed investigations, reused piles may be loaded up to a limit capacity, say 80% of the previous maximum imposed load. Pile capacity in London Clay has been shown by Wardle et al.[9.59] and Whitaker[4.11] to increase significantly for several years after installation, largely due to increase in shaft friction; similar results have been shown for piles in sand. However, it is not advisable to rely on any increase in the original imposed load on old piles, especially when combined with new piles, unless differential settlements can be accurately assessed and tolerated by the new superstructure. The use of rapid load tests[7.7,11.39] does not give sufficiently reliable data on the pile stiffness and displacement, particularly in clays, to ensure that the stiffness of the old and new foundations will be compatible. New under-reamed piles alongside old straight-shafted piles are unlikely to be compatible; existing under-reamed piles have been successfully reused, supplemented by straight-sided settlement-reducing piles. A potential difficulty for designers when dealing with structures in the public sector, apart from a lack of information, is the reconciliation of the structural codes used for the existing foundations with the requirement to apply Eurocode principles to the new foundations. The insurers for the completed building will have to be consulted to review the acceptability of combined old and new foundations.

While it is possible that the construction programme can be shortened by reuse of foundations, there is a risk that these studies will show reuse is not viable or that piles considered for reuse may have to be downgraded to a fraction of the original capacity, resulting in redesign causing delays. The insertion of fibre-optic instrumentation into a hole drilled in an existing pile can demonstrate pile performance before, during and after structural demolition to assist in determining reuse of piles.

Bauer has developed a technique for the extraction of unwanted piles which uses an 'annulus cutter' to debond the pile from the soil, before jacking out the pile.

## 9.12 UNEXPLODED ORDNANCE

Under the CDM Regulations 2007[9.55], the client for construction works has a duty to provide designers and contractors with specific information needed to identify hazards which may impact the design and construction. This includes the possibility of hazardous unexploded ordnance (UXO) being encountered on site—from aerial bombardment throughout the United Kingdom during two world wars and on abandoned military training grounds being redeveloped. Stone et al.[9.60], in CIRIA Report C681, provide detailed guidance on risk assessment and the implementation of a risk mitigation plan to ensure that the site can be worked on safely and that groundwork delays are minimised. It is essential that where UXO risk is to be assessed the client employs a contractor with expert experience in explosive ordnance disposal (EOD) able to detect unambiguously old ordnance and then render it harmless. Smith et al.[9.61] provide a practical illustration of how appropriate risk assessment can avoid extensive and unnecessary mitigation works while ensuring that where mitigation measures are carried out, they enable work to proceed safely.

For coverage of large areas of potential hazards, helicopters deploying multiple magnetic, electromagnetic and ground-probing radar sensors are used, flying low where feasible. Walking the potential UXO site with instruments should be avoided, but access towers around a small site can house instruments that sweep the site.

## REFERENCES

9.1 Anyaegbunam, A.J. Minimum foundation mass for vibration control, *Journal of Geotechnical and Geoenvironmental Engineering*. American Society of Civil Engineers, 137 (2), 2011, 190–195.

9.2 Hsieh, T.K. Foundation vibrations, *Proceedings of the Institution of Civil Engineers*, 22, June 1962, 211–226.

9.3 Prakash, S. and Sharma, H.D. *Pile Foundations in Engineering Practice*. John Wiley & Sons, Inc., New York, 1990.

9.4 Foundations for Dynamic Equipment. American Concrete Institute, Report ACI351.3R-04, 2004.

9.5 Bhatia, K.G. *Foundations for Industrial Machines: Handbook for Practising Engineers*. D-CAD Publishers, New Delhi, India, 2008.

9.6 Lizzi, F. Pali radice and reticulated pali radice, Micropiling, in *Underpinning*, S. Thorburn and J.F. Hutchison (ed.), Surrey University Press, Guildford, UK, 1985, 84–161.

9.7 Donnelly, I.J., Culshaw, M.G. and Bell, F.G. Longwall mining-induced fault reactivation and delayed subsidence ground movement in British coalfields, *Quarterly Journal of Engineering Geology, and Hydrogeology*, 41, August 2008, 301–314.

9.8 Price, D.G., Malkin, A.B. and Knill, J.L. Foundations of multi-storey blocks on the coal measures with special reference to old mine workings, *Quarterly Journal of Engineering Geology*, 1 (4), June 1969, 271–322.

9.9 Abandoned Mineworkings Manual. (Replacing SP32 'Construction over abandoned mine workings', 1983). Construction Industry Research and Information Association, London, UK, in preparation 2014.

9.10 Andersland, O.B. and Ladanyi, B. *Frozen Ground Engineering*, 2nd ed. ACSE Press and John Wiley & Sons, New York, 2004.

9.11 Tsytovich, N.A. Permafrost in the USSR as foundations for structures, *Proceedings of the 6th International Conference, ISSMFE*, Montreal, Quebec, Canada, Vol. 3, 1965, pp. 155–167.

9.12 Penner, E. and Irwin, W.W. Adfreezing of Leda Clay to anchored footing columns, *Canadian Geotechnical Journal*, 6 (3), August 1969, 327–337.

9.13 Zarling, J.P. Refrigerated foundations, in *Permafrost Foundations. State of the Practice*, E.S. Clarke (ed.), ASCE Technical Council on Cold Regions Engineering, Reston, VA, 2007, pp. 65–82.

9.14 Hambly, E.C. *Bridge Foundations and Substructures*. Building Research Establishment, HM stationary office, London, UK, 1979.

9.15 Highways Agency. *Use of Eurocodes for the Design of Highway Structures*. Interim Advice Note IAN 124/11. Department for Transport, London, UK, 2011.

9.16 Highways Agency. *The Design of Integral Bridges, Design Manual for Roads and Bridges*, Vol. 1 Section 3. BA 42/96. Department for Transport, London, UK, 2003.

9.17 Springman, S.M. and Bolton, M.D. The effect of surcharge loading adjacent to piles. Contractor Report. Transport and Road Research Laboratory, Wokingham, UK, 1990.

9.18 Springman, S.M., Ng, C.W.W. and Ellis, E.A. Centrifuge and analytical studies of full height bridge abutment on piled foundation subject to lateral loading, Project Report TRL98. Transport Research Laboratory, Wokingham, UK, 1995.

9.19 Tomlinson, M.J. and Boorman, R. *Foundation Design and Construction*. Pearson Education Ltd., Harlow, UK, 2001.

9.20 Highways Agency. *Design for Durability, Design Manual for Roads and Bridges*, Vol. 1 Section 3 BD 57/01. Department for Transport, London, UK, 2001.

9.21  Springman, S.M., Norrish, B.M. and Ng, C.W.W. Cyclic loading of sand behind integral bridge abutments. Project Report TRL146. Transport Research Laboratory, Wokingham, UK, 1996.

9.22  England, G.L., Tsang, N.C.M. and Bush, D.I. *Integral Bridges. A Fundamental Approach to the Time-Temperature Loading Problem.* Thomas Telford, London, UK, 2000.

9.23  Lehane, B.M. Lateral stiffness adjacent to deep integral bridge abutments, *Geotechnique*, 61 (7), 2011, 593–604.

9.24  Denton, S., Riches, O., Christie, T. and Kidd, A. Developments in integral bridge design, in *Bridge Design to Eurocodes – UK Implementation*, S. Denton (ed.), ICE Publishing, London, UK, 2011, 463–480.

9.25  Randolph, M.F. and Houlsby, G.T. The limiting pressure on a circular pile loaded laterally in cohesive soil, *Geotechnique*, 34 (4), 1984, 613–623.

9.26  De Beer, E. and Wallays, M. Forces induced by unsymmetrical surcharges on the soil around the pile, *Proceedings of the fifth European Conference on Soil Mechanics and Foundation Engineering*, Madrid, Spain, Vol. 1, 1972, 325–332.

9.27  Massarsch, K.R. and Broms, B.B. Pile driving in clay slopes, *Proceedings of the 10th International Conference, ISSMFE*, Stockholm, Sweden, Vol. 3, 1981, pp. 469–474.

9.28  Seaman, J.W. A guide to accommodating or avoiding soil-induced loading of piles foundations for highway bridges. Project Report TRL71. Transport Research Laboratory, Wokingham, UK, 1994.

9.29  Reid, W.M. and Buchanan, N.W. Bridge approach support piling, *Proceedings of the Conference on Piling and Ground Treatment*, Institution of Civil Engineers, London, UK, 1983, 267–274.

9.30  Beetstra, G.W., van Den Hoonard, J. and Vogelaar, L.J.J. Bridges and viaducts, in *The Netherlands Commemorative Volume*, E.H. de Leew (ed.), *Proceedings of the 11th International Conference*, ISSMFE, San Francisco, CA, 1985.

9.31  Bittner, R.B., Zhang, X. and Jensen, O.J. Design and construction of the Sutong bridge foundations, *Deep Foundations Institute, Marine Foundations Speciality Seminar*, New York, 2005, 19–27.

9.32  May, R.W.P., Ackers, J.C. and Kirby, A.M. Manual on scour at bridges and other hydraulic structures. CIRIA Report C551. Construction Industry Research and Information Association, London, UK, 2002.

9.33  Chin Fung, K. and McCabe, R. Penang Bridge Project: Foundations and reclamation work, *Proceedings of the Institution of Civil Engineers*, 88 (1), 1990, 531–549.

9.34  Stanley, R.G. Design and construction of the Sungai Perak Bridge, Malaysia, *Proceedings of the Institution of Civil Engineers*, 88 (1), 1990, 571–599.

9.35  Levesque, M. Les fondations du pont de Färo en Danemark, *Travaux*, November 1983, pp. 33–36.

9.36  Elazouni, A.M. and El-Razek, M.E.A. Adapting lift-slab technology to construct submerged pile caps, *Journal of Construction Engineering and Management*, American Society of Civil Engineers, 126, 2000, 149–157.

9.37  Waltham, A.C. and Fookes, P.G. Engineering classification of karst ground conditions, *Journal of Engineering Geology and Hydrology*, Geological Society of London, 36 (2), 2003, 101–118.

9.38  Uranowski, D.D., Dodds, S. and Stonecheck, P.E. Micropiles in karstic dolomite; similarities and differences of two case histories, in *Geo-Support 2004, Drilled Shafts, Deep Mixing, Remedial Methods and Speciality Foundation Systems*, Orlando, FL, J.P. Turner and P.W. Mayne (eds.), American Society of Civil Engineers Geotechnical Special Publication No 124, 2004, 674–681.

9.39  Fischer, J.A. A nuclear power plant on karst terrain? in *Sinkholes and the Engineering and Environmental Impacts of Karst*, B.F. Beck (ed.), American Society of Civil Engineers Geotechnical Special Publication No 122, 2003, 485–491.

9.40  Robertson, P.K. Evaluation of flow liquefaction and liquefied strength using the cone penetrometer test, *Journal of Geotechnical and Geoenvironmental Engineering*, American Society of Civil Engineers, 136 (6), 2010, 842–853.

9.41  Puri, V.K. and Prakash, S. Pile design in liquefying soil, *Proceedings of 14th World Conference on Earthquake Engineering*, Paper 301, Beijing, China, 2008.

9.42 Tabash, A. and Poulos, H.G. Design charts for seismic analysis of single piles in clay, *Proceedings of the Institution of Civil Engineers*, 160 (GE2), 2007, 85–96.

9.43 Bhattacharya, S. and Bolton, M. Errors in design leading to pile failures during seismic liquefaction, *Proceedings of fifth International Conference on Case Histories in Geotechnical Engineering*, Paper 12A-12, S. Prakash and V.K. Puri (eds.), New York, 2004.

9.44 Teyssandier, J.-P., Combault, J. and Morand, P. The Rion-Antirion bridge design and construction, in *Proceedings of the 12th Conference on Earthquake Engineering,* Paper 1115, Auckland, New Zealand, 2000.

9.45 Brandl, H. Energy foundations and other thermo-active ground structures, *Geotechnique*, 56 (2), 2006, 81–122.

9.46 Clarke, B.G., Agab, A. and Nicholson, D. Model specification to determine thermal conductivity of soils, *Proceedings of the Institution of Civil Engineers, Geotechnical Engineering*, 161 (3), 2008, 161–168.

9.47 Loveridge, F., Powrie, W. and Smith, P. A review of the design and construction aspects for bored thermal piles, *Ground Engineering*, 46 (3), 2013, 28–31.

9.48 Laloui, L. and Di Donna, A. Understanding the behaviour of energy geo-structures, *Proceedings of the Institution of Civil Engineers*, 164 (4), 2011, 184–191.

9.49 Bourne-Webb, P.J., Amatya, B., S Ogat, K., Amis, T., Davidson, C. and Payne, P. Energy pile test at Lambeth College, London, *Geotechnique*, 59 (3), 2009, 237–248.

9.50 Ito, T. and Matsui, T. The effects of piles in a row on slope stability, *Proceedings of the Ninth International Conference of the Soil Mechanics and Foundation Engineering*, Tokyo, Japan, 1977.

9.51 Carder, D.R. Improving the stability of slopes using a spaced piling technique. Transport Research Laboratory Report INS001, Wokingham, UK, 2009.

9.52 Ellis, E.A., Durrani, I.K. and Reddish, D.J. Numerical modelling of discrete pile rows for slope stability and generic guidance for design, *Geotechnique*, 60 (3), 2010, 185–195.

9.53 Viggiani, C. Ultimate lateral load on piles used to stabilise landslides, *Proceedings of 10th International Conference on Soil Mechanics and Foundation Engineering*, Stockholm, Sweden, Vol. 3, 1981, pp. 555–560.

9.54 Chmoulian, A.Y. Analysis of piled stabilisation of landslides, *Proceedings of the Institution of Civil Engineers, Geotechnical Engineering*, 157 (2), 2004, 55–56.

9.55 *Construction Design and Management Regulations*. Health and Safety Executive, London, UK, 2007.

9.56 Building Information Model (BIM) Protocol. Construction Industry Council, London, UK, 2013.

9.57 Chapman, T., Anderson, S. and Windle, J. Reuse of foundations. Construction Industry Research Information Association (CIRIA), Report C653, London, UK, 2007.

9.58 Butcher, A.P., Powell, J.J.M. and Skinner, H. (ed). *The 'RuFUS Project Handbook and Conference Proceedings on the Re-use of Foundations*, Building Research Establishment, CRC, London, UK, 2006.

9.59 Wardle, I., Price, G. and Treeman, T.J. Effect of time and maintained load on the ultimate capacity of piles in stiff clay, in *Piling: European Practice and Worldwide Trends*, M.J. Sands (ed.) Thomas Telford Ltd., London, 1993, 92–99.

9.60 Stone, K., Murray, A., Cooke, S., Foran, J. and Gooderham, L. Unexploded ordnance (UXO). A guide for the construction industry. Construction Industry Research Information Association, Report C681, London, UK, 2009.

9.61 Smith, P., Lawrence, U., Terry, S. and Cooke, S. Unexploded ordnance risk assessment on Crossrail project in London – Pre-empting best practice, *Proceedings of Institution of Civil Engineers, Geotechnical Engineering*, 166 (4), 2013, 333–342.

# Chapter 10

# Durability of piled foundations

## 10.1 GENERAL

In all situations, consideration must be given to the possibility of the deterioration of piled foundations due to aggressive substances in soils, in rocks, in groundwaters, in the sea and in river waters. Piles in river or marine structures are also exposed to potentially aggressive conditions in the atmosphere and they may be subjected to abrasion from shifting sand or shingle, or damage from floating ice or driftwood.

In considering schemes for protecting piles against deterioration due to these influences, the main requirement is for detailed information at the site investigation stage on the environmental conditions. In particular, adequate information is required on the range of fluctuation of river or sea levels and of the groundwater table. In the latter case, the highest levels are required when considering the likely severity of sulphate attack on concrete piles or the corrosion of steel piles, and the lowest possible levels are of considerable importance in relation to the decay of timber piles. The possibility of major changes in groundwater levels due to, say, drainage schemes, irrigation or the impoundment of water must be considered.

In normal soil conditions, it is usually sufficient to limit chemical analyses of soil or groundwater samples to the determination of pH values, water-soluble sulphate content and chloride content. Where the sulphate content exceeds 0.24% in soils, it is advisable to determine the water-soluble sulphate content, expressing this in mg of $SO_4$ per litre of water extracted. For brownfield sites, full chemical analyses are required to identify potentially aggressive substances[2.8]. Methods of investigating and assessing brownfield sites are given by Rudland et al.[10.1], drawing attention to the health and safety precautions necessary, the need to employ specialist personnel and care in selecting representative samples. (See also Statutory Guidance[11.1].)

Bacterial action can be an influence in the corrosion of steel piles. Samples of soil and groundwater should be obtained in sterilised containers, which are then sealed for transportation to the bacteriological laboratory for later analyses. Where steel piles are used for foundations in disturbed soils or fill material on land, an electrical resistivity survey is helpful in assessing the risk of corrosion and in the design of schemes for cathodic protection (see Section 10.4.2).

Investigations for marine or river structures should include a survey of possible sources of pollution which might encourage bacteriological corrosion, such as contaminated tidal mud flats, discharges of untreated sewage or industrial effluents, dumping grounds for industrial or household refuse and floating rubbish discharged from ships or harbour structures. The pattern of sea or river currents should be studied and water samples taken at various stages of spring and neap tides or during dry weather  and at flood stages in rivers. Particular attention should be paid to sampling water from currents originating at the areas of contamination previously identified. Chemical and bacteriological analyses should be made on the full range of samples to assess the daily or seasonal variation in potentially aggressive substances. Other items for study include the presence and activity of organisms such as weeds and barnacles, and molluscan or crustacean borers (see Section 10.2.2).

## 10.2 DURABILITY AND PROTECTION OF TIMBER PILES

### 10.2.1 Timber piles in land structures

Timber piles permanently below groundwater level have an indefinite life. There are numerous examples of stumps of timber piles that are more than 2000 years old being found in excavations below the water table. While timber does not decay from fungal attack if the moisture content is kept below 20%, it is impossible to maintain it in this dry condition when buried in the ground above water level. Hence, damp timber which does not have natural durability is subject to decay by fungal attack, resulting in its complete disintegration. Figure 10.1 shows an example of the decay of timber piles above the water table. Figure 10.1a shows the cavities left by the complete decay of the timber. The timber capping beams have also decayed, allowing the stone lintels to sink down onto the ground surface. Figure 10.1b is a view down a cavity which is partly filled by soil debris and fragments of decayed timber. The piles were driven into clay fill in the early nineteenth century. Preservative treatment can, however, give a useful life to timber piles in the zone above groundwater level. If treatment is applied to properly air-seasoned wood at the correct moisture content for the impregnation of the preservative, a life of several decades may be achieved.

The durability of the various grades of timber in terms of their approximate life when in contact with the ground and water has been classified in several standards in similar terms: the Building Research Establishment Digest 429[10.2]; and BS EN 350-2 Durability of wood (Table 10.1) and in BS 8417 Preservation of Wood.

(a)                                             (b)

Figure 10.1  Decay of timber piles above groundwater level. (a) cavities left by complete decay of piles and timber capping sills; (b) view down cavity left in clay after complete decay of timber pile. (Crown copyright reserved. Reproduced with permission of BRE, Watford, UK.)

*Table 10.1* Natural durability classifications of heartwood of untreated timbers in contact with the ground

| BS EN 350–2 description | Class | BRE Digest 429 description | BRE mean life (years) |
|---|---|---|---|
| Very durable | 1 | Very durable | >25 |
| Durable | 2 | Durable | 15–25 |
| Moderately durable | 3 | Moderately durable | 10–15 |
| Slightly durable | 4 | Non-durable | 5–10 |
| Not durable | 5 | Perishable | Up to 5 |

The natural durability and *treatability* (an assessment of the take-up of preservative) depends on the structure of the wood and the method of treatment. The heartwood of some timbers suitable for piles as given in Table 2.1 are summarised as follows:

| Timber species | Durability | Class | Treatability | Class |
|---|---|---|---|---|
| Sitka spruce | Non-durable | 4 | Difficult to treat | 3 |
| Western red cedar | Durable | 2 | Difficult to treat | 3 |
| British pine | Non-durable | 4 | Moderately easy to treat | 2 |
| Douglas fir (United States) | Non-durable | 4 | Difficult to treat | 3 |
| All the tropical hardwoods | Very durable | 1 | Extremely difficult to treat | 4 |

The natural durability refers to resistance to fungal attack only and durability does not imply total resistance. The classification for resistance to marine borers uses a different system.

Precautions against fungal attack must be commenced at the time that the timber is felled. It should be carted away from the forest as quickly as possible and then stacked clear of the ground on firm, well-drained and elevated ground from which all surface soils which might harbour organisms have been stripped. The timber stacks should have spaces between the baulks to encourage the circulation of air and the drying of the timber to the moisture content suitable for the application of the preservative treatment.

'Use classes' (identifying biological hazards) are provided in BS 8417: 'Class 4' for timber in contact with the ground or fresh water likely to suffer fungal attack and 'Class 5' for timber in salt water subject to borers and fungi. The need for preservative treatment is also classified: 'desirable' with Service Factor C or 'essential' Factor D depending on the natural durability as mentioned earlier. It is noted that durability Class 1 cannot be relied upon to give more than 15 years' service in seawater.

Many of the timber preservatives which were available in the past have been prohibited or their use restricted under European Directives and US regulations. The UK *REACH Enforcement Regulations 2008*[10.3], dealing with all forms of pesticides including timber preservatives such as creosote and chromated copper arsenate (CCA), has introduced limits for constituents and their use. However, these regulations do not apply where previously creosote-treated timber is available and timber already in use prior to 2002; wood treated after 2002 can only be marketed for 'industrial' use, which could include piles. All existing approvals for the use of CCA have been withdrawn, although timber treated and in use prior to 2004 is not affected; CCA remains prohibited for use in seawater. As a result, alternative preservatives such as copper azole compounds have been developed to treat Use Class 4 timber. BS 8417 provides data on preservative requirements and penetration depths for Use Classes 4 and 5 and Service Factor D for service life up to 30 years in fresh water and seawater. The Wood Protection Association (WPA) Manual[10.4] gives detailed guidance on the preparation of timber, air seasoning and treatment. In the United States, the specifications of the American Wood Preservers' Institute are followed. Biological deterioration including termite attack is much more severe in tropical countries and the loadings or the selection of resistant species for these conditions should be specified in consultation with a specialist authority in the country under consideration.

As can be seen from the above tables, the use, durability and treatability classes are not necessarily related when considering pile selection. Class 3 and 4 woods can only be treated to a limited depth (3–6 mm) under sustained pressure, whereas Class 2 woods are relatively easy to treat to depths of 18 mm. Round timbers are preferred for piles where the sapwood can be thoroughly impregnated (e.g. Scots pine to depth of 75 mm) to resist fungal attack for many years. This receptive sapwood is removed when squaring timbers.

When bolt holes are drilled or other incisions made for lifting hooks after the main impregnation treatment, preservative should be poured into the holes or painted on scars. The exposed end grain should be given two heavy coats of the preservative prior to attaching the pile shoe or the driving ring (Figure 2.2).

Some hardwoods, for example ekki (botanical name *Lophira alata*), greenheart (*Ocotea rodiei*) and jarrah (*Eucalyptus marginata*), can be used without preservative treatment, but in these cases, it is usual to specify that no sapwood is left on the prepared timber. Expert advice should be sought on the removal of sapwood or whether a preservative should be used to treat these sapwoods as a precautionary measure. However, as timber used for piling is normally required to have large cross-sectional dimensions, it is generally not practicable to reduce the pile section by removing the sapwood.

The adoption of preservative treatment does not give indefinite life to the timber above groundwater level, and it may be preferable to adopt a form of composite pile having a concrete upper section and timber below the waterline, as shown in Figure 2.1a.

## 10.2.2 Timber piles in river and marine structures

The moisture and oxygen in the atmospheric zone of timber marine piles above the waterline creates a favourable environment for fungal growth, which usually starts in the centre portion where preservatives have not penetrated. Fungal activity occurs in the splash zone but is limited due to poor oxygen supply. Marine borers do not attack wood in these zones. Brown rot decay is the most common type of fungal decay in coniferous wood species, and in the early stages of attack the wood will have lost weight and, while visually appearing sound, will have suffered considerable loss of elasticity. Fungal attack does not occur below a maintained water table and immersion in salt water protects against fungal decay.

The most destructive agency which can occur in piles fully immersed in brackish or saline waters in estuaries or in the sea is attack by molluscan or crustacean borers. Conditions in the tidal zone are also likely to be favourable for attack by borers where adequate oxygen and salt water are present, but crustacean borers can often attack near an exposed mud line. Below the mud line, adequate oxygen is not available for the survival of marine borers. These organisms burrow into the timber, forming networks of holes that eventually result in the complete destruction of the piles. Timber jetties in tropical waters have been destroyed in this way in a matter of months.

The main types of marine boring organisms are as follows:

| | |
|---|---|
| Molluscan borers | Teredo (shipworm) |
| | Bankia |
| | Martesia (in tropical waters only) |
| | Xylophaga dorsalis |
| Crustacean borers | Limnoria (gribble or sea louse) |
| | Chelura |
| | Sphaeroma |

The young molluscan borers enter the timber through minute holes in the surface or through incisions. They then grow to a considerable size (*Bankia* can grow to a diameter of 25 mm and to nearly 2 m long) and destroy the wood as they grow (Figure 10.2a). The crustaceans work on the surface of the timber, forming a network of branching and interlacing holes (Figure 10.2b). Their activity depends on factors such as the salinity, temperature, pollution level, dissolved oxygen content and current velocity of the water. A salinity of more than 15 parts per 1000 (the normal salinity of seawater is between 30 and 35 parts per 1000) is necessary for the survival of most species of borer, but *Sphaeroma* have been found in nearly fresh tropical waters in South America, South Africa, India, Ceylon, New Zealand and Australia. Attack by *Chelura* is usually dependent on the presence of *Limnoria*. *Limnoria* cannot survive in fresh water.

(a)  (b)

*Figure 10.2* Attack on timber piles by marine borers: (a) Attack by *Teredo* and (b) attack by *Limnoria*. (Crown copyright reproduced with permission of BRE. Walford, UK.)

Chellis[10.5] states that *Teredo* and *Limnoria* do not attack in current velocities higher than 0.7 m/s (1.4 knots) and 0.9 m/s (1.8 knots), respectively. Although activity from some species may be marked in tropical waters, borers have been found above the Arctic Circle. They show cyclic activity rising to a peak in some years, and not infrequently dying away completely. Conversely, previously trouble-free areas can become infested with borers brought in by ships or driftwood.

No species of timber is absolutely free from borer attack, but certain species are highly resistant and in many conditions of exposure they may be considered to have practical immunity[10.6]. The more-resistant species (now classified as *D* in BS 350-2) greenheart, pynkadou, turpentine, totara and jarrah are suitable for conditions of heavy attack by *Limnoria* and *Teredo* in temperate and topical waters. 'Moderately durable' woods (*M*) will resist moderate attack by *Limnoria*. 100% coal-tar creosote has given reliable service as a timber preservative for over 150 years, and in British waters, any timber which is efficiently impregnated with creosote should be practically immune to borer attack. Hence, BS 8417 includes Table 6 for treatment using creosote for Use Classes 4 and 5 but notes that the UK Regulations 2008[10.3] restrict its use. *Limnoria tripunctata* are tolerant to creosote but the species can be effectively controlled, where authorised, by the addition of copper pentachlorophenate to the creosote.

The WPA Manual[10.4] lists the range of timbers, in addition to the common species noted above, which have heartwood resistant to borer attack and best suited for marine work; their properties, durability, preservation and uses are described in the BRE Digest 429. However, the sapwood of these timbers is liable to be attacked by borers, and if it is impossible to ensure the removal of all sapwood, the timber should be treated as a precautionary measure. Greenheart fenders in Milford Haven were attacked in the sapwood by *Teredo*, causing about 10 mm of damage in 5 years.

The methods of preparing, air seasoning and preserving timber against borer attack are the same as those described earlier for fungal decay in Section 10.2.1. However, great care is necessary to avoid making incisions through which borers can enter the untreated wood in the interior of the pile. The timber should be handled by slings rather than hooks or dogs after treatment, and purpose-made devices should be used to give pressure impregnation of the bolt holes after drilling.

Other methods of protecting timber piles against attack by borers include sleeving with non-ferrous metal or precast concrete tubes, encasing the pile in concrete and applying sprayed concrete (*gunite*). These measures will also give some protection from abrasion by seabed shingle, but non-ferrous metal is expensive and it may be preferable to use sacrificial timber strapped around the main bearing piles or to accept the cost of periodical renewal.

Reliable methods of repairing decayed marine timber piles to provide substantial recovery of original strength are not available, not least because of the difficulty in gaining access to the critical zones. Experimental techniques which first remove the decayed material, treat the remaining wood with preservative and infill the void with epoxy resin mortar followed by wrapping with glass fibre have shown some small-scale success. Voids left by rotting timber piles below the Royal Scottish Academy in Edinburgh were successfully treated by Keller Ground Engineering using their *Soilfrac* process and cement injection through horizontal tubes à manchette 2 m below the pile cap stonework.

In tropical and subtropical countries, timber piles can be destroyed by termites above the waterline unless a resistance species is used or preservative applied. Also, the end grain at the heads of piles is particularly susceptible to attack by fungi or beetles when in a damp condition. The pile heads can be protected by heavy coats of hot-applied creosote followed by capping with metal sheeting, bituminous felt or glass fibre set in coal tar pitch.

Some species of wood corrode iron fastenings by the secretion of organic acids. Either non-ferrous fastenings should be used or steel components should be heavily coated with tar or sheathed in plastics. Stainless steel fastenings can be used if the type of steel is resistant to corrosion by seawater.

## 10.3  DURABILITY AND PROTECTION OF CONCRETE PILES

### 10.3.1  Concrete piles in land structures

Properly designed and mixed concrete[2.30] compacted to a dense impermeable mass is one of the most permanent of all constructional materials and gives little cause for concern about its long-term durability in a non-aggressive environment. However, concrete can be attacked by sulphates and sulphuric acid occurring naturally in soils, by corrosive chemicals which may be present in industrial waste in fill materials, and by organic acids and carbon dioxide present in groundwater as a result of decaying vegetable matter[10.7]. Attack by sulphates is a disruptive process, whereas the action of organic acids or dissolved carbon dioxide is one of leaching. Attack by sulphuric acid combines features of both processes.

The naturally occurring sulphates in soils are those of calcium, magnesium, sodium and potassium. The basic mechanism of attack by sulphates in the ground is a reaction with hydrated calcium aluminate in the cement paste to form calcium sulphoaluminate. The reaction is accompanied by an increase in molecular volume of the minerals, resulting in the expansion and finally the disintegration of the hardened concrete. Other reactions can also occur, and in the case of magnesium sulphate, which is one of the most aggressive of the naturally occurring sulphates, the magnesium ions attack the silicate minerals in the cement in addition to the sulphate reaction. Ammonium sulphate, which attacks Portland cement very severely, does not occur naturally. However, it is used as a fertiliser and may enter the ground in quite significant concentrations, particularly in storage areas on farms or in the factories producing the fertiliser. Ammonium sulphate is also a by-product of coal gas production and it can be found on sites of abandoned gasworks. Because calcium sulphate is relatively insoluble in water, it cannot be present in sufficiently high concentrations to cause severe attack. However, other soluble sulphates can exist in concentrations that are much higher than that possible with calcium sulphate. This is particularly the case where there is a fluctuating water table or flow of groundwater across a sloping site. The flow of groundwater brings fresh sulphates to continue and accelerate the chemical reaction. High concentrations of sulphates can occur in some peats and within the root mass of well-grown trees and hedgerows due to the movement and subsequent evaporation of sulphate-bearing groundwater drawn from the surrounding ground by root action. The severity of attack by soluble sulphates must be assessed by determining the soluble sulphate content and the proportions of the various cations present in an aqueous extract of the soil. These determinations must be made in all cases where the concentration of sulphate in a soil sample exceeds 0.5%.

The thaumasite form of sulphate attack (TSA) which consumes the binding calcium silicate hydrates in Portland cement, thereby weakening the concrete, has been investigated extensively[10.8,10.9]. The reaction requires the presence of sulphates, calcium silicate, carbonate and flowing groundwater; it is more vigorous at temperatures below 15°C. Carbonation of concrete due to atmospheric carbon dioxide acting on the calcium hydroxide in the concrete matrix causes a reduction in the pH rendering the concrete susceptible to sulphate reactions forming thaumasite. In well-compacted concrete, the carbonation is a slow process, and a thin layer will provide resistance to sulphate attack, but not to acids. Recent research by Brueckner[10.10] indicated that a reduction in pile skin friction due to thaumasite

*Figure 10.3* Disintegration of concrete in bored and cast-in-place pile due to attack by sulphuric acid leaking into fill from industrial processes.

attack on concrete in clay soils is not expected, despite the fact that the surface roughness is changed. Generally, TSA is not a problem in the United Kingdom, provided that the current specifications[2.30] for concrete design and the recommendations given in BRE Special Digest 1: 2005[2.8] (SD1) are adhered to.

Free sulphuric acid may be formed in natural soil or groundwater as a result of the oxidation of pyrites in some peats or in ironstone or alum shales. Sulphuric acid can also be present in industrial waste materials which have been contaminated by leakages from copper and zinc smelting works and from dyeing processes. The acid has an effect on the cement in hardened concrete that is similar to that of sulphate attack, but the degradation may not result in significant expansion. Figure 10.3 shows the disintegration of the concrete in the shaft of a bored and cast-in-place pile caused by the seepage of sulphuric acid into porous fill material.

The distribution of sulphates in various ground conditions in Britain is described in detail by Forster et al.[10.11]. Sulphates occurring naturally in soils are generally confined to the Mercia Mudstone, which is rich in gypsum, and to the Lias, London, Oxford, Kimmeridge and Weald Clays. They are also found in glacial drift associated with these formations. They may be present in the form of gypsum plaster in brick rubble fill.

The sulphate content of the groundwater gives the best indication of the likely severity of sulphate attack, particularly that resulting from soluble sulphates. Where the water samples are taken from boreholes, care should be taken to ensure that the sample is not diluted by the water added to assist the drilling. If possible, the groundwater should be sampled after a long period of dry weather. Groundwater flow across a sloping site through sulphate-bearing ground results in the highest concentration on the downhill side of the site, and the flow may continue into permeable soil deposits which are not naturally sulphate bearing. Methods of analysis to determine the sulphate content and pH value of soils and groundwaters are set out in BS 1377-3:1990 (under review by ISO) and by Bowley[10.12] in BRE Report 279.

A dense, well-compacted concrete provides the best protection against the attack by sulphates on concrete piles, pile caps and ground beams. The low permeability of dense concrete prevents or greatly restricts the entry of the sulphates into the pore spaces of the concrete. For this reason, high-strength precast concrete piles are the most favourable type to use. However, for the reasons explained in Chapter 2, precast concrete piles are not suitable for all site conditions and the mixes used for the alternatives of bored and cast-in-place or driven and cast-in-place piles must be designed to achieve the required degree of impermeability and resistance to aggressive action.

In British practice, recommendations for the types of cement and the mix proportions to resist chemical attack are given in SD1[2.8] and are compatible with the exposure subdivisions in BS 8500-1. Five classes of severity of attack ('design sulphate' [DS] classes 1 to 5) for natural ground and brownfield sites are defined, from which are derived the 'Aggressive Chemical Environment for Concrete' (ACEC) classes (AC1 to 5), which are subject to certain conditions (e.g. pH should be greater than 2.5). The AC classes provide for adjustment from one DS class to another depending on the conditions of exposure, the pH and mobility of groundwater, and other environmental conditions. For a given AC class a 'design chemical' (DC) class is derived for the intended working life, either 50 or 100 years, together with recommended 'additional protective measures' specific to highly aggressive ground types. Concrete mixes are then tabulated to suit the DC class giving a wide selection of free-water/cement ratios and aggregate sizes down to 10 mm and the appropriate cement and cement combinations in accordance with BS EN 197-1 and BS 8500-2.

Concrete-incorporating ground granulated blast-furnace slag (ggbs) or pulverised fuel ash (pfa – now referred to as *fly ash* in codes) is recommended in place of sulphate-resisting cement where thaumasite attack may occur in United Kingdom. Table D3 in SD1 and A6 in BS 8500-1 provide for the use of Portland cements containing ggbs and for a variety of Portland cement–pozzolan combinations to give enhanced sulphate-resisting properties.

The workability of the SD1 in situ concrete mixes may, in some cases, be too low for placing in bored and driven small-diameter cast-in-place piles. Slightly modified mixes are given for certain precast products, including the manufacture of surface-carbonated precast concrete suitable for precast piles (see Section 10.3.2).

Mixes suitable for concrete in pile caps, ground beams and blinding concrete depend on the size, shape and amount of reinforcement of the members which govern the workability requirements. Footnotes to SD1 Table D1 provide for modifications to the DC class depending on the size of a structural member.

Generally, no additional protection measures (APMs as given in SD1) are necessary where the groundwater is considered 'static', but other conditions may override this (e.g. thickness of concrete section). When in doubt, the 'mobile' groundwater condition should be used. For example, it would be unwise to assume a static groundwater table at a shallow depth for cast-in-place concrete piles where the concrete may be weaker than in the body of the pile due to accumulation of laitance. Weak concrete used as a blinding layer beneath pile caps is also vulnerable to sulphate attack when the resulting expansion of the blinding concrete could lift the cap; hence, the quality of blinding concrete should match the structural quality.

Pile caps and ground beams can be protected on the underside by a layer of heavy-gauge polyethylene sheeting (designated APM3) laid on a sand carpet or on blinding concrete. The vertical sides can be protected after removing the formwork by applying hot bitumen spray coats, bituminous paint, trowelled-on mastic asphalt or adhesive plastics sheeting. The recommendation for placing a membrane between floors and fill, or hardcore containing sulphates, should be considered for the undersides of slender pile capping beams or shallow pile caps.

Coatings of tar or bitumen on the surface of precast concrete piles do not give adequate protection against sulphate attack since they are readily stripped off by abrasion as the piles are driven down in all but the softer soils. The addition of a sacrificial layer of concrete (APM4) to friction piles to improve resistance to expansion due to sulphate attack is not needed provided the concrete is designed as recommended.

The use of special cements, calcium aluminate cement (BS EN 14647), also referred to as high-alumina cement (HAC), or supersulphated cement (BS EN 15743), for high sulphate concentrations is referred to in SD1. The latter cement is attacked by ammonium sulphate to which HAC alone is resistant. However, neither HAC nor supersulphated cement is favoured for UK piling work. In any case, the use of HAC in structural concrete is restricted in the United Kingdom and some other countries due to the risk of internal crystalline rearrangement ('chemical conversion'[10.7]), except where close mix control is provided and the structural properties can be reliably predicted. Another drawback to using HAC is the practical difficulty of placing concrete in pile shafts due to its rapid setting. Manufacturers are investigating HAC blends which could address some of these problems. Sulphate-resisting Portland cement which may be useful in pile caps is covered in BS EN 197-1.

The leaching of concrete exposed to flowing river or groundwater containing organic acids or dissolved carbon dioxide was mentioned at the beginning of this Section. Organic acids are present in run-off water from moorlands and in groundwater in peaty and lignitic soils. The recommendations for concrete exposed to acid attack as determined by the pH value of the soil or groundwater are covered by the ACEC Tables in SD1. Good-quality concrete, made with any of the tabulated cements and nondegradable aggregates, is essential.

The deterioration of concrete due to alkali–silica reaction (ASR) is the result of a reaction between the hydroxyl ions in the cement and reactive forms of silica in the aggregate (e.g. chert) producing an expanding gel in the concrete over long periods. As a result of the comprehensive guidance given in BRE Digest 330[10.13] since the early 1980s, no verifiable deleterious effect of ASR has been reported in the United Kingdom. BS 8500-2 now includes, at Annex D, the requirements for ensuring that ASR risk has been minimised.

## 10.3.2 Concrete piles in marine structures

Precautions against the aggressive action by seawater on concrete need only be considered in respect of precast concrete piles. In situ concrete is used only as a hearting to steel tubes or cylindrical precast concrete shell piles, where the tube or shell acts as the protective element. A rich concrete, well compacted to form a dense impermeable mass, is highly resistant to aggressive action and ASR, and, provided that a cover of at least 50 mm is given to all reinforcing steel, precast concrete piles should have satisfactory durability over the normal service life of the structures they support.

When the disintegration of reinforced concrete in seawater does occur, it is usually most severe in the *splash zone* and is the result of porous or cracked concrete caused by faulty design or poor construction. Evaporation of the seawater in the porous or cracked zone is followed by the crystallisation of the salts, and the resulting expansive action causes spalling of the concrete and the consequent exposure of the reinforcing steel to corrosion by air and water. The expansive reaction that occurs when corrosion products are formed on the steel accelerates the disintegration of the concrete. Freezing of seawater in porous or cracked concrete can cause similar spalling. However, where concrete piles are wholly immersed in seawater, there is no degradation of properly made and well-compacted concrete. Attention must be paid to the source of aggregate: concrete made with certain limestones in the Arabian Gulf has been attacked by rock boring molluscs.

In an extensive review of literature and the inspection of structures which had been in the sea for 70 years, Browne and Domone[10.14] found no disintegration in permanently immersed reinforced concrete structures even though severe damage had occurred in the splash zone. They concluded that corrosion of the steel cannot occur with permanent immersion because the chloride present is restricted to a uniform low level and the availability of oxygen is low.

Although seawater typically has a sulphate content of about 230 parts per 100,000, the presence of sodium chloride has an inhibiting or retarding effect on the expansion caused by its reaction with ordinary Portland cement. The latter material is, therefore, quite satisfactory for the manufacture of precast concrete piles for marine conditions, but to avoid disintegration in the splash zone, the concrete should have a minimum cement content of 360 kg/m³ and a maximum water/cement ratio of 0.45 by weight. SD1 does not provide recommendations for concrete exposed to seawater, but reference should be made to BS 6349-1 on marine structures and BS 8500-1 for exposure classifications as indicated previously in Tables 2.3 and 2.4. Air entrainment of concrete in accordance with BS 8500 as a safeguard against frost attack on piles above the waterline is unnecessary if the water/cement ratio is less than 0.45.

The concrete in precast piles should be moist cured for 7 days after the removal of the formwork (with a further 10-day exposure to air in order to be classified as 'surface carbonated'). Great care should be taken in handling the piles to avoid the formation of transverse cracks which would expose the steel to corrosion in the splash zone. Coatings on precast concrete piles to protect them against deterioration in the splash zone are of little value since they are soon removed by the erosive action of waves and by abrasion from floating debris or ice.

The repair and protection of degraded concrete in the splash zone can be effected by wrapping with polyurethane-impregnated fabric and covered with a sheath of high-density polyethylene (HDPE), as provided by commercial repair companies.

SD1 provides a wide-ranging and flexible approach to the protective measures applicable to concrete in the ground, and the various comments and qualifications to the recommendations cannot be fully covered in this text. It is important to read the Digest as a whole and to follow the step-by-step approach to determine the appropriate concrete quality for a particular assessment of ground conditions. Case studies highlighting critical issues with buried concrete foundations are provided by Henderson et al.[10.15] in the CIRIA Report C569.

## 10.4 DURABILITY AND PROTECTION OF STEEL PILES

### 10.4.1 Steel piles for land structures

Corrosion of iron or steel in the electrolyte provided by water or moist soil is an electrochemical phenomenon in which some areas of the metal surface act as anodes and other areas act as cathodes. Pitting occurs in anodic areas, with rust as the corrosion product in cathodic areas. Air and water are normally essential to sustain corrosion but bacterial corrosion can take place in the absence of oxygen, that is in anaerobic conditions. Anaerobic corrosion is caused by the action of sulphate-reducing bacteria which thrive below the sea- or riverbed in polluted waters, particularly in relatively impermeable silts and clays.

The comprehensive investigations of the corrosion rates of steel sheet piles and bearing piles in soils carried out by Romanoff[10.16,10.17] and Morley[10.18] in the 1970s laid the foundations for the codes and standards which are now in use for the protection of steel piles in the ground. It was established that the rate of corrosion is influenced by the nature of the soil – from permeable sands to relatively impermeable clays to uncompacted fills – and the lower the resistivity of the ground, the higher the corrosion rate. Similarly, the lower the pH of the soil, the greater the potential for corrosion. Piles driven into 'undisturbed' natural soils showed little sign of corrosion at resistivities ranging from 300 to 50,000 Ω-cm, and the minor pitting and loss of mill scale observed was considered negligible in terms of serviceability. These soils are likely to be poorly aerated and not capable of producing damaging iron oxides, but the occurrence of anaerobic sulphate-reducing bacteria may be a cause of corrosion, although there is little evidence of this problem in UK natural soils.

The Romanoff studies also looked at long-term corrosion in sheet piles driven in fills. The more limited scope of this investigation only produced one significant case of severe corrosion of piles driven in clinker, with pitting up to 6 mm deep. However, his conclusions regarding the galvanic actions taking place on the surface of the steel are now well established: corrosion pitting occurs at the anode; differences then occur in the surface potential between the anode and cathode, causing the anode to move to new locations, in turn resulting in general corrosion.

Morley's investigations of piles extracted from the United Kingdom showed losses below the natural soil line varying from nothing to 0.03 mm per year with a mean of 0.01 mm per year. Where piles in land structures are extended above ground, mild steel thickness losses of 0.2 mm per year were measured over a 10-year period in a coastal environment. For steel piles immersed in fresh water, Morley reported a corrosion rate of 0.05 mm per year, except at the waterline in canals where the rate was as high as 0.34 mm per year. This locally higher corrosion zone may be due to abrasion by floating debris or to cell action between parts of the structure in different conditions of oxygen availability. The pH range of fresh water had little effect on corrosion.

In potentially susceptible soils, the ground investigation should determine the resistivity and pH values, chloride and sulphate contamination levels, the depth of water table and information on the potential for damaging anaerobic bacteria to give guidance on the need for protection of the steel. Soil with a resistivity below 300 Ω-cm and pH less than 4 will be susceptible to severe corrosion, but above 10,000 Ω-cm, the risk is low.

The guidance for the 25-year maximum corrosion rates given in Table 4.1 of the UK National Annex of EC3-5 ranges from 0.012 mm per side per year for piles in undisturbed natural soils to 0.08 mm per side per year in non-compacted and aggressive fills (say, resistivity <1000 Ω-cm). These values are based on experience and the research referenced above and do not require special assessment of electrochemical reactions. For longer working life, the values in EC3-5 are extrapolated. Atmospheric corrosion loss may be taken as 0.01 mm per side per year. Localised conditions in coastal areas and varying groundwater may give rise to more aggressive conditions requiring additional allowance for corrosion; if external protection is to be provided, these allowances may not be needed. For example, paint treatment[10.19] would be a suitable precautionary measure for the exposed steel and for aesthetic reasons, provided that it is accessible for maintenance. Where the water table is shallow (and subject to construction constraints), the concrete pile cap can be extended down to a depth of 0.6 m below water level to protect the steel of the piles.

If protection is deemed necessary, the organic coatings recommended in Section 10.4.2 can be used for piers and jetties in fresh water with the nominal coating thickness of 400 μm extending the time to the first maintenance period to beyond 20 years. An alternative for shorter maintenance periods, in both immersed and atmospheric exposures[10.20], is a polyamine-cured epoxy with dry film thickness of 300 μm. The coatings must be applied over blast-cleaned steel. Isocyanate-cured pitch epoxy and coal tar epoxy coatings are no longer available for environmental health and safety reasons. Hot dip galvanising is used as protection for helical plate screw piles.

Paint coatings are not generally satisfactory for protection against bacterial corrosion. Any pinholes in the coating or areas removed by abrasion serve as points of attack by the organisms. Cathodic protection (see Section 10.4.2) is effective but higher current densities are required than those needed to combat normal corrosion in aerobic conditions.

Where steel piles are buried in fill or disturbed natural soil, the thickness of metal in a bearing pile should be such that the steel section will not be overstressed due to wastage of the metal by corrosion over the period of useful life of the structure. Taking a maximum loss of 2.0 mm per face over 25 years as EC3-5, a steel H-pile in aggressive fill with web and flange thicknesses of 15.5 mm will lose 26% of its thickness over this period, although there may be localised areas of deeper pitting. Marsh and Chao[10.21] have refined the contamination guidelines so that more accurate long-term corrosion allowances can be made. In stratified fills, differential aeration can set up damaging macro cells with resistivity below 1000 $\Omega$-cm. Where values of soil resistivities are not available, corrosion rates can be predicted from soluble salt concentrations; for example, concentrations above 1000 mg/kg are likely to produce high corrosion rates.

Protection coating of piles in severely contaminated ground should resist abrasion, impact and acidic attack using, for example, a polyamide-cured epoxy system with increased chemical resistance and a nominal dry film thickness of 400 μm onto blast-cleaned surfaces[10.20]. Protection should extend to around 0.6 m below water table. Filling the shafts of hollow piles with concrete capable of carrying the full applied load will give long-term support.

## 10.4.2 Steel piles for marine structures

Steel piles supporting jetties, offshore platforms and other river or marine structures must be considered for protection against corrosion in six separate zones as follows:

1. *Atmospheric zone*: Exposed to the damp conditions of the atmosphere above the highest water levels or to airborne salt spray
2. *Splash zone*: Above mean high water level and exposed to waves and spray and wash from ships
3. *Tidal zone*: Between mean high and mean low water spring levels (MLWS)
4. *Intertidal low water zone*: Between the lowest astronomical tide (LAT) and MLWS
5. *Continuous immersion zone*: From lower limit of low water to seabed
6. *Embedded zone*: Below the soil line

Design thicknesses to allow for loss of steel due to corrosion and methods of protection should take into account the variation in type and rate of corrosion over these zones, particularly in the low water and splash zones. EC3-5 guidance in Table 4.2 of the National Annex on loss of thickness due to corrosion for a working life of 25 years ranges from 0.022 mm per side per year in fresh water to 0.076 mm per side per year in temperate seawater (excluding an allowance for accelerated low water corrosion (ALWC); see below).

Breakell et al. in the CIRIA Report C634[10.22] provide a summary of marine corrosion rates in steel piles researched and published by a range of worldwide sources, stating both average and upper limits of thickness loss in exposed, unprotected structural steel in temperate climates as follows:

| Zone | Range of average corrosion rate (mm/side/year) | Range of upper-limit corrosion rate (mm/side/year) |
|---|---|---|
| Atmospheric zone | 0.02–0.04 | 0.10–0.41 |
| Splash zone | 0.08–0.42 | 0.17–0.30 |
| Tidal zone | 0.04–0.10 | 0.10–0.18 |
| Intertidal zone | 0.08–0.20 | 0.17–0.34 |
| Continuous immersion zone | 0.04–0.13 | 0.13–0.20 |
| Embedded zone | 0.03–0.08 | 0.02–0.10 |

This report deals specifically with the increased corrosion encountered in recent years in a narrow horizontal band in the intertidal zone, approximately 0.5 m below the MLWS and the LAT, referred to as ALWC or 'concentrated corrosion', where the rate of loss ranges from 0.3 mm to, exceptionally, 4.0 mm per side per year[10.23]. This is considerably higher than traditional reported values[10.24] and the guidance in EC3-5. A type of ALWC has also been reported in the continuous immersion zone and near seabed, but no quantitative data are currently available. ALWC is a localised form of microbiologically influenced corrosion which is initiated by sulphate-reducing bacteria producing hydrogen sulphide causing anaerobic corrosion of the steel. The hydrogen sulphide is then converted by sulphate-oxidising bacteria into sulphuric acid in aerobic tidal conditions with the combined process producing severe pitting and rapid corrosion. Stratified layers of corrosion products of black iron sulphide sludge under a soft outer orange deposit (including rust) and other biofilms cover the shiny pitting caused by the acid attack, all of which may be masked by a covering of marine growth and barnacles. As reported by PIANC[10.23], these chemical processes and the microbiology produced by the differential aeration of the active area are now well recognised, but the external environmental causes of ALWC are not fully understood. Potential causes include bacterial infection from dredged harbour mud, discharges of high-level dissolved organic waste from factories, discharges of waste ballast water and changes in the type of antifouling marine coatings. Some correlation with the total organic carbon content of seawater has been observed. Abrasion from fenders and floating debris also affects the rate and mechanisms of corrosion in the low water zone.

As indicated in Figure 10.4, the most severe conditions of general corrosion are experienced in the splash zone, where Hedborg[10.25] quotes corrosion rates of 0.13–0.25 mm per year in the Panama Canal zone and the Hawaiian Islands and a rate of 0.88 mm per year which has been observed on a platform at Cook Inlet, Alaska. In the narrow ALWC zone, the loss rate can be much higher than shown in the figure.

The presence of marine growth has a considerable influence on protective measures. There is no growth within the atmospheric and splash zones, but in the intertidal and continuously immersed zones, heavy growths of barnacles and weeds can develop, which damage paint treatment and prevent its renewal. However, the growth can shield the steel from exposure to oxygen and in this way reduce the rate of corrosion, counterbalanced by the removal of the growth by abrasion and wash from ships, particularly those with bow-thrust propellers. Macro cells (where the anode and cathode sites are well separated and where the tidal zone is cathodic to the low water zone) may also limit the corrosion rate to a level similar to that of the immersion zone.

Piles forming the main supporting structures in important marine jetties or in offshore platforms exposed to a marine environment frequently require elaborate and relatively expensive treatment to ensure a long life. Recommendations by steel pile manufacturers for protection of new marine structures may be summarised as follows:

*Atmospheric zone and splash zone*: Organic coatings or high-quality concrete encasement, well compacted with appropriate cover, extending 1 m below mean high water level. Coatings should have a minimum 400 μm dry film thickness to give an estimated 20-year life and also extend 1 m below high water.

*Tidal and intertidal zone*: Bare steel with appropriate corrosion allowance, high-yield steel or cathodic protection. Because of uncertainty in driving depths, it may be necessary to extend the coating from the splash zone into the intertidal zone. Special attention must be given to potential ALWC by plating or increased section.

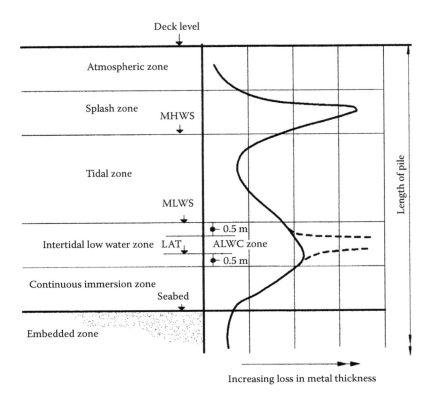

*Figure 10.4* Schematic of corrosion zones and indicative corrosion rates in unprotected marine steel piles. (After Breakell, J.E. et al., Management of accelerated low water corrosion in steel maritime structures, Construction Industry Research and Information Association (CIRIA), Report C634, London, UK, 2005.)

*Continuous immersion zone*: Bare steel with appropriate corrosion allowance or cathodic protection.

*Embedded zone*: No protection necessary.

*Painting/coating*: Long-term evidence shows that the life of protective paint coatings from time of application to first maintenance period is unlikely to exceed 12–15 years, using zinc silicate with layers of epoxy coal-tar paint (now largely prohibited). The need for continual maintenance by periodic cleaning and painting on exposed steelwork was costly and time consuming, but as noted below, new formulations of epoxy and polyester coatings can extend the maintenance period to over 25 years. It is still appropriate to balance the cost of painting against the alternative of increasing steel thickness or the use of high-tensile steel at mild steel stresses. This provides an additional corrosion loss of 30% without the loss of load-bearing capacity at an additional steel cost of about 7%. In some cases, steel thicknesses may be determined by the higher stresses caused during driving, giving a reserve available for the lower stresses under service conditions. Also, the maximum stresses for working conditions in marine structures may be at or near the soil line where corrosion losses are at the minimum rate.

Where the steel in the atmospheric zone is to be protected by paint, the first essential is to obtain thorough cleaning of the metal by sand or grit blasting to produce a white metal or near white metal condition. Well-tested, high-build, organic coatings, shop applied under dry conditions to well-prepared surfaces, are more durable[10.19] than the zinc and the old coal-tar-based products. These new coatings include glass flake epoxy in a single coat of 500 μm (effective against ALWC) and rapid-cure glass flake polyester applied in a single spray or brush coat up to 600 μm. Vinyl copolymer multi-coat systems, which are tolerant of surface conditions, are useful in tidal maintenance work. Specialist equipment is required for the application of hot-applied, solvent-free epoxies and can take weeks to cure, but they perform well in ALWC conditions. High-build, solvent-free epoxies which can be applied and cured under water are also available; two coats are desirable to reduce pinholes. Coatings in the tidal zone also potentially reduce the galvanic area of the macro cell. The problem of abrasion allowing potential bacterial attack remains.

The particular case of protecting the steel members of offshore platforms in the atmospheric zone has been successfully addressed by the application of metal coatings to the large-diameter tubular sections of jacket legs in the fabrication yard, using automatic spray systems which produce low porosity in the coating and avoid the need for overcoating with a sealant.

*Concrete encasement* of piles in the atmospheric, splash and tidal zones will act as a corrosion barrier, but where the splash zone is only partially covered, increased corrosion may occur at the steel–concrete junction due to electrochemical effects. The enlarged area of the encasement will attract greater current and wind forces.

*Plating*: Protection of steel over the length in the splash zone is achieved either by increasing the thickness or quality of the steel as previously mentioned or by providing profiled cover plates of steel to the same specification as the piles, bracings or jacket members. Corrosion-resistant material such as rolled Monel metal (an alloy containing nickel and copper with small amounts of other metals) is particularly effective for plating, but it is costly. Retrofitting of encasement and welded plating will rely on the expertise of divers and may require specially designed chambers in which to work.

*Cathodic protection* of the bare or painted steel below the splash zone is achieved by measures which utilise the characteristic electrochemical potential possessed by all metals (see BS EN 12473). The metals which are higher in the electromotive series act as anodes to the metals lower in the series which form the cathodes. Thus, if a steel structure is connected electrically to a zinc anode, the potential difference between the metals sets up a current so that the whole area of the steel becomes cathodic and does not corrode. The two methods of cathodic protection used in marine structures are the *sacrificial* (galvanic) anode system and the *impressed-current* (or power-supplied) system[10.26]. In the former, large masses of metal such as magnesium, aluminium or zinc, which are higher in the electromotive series than steel and attached to the structure, are used as the corroding anodes. In the impressed-current system, the anodes are inert (non-wasting) and consist of a variety of materials such as mixed metal oxides, lead–silver or other noble metals. They are not attached to the pile but suspended in seawater adjacent to the piles and supplied with direct current from a generator or rectifier with the negative return cable from the cathode structure being protected. Electrical connections should be well-insulated, low-resistance conductors. Suitable compatible coating of the pile will reduce the current demand.

The sacrificial anode system, with a design life of 10 years, is generally preferred for marine structures since it does not require the use of cables which are liable to be damaged by vessels or objects dropped or lowered into the water from the structures. In depths of water of up to 60 m, the wasted anodes can be replaced by divers at a reasonable cost, and for deep-water applications, diver-less sacrificial systems can be designed. The length of the anode determines the area of structure which can be protected, and again, suitable coatings will reduce the area requiring protection. The choice between sacrificial anode systems, with or without a coated structure, and power-supplied systems is a matter of economics, taking into account the capital costs of installation, the current consumption, the costs of maintenance and the intended life of the structure.

There are limitations to cathodic protection: it is ineffective in the splash zone and hydrogen evolution at the steel surface may cause hydrogen embrittlement in high-strength steels. Monitoring of the system is by measuring the potential of the steel against a standard reference electrode such as silver/silver chloride/seawater cell.

Other means of repairing tubular piles include the following:

- Layers of epoxy or polyester glass fibre tape wrappings in the splash zone: using divers below water level. These surface-tolerant proprietary products are effective where thorough cleaning is not possible.
- Clamped sleeves over the damage, with the annulus grouted as described in BS EN ISO 19902, Clause 15.3.6; applicable also to tubular bracing members.

Methods of dealing with ALWC are considered in the CIRIA Report C634[10.22] and the PIANC report[10.23].

## REFERENCES

10.1   Rudland, D.J., Lancefield, R.M. and Mayell, P.N. Contaminated land risk assessment. A guide to good practice. Construction Industry Research and Information Association (CIRIA), Report C552, London, UK, 2001.

10.2   Timbers: The natural durability and resistance to preservative treatment. Building Research Establishment, (BRE) Digest 429, Construction Research Communications Ltd, Watford, UK, 1998.

10.3   The REACH Enforcement Regulations 2008 SI 2008/2852. Health and Safety Executive, London, UK, 2008.

10.4   Industrial Wood Preservation, Specification and Practice, 1st ed. The Wood Protection Association, Castleford, UK, 2007.

10.5   Chellis, R.D. Pile foundations, in Foundation Engineering, G.A. Leonards (ed.), McGraw-Hill, Whitby, ON, Canada, 1962, pp. 723–731.

10.6   Deterioration of structures of timber, metal, and concrete exposed to the action of seawater, 21st Report of the Sea Action Committee of the Institution of Civil Engineers, London, UK, 1965.

10.7   Hewlett, P.C. (ed). Lea's Chemistry of Cement and Concrete, 4th ed. Arnold, London, UK, 1998.

10.8   Thaumasite in cementitious materials, Proceedings of First International Conference, AP147, BRE-CRC Ltd., Watford, UK, 2002.

10.9   Halliwell, M.A. and Crammond, N.J. Avoiding the thaumasite form of sulphate attack, two-year report, BRE-CRC Ltd., Watford, UK, 2000.

10.10  Brueckner, R. Accelerating the thaumasite form of sulfate attack and investigation of its effects on skin friction, PhD thesis, University of Birmingham, Birmingham, UK, 2007.

10.11  Forster, A., Culshaw, M.G. and Bell, F.G. Regional distribution of sulphate in rocks and soils of Britain, in Engineering Geology of Construction, M. Eddleston, S. Walthall, J.C. Cripps and M.G. Culshaw (ed.), The Geological Society of London, London, UK, 1995, pp. 95–104.

10.12  Bowley, M.J. Sulphate and acid attack in the ground: Recommended procedures for analysis. Building Research Establishment, (BRE) Report BR279, Construction Research Communications Ltd, Watford, UK, 1995.

10.13  Alkali-silica reaction in concrete. BRE Digest 330. IHS BRE Press, Bracknell, UK, 2004.

10.14  Browne, R.D. and Domone, P.L.J. The long term performance of concrete in the marine environment, *Proceedings of the Conference on Offshore Structures*, Institution of Civil Engineers, London, UK, 1974, pp. 31–41.

10.15  Henderson, N.A., Baldwin, N.J.R., McKibbins, L.D., Winsor, D.S. and Shanghavi, H.B. Concrete technology for cast in-situ foundations. Construction Industry Research and Information Association (CIRIA), Report C569, London, UK, 2002.

10.16  Romanoff, M. Corrosion of steel pilings in soils, National Bureau of Standards, NBS Monograph 58, US Department of Commerce, Washington, DC, October 1962.

10.17  Romanoff, M. Performance of steel pilings in soils, *Proceedings of the 25th Conference of National Association of Corrosion Engineers,* Houston, TX, 1969, pp. 14–22.

10.18  Morley, J. The corrosion and protection of steel piling *British Steel Corporation,* Teesside Laboratories, 1979.

10.19  A corrosion protection guide – For steelwork exposed to atmospheric environments. Corus Construction and Industrial, Scunthorpe, UK, 2004.

10.20  *Piling Handbook*. Arcelor RPS, Luxembourg and Scunthorpe, UK, 2005.

10.21  Marsh, E. and Chao, W.T. *The Durability of Steel in Fill Soils and Contaminated Land*. Corus Research, Development and Technology, Rotherham, UK, 2004.

10.22  Breakell, J.E., Siegwart, M., Foster, K., Marshall, D., Hodgson, M., Cottis, R. and Lyon, S. Management of accelerated low water corrosion in steel maritime structures. Construction Industry Research and Information Association (CIRIA), Report C634, London, UK, 2005.

10.23  Accelerated Low Water Corrosion. Report of Working Group 44 of the Maritime Navigation Commission. International Navigation Association of PIANC General Secretariat, Brussels, Belgium, 2005.

10.24  Morley, J. and Bruce, D.W. Survey of steel piling performance in marine environments, final report. Commission of the European Communities, Document EUR 8492 EN, 1983.

10.25  Hedborg, C.E. Corrosion in the offshore environment, *Proceedings of the Offshore Technology Conference*, Houston, TX, 1974, Paper No. OTC, 1949, pp. 155–168.

10.26  Baeckmann, W.V., Schwenk, W. and Prinz, W. (eds.). *Handbook of Cathodic Corrosion Protection*, 3rd ed. Gulf Professional Publishing, imprint of Elsevier Ltd., London, 1997.

# Chapter 11

# Ground investigations, piling contracts and pile testing

The importance of a thorough ground investigation as an essential preliminary to piling operations cannot be overemphasised. Accurate and detailed descriptions of soil and rock strata and an adequate programme of field and laboratory tests are necessary for the engineer to design the piling system in the most favourable conditions.

Detailed descriptions of the ground conditions are also essential if the piling contractor is to select the most appropriate equipment for pile installation, while giving prior warning of possible difficulties when driving or drilling through obstructions in the ground.

The employer, through the appointed engineer/project manager, must have assurance that the piles have been correctly designed and installed in a sound manner without defects which might impair their bearing capacity. To this end, piling contracts must define clearly the responsibilities of the various parties, and the installation of piles must be controlled at all stages of the operations. It will have become evident from the earlier chapters of this book that load testing cannot be dispensed with as a means of checking that the correct assumptions have been made in design and that the deflections under the applied load conform, within tolerable limits, to those predicted. Load testing is also one of the most effective means of checking that the piles have been soundly constructed. Pile design using the results of preliminary load tests for a specific site is emphasised in EC7-1, Clause 7. This approach is not routinely adopted in the United Kingdom where pile testing is used to check design based on calculation and to confirm the suitability of the construction method.

## 11.1 GROUND INVESTIGATIONS

### 11.1.1 Planning the investigation

The Codes of Practice which served to define ground investigations, reporting and laboratory testing, BS 5930 and BS 1377, are subject to the ongoing changes resulting from the adoption by British Standards Institute (BSI) of the structural Eurocodes and the new BS EN ISO standards and are being progressively withdrawn.

The objective of an investigation in respect of piling works is to produce geotechnical parameters by 'theory, correlation or empiricism from test results' which can then be applied either directly or as characteristic values to the design process. Theoretical parameters related to soil mechanics are usually obtained from laboratory testing, correlations from well-established field tests and empirical values from practical experience and experimentation as demonstrated in the design Chapters 4 through 6.

The project designer (preferably with a geotechnical specialist) will initiate the desk study required by Clause 3.2.2 of EC7-1 and produce the broad conceptual ground model to define the appropriate scale of fieldwork and laboratory testing for the project. The investigation proper will usually be carried out in two phases – preliminary work to check the site suitability followed by detailed investigations to produce the geotechnical parameters for the foundation design. The results of the investigations are firstly produced in a *factual report*, designated the 'ground investigation' report (GIR), to provide a full description of all ground conditions relevant to the site and the proposed works. The GIR is then produced as part of the 'interpretative geotechnical design report' (GDR), mandatory under EC7-1 Clause 2.8 and EC7-2 Clause 6, for all geotechnical design, large or small, and must provide all the data used in the study, the design assumptions, methods of calculations and the checks on safety and serviceability. It must also include a plan (the scope of which will depend on the 'Geotechnical Category') for 'adequate supervision' and monitoring of the works by experienced personnel[1.9]. A third phase would be the comparison of actual ground conditions with those given in the report.

Guidance on investigating and reporting on contaminated land is provided in BS 10175 and the Statutory Guidance[11.1] from DEFRA (for England only) on the definition of contaminated land, remediation and liabilities as provided under the Environment Protection Act 1990 Part 2A. The guidance is legally binding on enforcing authorities, the objectives being to quantify the risks to human health and damage to materials in buildings. Relevant information from the formal Contaminated Land Report[11.2] should be incorporated into the GIR. CIRIA Report RP961[11.3] provides guidance on investigating and dealing with asbestos contamination in soil.

At the time when a ground investigation is planned, it is not always certain that piled foundations will be necessary. Therefore, the programme for the site work should follow the usual pattern for a ground investigation with boreholes that are sufficient in number to give proper coverage of the site both laterally and in depth. Borehole spacing recommended in Annex B of EC7-2 is a grid of 15–40 m for high-rise and industrial structures, less than 60 m for large-area structures and 20–200 m for linear structures. For small sites, it would be appropriate to have a minimum of three investigation points and, for bridges and machinery foundations, two to six holes per foundation.

If it becomes evident from the initial boreholes that piling is required or is an economical alternative to the use of shallow spread foundations, then special attention should be given to ascertaining the level and characteristics of a suitable stratum in which the piles can take their bearing. Where loaded areas are large in extent, thus requiring piles to be arranged in large groups rather than in isolated small clusters, the borings should be drilled to a depth of 1.5 times the width of the group below the intended base level of the piles or 1.5 times the width of the equivalent raft below the base of the raft (Figure 11.1). This depth

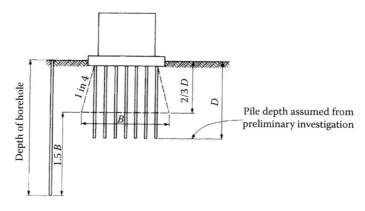

Figure 11.1 Required depth of boreholes for pile groups in compressible soils.

of exploration is necessary to obtain information on the compressibility of the soil or rock strata with depth, thus enabling calculations to be made of the settlement of the pile groups in the manner described in Sections 5.2 and 5.3. If the piles can be founded on a strong and relatively incompressible rock formation, the drilling need not be taken deeper than a few metres below *rockhead* (the buried interface between overburden or superficial sediments and rock), to check that there are no layers or lenses of weak weathered rock which might impair the base resistance of individual piles. Before permission is given for the drilling depth to be curtailed in this manner, there must be reliable geological evidence that the bearing stratum is not underlain by weak compressible rocks which might deform under pressures transmitted from heavily loaded pile groups and that large boulders have not been mistaken for bedrock. Rockhead contours formed due to erosion prior to the deposition of the overburden may be unrelated to current topographical surface, for example karstic conditions.

Particular care is necessary in interpreting borehole information where the site is underlain by weathered rocks or by alternating strong and weak rock formations dipping across the site. Without an adequate number of cored boreholes and their interpretation by a geologist, wrong assumptions may be made concerning the required penetration depth of end-bearing piles. Two typical cases of misinterpretation are shown in Figure 11.2.

Where piles are end bearing on a rock formation, it may be desirable, for economic reasons, to obtain a detailed profile of the interface between the bearing stratum and the overburden, so enabling reliable predictions to be made of the required pile lengths over the site. Cased light cable percussive rig borings followed by rotary core drilling to prove the rock conditions can be costly when drilled in large numbers at the close spacing required to establish a detailed profile. Geophysical exploration by seismic refraction on land and by continuous seismic profiling at sea are economical methods of establishing bedrock profiles over large site areas. With the improvements in geophysical data processing, less intrusive techniques, such as electrical resistivity and tomography, provide good resolution of stratification and relatively shallow bedrock profiles. Reynolds[11.4] describes common geophysical methods and their application to geotechnical investigations.

Geophysical methods are not usually economical for small site areas, but where the overburden is soft or loose, either uncased wash probings or continuous dynamic probing tests and cone penetration tests (see Section 11.1.4) are cheap and reliable methods of interpolating between widely spaced cable percussion boreholes.

Information on groundwater conditions is vital to the successful installation of driven and cast-in-place and bored and cast-in-place piles. The problems of installing these pile types in water-bearing soils and rocks are discussed in Sections 3.4.8 and 3.4.9. Standpipes or piezometers should be installed in selected boreholes for long-term observations of the fluctuation in groundwater levels, well before construction operations.

Trial pits and trenches are often a useful adjunct to borehole exploration for a piling project. Shallow trial pits are excavated in filled ground to locate obstructions to piling such as buried timber or blocks of concrete. The Health and Safety Executive Information Sheet (No. 8) states that no one should enter an unsupported excavation; safely battered side slopes may be acceptable. Deep trial pits, properly shored, may be required for the direct inspection of a rock formation by a geologist or to conduct plate bearing tests to determine the modulus of deformation of the ground at the intended pile base level (see Sections 4.7 and 5.5). It may be more convenient and economical to make these tests at the preliminary test piling stage.

The production of the GDR requires regular communication between the ground investigation contractor responsible for the primary data, the geotechnical specialist assessing the derived values and the structural designer. The GDR should contain recommendations

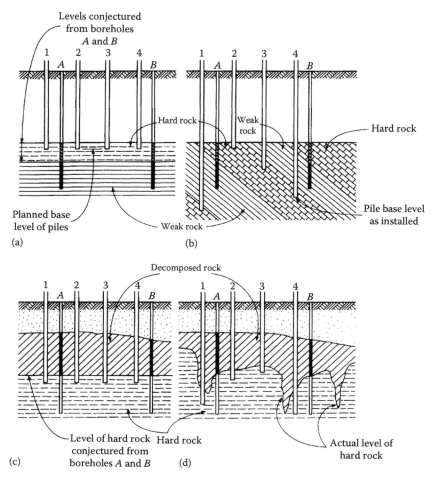

*Figure 11.2* Misinterpretation of borehole information. (a) Horizontal stratification interpreted by interpolation between boreholes A and B. Piles 1–4 planned to have uniform base level. (b) Actual stratification revealed by drilling boreholes for piles 1, 3 and 4, showing base level required by dipping strata. (c) Uniform level of interface between decomposed rock and hard rock interpreted by interpolation between boreholes A and B. Piles 1–4 planned to have uniform base level. (d) Actual profile of hard rock surface.

for inclusion in the contract conditions and specification to deal with geotechnical uncertainties that cannot be mitigated by the detailed design, including comments as to when additional investigation may be needed. The ground investigations recommended in EC7 to produce the GDR may not provide the data needed for inclusion in commercial foundation analysis programmes; this must be addressed at the planning stage to avoid misapplication of software. Generally, the default values in software packages may be selected to provide conservatively low estimates of geotechnical parameters.

## 11.1.2 Boring in soil

Cased cable percussion borings (by *shell and auger*) give the most reliable information for piling work. Operation of the boring tools from the winch rope gives a good indication of the state of compaction of the soil strata. If the casing is allowed to follow down with the boring and drilling, and water to aid drilling is used sparingly, reliable information can be obtained on groundwater conditions, but where groundwater fluctuates seasonally and tidally, standpipe readings over a period are essential. Such information cannot be obtained from wash borings or by drilling in uncased holes supported by bentonite slurry. Borings by continuous flight auger are satisfactory provided that there is a hollow drill stem down which sample tubes can be driven below the bottom of the boring and measurements of the groundwater level obtained. The spoil from the auger blades is not satisfactory for describing strata.

Information on the size of buried boulders is essential for a proper assessment of the difficulties of driving piles or constructing bored piles past these obstructions. To ensure that boulders are not confused with rockhead, it is usual practice for the boring rig to be replaced with a core drill to obtain information on the size and nature of boulders; coring then continues to provide information on the soil overburden and to determine rockhead. Norbury[11.5] gives recommendations for field descriptions of the boundary conditions which should be used by the site geologist; he comments on some anomalies in the new standards in respect of soil and rock descriptions.

Investigation of glacial tills for piled foundations requires particular care. For example, in addition to assessing the potential for random boulders, it is necessary to identify mixed sequence of strata, laminations of silty clay, perched water tables and infilled buried channels and also to provide samples for testing. The presence of soft clays if they are to be subjected to lateral loads from the pile shaft and compressible clays below the pile toe should be investigated.

The UK practice of providing 'undistorted' samples of fine-grained soils from cable percussion boreholes by means of 100 mm open-drive thick-wall sample tubes (U100) is no longer compliant with BS EN 22475-1 (sampling methods). Such tubes cannot be considered as *Category A* samplers and cannot be used to provide EC7-2 quality *Class 1* samples for laboratory shear strength and compressibility testing – although they may be suitable as *Class 2* samples for other laboratory testing. This is important when using 'derived values' from triaxial tests to determine characteristic values of geotechnical parameters for application to pile design (EC7-1 Clause 2.4.5.2). Thin-wall sample tubes which are pushed into the soil by either a cable rig or rotary drill to obtain better quality undisturbed samples in soft clays can be considered Class 1 samples; the piston sampler is similar. Core drilling using triple core barrels to obtain continuous Class 1 samples in over-consolidated clay and some coarser soils is also practical as described by Binns[11.6]. Baldwin and Gosling[11.7,11.8] clarify the sampling categories given in EC7-2 and BS EN ISO 22475-1 and demonstrate the differences in triaxial tests on U100 samples and thin-wall UT100 samples.

## 11.1.3 Drilling in rock

Weak rocks can be drilled by percussion equipment, but this technique is useful only to determine the level of the interface between the rock formation and the soil overburden. Little useful information is given on the characteristics and structure of the rock layers because they are reduced to a gritty slurry by the drilling tools, and drilling should be stopped as soon as it is evident that a rock formation has been reached. Some indication of the strength of weak rocks can be obtained from standard penetration tests (SPTs) (see Section 11.1.4). Percussion boring can provide reliable information from rocks which have been weathered to a stiff or hard clayey consistency such as weathered chalk, marl or shale. Hammering sample tubes into shattered rock will not produce useable samples for laboratory tests and frequently lead to confusion and error in determination of rockhead.

The improved triple-barrel core drilling equipment noted above is now the preferred method of sampling weak and weathered rock, supplemented by in situ testing.

There are three main types of rotary coring: *conventional*, with 3 m long double- or triple-core barrels, *wireline* coring where the core barrel is latched on the bottom of the casing (which also acts as the drill rod) and the core run retrieved by winch, and *sonic* coring. In the conventional and wireline systems, the core diameter must be large enough to ensure complete or virtually complete recovery of weak or heavily jointed rocks to allow reliable assessment to be made of bearing pressure. The percentage core recovery achieved and the rock quality designation (RQD)[11.5] should be recorded. All cores should be stored in secure, correctly sized core boxes, and selected cores should be promptly coated in wax or sealed in aluminium foil and cling film to preserve in situ moisture content. Generally, the larger the core size, the better will be the core recovery. Drilling to recover large diameter cores, say up to the ZF size (165 mm core diameter), can be expensive, but the costs are amply repaid if claims by contractors for the extra costs of installing piles in 'unforeseen' rock conditions can be avoided. Also, by a careful inspection and testing of the cores to assess the effects of the joint pattern on deformability and to observe the thickness of any pockets or layers of weathered material, the required depth of the rock socket (see Section 4.7.3) can be reliably determined. It must be remembered that drilling for piles in rock by chiselling and baling or by the operation of a rotary rock bucket (Figure 3.28) will form a weak slurry at the base of the pile borehole which may make it impossible to ascertain the depth to a sound stratum for end-bearing piles. Whereas if there has been full core recovery from an adequate number of boreholes together with sufficient testing of core specimens, the required base level of the piles can be determined in advance of the piling operations. Developments in instrumentation to record operating parameters similar to that used on piling rigs, together with electronic logging and immediate core scanning, are improving the quality of data from rotary cored holes.

Sonic coring can rapidly produce a continuous core to considerable depths, and, while in situ tests can be performed, it is mainly used for environmental investigations and cross-hole seismic tomography. One hundred percent core recovery is possible, but samples may be disturbed to some extent. The technique operates on the principle of controlled high-frequency vibration, with or without rotation in casing sizes up to 300 mm, and bores can be drilled dry or with fluid flush.

Investigation of chalk for piled foundations requires attention to defining the *marker beds* (marl and flints), variability of the chalk with depth, possible fissures and dissolution cavities, leading to determination of the *grades* as given in the revised engineering classification of chalk[4.58] (also see Appendix A). Exploration should continue for at least 5 m below the tip of the longest pile anticipated. Percussion boring can cause disturbance and is best used in low and medium density chalks. Rotary drilling in most grades will produce cores, but even with high-quality large diameter cores, identification of the fracture size is difficult.

## 11.1.4 In situ and laboratory testing in soils and rocks

The following are summaries of the tests described in Clause 4 of EC7-2 and the 13 parts of BS EN ISO 22476 on field testing (due for implementation by 2015).

*Field vane tests*, mainly used to determine the undrained in situ shear strength of soft to very soft, sensitive, fine-grained soils, have little application to piling operations. Shaft friction in such soils can contribute only a small proportion of the total pile resistance, and it is of no great significance if laboratory tests for shearing strength on Class 1/2 'undisturbed' soil samples indicate shearing values that are somewhat lower than the indicated in situ strengths. The lateral resistance of piles is particularly sensitive to the shearing strength of clays at shallow depths.

The *standard penetration test* (SPT) as described by Clayton[11.9] is the most useful all-round test for piling investigations which, in clays, silts and sands, is performed with an open-ended tube and in gravels and weak rocks can be made by plugging the standard tube with a cone end. The blow counts (*N-values*, blows per 300 mm of penetration) for the SPT have been correlated with the angle of shearing resistance of coarse-grained soils (Figure 4.10) by Peck et al.[4.24]

Relative density descriptors for coarse-grained soils based on SPTs have been modified as follows (after Norbury[11.5]):

| BS 5930+A2 2010 | | BS EN ISO 14688-2 Table 4 | BS EN 1997-2 (Annex F) |
|---|---|---|---|
| SPT N | Term | Density index ID (%)$^a$ | SPT N I$_{(60)}$ |
| <4 | Very loose | 0–15 | 0–3 |
| 4–10 | Loose | 15–35 | 3–8 |
| 10–30 | Medium-dense | 35–65 | 8–25 |
| 30–50 | Dense | 65–85 | 25–42 |
| >50 | Very dense | 85–100 | 42–58 |

$^a$ Or relative density.

The BS 5930 description is shown on field logs and is uncorrected. The SPT *N*-value using EC7-2 Annex F is the *N*-value corrected for hammer efficiency and overburden pressure as provided in BS EN ISO 22475-3, that is, $N_{1(60)}$ (at 60% of the free-fall energy and 1 bar pressure).

For fine-grained soils, the traditional Terzaghi and Peck correlations for soil consistency and strength have been modified in BS EN ISO 14688-2 (classification of soil) as follows:

| N-value uncorrected (blows/300 mm) | Strength classifier | Shear strength (kN/m$^2$) |
|---|---|---|
| <2 | Extremely low | <10 |
| <4 | Very low | 10–20 |
| 4–8 | Low | 20–40 |
| 8–15 | Medium | 40–75 |
| 15–30 | High | 75–150 |
| Over 30 | Very high | 150–300 |
| No value given | Extremely high | >300 |

This table is similar to the empirical relationship described by Stroud[5.7] between the SPT and the undrained shear strength of stiff over-consolidated clays as shown in Figure 5.20. The cone-ended SPT can also be made in weak rocks and hard clays. Useful correlations have been established between the *N*-values of stiff to hard clays and the modulus of volume compressibility (Figure 5.20). The test should also be made if percussion borings are carried down below rockhead.

The SPT is liable to give erroneous results if the drilling operations cause loosening of the soil below the base of the borehole. This can occur if the borehole is not kept filled with water up to ground level, or above ground level, to overcome the head of ground-water causing 'blowing' of a granular soil. Careful manipulation of the 'shell' or baler is also necessary to avoid loosening the soil by sucking or surging it through the clack valve on the baler. It is particularly necessary to avoid misinterpretation of SPT data on piling investigations since denser conditions than indicated by the test may make it impossible

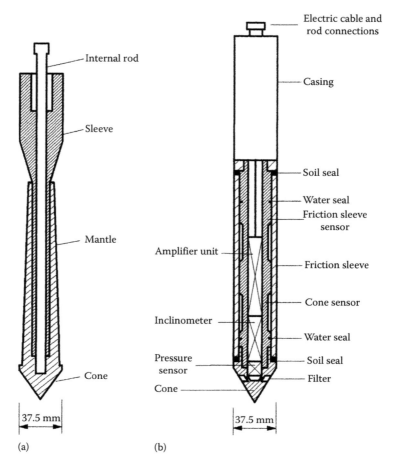

*Figure 11.3* Types of 60° cone for static CPT. (a) Mechanical cone. (b) Fugro piezocone. (Courtesy of Fugro Engineering Services Ltd, Wallingford, England.)

to drive piles to the required penetration level. The SPT cannot be performed satisfactorily at deep-sea locations for use in pile design; the cone penetration test is now widely used (see the following text).

The application of the *cone penetration test* (CPT)[11.10] results to the design of individual piles is described in Section 4.3.6 and to the design of pile groups in Section 5.3. The original *Dutch* cone (Figure 11.3a) is still in use with mechanical driving, now designated CPTM. Empirical correlations have been established between the static cone resistance and the angle of shearing resistance of coarse-grained soils (see Figure 4.11). The CPT also gives useful information on the resistance to the driving of piles over the full depth to the design penetration level.

The more commonly used *electrical* cone has electrical-resistance strain gauges mounted behind the cone and inside the sleeve, giving continuous readings of penetration resistance by means of electrical signals recorded on data loggers at the surface. The test is rapid and on land is carried out by a custom built, ballasted truck rig which pushes the cone into the ground at a continuous rate with a hydraulic piston. The 60° cone area is usually 10 cm² (as the original Dutch cone) equivalent to a diameter of 35.7 mm, but cones up to 20 cm² in area are also used. In addition to providing the basic cone resistance parameter ($q_c$) and side friction ($f_s$), the parameters and charts developed by Robertson[11.11] for 'normalised' cone resistance and 'normalised' friction ratio can be useful in identifying the soil and soil performance. The CPT does not provide a soil sample; hence, most operators provide samples from an adjacent hole by using the CPT installation equipment to push-in a 25 mm diameter sampling tube. The *piezocone* (CPTU) in Figure 11.3b is a cone penetrometer which also provides measurement of the pore-water pressure at one or more locations on the penetrometer surface. This penetrometer can be used in deep-sea ground investigations, usually with a 15 cm² cone, allowing more space for sensors giving data on deformation and consolidation parameters. The continuous profile procedure can be speeded up using the drill–push technique developed by Fugro Seacore. The *seismic cone* provides information which links the CPT net cone resistance to shear wave velocity and soil modulus[11.11].

The updated requirements in BS EN ISO 22476-1 for the electrical cone procedures will resolve the differences in interpreting the $q_c$ values from different operators and are compatible with EC7-1 design procedures. Under the new standard, the CPT/CPTU procedures have to be related to end use.

The *continuous dynamic probing test* is a useful means of logging the stratification of layered soils such as interbedded sands, silts and clays. The number of blows with different weight hammers to penetrate 10 mm is designated $N_{10}$ and $N_{20}$ for 20 mm penetration. The $N_{10}$ values can be converted to unit cone resistance ($r_d$) or dynamic cone resistance ($q_d$) using pile driving formulae as given in Annex E of BS EN ISO 22476. But note the correlations in EC7-2 Annex G have not been proven in UK conditions.

*Pressuremeter* (PMT) or *dilatometer tests* provide approximate determinations of the deformation modulus of soil and rocks by expanding a cylindrical rubber membrane against the walls of the borehole test section and measuring the increase in diameter of the cylinder over an increasing range of cell pressures. Apparatus developed for this purpose includes the following:

- Ménard pressuremeter (MPM): Provides data for direct application to pile design[11.12]. Used in stiff clays and weathered rock in a preformed pocket in a borehole (BS EN ISO 22476-4).
- Cambridge self-boring PMT: A plug of soil is removed by a drill bit at the base of the device to accommodate the PMT membrane unit. Depths around 60 m in sands and clays; minimal ground disturbance.
- High-pressure dilatometer (HPD): A similar device for use in cored boreholes in stiff soils and weak rock. Depths >200 m feasible.
- Marchetti dilatometer[11.13] (DMT): A spade-shaped device which is pushed or hammered into soft to firm clays and silts.

The PMT should be distinguished from a *borehole jack* which applies forces to the sides of boreholes by forcing apart circular plates, imposing different boundary conditions on the test.

As noted in Section 6.3.7, the PMT is useful in determining the ultimate resistance to lateral loads on piles and the calculation of deflections for a given load. Because the PMT only shears a soil or rock (there is no compression of the elastic soil or rock), the slope of the pressure/volume change curve in Figure 6.34 gives the shear modulus G. This can be converted to Young's modulus from Equation 6.49. The frequency of the datapoints on the load/unload loop are such that the change in strain can be accurately measured for each successive point from a selected zero—with the smallest increment being around 0.01% radial strain. G calculated in this way more accurately reflects actual strain produced in the ground by structures and is greater than $G$ obtained from slopes of lines through the loops. The undrained strength of clays obtained from the PMT is usually greater than that obtained from triaxial tests, possibly due to sample disturbance, but mainly because the soil is tested differently. When using the PMT to obtain $E$ values for pile group settlements using the methods described in Chapter 5, it is necessary to take into account the drainage conditions in the period of loading. Clarke[6.24] covers the practical operation and interpretation of results of the PMT in detail. Two hundred and forty PMT tests using the HPD to determine in situ deformation properties in Triassic and Carboniferous strata were carried out in 75 mm diameter cored holes to depths up to 50 m for the piling to the viaduct piers on the Second Severn Crossing[11.14].

The *cone PMT* is a device which incorporates a 15 cm² cone below the friction sleeve of the expanding PMT module and pushed into the soil using standard cone rods. A piezocone may replace the cone to assist in identifying soil types.

*Plate bearing tests* can be used to obtain both the ultimate resistance and deformation characteristics of soils and rocks. When used for piling investigations, these tests are generally made at an appreciable depth below the ground surface, and rather than adopting costly methods of excavating and timbering pits down to the required level, it is usually more economical to drill cased holes 1–1.5 m in diameter by power auger. The holes are lined with casing, the soil at the base carefully trimmed and the rigid plate (300–600 mm diameter) accurately levelled on a bed of cement mortar or plaster of Paris[11.15]. A flat jack may be inserted below the plate to help even out stresses. The load is transmitted to the plate through a tubular or box-section strut and is applied by a hydraulic jack bearing against a reaction girder at the surface as described for pile loading tests (see Section 11.4.1). The deformation of the soil or rock can be measured by inserting a borehole extensometer, containing electrolevels at various depths, into a borehole below the centre of the plate. This procedure[11.16] is helpful in obtaining the modulus of deformation of layered soils and rocks. Loading tests of this type were carried out on 500 mm diameter plates in 600 mm holes 20 m deep drilled offshore from a jack-up platform to determine deformation properties in Triassic rocks for the main span foundations for the Second Severn River crossing[11.14].

Small-diameter plate loading tests can be made using a 143 mm plate in a 150 mm borehole, but it is impractical to trim the bottom of the hole or to ensure even bedding of the plate. However, these tests can be useful means of obtaining the ultimate resistance of stiff to hard stony soils[11.17] or weak rocks[11.18]. They do not give reliable values of the deformation modulus.

Evaluation of the above-mentioned field tests to provide derived design parameters should be carried out as specified in the Annexes to EC7-2, summarised in the table as follows:

| EC7-2 annex | Field test | Parameter |
|---|---|---|
| D | CPT | $\phi$ $E'$ $E_{oed}$ |
| E3 | PMT test (MPM) | $Q$ |
| F | SPT | $\phi$ $I_D$ |
| G | Dynamic probing test | $\phi$ $I_D E_{oed}$ |
| I | Field vane test | $c_u^a$ |
| J | Flat dilatometer test (DMT) | $E_{oed}$ |
| K | Plate loading test | $c_u$ $E'$ $k_s$ |

$Q$ is the ultimate compressive resistance of a single pile; $k_s$ is the coefficient of subgrade reaction; $I_D$ is the density index.
a  Requires correction using Atterberg limits.

Simple forms of *in situ permeability test* can give useful information for assessing problems of placing concrete in bored and cast-in-place piles in water-bearing ground. The *falling-head test* consists of filling the borehole with water and measuring the time required for the level to drop over a prescribed distance. In the *constant-head test*, water is poured or pumped into the borehole, and the quantity required to maintain the head at a constant level above standing groundwater level is recorded. *Pumping-in tests* made through packers in a borehole or pumping-out tests with observations of the surrounding drawdown are too elaborate for most piling investigations. Basic groundwater information can often be obtained by baling the borehole dry and observing the rate at which the water rises to its standing level. BS EN ISO 22282 replaces the tests in BS 5930 for all permeability testing, including large-scale pumping tests, in accordance with EC7-1 and EC7-2.

The *point load test*[11.19] is a quick and cheap method of obtaining an indirect measurement of the compression strength of a rock core specimen. It is particularly useful in closely jointed rocks where the core is not long enough to perform uniaxial compression tests in the laboratory. The equipment is easily portable and suitable for use in the field. The tests are made in the axial and diametrical directions on cores or block samples. The failure load to break the specimen is designated as the point load strength $(I_s)$ which is then corrected to the value of point load strength which would have been derived from a diametral test on a 50 mm diameter core using a standard correction (Table 11.1) to obtain $I_{s(50)}$.

EC7-1 Clause 2.4.3(6) requires *calibration factors* to be applied to certain field and laboratory tests given in EC7-2, in order to convert them into values which 'represent the behaviour of the soil or rock in the ground' (see Section 4.3.1). This is before applying the *correlation factors* required in EC7-1 Clause 7.6.2.3 for parameters based on multiple tests. It is also necessary to consider other relevant data from published material and local experience when establishing the derived characteristic value from empirical correlations.

*Laboratory testing* of soil is described in EC7-2 Clause 5, cross-referenced to BS EN ISO 14689 and BS 1377 (being replaced by ISO 17892). Methods of determining shear strength parameters $\phi$, $c_u$ and $E$ ('theoretically derived values') from Class 1 samples include the various types of the triaxial test and shear box test. The improved sampling and triaxial compression testing techniques now available are likely to result in undisturbed shear strength values higher than those obtained in earlier practice, particularly in very stiff to hard clays. These higher values may require modification of correlations established between the shear strength of clays and shaft friction and end-bearing resistance of piles.

Laboratory tests on rock cores should include the determination of the unconfined compression strength of the material, either directly in the laboratory or indirectly in the field or laboratory by means of point load strength tests. Young's modulus values of rock cores can

*Table 11.1* Empirical relationship between uniaxial compression strength ($q_{comp}$) and point load strength ($I_{s(50)}$) of some weak rocks

| Rock description | Average $q_{comp}$ (MN/m²) | $q_{comp}/I_{s(50)}$ |
|---|---|---|
| Jurassic limestone | 58 | 22 |
| Magnesium limestone | 37 | 25 |
| Upper chalk (Humberside) | 3–8 | 18 |
| Mudstone/siltstone (Coal Measures) | 11 | 23 |

be obtained by triaxial compression testing using the transducer equipment or strain gauges stuck on the specimen for small strain measurements.

### 11.1.5 Offshore investigations

Offshore investigations for deep-water structures and oil production platforms are highly specialised, and although the basic procedures contained in the British Standards and Eurocodes for geotechnical investigations should be adhered to for UK waters, there are many additional statutes and regulations which apply to such work and are outside the scope of this text. Draft 'guidance notes' on site investigations[11.20] have been prepared by the Society for Underwater Technology to provide a basic framework for offshore investigations, particularly for renewable energy projects, citing practices and regulations for the United Kingdom, for other European regulatory bodies and under API regulations.

The range of design issues which has to be investigated for piled offshore structures is more extensive than that required for land-based buildings. In addition to bearing capacity and settlement, it is usually necessary to consider cyclic displacements, foundation stiffness, stability of footings for jack-up rigs, liquefaction of soils and scour and erosion potential. The ISSMGE[11.21] comprehensive review of offshore and nearshore investigations describes current good practice for obtaining data on soil and rock properties using a variety of techniques for marine structures ranging from jetties to deep offshore platforms. BS EN ISO 19901-8 dealing with marine soil investigations, due to be published in 2015, will expand on this review. BS EN ISO 19901-4 deals only with 'geotechnical considerations' for shallow foundations offshore except for a note in Annex B regarding drilled and grouted piles in carbonate sands. (Note that ISO 19901-4 applies material factors to the soil strength, whereas ISO 19902 uses the LRDF method of applying a resistance factor to the foundation capacity.)

Depending on the site variability, at least one borehole is required for a small inshore platform, with samples and CPTs taken alternately every metre to a depth of several pile diameters below the likely pile penetration. For larger platforms in deeper waters, several boreholes will be needed with continuous CPTs in a dedicated hole and sampling at 0.5 m intervals and PMT and vane tests in other holes. Depths will vary from 70 m in sands and over-consolidated clays to over 150 m in normally consolidated clays. For inshore wind farms with monopile foundations, one borehole may be drilled and cored at each location or, in reasonably uniform soil conditions, one for say five adjacent monopiles. Vertically anchored structures such as piles for semi-submersible production platforms will require investigations of the seabed and geology over the full anchor spread area using geophysical and geotechnical techniques.

Plant and equipment deployed for investigations include dynamically positioned vessels with heave compensation for deep coring through a moon pool; jack-up rigs and anchored barges inshore; remote-controlled, seabed devices for sampling, coring, and CPTs and vessels for geophysical surveys. Positioning of offshore boreholes is provided by differential GPS giving $x$ and $y$ coordinates to 1–3 m accuracy and the surface level of the vessel to give borehole penetration depth. Acoustic devices using transponders are used for underwater positioning. Figure 11.4 shows a remotely controlled seafloor drill capable of wireline coring and CPT up to 150 m deep in water depths of 3000 m.

Laboratory testing for offshore structures will generally follow the same standards and codes used for onshore developments. Particular attention has to be given to ensure high-quality samples are retrieved from sensitive soils (by piston samplers or vibrocorers), then carefully stored and transported.

*Figure 11.4* Remotely controlled seafloor drilling rig for sampling and in situ testing. (Designed and built by Gregg Marine. Moss Landling, CA, and reproduced with permission.)

## 11.2 PILING CONTRACTS AND SPECIFICATIONS

### 11.2.1 Contract procedure

The Conditions of Contract document provides the procedures to allocate between the parties the legal and financial risks and obligations involved in managing and undertaking the project. It ranks alongside the tender, drawings, specifications and pricing schedules (and in some cases ranks above these). Employing authorities, whether public sector or generally, procure construction work by inviting tenders (offers) from interested parties based on enquiry documents defining the works and one of the following procedures: a prequalification process, an open tender, selected tender or negotiated tender. A binding contract between employer and contractor is formed by acceptance of the contractor's offer, subject to defined legal requirements[11.22], whether the procurement process was based on a standard form of contract, an exchange of letters or concluded wholly or partially orally.

The terms and conditions of a contract should fairly allocate the risks between the parties on the basis that risks and responsibilities are placed with the party best able to influence them. If these are not clear, then arguments over acceptance of the works and payment can ensue with the possibility of expensive legal action to resolve disputes. The problem of late payments and late release of retention money to subcontractors has been addressed to a degree in the new *Construction Act 2009*[11.23], which requires payment dates to be stated in the contract and prohibits 'pay when certified' contract clauses; also, if a compliant dispute resolution clause is not included, then statutory provisions will apply.

Piling rarely forms a high proportion of the total cost of a project on land in the United Kingdom, and it is usual for the works to be carried out as a subcontract to the main general contract. On large projects where it is necessary to let separate contracts for advance works, it may be advisable for the piled foundations to be treated separately. Piling can be a significant part of the cost for marine construction and is usually undertaken directly by the main contractor, subcontracting being limited to specialist services such as drilling and grouting or the construction of anchorages to tension piles.

The traditional, well-proven form of civil engineering contract in which an independent engineer is appointed by the employer to design and supervise the works is provided by the Civil Engineering Contractors' Association (CECA) in the new *Infrastructure Conditions of Contract* (ICC)[11.24]. However, for major works, this form is being increasingly replaced by the *New Engineering and Construction Contract* (NEC3)[1.10] in which the traditional engineer's role has been removed and replaced with a *project manager*, responsible directly to the employer, with authority to issue instructions and certify payments, and a separate *supervisor* with duties for testing and inspecting the works. This is a complex document with 11 separate sections and more than 20 'options' for the employer to select and build up a contract to suit individual requirements. It allows for design responsibility to be carried by either the employer or contractor depending on which party has the competency. Specialist subcontractors may be required to enter into 'back-to-back' agreements with the employer and other parties which frequently place additional risk-taking burdens on the subcontractor.

For building contracts, the main contract conditions are set out in the *Standard Building Contract*[11.25] prepared by the Joint Contracts Tribunal (the *JCT* forms), generally with design and contract control by an architect or by a contract administrator (CA) with the architect as designer. Building and construction works on behalf of government departments are usually carried out under a form of contract designated *GC/Works/1*. Here, the employer, advisers and designers are termed the 'authority', and the supervisory duties of the engineer are delegated to the 'superintending officer'.

The revised ICE Ground Investigation Specification[11.26] has been aligned with the NEC3 contract terminology, replacing the engineer with an 'investigation supervisor or ground practitioner' with 'suitable experience' to procure and supervise the investigation. The ICC *Ground Investigation Contract*[11.27] is the traditional form providing for an engineer or geotechnical advisor acting for the employer to design and supervise the work.

The standard main contracts mentioned earlier operate with compatible forms of subcontract which will govern most piling works. The NEC3 contract states that the contractor is responsible for the works as if he had not subcontracted; the piling subcontractor therefore needs to be aware that risks may be transferred through the NEC3 subcontract. The CECA *Blue Form* of subcontract can be used with the ICC form, and the JCT contracts have specialist forms for domestic (direct) and nominated subcontractors. Additionally, ad hoc and bespoke forms of subcontract prepared by main contractors are increasingly used to change the liabilities and risk-sharing obligations given in standard forms. Collateral warranties, which are separate from, but operate alongside, the works contract, are now frequently requested by the employer, developer or project funders. They provide for liability to a beneficiary (who may not be the same person as the principal works owner) in respect of defective performance over a period of time, possibly longer than the statutory defects period, and must be treated with caution by specialist contractors. Similar long-term benefits may be conferred on identified, non-contracting parties through a clause in the construction contract listing the contractor's obligations/liabilities to such parties under the Third Parties Act which may be passed on to the piling contractor.

The authority of the person fulfilling the role of the CA (whether project manager, engineer or architect) to act as the interface between the employer and contractor derives from the specific contract under which he was appointed. These terms are different for the different forms of contract mentioned earlier and can include broad powers to act.

The open tender process, which allows for any interested party to make an offer, is not recommended for piling works. Prequalification and selection will demonstrate to the employer or contractor that the tenderer has the necessary skills and financial standing to undertake the works. Once the tender list is established, there are two basic methods of obtaining offers for piling works, either as main contractors or subcontractors:

Employer's design

The employer's CA invites tenders from specialist contractors to undertake piling in accordance with detailed designs, pile layout, specification and relevant contract data prepared and provided on behalf of the employer by his designer. The type of non-proprietary pile (or alternative), the applied loads and acceptable settlement under test load must be stated, and also the diameter, penetration depth and pile material. The contractor will provide information (e.g. as Table B1.1 in the ICE SPERW [2.5]) and a price for the specified work.

Benefits: The designer may be more objective in selecting the best overall piling system, particularly for difficult sites requiring significant engineering input. Responsibility of each party is clearly defined. Full-time engineer supervision is recommended.

Drawbacks: The knowledge and experience of the piling contractor may not be fully utilised; his involvement is limited to selecting the most efficient type of plant and to installing the piles in a sound manner complying with the specification.

Contractor's design

The CA invites tenders from specialist contractors to undertake piling in accordance with designs prepared by the specialist. The employer's designer must provide pile layout and loads, information on ground conditions, a specification and any site constraints and specify the requirements for performance under loading tests. The tenderer must submit a design based on his choice of pile, a method statement and a quality control plan and guarantee the successful performance of the piles. Overall responsibility for the project substructure works (e.g. pile caps) should be stated.

Benefits: Widest choice of piling systems; utilises the experience of the specialist to the optimum extent. Tender based on in-house experience and established performance.

Drawbacks: Possible unrealistic performance specification. Independent check on the design may be necessary.

Contractor's design (alternative)

The CA provides a layout drawing of columns and walls with the loadings and general site information. The tenderer must supply the pile layout and the specification and demonstrate how compliance will be achieved using his design and construction method, together with a guarantee of performance.

*Benefits*: Appropriate for a selected tender list or negotiated tender. Allows application of new types of piles, designed to optimise bearing capacity and minimise concrete usage and spoil disposal; can provide the best value. The contractor explicitly assumes responsibility for all aspects of the piling work, including unforeseen adverse ground conditions and the possibility of having to increase the penetration depth or increase the number of piles or even to abandon a particular system.

*Drawbacks*: Ground investigation and preliminary pile testing may have to be carried out by the tenderer, with possible delays to the start of the work. The employer's designer will need to check the different proposals offered. Insurers providing the main contractor with the essential warranty for his work and the employer with cover for the structure may need to be involved in the tender process. The employer may have critical information not released to the tenderer.

Other matters which the piling contractor should ensure are included in the contract terms and conditions are as follows:

Responsibility for the *design of the piling works* should be unambiguously stated. If a separate designer is appointed for the piles or if the designer is the piling contractor's in-house designer, then for an NEC3 contract, the design must be accepted by the project manager before work commences. The design may be rejected if it 'does not comply with either the Works Information or the applicable law'. If the project manager supplies the pile design, it should be stated that this is the approved design for the works.

*Ground investigations* undertaken on behalf of the employer before inviting tenders for the piling should be provided and include the GDR if the contractor is to be responsible for the design of the piles. In SPERW[2.5] (Clause B1.7), the employer/engineer is not liable for the opinions and conclusions provided in the GIR and GDR. Site information under a NEC3 contract may only be available as 'reference' data—with financial implications for contractual compensation events. There is no obligation on the project manager to provide 'additional information' which the piling contractor may need before tendering. It may therefore fall to the subcontractor to fill in gaps in the data.

The *facilities and attendances* to be provided by the main contractor, or those to be included in the piling contract, should be stated. These include such items as access roads, storage areas, fencing, watching, lighting and the supply of electrical power and water. Hardstandings (working platforms)[3.20] for large piling plant may need to be of substantial construction, and the contract should state if the main contractor will make the site stable and at what level in relation to the pile commencing surface and cut-off level.

*Underground services and obstructions* can be a contentious item. Under ICC contracts, it is normally the engineer's responsibility to locate all known buried services and other man-made obstructions to pile installation. The employer has the right to expect that the contractor will not push on blindly with the piling work with complete disregard for the safety of the operatives or the consequences of damage which can be severe. The NEC3 contract data contain a 'risk register' which should explicitly refer to services and potential obstructions.

A *Quality Management System* (conforming to BS EN ISO 9001) operated by the piling contractor and a project quality plan should be provided as a means of assuring the employer that the required standards for the particular works have been met through traceable documentation. The plan should include the 'pile installation plan' as stated in EC7 Clause 7.6(1). The system and the plan may be subject to audit and certification either by an independent third party or by the CA. Self-certification by the contractor to assure compliance with the specification may be acceptable—except for laboratory testing. Surveillance and intervention by the CA will be in addition to the contractor's demonstration of conformance under his plan.

*Risk assessments* to identify the hazards (risk events), probability of a risk event occurring and the consequences of ensuing injury, damage and loss, and any general uncertainties are the responsibility of the employer as part of the project feasibility study. There will be an obligation on the piling contractor to advise the CA of potential hazards involved in the particular method proposed, such as the potential for methane in Coal Measures and contaminated land. This information will be provided in the NEC3 risk register for the contract together with the actions to be taken to avoid or reduce the risks.

Compliance with the statutory *Construction (Design and Management) Regulations 2007* (CDM)[9.55], the *Health and Safety at Work Act 1974* and the *Environment Protection Acts* (EPA) involves all parties to the contract even if not expressly stated. Wide legal duties are now placed on virtually everyone involved in construction as explained in the CDM Approved Code of Practice. The piling contractor may be the first and only party on the site initially and, on a 'notifiable' project, may be appointed by the employer to undertake the statutory duties of the *CDM coordinator* and the other supervisory roles as defined. If no appointment is made, the duties fall to the employer by default. The statutory regulations under the EPA control the disposal of arisings from bored piles and waste drilling fluids, and the health and safety regulations cover all aspects of construction from protective clothing, lifting and hoisting appliances to access into excavations and welfare facilities.

Increasingly, tender information is provided electronically, and online bidding for contracts is approved in European Directive 2004/18/EC, but 'non-quantifiable elements should not be the object of electronic auctions'. It is submitted that foundation design and construction comes within this exclusion. In any event, contract liability may be limited by the tenderer or the employer to the documents provided in hard copy.

## 11.2.2 Piling specifications

The ICE SPERW[2.5] details items which should be included in the project specification for a piling contract. These include stating responsibility for design, performance criteria to be applied, requirement for additional ground investigation, and routine matters on site location, personnel, etc. Materials and workmanship shall be in accordance with the appropriate British Standards which should be quoted for the various work classifications. In addition to the SPERW, guidance on preparing appropriate clauses is given in the Department for Transport's Specification for Highway Works[11.28]. Some matters which require particular attention are listed as follows.

*Setting out:* The responsibility for setting out rests with the piling contractor if he is also the main contractor. The CA has no responsibility in the matter but should check the positions of the piles from time to time, since if these are inaccurately placed, the remedial work can be very costly. Problems can arise when a piling subcontractor does the setting out from a main contractor's grid-lines. If these are inaccurate or obscured, then there can be major errors in pile positions, and the main contractor may decline to accept the responsibility for the cost of the replacement piling. GPS surveying techniques may provide a reasonably accurate check of setting out under ideal conditions.

*Ground heave:* In the case of the employer-designed project, having specified the type and principal dimensions of the pile, the employer (engineer, project manager or other designer) would normally be liable for the effects of ground heave, as described in Section 5.7. However, for a contractor-designed piling project, the matter is not so clear. Unless contract responsibility explicitly lies with the piling contractor (as in the alternative contractor design mentioned earlier), it is difficult to assess liability for ground heave, either for remedial work to risen piles or for repairing damage to surrounding structures. It is suggested that where the piling contractor decides on the type and dimensions of the pile, he should have experience of ground heave effects and accept full responsibility for the site operations. If pre-boring or other measures are considered insufficient to prevent ground heave at tender stage, piling alone may not be the solution for the site foundations.

*Surcharge*: The piling contractor must be advised if additional temporary loads or excavations are planned adjacent to the piles.

*Loss of ground due to boring*: The consequences of a loss of ground while boring for piles were described in Section 5.7. The responsibilities for these are similar to those for ground heave.

*Noise and vibration*: The contractor is responsible for selecting the plant for installing piles and is therefore responsible for the effects of noise and vibration (see Section 3.1.7). The current statutory and local authority regulations limiting noise emissions should be stated in the conditions of contract.

*Piling programme*: If the CA wishes to install the piles for the various foundations in a particular sequence to suit the main construction programme, the sequence should be stated in the specification or Works Information, since it may not be the most economical one for the piling contractor to follow.

*'Set' of driven piles*: This should not be stated in precise terms in specifications for driven or driven and cast-in-place piles. The set for a particular site and applied load cannot be established until preliminary piles have been driven and the driving records checked against the ground conditions assumed in design.

*Tolerances*: Tolerances in plan position, vertical deviation from the required rake and deviation in level of the pile head, should be specified. Suitable values for tolerances are given in Section 3.4.13.

*Monitoring* of piling is mandatory under EC7-1 Clause 7.9(1) and the BS EN standards for execution of special geotechnical works, in accordance with a pile installation plan or project quality plan which is consistent with the design.

*Piling records*: The CA and the piling contractor should agree the form in which records should be submitted (see Section 11.3).

*Cutting down pile heads*: The specification should define whether it is the main contractor's or the piling contractor's responsibility to remove excess lengths of pile projecting above the nominal cut-off level. The responsibility for cutting away concrete to expose reinforcement and trimming and preparing the heads of steel piles should also be stated.

*Method of measurement*: The method of measuring pile lengths as installed should be based on an appropriate standard, for example as given in the Civil Engineering Standard Method of Measurement (CESMM4)[11.29] or in the ICE SPERW. Care is needed to define the length of pile to be measured (i.e. from cut-off level to pile toe or 'commencing surface' to toe). The standard method provides 'ancillary' bill items for extensions of preformed piles (timber, steel and concrete), but credits for short piles installed are generally excluded. For employer-designed piles, the liability for extensions or reductions in length due to unforeseen conditions would normally lie with the employer.

*Removal of spoil*: The respective responsibilities for the removal of spoil from bored piles, the removal of cut-off lengths of pile, trimming off laitance and ground raised by ground heave should be defined. The disposal of used bentonite slurry is usually by tanker, but statutory regulations now prohibit placing fluids in landfill; hence, flocculation and dewatering, preferably on site, should be specified.

## 11.3 CONTROL OF PILE INSTALLATION

### 11.3.1 Driven piles

Control of driven pile operations commences with the inspection and testing of the pre-fabricated piles before they are driven. Thus, timber piles should be inspected for quality, straightness and the application of preservative. The operations of casting precast concrete piles on site or in the factory should be inspected regularly, and cubes or cylinders of the concrete should be made daily for compression testing at the appropriate age. Materials used for concrete production should be tested for compliance with the relevant standards. In the case of steel piles, tests should be made for dimensional tolerances, and full documentation of the quality of the steel in the form of manufacturers' test certificates should be supplied with each consignment. Welding tests should be made for piles fabricated in the factory or on site. Full radiographic inspection of welds may be necessary only for marine piles, where the exposure conditions are severe (Section 10.4.2). The coating treatments should be checked for film thickness, continuity, and adhesion. Degaussing may be needed to counter magnetisation of the pile heads caused by driving, as this can be detrimental to the quality of welds made for pile extensions.

The ICE SPERW[2.5] lists the information which should be recorded for each type of pile; Table 11.2 is a typical compliant form. A separate record should be provided for each pile, and records should be signed by the piling contractor's and employer's representatives and submitted daily. Records to comply with EC7, Clause 7.9, are similar to Table 11.2, but should be provided in two parts according to BS EN 12699 for each displacement pile driven. Part 1 should give general information on the contract and type of pile, methods, and quality of materials; Part 2 should give 'particular information' as tabulated in Clause 10 of this standard for each pile. 'As-built' records of piles have to be submitted to the employer for retention under the CDM[9.55] regulations and, possibly in the near future, under the BIM[9.56] protocols.

While it is essential for the toe level and final set of every pile to be recorded, BS EN 12699 does not mandate a full record of sets during driving. There are, however, advantages in providing a log of the blow count against penetration over the full depth for every pile driven. If, for example, piles are to be driven to end bearing on a hard stratum, it may be sufficient to record the sets in blows for each 25 mm of penetration after the pile has reached the hard stratum. On the other hand, where piles are supported by shaft friction, say in a stratum of firm to stiff clay or in a granular soil overlain by weak soils, it is essential to record for every pile the level at which the bearing stratum is encountered and to check that the required length of shaft to be supported is achieved. For this purpose, the blows required for each 500 mm or each 250 mm of penetration must be recorded over the full depth of driving of each pile, until the final metre or so when the sets are recorded in blows for each 25 mm.

The pile driving analysers mentioned in Section 7.3 will record blow counts electronically at intervals selected by the monitoring technician. These data, together with driving resistance, transferred energy, and stresses in the pile at the selected depths as measured by accelerometers and strain transducers (in up to eight channels), can be transmitted wirelessly or via the Internet to the design engineer in real time. The printouts can be scrutinised to assess the cause of any problems, such as pile breakage occurring during driving, or determine the need for re-driving or testing.

If the methods of Chapter 4 have been used for calculating the penetration depth of friction piles, the depth into the bearing stratum should, theoretically, be the only criterion, and final sets should be irrelevant. However, because of natural variations in soil properties,

Table 11.2 Daily pile record for driven pile

| DAILY PILE RECORD FOR DRIVEN PILES | | | | | | | | | | | |
|---|---|---|---|---|---|---|---|---|---|---|---|
| PILE RECORDS TO BE SUBMITTED TO OFFICE DAILY | | | | | | | | | | | |
| A SEPARATE SHEET TO BE USED FOR EACH PILE | | | | | | | | | | | |
| **BLOCK NUMBER** | | | | **DRAWING NUMBER** | | | | | | | |
| **SUB CONTRACTOR** | | | | **PILE TYPE** | | | | | | | |
| **1.GENERAL** | PILE REF. NO. | | | PILE DIA. | | | | LEVEL OF TOE | | | |
| | GROUND LEVEL | | | CUT OFF LEVEL | | | | CONCRETED LEVEL | | | |
| **2.DRIVING** | DATE STARTED | | | DATE COMPLETED | | | | AIR TEMPERATURE | | | |
| | ERROR IN POSITION ON PLAN | | | ERROR IN PLUMB | | | | DEPTH DRIVEN | | | |
| **3. OBSTRUCTIONS** | TYPE | | | DEPTH ENCOUNTERED | | | | PENETRATION TIME | | | |
| **4.*STEEL** MAIN | NO. OF BARS | | | DIAMETER | | | | LENGTH | | | |
| LINKS OR HELIX | CENTRES OF BARS/PITCH | | | DIAMETER | | | | COVER TO ALL STEEL | | | |
| **5.CONCRETE** | DATE STARTED | | | TYPE OF CEMENT | | | | CONCRETE TEMPERATURE | | | |
| CAST-IN-SITU WORK ONLY | MIX | | | SLUMP | | | | SUPPLIER | | | |
| **6.*HAMMER** | TYPE | | MASS | CAST-IN-SITU CONCRETE PILES ONLY | | | | | | | |
| | | | | ESTIMATED QUANTITY OF CONCRETE IN BULB | | | | | | | |
| **7.FINAL SET** | NO. OF BLOWS | | | DROP OF HAMMER | | | | MOVEMENT OF PILE | | | |
| **8.PRECAST CONCRETE PILES ONLY** | COMPONENTS OF PILE | | | **A** | | **B** | | **C** | | **D** | |
| | LENGTH OF COMPONENT | | | | | | | | | | |
| | REF. NO./DATE CAST | | | | | | | | | | |
| | O.D. LEVEL OF JOINT | | | | | | | | | | |
| **9.PILE DRIVING LOG** | DEPTH DRIVEN | NO. OF BLOWS | HAMMER DROP | DEPTH DRIVEN | NO. OF BLOWS | HAMMER DROP | DEPTH DRIVEN | NO. OF BLOWS | HAMMER DROP | DEPTH DRIVEN | NO. OF BLOWS | HAMMER DROP |
| | | | | | | | | | | | |
| | | | | | | | | | | | |
| | | | | | | | | | | | |
| | | | | | | | | | | | |
| OTHER REMARKS MAY BE RECORDED ON REVERSE SIDE OF THIS SHEET. | | | | | | | | | | | |
| | | | | | | | | | | | |
| | | | | | | | | | | | |
| | | | | | | | | | | | |
| | | | | | | | | | | | |
| **NOTE** * IF THERE ARE NO CHANGES TO BE RECORDED ITEMS 4 & 6 NEED BE COMPLETED FOR THE FIRST PILE ONLY IN EACH BLOCK. | | | | SIGNED | | | | | CONTRACT SITE ENGINEER | | |

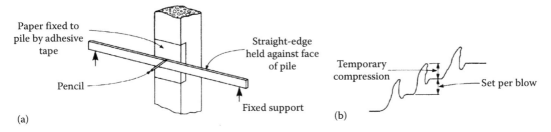

*Figure 11.5* Measuring set and temporary compression on driven pile. Arrangement of straight edge and paper card. Pencil trace showing set and temporary compression.

piles with identical lengths in the bearing stratum will not necessarily have identical ultimate loads. By driving to a minimum depth into the bearing stratum and to a constant final set (or to within a specified range of set), the variations in the soil properties can be accommodated.

A minimum penetration is necessary because random compact layers in the soil may result in localised areas of high driving resistance. The driving records within these layers should be compared with the ground investigation data, so that suitable termination levels can be established. The establishment of criteria for controlling the termination of piles driven into layered soils is described in Section 4.5.

It is advisable to conduct re-driving tests on preliminary piles and on random working piles. These tests are a check on the effects of heave and on possible weakening in resistance due to pore pressure changes. Re-driving can commence within a few hours in the case of granular soils, after 12 h for silts and after 24 h or more for clays. If the re-driving shows a reduction in resistance after about 20 blows, driving should continue until the original final set is regained. Careful monitoring is essential when re-driving a friction working pile in stiff clay.

The temporary compression at various intervals of pile driving is irrelevant if applied loads have been obtained by the methods described in Chapter 4. However, if pile driving formulae are adopted, the temporary compression values must be taken at intervals after the pile enters the bearing stratum. Figure 11.5 shows a simple field measurement, but pile driving programmes will provide accurate readings of such compression.

Other items to be recorded include any obstructions to driving or damage to the pile and deviations in alignment which might indicate breakage below the ground surface. Methods of checking the alignment of steel tubular and H-piles are described in Section 2.2.4. Hollow precast concrete piles can be checked for alignment in a similar way to steel tubes. It would be advantageous if manufacturers of jointed precast concrete piles were to provide a central hole in each unit, or in a proportion of the piles cast, down which an inclinometer could be lowered on the completion of driving.

## 11.3.2 Driven and cast-in-place piles

Table 11.2 is a suitable form of record. Generally, the procedure for recording driving resistances and sets is similar to that described in the preceding section, but in the case of proprietary piles, the piling contractor decides the criteria for the final set.

The concrete mix[2.30] should be designed to produce the required strength and workability properties. The concrete supply to the site should be checked regularly for compliance and test cubes or cylinders taken daily for compression tests. In addition, the following checks should be recorded:

- The quantity of concrete placed in the shaft of each pile as assurance against the possible collapse of the soil during the withdrawal of the tube.
- The level of the concrete as each batch is placed to give an indication of possible 'necking' of the shaft.
- The volume of concrete in an enlarged base as a check on the design assumptions for the diameter of base.
- The level of the reinforcing cage after withdrawing the drive tube on every pile driven (a safeguard against the cage being lifted with the tube).
- Thin shell piles should be inspected before placing the concrete by shining a light down the hole (to reveal any torn or buckled shells).

### 11.3.3 Bored and cast-in-place piles

The record in Table 11.3 gives information required in SPERW and complies generally with EC7-1 Clause 7.9, but as for displacement piles, records have to be provided in two parts according to BS EN 1536 Annex B for each bored pile. Records for CFA piles should include the pitch of the screw, and the factors included on data loggers used to monitor construction, for example the penetration per revolution, torque of drilling motor and pumping pressure of grout or concrete (see pile log in Figure 2.32). Clause 9.2.5 of BS EN 1536 states that 'ground behaviour' during excavation shall be observed, and any changes which may be important shall be communicated to the designer. Reference should be made to the comprehensive set of tables which detail the information and frequencies required under this standard.

If the boreholes are free of water, the conditions at the base of small-diameter piles in dry boreholes can be checked by shining a light down to the bottom before placing the concrete. In critical cases (and subject to the specification on safety assessment), large-diameter piles may be inspected from a safety cage of the type shown in Figure 11.6, following the safety procedures described in BS 8008. The presence of cuttings or cake at the base of a pile bored under a support slurry should be checked with a weighted dip line as a minimum precaution. Kort et al.[11.30] describe the use of a multidirectional head sonar calliper to determine the bore profile and verticality in these conditions.

The procedures and problems in placing concrete in pile boreholes are described in Sections 3.4.5 and 3.4.6. The controls required for the design, mixing and placing are these mentioned for driven and cast-in-place earlier, with additional account taken of exposures to aggressive ground or water.

### 11.4 LOAD TESTING OF PILES

EC7 provides for pile design to be based on static loading tests, dynamic impact tests and pile driving formulae (subject to determination of ground stratification). Pile design in the United Kingdom generally relies on proven calculation methods using selected soil parameters with loading tests undertaken where appropriate to verify the calculations and check the construction method.

However, there are conditions given in EC7 at Clause 7.5.1(1)P when pile tests are mandatory. There is no specific guidance on the number of piles to be tested for design purposes or to check designs and what type of test should be used. In complex ground conditions and

Table 11.3 Record for bored pile

## DAILY PILE RECORD FOR LARGE-AND SMALL-DIAMETER BORED PILES
### PILE RECORDS TO BE SUBMITTED TO OFFICE DAILY
### A SEPARATE SHEET TO BE USED FOR EACH PILE

| BLOCK NUMBER | | DRAWING NUMBER | | / | / |
|---|---|---|---|---|---|

| 1. General | PILE REF. NO. | | PILE DIA. | | LEVEL OF BASE | |
|---|---|---|---|---|---|---|
| | | | UNDERREAM DIA. | | | |
| | GROUND LEVEL | | CUT OFF LEVEL | | CONCRETED LEVEL | |
| 2. Drilling | DATE STARTED | | DATE COMPLETED | | AIR TEMP | |
| | ERROR IN POSITION ON PLAN | | ERROR IN PLUMB | | DEPTH BORED | |
| 3.* Obstructions Natural Unnatural | TYPE | | DEPTH ENCOUNTERED | | PENETRATION TIME | |
| | TYPE | | DEPTH ENCOUNTERED | | PENETRATION TIME | |
| 4. *Steel main steel links or helix | NO. OF BARS | | DIAMETER | | LENGTH | |
| | CENTRES OF BARS/PITCH | | DIAMETER | | COVER TO ALL STEEL | |
| 5. Concrete | DATE STARTED | | DATE COMPLETED | | CONCRETE TEMP. | QUANTITY ACTUAL: THEORETICAL: |
| | MIX | | SLUMP | | SUPPLIER | |

| 6. Borehole log and rock excavation | DEPTH OF SOIL | DESCRIPTION OF SOIL | DEPTH OF ROCK | DESCRIPTION OF ROCK | DEPTH OF ROCK AUGERED | DEPTH OF ROCK CHISELLED |
|---|---|---|---|---|---|---|
| | | | | | | |

| 7. *Casing | DEPTH OF TEMPORARY CASING | | DEPTH OF PERMANENT CASING | | REASON FOR USE OF PERMANENT CASING | |
|---|---|---|---|---|---|---|
| 8. *Water | DEPTH ENCOUNTERED | | DETAILS OF STRONG FLOW | | DETAILS OF REMEDIAL MEASURES | |
| | DEPTH TO STRONG FLOW | | | | | |

Note : * If there are no changes to be recorded, items 3, 4, 7 and 8 need be completed for the *first pile only* in each block.

Remarks

SIGNED                                                    CONTRACT SITE ENGINEER

*Figure 11.6* Safety cage used for inspection of pile boreholes.

where risk is high, it is suggested that as a design check, at least one preliminary fully instrumented static pile test be undertaken for every 250 working piles and 1%–2% of the working piles proof loaded. For new piling techniques and to satisfy EC7, this frequency may need to be increased. Short-term testing on a single pile in a group may not be representative of the combined resistance of the group (but see Section 4.9.4). There is no objection to preselecting working piles for testing, as opposed to random selection after completion, provided that the preselected piles have been shown to have been constructed in the same way as others.

## 11.4.1 Compression tests

Two principal types of test are used for compressive loading on piles. The first of these is the *constant rate of penetration* (CRP) test, in which the compressive force is progressively increased to cause the pile to penetrate the soil at a constant rate until failure occurs. The second type of test is the *maintained load* (ML) test in which the load is increased in stages to some multiple, say 1.5 times or twice the applied load with the time/settlement curve recorded at each stage of loading and unloading. The ML test may also be taken to failure by progressively increasing the load in stages.

EC7-1 Clause 7.5 outlines procedures for static and dynamic load tests, trial piles and testing working piles. BS EN 1536 refers to EC7-1 requirements giving recommendations for CRP, ML and dynamic and integrity testing. BS EN 12699 is less prescriptive for tests on displacement piles, but requires testing to be in accordance with the relevant parts of

EC7 and the specifications. In all cases, sufficient time must be allowed for the pile material to achieve the required strength and, ideally, for pore-water pressures to regain their initial values (or pore pressures monitored to assess the effects on the test).

The CRP method is essentially a test to determine the ultimate load on a pile and is therefore applied only to preliminary test piles or research-type investigations. The method has the advantage of speed in execution, and because there is no time for consolidation or creep settlement of the ground, the load/settlement curve is easy to interpret. Penetration rates of 0.75 mm/min are suitable for friction piles in clay and 1.5 mm/min for piles end bearing in a granular soil and are compatible with Clause 7.5. The CRP test is not suitable for checking compliance with the specification requirements for the maximum settlement at given stages of loading. The London District Surveyors Association considers that CRP tests are not appropriate for piles in London Clay[4.13].

The ML test is best suited for proof loading tests on working piles. Clause B15 of SPERW dealing with static loading tests defines the *specified working load* (SWL) as 'the specified load on the head of a pile as stated in the relevant particular specification'. This is differentiated from the *design verification load* (DVL) which is defined as 'a load which will be substituted for the specified working load for the purpose of a test and which may be applied to an isolated or singly loaded pile at the time of testing the given conditions of the site' The DVL takes into account special conditions which may not apply to all piles on the site such as negative skin friction or variations in pile head casting level. A proof load test on working piles should normally be the sum of the DVL plus 50% of the SWL (or as specified), applied in the sequence shown in Table 11.4 for multi-cyclic pile tests. A footnote to EC7-1 Clause 7.5.2.1 refers to an earlier, and slightly different, loading sequence.

Following each load increment, the load is held for the periods shown and until the rate of settlement is reducing. This is dependent on the pile head displacement achieved: for example, for less than 10 mm displacement, the rate should be $\leq 0.1$ mm/h and for greater than 24 mm displacement, $\leq 0.24$ mm/h.

*Table 11.4* Loading sequence for a proof load multi-cyclic test

| Load | Minimum time of holding load |
| --- | --- |
| 25% DVL | 30 min |
| 50% DVL | 30 min |
| 75% DVL | 30 min |
| 100% DVL | 6 h |
| 75% DVL | 10 min |
| 50% DVL | 10 min |
| 25% DVL | 10 min |
| 0 | 1 h |
| 100% DVL | 1 h |
| 100% DVL + 25% SWL | 1 h |
| 100% DVL + 50% SWL | 6 h |
| 100% DVL + 25% SWL | 10 min |
| 100% DVL | 10 min |
| 75% DVL | 10 min |
| 50% DVL | 10 min |
| 25% DVL | 10 min |
| 0 | 1 h |

Source:  Courtesy of Thomas Telford Limited, London, UK

To obtain the ultimate load on a preliminary test pile, it is useful to adopt the ML method for up to twice the applied load and then to continue loading to failure at a CRP. A further modification of the ML test consists of returning the load to zero after each increment. This form of test is necessary if the net settlement curve is used as the basis of defining the failure load (Section 11.4.2). An ISSMGE committee has recognised the need for improved standardisation of ML and CRP testing and has produced a recommendation document[11.31] for the execution and interpretation of axial static pile loads, but this is not yet a UK or European standard.

CRP and ML tests use the same type of loading arrangements and pile preparation. A square cap is cast onto the head of a concrete pile with its underside clear of the ground surface. Steel piles are trimmed square to their axis, and a steel plate is welded to the head, stiffened as necessary by gussets. Suitable loading arrangements for applying the load to the pile by a hydraulic jack using as the reaction, either kentledge, tension piles or cable anchors, are shown in Figures 11.7 through 11.9, respectively. The clearances between the pile and

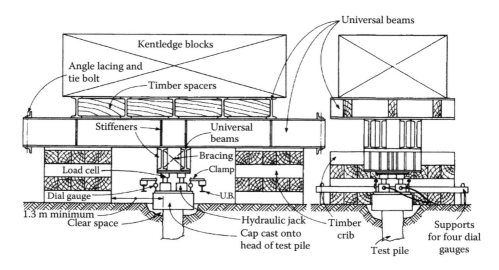

Figure 11.7 Testing rig for compressive test on pile using kentledge for reaction.

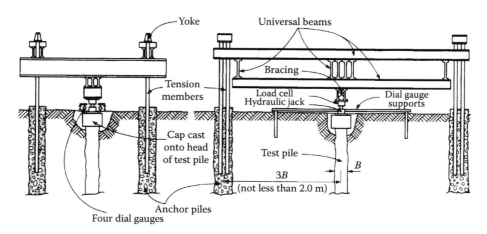

Figure 11.8 Testing rig for compressive test on pile using tension piles for reaction.

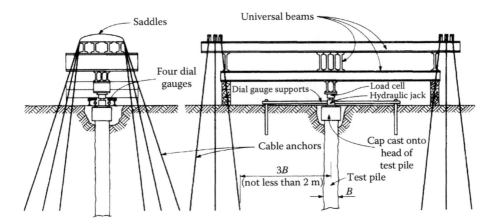

*Figure 11.9* Testing rig for compressive test on pile using cable anchors for reaction.

the reaction support systems are those recommended in SPERW Clause B15.9.5. These are necessary to avoid the induced horizontal pressures from the supports having an appreciable effect on the shaft friction and base load of the test pile. It is uneconomical to space the supports so widely apart that all effects are eliminated, and if necessary, the contribution of these surcharge effects should be calculated and allowed for in the interpretation of the test results.

Where piles are installed through fill or soft clay, these materials give positive support in shaft friction to the test pile, whereas they may add to the applied load in negative shaft friction on the permanent piles. It may therefore be desirable to sleeve the pile through these layers by using a double-sleeve arrangement. Alternatively, the outer casing can be withdrawn, after filling the annular space between it and the steel tube encasing the test pile with a bentonite slurry.

It is inadvisable to test raking piles by a reaction from kentledge or tension piles since the horizontal component of the jacking force cannot be satisfactorily restrained by the jacking system. Cable anchors inclined in the same direction as the raking piles can be used with a suitable stabilising crib. Separate vertical piles with similar dimensions to the proposed rakers could be installed and tested if conditions mean that comparisons are feasible. *Statnamic* testing (see the following text) is now the preferred method for determining the ultimate loads on raking piles.

The combined weight of the kentledge and reaction girders, or the calculated resistance capacity of tension piles or cables, must be greater than the maximum jacking force required to achieve ultimate loading. In the case of kentledge loading, the combined weight should be about 20% greater than this force. Cable anchorages or tension piles should have an ample safety factor against uplift. The former can be tested by stressing the anchors after grouting. If there is any doubt about the uplift capacity of tension piles, a test should be made to check the design assumptions. Increased capacity of tension piles in clays can be obtained by under-reaming (Section 6.2).

The reaction girders and load-spreading members should be so arranged that eccentric loads caused by any lateral movement of the pile head will not cause dangerous sidesway or buckling of the girders. Connections should be bolted so that they will not become dislodged if there is a sudden rebound of load due to the failure of the pile shaft or of the jack. Similarly, the kentledge stack should not be arranged in such a way that it may topple over.

*Figure 11.10* Patented arrangement for a 5800 tonne static load test.

The reaction girders, anchorages and jacking arrangements for a 5800 tonnes static load test in Taipei are shown in Figure 11.10.

Restraint by a pair of anchors from a single pile to each end of the reaction girder is not a good practice as it can cause dangerous sidesway of a deep girder. The piles or anchor cables should be placed in pairs at each end of the girders, as shown in Figures 11.8 and 11.9. Permanent piles can be used as anchorages for ML tests on working piles, but it is unwise to use end-bearing piles for this purpose when the shaft friction will be low and the pile may be lifted off its seating. When using tension piles, special threaded anchor bars extending above the pile head should be cast into the piles for attachment to the reaction girders. It is inadvisable to weld such bars to the projecting reinforcing bars because of the difficulty in forming satisfactory welds to resist the high tensile forces involved.

The hydraulic jack should have a nominal capacity which exceeds by 20% or more the maximum test load to be applied to the pile. This will minimise the risks of any leakage of oil through the seals when reaching maximum load. The ram of the jack should have a long travel (15% of the pile width) where piles are being loaded near to the failure condition. This is to avoid having to release oil pressure and repack with steel plates above the ram as the pile is pushed into the ground.

The load is best applied through an accurate servo-hydraulic jacking system and measured by load cells as shown in Figure 11.11, rather than relying on manually operated jacks and observing pressure gauges. Measurement of the load can then be carried out remotely by the strain gauge load cell arrangement and the settlement by displacement transducers (reading to 0.01 mm), with the records logged immediately on a computer, giving a fully automated, safe system. For high test loads, load columns capable of loads up to 1000 tonnes capacity each are used. The computer programs TIMESET® and CEMSOLVE® (see Appendix C) were developed by Cementation Skanska to monitor and predict pile test performance as reported by Fleming[11.32]. Precise levelling checks should be carried out on datum beams. A load pacer can be added for CRP tests. The data are then reproduced in the format of the test report and can be used to analyse the pile behaviour throughout the whole range of loading.

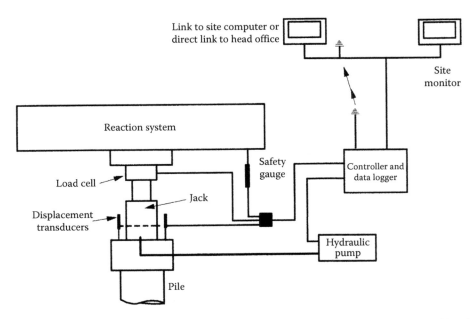

Link to site computer or direct link to head office

Site monitor

Reaction system

Safety gauge

Controller and data logger

Load cell

Jack

Displacement transducers

Hydraulic pump

Pile

*Figure 11.11* Schematic of typical automated static load test. (Courtesy of Cementation Skanska Ltd, Rickmansworth, UK.)

The traditional method of measuring settlement using dial gauges on reference points on the pile head is covered in SPERW and shown in Figures 11.7 through 11.9. In all cases, and especially where access is needed for technicians to carry out measurements at the pile while the test in underway, kentledge support must be carefully designed to allow sufficient space for technicians to work safely.

Where piles have been designed by the methods described in Chapter 4, it is very helpful to provide devices whereby the shaft and base loads can be evaluated separately. The *Osterberg load cell (O-cell)* as provided and operated by Fugro Loadtest Ltd. was originally used to assess the shaft friction and end bearing of rock sockets[4.63] but is now applied to a range of piles and test loads. It comprises a sacrificial hydraulic flat jack mounted between bearing plates and installed within the pile in order to load the pile from the base rather than the pile top. The cell assembly is attached to the bored pile reinforcement cage and cast into the pile to provide the reaction to the jacking force (Figures 11.12 and 11.13). Cells can be placed at levels up the shaft, but in order to maximise the mobilised load, they are usually placed where there is equal resistance above and below the O-cell. In a bored pile, once the concrete has been placed and reached adequate strength, the O-cell is pressurised, applying load upwards against the upper shaft friction and downwards against base resistance and lower frictional capacity. These are known as *bidirectional* tests and methods of installation and analysis are given by England[11.33]. The cells can be used in CFA piles and barrettes, and especially constructed cells can be pre-installed in driven piles, prestressed concrete piles and tubular steel piles. The test can be a substitute for tension tests which attempt to pull out the pile and can also be applied to offshore piles where the concrete cut-off level is low. The load increments can follow the ICE sequence as presented previously, but it is usual to apply more increments at 8–10 min intervals. The test continues until either the base or the shaft reaches the ultimate resistance, but unless the shaft and base have similar values, the full value of pile resistance may not be determined. To avoid potential overestimate of the pile head

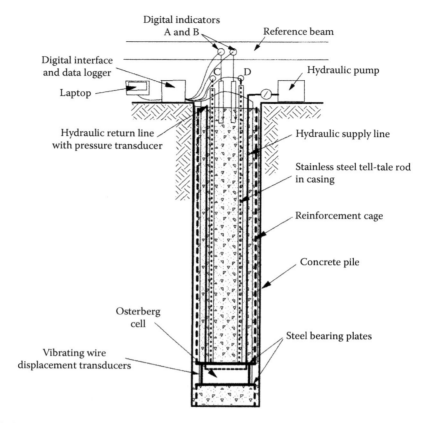

*Figure 11.12* Schematic of the Osterberg cell test in bored pile. Gauges A and B measure upward movement of top of shaft; C and D measure shaft compression. Vibrating wire transducers measure the expansion of O-cell. (Courtesy of Fugro Loadtest, Sunbury-upon-Thames, UK.)

stiffness at low displacements, it is advisable to limit the test to piles with length/diameter ratio, $L/d < 50$, and maximum length of 40 m.

The benefit of this method of testing is that very high test loads are achievable without the need for costly, large kentledge frames. For example, one of the 3.5 m diameter drilled shafts 36.3 m deep for the 1.22 mile cable-stayed bridge over the Mississippi River at St Louis, Missouri, was loaded to 321 MN with a bidirectional load of over 180 MN using four 860 mm diameter O-cells. The test is generally applied to preliminary piles, but for deep, large-diameter piles this may not be economic. In such cases the cell can be grouted post-test to restore the structural integrity and the pile incorporated into the structure.

*Dynamic load tests* and high strain integrity testing have developed significantly in recent years to give real-time calculations of bearing capacity of driven piles during driving. The test uses the short duration of the pile hammer impact (typically 5–20 ms) and instrumentation attached to the pile above ground to measure the resulting axial strain and acceleration of the pile. Pile diving analysers and computer software, such as PDA and CAPWAP® from Pile Dynamics, Inc. (see Section 7.3 and Appendix C), process the data acquired from the impact and the propagation of the wave in the pile and surrounding soil to give estimates of the static bearing capacity on completion of driving. The complementary program, iCAP® applicable to uniform piles, produces a load/settlement calculation in real time during driving without having to allow a 'soil damping' factor as in CAPWAP. The test is frequently

*Figure 11.13* O-cell installation in 1500 mm test pile at Farringdon Station redevelopment, London. Maximum load mobilised was 80 MN. (Courtesy of Fugro Loadtest, Sunbury-upon-Thames, UK.)

carried out during re-striking of piles, with account being taken of time effects. EC7-1 refers to the dynamic test procedures in ASTM Designation D4945-08 and requires that the method be calibrated against static load tests on the same type of pile, of similar length and cross section, and in comparable ground conditions. Clause B14 of SPERW requires that dynamic testing of cast-in-place piles be delayed for 4 days after casting.

The SIMBAT® dynamic test for bored cast-in-place piles applies a series of blows (5–10) to the pile and measures strain, acceleration, and displacement of the pile to produce the static load/settlement curve. The analysis is again based on wave propagation in the pile cylinder, to determine firstly the dynamic soil reaction then the static reaction; as with other dynamic tests, it is essential to achieve sufficient displacement to assess ultimate pile resistance. The analysing software corrects the acceleration data using the input from a high-speed theodolite to give displacement and can model separate shaft and end-bearing resistance; no damping factor is needed. Long[11.34] recommends that the ratio of the drop weight to pile weight should be 0.5 and to the applied load, 0.015; this should allow proof testing up to 1.5 SWL. Testconsult offers a range of systems including a portable rig for minipiles (see Figure 11.14) and free-fall drop weights up to 30 tonne handled from a crane. Piles of 2 m diameter have been successfully tested at 30 MN.

In the *Statnamic* rapid load test developed by Bermingham Foundations Solutions[11.35], loads ranging from 0.1 to 50 MN are generated by rapidly propelling a reaction mass upward off the foundation producing an equal and opposite reaction on the pile (Figure 11.15). The burning of a special fuel inside a combustion chamber provides the explosive force to lift

Figure 11.14 SIMBAT® test using a minirig for small-diameter dynamic tests. (Courtesy of Testconsult, Warrington, UK.)

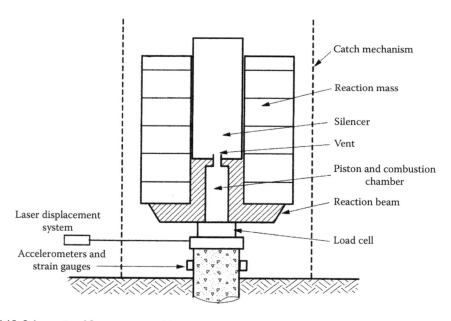

Figure 11.15 Schematic of Statnamic rapid load test set-up on a pile.

weights (5%–10% of the required test load) mounted on the pile head to a height of about 2.5 m at accelerations of up to 20 g, the impulse duration being 50–200 ms. The reaction mass is safely caught by hydraulic or mechanical latching. The load on the pile is measured by a dynamic load cell and the displacement of the pile by a laser beam and photovoltaic sensor; hence, the test is effective on non-uniform piles. The displacement should be at least 10% of the pile diameter to give ultimate capacity, requiring a greater explosive load in fine-grained soil. Accelerometers and strain gauges can be attached at the pile head and also cast into the pile toe which, together with a powerful data acquisition system, provides measurement of load distribution in the pile and the pile bearing capacity. Damping and inertia effects must be allowed for. The test is a recognised method under ASTM Designation D7383-10 and is used on preliminary and working piles.

Because of the time effects of short duration dynamic tests, they cannot be used to assess creep or consolidation and must be considered undrained tests. When calibrating dynamic tests against static load tests, it is fundamental that the pile types and soil conditions are the same and that time-related effects are considered.

Further guidance on the procedure for pile load testing is given by the Federation of Piling Specialists[11.36].

## 11.4.2 Interpretation of compression test records

A typical load/settlement curve for the CRP test and a load–time–settlement curve for the ML test are shown in Figure 11.16. The ultimate or failure load condition can be interpreted in several different ways. There is no doubt that failure in the soil mechanics sense, as stated by Terzaghi, occurs when the pile plunges down into the ground without any further increase in load. From the point of view of the structural designer, the pile has failed when its settlement has reached the stage that unacceptable distortion and cracking is caused to the structure which it supports. The latter movement can be much less than that resulting from ultimate failure in shear of the supporting soil. In the case of high loading on long slender piles, the elastic shortening under the test load will produce increased pile head settlements.

With reference to Figure 11.16, some of the recognised criteria for defining failure loads are listed as follows:

1. The load at which settlement continues to increase without any further increase of load (point A)
2. The load causing a gross settlement of 10% of the least pile width (point B)
3. The load beyond which there is an increase in gross settlement disproportionate to the increase in load (point C)
4. The load beyond which there is an increase in net settlement disproportionate to the increase of load (point D)
5. The load that produces a plastic yielding or net settlement of 6 mm (point E)
6. The load indicated by the intersection of tangent lines drawn through the initial, flatter portion of the gross settlement curve and the steeper portion of the same curve (point F)
7. The load at which the slope of the net settlement is equal to 0.25 mm per MN of test load

EC7-1, Clause 7.6.2.2, prescribes a method for assessing design pile loads from a series of static load tests as described in Section 4.1.4. With experience, the load/settlement curve from a compression test can be used to interpret the mode of failure of a pile. A defective pile shaft is also indicated by the shape of the curve. Some typical load/settlement curves and their interpretation are shown in Figure 11.17; Figures 11.17e and f demonstrates the value of loading tests in detecting defects in piles.

A method of analysing the results of either CRP or ML tests to obtain an indication of the ultimate load in conditions where the maximum applied test load does not reach the ultimate pile resistance is described by Chin[11.37] and included in SPERW. The settlement $\Delta$ at

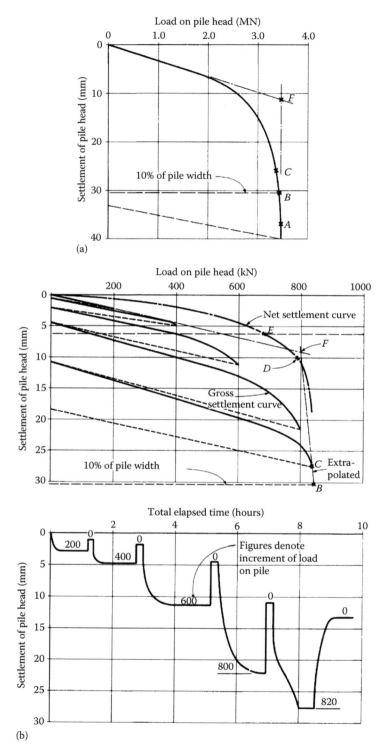

*Figure 11.16* Compression load tests on 305 × 305 mm pile: (a) Load/settlement curve for CRP test for pile on dense gravel. (b) Load/settlement and time/settlement curves for ML test for pile on stiff clay.

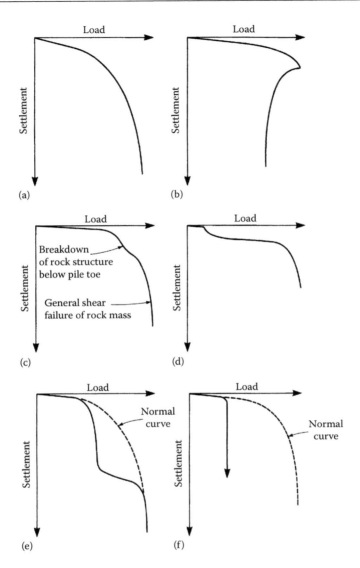

*Figure 11.17* Typical load/settlement curves for compressive load tests. (a) Friction pile in soft, firm clay or loose sand. (b) Friction pile in stiff clay. (c) Pile bearing on weak porous rock. (d) Pile lifted off seating on hard rock due to soil heave and pushed down by test load to new bearing on rock. (e) Gap in pile shaft closed up by test load. (f) Weak concrete in pile shaft sheared completely by test load.

each loading stage $P$ is divided by the load $P$ at that stage and plotted against $\Delta/P$ as shown in Figure 11.18. For an undamaged pile, a straight line plot is produced. For an end-bearing pile, the plot is a single line (Figure 11.18a). A combined friction and end-bearing pile produces two straight lines which intersect (Figure 11.18b). The inverse slope of the line gives the ultimate load in each case. However, if either the frictional resistance or base resistance is predominant, then the separation of the resistances may not be clearly defined. Chin also describes how a broken pile is detected by a curved plot (Figure 11.18c).

The Osterberg test separates the resistances of pile shaft and base which have to be combined and analysed to reconstruct a characteristic top-loaded settlement diagram.

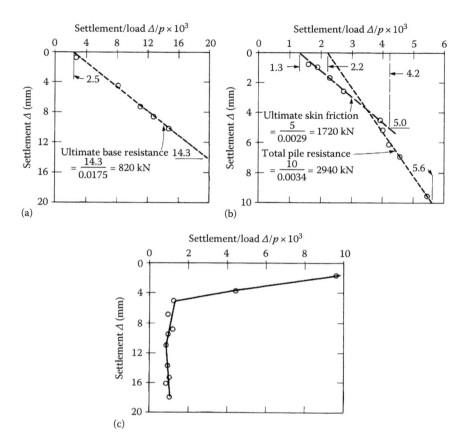

Figure 11.18 Analysis of load/settlement curves from pile loading test. (a) End-bearing pile. (b) Friction and end-bearing pile. (c) Broken pile. (After Chin, F.K., *Geotech. Eng.*, 9, 85, 1978.)

Provided that both elements are displaced sufficiently, the resulting combinations can give the characteristic behaviour of the pile up to its ultimate capacity. Figure 11.19 shows a simplified example of bidirectional displacement curves, where the upward movement is greatest, and a simple conversion to the equivalent pile head settlement, using the summation of the measured results using a hyperbolic extrapolation of the downward movement. The magnitude of mobilised shaft friction and/or end bearing may dictate which method of analysis would be appropriate to determine the pile top settlement. For example, other methods of interpretation include summation of modelled results as in the CEMSOLVE/CEMSET® programs[11.38], the Chin[11.37] method and finite element analysis[11.32]. Consideration must also be given to the elastic compression of the pile when comparing top-loaded results. It is important for the geotechnical designer to see the actual results of the extensometer and transducer read-outs.

*Dynamic tests* are analysed using the stress wave theory and continuous pile model as in CAPWAP with appropriate dynamic and static soil resistances modelled to match the measured behaviour as described in Section 7.3. However, as the pile toe resistance may not be fully mobilised by the energy which can be safely applied to avoid overstressing and damage to the pile, the ultimate pile resistance may not be determined without 'forcing' the programme iterations to produce the required 'matching'. In his review of a large database, Long[11.34] found that dynamic tests on CFA piles up to 600 mm underestimate settlement

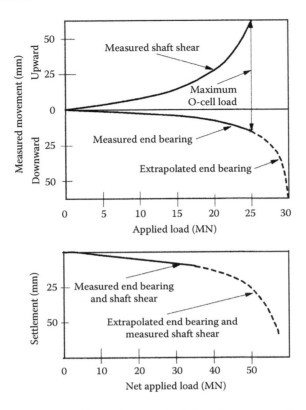

*Figure 11.19* Example of construction of equivalent top-loaded settlement curve from measured Osterberg cell test results. (Courtesy of Fugro Loadtest, Sunbury-upon-Thames, UK.)

(an average of 3 mm), possibly due to inadequate applied energy. Good correlations were achieved with static tests in rock and coarse-grained soil but poor in clays.

*Rapid load tests* can be analysed using stress wave procedures but more usually are interpreted by the simpler 'unloading point method' (UPM) to determine the static resistance of the pile. The UPM first identifies the point where the pile has zero velocity (the unloading point) and assumes that the pile resistance at this point is equivalent to the static pile resistance. The damping effects are extracted, and if the pile is instrumented at the head and toe, the inertial forces can be more accurately determined and improved static resistance derived. The UPM works well in granular soil, but in fine-grained soil and for long piles (>40 m), a non-linear approach which allows for changes in the damping effects and includes soil-dependent parameters is desirable as noted below.

A detailed research report by Paikowsky[11.39] on static and dynamic testing concluded that end bearing and shaft resistances from the O-cell test generally compare well with the conventional top-loaded static test. He found that a reduction factor should be applied to the UPM calculation for the equivalent static ultimate capacity derived from rapid loading tests to allow for rate effects; this ranges from 0.91 in sands to 0.65 in clays. Brown and Powell[11.40] describe two case studies comparing static loading (ML and CRP tests) with Statnamic tests on CFA piles in stiff London Clay and glacial till. They suggest that an analysis which incorporates a soil-dependent rate parameter to vary the damping effects with pile depth, such as liquid limit and plasticity index, provides an improved framework for the selection of the UPM reduction factor. The 'improved UPM' gave the best predictions in the medium plasticity till, being 9%–17% of the static test result equating to a reduction (or correction) factor of 0.56. The researchers' modifications using PI corrections for the London Clay were within the range of 2%–15%, which would equate to an average reduction factor of 0.39, whereas the UPM prediction ranged from 10% to 85%. However, they conclude that more data are required before realistic ultimate pile capacity and settlement can be determined from the Statnamic test in high plasticity clay.

### 11.4.3 Uplift tests

Uplift or tension tests on piles can be made at a continuous rate of uplift (CRU) or an incremental loading basis (ML). Where uplift loads are intermittent or cyclic in character, as in wave loading on a marine structure, it is good practice to adopt repetitive loading on the test pile. The desirable maximum load for repeated application cannot be readily determined in advance of the load testing programme since the relationship between the ultimate load for a single application and that for repeated application is not known. Ideally, a single pile should be subjected to a CRU test to obtain the ultimate load for a single application. Then two further piles should be tested: one cycled at an uplift load of, say, 50% of the single-application ultimate load and the second at 75% of this value. At least 25 load repetitions should be applied. If the uplift continues to increase at an increasing rate after each repetition, the cycling should be continued without increasing the load until failure in uplift occurs. Alternatively, an incremental uplift test can be made with, say, 10 repetitions of the load at each increment.

A typical load–time–uplift curve for an ML test is shown in Figure 11.20. The criteria for evaluating the failure load are similar to those described in Section 11.4.2.

EC7 Clause 7.6.3.2 prescribes a method of deriving the design tensile capacity, $R_{td}$, of a single pile from tension tests as discussed in Section 6.2.2.

A loading rig for an uplift test is shown in Figure 11.21. The methods used for measuring the jacking force and the movement of the pile head are the same as those used for compressive tests. It is particularly important to space the ground beams or bearers at an ample distance from the test pile. If they are too close, the lateral pressure on the pile induced by the load on the ground surface will increase the shaft friction on the pile shaft.

### 11.4.4 Lateral loading tests

Lateral loading tests are made by pulling a pair of piles together or jacking them apart. If the expected movements are large, for example when obtaining the load–deflection characteristics of breasting dolphin piles, a *Tirfor* or block and tackle can be employed to pull the piles together and a graduated staff used to measure the horizontal movement, as shown in Figure 11.22. Where the lateral loads on piles are of a repetitive character, as in wave loading or traffic loads on a bridge, it is desirable to make cyclic loading tests. This involves alternately pushing and pulling a pair of piles, using a rig of the type shown in Figure 11.23. Instead of a pair of piles, a single pile can be pushed or pulled against a thrust block (Figure 11.24). Where pushing methods are used, restraining devices should be provided to ensure that the jack and strut assembly does not buckle during the application of load.

The lateral movement of the pile heads may be measured by dial gauges mounted on a frame supported independently of the test piles. As with the axial load tests, the use of electronic strain gauge load cells and extensometers allows for a high degree of remote monitoring downloaded to data acquisition systems for rapid on-site analysis. Laser displacement devices are useful for marine testing, avoiding the need for an over-water support frame, thereby reducing the problem of oscillations.

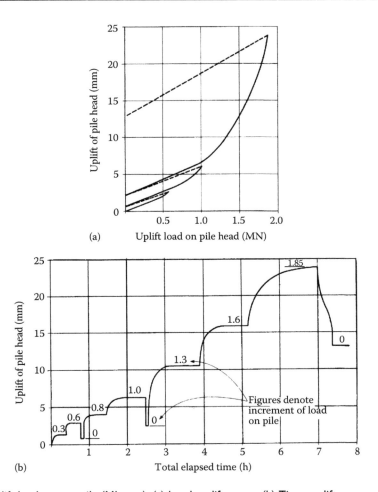

Figure 11.20 Uplift load on test pile (ML test). (a) Load–uplift curve. (b) Time–uplift curve.

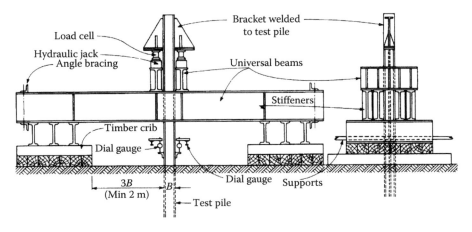

Figure 11.21 Testing rig for uplift on H-section pile using ground reaction.

Figure 11.22 Lateral loading test on two steel tubular piles forming part of a breasting dolphin.

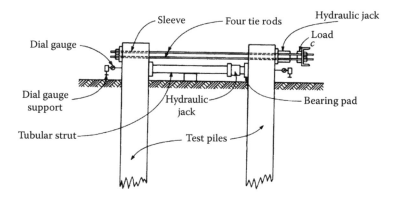

Figure 11.23 Testing rig for push and pull lateral loading test on a pair of piles.

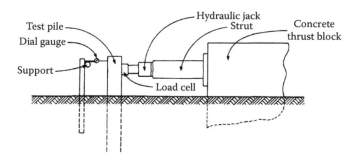

Figure 11.24 Testing rig for lateral loading test on single pile.

Downhole inclinometers and electrolevels can be mounted in a probe and lowered down a sleeve cast centrally into a pile or attached inside a steel pile to measure the slope at the pile head and down the pile, checking the assumptions made on the point of fixity as described in Chapter 6. Strain gauges can provide information on bending moments in the compressive and tensile zones. When testing piles in marine structures, it is helpful to make two separate tests by applying the load at the pile head and just above low water of spring tides. This provides two sets of curves relating deflections to bending moments. Lateral load tests are not normally continued to failure, but should simulate the design loading. Typical load/deflection curves for cyclic tests are shown in Figure 11.25.

Full-scale lateral Statnamic rapid pile tests are used in both on- and offshore applications with test loads between 1.5 and 200 kN. The equipment is mounted on a sled (or barge as Figure 11.26) and the load transmitted to the foundation through a hemispherical bearing to overcome potential rotations; instrumentation and analyses are previously discussed.

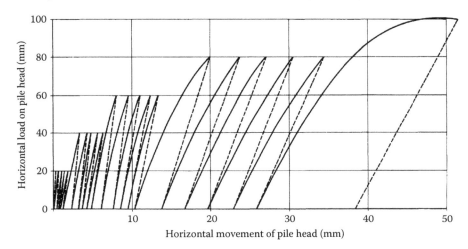

Figure 11.25 Load/deflection curve for cyclic horizontal loading test on pile (some load cycles omitted for clarity).

Figure 11.26 1100 tonne lateral Statnamic test on a pair of drilled piers for bridge foundations.

Reese and Van Impe[6.13] provide information on a variety of instrumentation and interpretation methods for lateral testing. The ICE SPERW does not comment on procedures, but ASTM Designation D3966-07 describes appropriate lateral test methods. EC7-1 Clause 7.7 does not prescribe a method of deriving the ULS design transverse load, $F_{trd}$.

## 11.5 TESTS FOR THE STRUCTURAL INTEGRITY OF PILES

From time to time, doubts are raised about the soundness of pile shafts. Excavations for pile caps may show defective conditions of the type illustrated in Figures 3.42 and 3.43, and questions are immediately asked about the likelihood of similar defects at greater depths and in other piles on the site. Where preformed piles, such as precast concrete or steel tubular sections, are used, defects can readily be explored by lowering inclinometers down guides fixed to the interior (see Section 2.2.4) or by inserting a light or TV camera down the interior of a hollow pile. Relatively inexpensive non-destructive testing of the structural integrity of piles can be undertaken using a variety of methods, including low-strain impacts (0.5–2 ms duration), the high-strain tests mentioned in Section 11.4.1 and cross-hole tomography, all as classified and described by Turner[11.41] in CIRIA Report 144.

The low-strain seismic method of dropping a weight onto the pile head and observing the time of the seismic reflections is quite widely used and has been shown by experience to give reliable results if the $L/d$ ratio is <30 and when operated and interpreted by specialists. Reliability decreases for high $L/d$ ratios in stiff soils and in jointed precast concrete piles. *Parallel seismic testing*, where a tube is grouted into the ground adjacent to piles under structures which are to be redeveloped, is a useful method for checking potential reuse of the piles. An acoustic receiver is lowered down the water-filled tube to record the time for the stress wave from a hammer blow on the foundation to the receiver. The *dynamic response* method consists of mounting a vibrating unit on the pile head and interpreting the oscillograph of the response from the pile.

Integrity testing may be applied as a routine feature of the piling contract or may be needed to resolve anomalies in the pile installation or to check reuse of an existing foundation. The main advantage of specifying integrity testing of all or randomly selected piles while pile installation is underway is that it encourages the piling contractor to keep a careful check on all the site operations. The designer will need to consider the percentage defects which can be tolerated and still provide safe foundations; if this is zero, then all piles should be tested. On a large site (say >30 piles), the first piles should all be tested, and depending on the results, random sampling may be appropriate—reverting to 100% testing if defects are located. Satisfactory evidence should be provided by the specialist performing the tests that a particular method of non-destructive testing or integrity testing will be appropriate to the site and type of pile. The methods do not replace the need for full-time supervision of the piling work by an experienced engineer or inspector, and the results of low-strain tests should not be the sole reason for acceptance or rejection of a pile.

The limitations of integrity testing were demonstrated by experiences of a field trial competition in the Netherlands[7.6]. Somewhat better results from a more recent comparative blind testing were reported by Iskander et al.[11.42] in 2003 for pulse echo and impulse response methods. Defects as small as 6% of the cross-sectional area of bored piles in varved clay were correctly identified. Cross-hole tomography was not as effective but was able to identify the pile lengths and lateral locations of the defects.

Integrity testing will indicate if a pile is badly broken but will not reveal hair cracks. Where possible defects cannot be readily interpreted from non-destructive testing, it may be necessary to check the anomalies revealed by another method. Intrusive drilling may be used to resolve the situation, either by open-hole methods using a percussion drill or rotary rock roller bit or by rotary coring. It is difficult to keep the drill hole within the confines of the shaft of a small-diameter pile, but drilling may be feasible in piles of medium to large diameters. If it is possible to flush an open-hole clear of dirty water, an inspection can be made by CCTV camera to look for cavities or honeycombed concrete. Heavy water losses when the drill hole is filled with water also indicate defective concrete. A cored hole provides a better indication of concrete soundness, and compression tests can be made on the cores, but the method is more costly than open-hole drilling. It should be noted that cores are only likely to be obtained from sound concrete, and any defective zones may not be recovered for testing. Calliper logging down a drill hole gives an indication of overbreak caused by weak concrete or cavities. A thin cable embedded in the shaft of a precast pile can provide a simple check for electrical continuity after driving. Strain gauges installed in bored piles are useful in monitoring performance of pile-supported rafts.

Pairs of ducts can be attached to the reinforcement cage of bored piles and concreted in, allowing various logging devices to be used to scan the concrete between the ducts for defects. These include sonic pulse measurements, gamma-ray logging and neutron emissions. The latter methods are believed to be reliable indicators of density changes and water content respectively, but are costly since they involve the use of skilled technicians and the transportation to site and operation of nuclear testing devices under strict safety precautions. The results of these in-pile techniques can be affected by bonding of the ducts to the concrete and their distance apart. Turner[11.41] gives a useful summary of the conditions suitable for integrity testing, and Hertlein and Davis[11.43] give guidance on specifying and interpreting low-strain and cross-hole sonic testing.

Excavation or extraction of a pile is rarely economical as a means of checking integrity as they are frequently installed in soft or loose ground, making excavation difficult and costly particularly below the water table. Over-coring as noted in Section 9.10 may assist.

## REFERENCES

11.1 *Contaminated Land. Statutory Guidance.* Department for Environment, Food and Rural Affairs, PB13735, London, UK, 2012.

11.2 Model procedures for the management of land contamination. Contaminated Land Report 11, Environment Agency, Bristol, UK, 2004.

11.3 Guide to managing and understanding the risks of asbestos in soil and on brownfield sites, RP961. Construction Industry Research and Information Association (CIRIA), Report RP961, 2013.

11.4 Reynolds, J.M. *An Introduction to Applied and Environmental Geophysics*, 2nd ed. John Wiley & Sons, Chichester, England, 2011.

11.5 Norbury, D. *Soil and Rock Description in Engineering Practice*. Whittles Publishing, Caithness, Scotland, 2010.

11.6 Binns, A. Rotary coring in soils and soft rock for geotechnical engineering, *Geotechnical Engineering*, 131 (2), 1998, 63–74.

11.7 Baldwin, M.J. and Gosling, D. BS EN ISO 22475-1: Implications for geotechnical sampling in the UK, *Ground Engineering*, 42 (8), 2009, 28–31.

11.8 Gosling, D. and Baldwin, M.J. Development of a thin wall open drive tube sampler (UT100), *Ground Engineering*, 43 (3), 2010, 37–39.

11.9 Clayton, C.R.I. The Standard penetration test (SPT): Methods and use. Construction Industry Research and Information Association (CIRIA), Report R143, 1995.

11.10 Lunne, T., Powell, J.J.M. and Robertson, P.K. *Cone Penetration Testing: In Geotechnical Practice*. Spon Press, London, UK, 1997.

11.11 Robertson, P.K. Interpretation of cone penetrometer tests – A unified approach, *Canadian Geotechnical Journal*, 46, 2009, 1337–1355.

11.12 Gambin, M. and Frank, R. Direct design rules for piles using Ménard pressuremeter test, *Contemporary Topics in In-Situ Testing, Analysis and Reliability of Foundations, GeoFlorida 2009*, American Society of Civil Engineers, Reston, VA, 2009, pp. 110–118.

11.13 Marchetti, S., Monaco, P., Totani, G. and Calabrese, M. The Flat Dilatometer Test (DMT) in soil investigation. ISSMGE TC6 Report, Bali, Indonesia, 2001, 41.

11.14 Maddison, J.D., Chambers, S., Thomas, A. and Jones, D.B. The determination of deformation and shear strength characteristics of Trias and Carboniferous strata from in-situ and laboratory testing for the Second Severn Crossing, in *Advances in Site Investigation Practice*, C. Craig (ed.), Institution of Civil Engineers, London, UK, 1996, 598–609.

11.15 Marsland, A. Large in-situ tests to measure the properties of stiff fissured clays, Building Research Establishment, Current Paper CP1/73, Department of the Environment, January 1973.

11.16 Marsland, A. and Easton, B.J. Measurements of the displacement in the ground below loaded plates in deep boreholes, *Proceedings of the Symposium on Field Instrumentation*, British Geotechnical Society, London, UK, 1973, 304–317.

11.17 McKinlay, D.G., Tomlinson, M.J. and Anderson, W.F. Observations on the undrained strength of a glacial till, *Geotechnique*, 24 (4), 1974, 503–516.

11.18 Butler, F.G. and Lord, L.A. Discussion, *Proceedings of the Conference on In-Situ Investigations in Soils and Rocks*, British Geotechnical Society, London, UK, 1970, 51–53.

11.19 International Society of Rock Mechanics. Point load test – Suggested method for determining point loads strength, *International Journal of Rock Mechanics and Mining Sciences*, 22 (2), 1985, 53–60.

11.20 Guidance Notes for Site Investigations for Offshore Renewable Energy Projects, Society for Underwater Technology, London, UK, 2005.

11.21 Danson, E. (ed.). *Geotechnical & Geophysical Investigations for Offshore and Nearshore Developments*. International Society for Soil Mechanics and Geotechnical Engineering, London, UK, 2005.

11.22 Uff, J. *Construction Law*, 11th ed. Sweet & Maxwell, London, UK, 2013.

11.23 Local Democracy, Economic Development and Construction Act 2009. Part 8 Construction. Department for Business Innovation and Skills, London, UK.

11.24 *ICC Infrastructure Conditions of Contract 2011*. Civil Engineers' Contractors Association, London, UK, 2011.

11.25 *JCT Standard Building Contract*. RIBA Publishing, London, UK, 2011.

11.26 *ICE Specification for Ground Investigation*, 2nd ed. ICE Publishing, London, UK, 2011.

11.27 *ICC Infrastructure Conditions of Contract 2011*, Ground Investigation version. Civil Engineers' Contractors Association, London, UK, 2011.

11.28 Notes for Guidance on the Specification for Highway Works. Series NG1600 Piling and embedded retaining walls. Department for Transport. Updated 2003.

11.29 *Civil Engineering Standard Method of Measurement*, 4th ed. (CESSM4), Institution of Civil Engineers, London, UK, 2012.

11.30 Kort, D.A., Hayes, J.A. and Hayes, J.S. Sonar calipering of slurry constructed drilled shafts – Providing quality assurance and quality control in deep foundations, *Proceedings of 32nd Conference on Deep Foundations*, Deep Foundation Institute, Hawthorne, NJ, 2007.

11.31 De Cock, F., Legrand, C. and Huybrechts, N. Axial static pile test in compression or in tension – Recommendations from ERTC – Piles, ISSMGE Subcommittee, *Proceedings of the 13th European Conference on Soil Mechanics and Geotechnical Engineering*, Prague, Czech Republic, Vol. 3, 2003, 717–741.

11.32 Fleming, W.G.K. Static load tests of piles and their application, *Institution of Civil Engineers, Ground Engineering Board*, Notes for meeting on May 22, 1991.

11.33 England, M. Review of methods of analysis of tests results from bi-directional static load tests, in *Deep Foundations and Augered Piles*, W.F. Van Impe and P.O. Van Impe (ed.), Taylor & Francis Group, London, UK, 2009, 235–239.

11.34  Long, M. Comparing dynamic and static test results of bored piles, *Geotechnical Engineering*, 160, 2007, 43–49.

11.35  Bermingham, P. and Janes, M. An innovative approach to load testing of high capacity piles, *Proceedings of the International Conference on Piling and Deep Foundations*, London, UK, 1989.

11.36  *Handbook on Pile Load Testing*. Federation of Piling Specialists, London, UK, 2006.

11.37  Chin, F.K. Diagnosis of pile condition, *Geotechnical Engineering*, 9, 1978, 85–104.

11.38  Fleming, W.G.K. A new method of single pile settlement prediction and analysis, *Geotechnique*, 42 (3), 1992, 411–425.

11.39  Paikowsky, S.G. Innovative load testing systems. National Cooperative Highway Research Program. Final Report, NCHRP 84 Project 21-08, 2006.

11.40  Brown, M.J. and Powell, J.J.M. Comparison of rapid load test analysis techniques in clay soils, *Journal of Geotechnical and Geoenvironmental Engineering*, 139 (1), 2013, 152–161.

11.41  Turner, M.J. Integrity testing in piling practice. Construction Industry Research and Information Association (CIRIA), Report 144, 1997.

11.42  Iskander, M., Roy, D., Shelly, S. and Ealy, C. Drilled shaft defects: Detection, and effects on capacity in varved clay, *Journal of Geotechnical and Geoenvironmental Engineering*, 129, December 2003, 1128–1137.

11.43  Hertlein, B. and Davis, A. *Non-Destructive Testing of Deep Foundations*. John Wiley & Sons, Ltd., Chichester, England, 2006.

# Appendix A: Properties of materials

## A.1 COARSE-GRAINED SOILS

| | Density when drained above groundwater level (Mg/m³) | Density when submerged below groundwater level (Mg/m³) | Angle of shearing resistance $\phi$ (degrees) |
|---|---|---|---|
| Loose gravel with low sand content | 1.6–1.9 | 0.9 | 28–30 |
| Medium-dense gravel with low sand content | 1.8–2.0 | 1.0 | 30–36 |
| Dense to very dense gravel with low sand content | 1.9–2.1 | 1.1 | 36–45 |
| Loose well-graded sandy gravel | 1.8–2.0 | 1.0 | 28–30 |
| Medium-dense well-graded sandy gravel | 1.9–2.1 | 1.1 | 30–36 |
| Dense well-graded sandy gravel | 2.0–2.2 | 1.2 | 36–45 |
| Loose clayey sandy gravel | 1.8–2.0 | 1.0 | 28–30 |
| Medium-dense clayey sandy gravel | 1.9–2.0 | 1.1 | 30–35 |
| Dense to very dense clayey sandy gravel | 2.1–2.2 | 1.2 | 35–40 |
| Loose coarse to fine sand | 1.7–2.0 | 1.0 | 28–30 |
| Medium-dense coarse to fine sand | 2.0–2.1 | 1.1 | 30–35 |
| Dense to very dense coarse to fine sand | 2.1–2.2 | 1.2 | 35–40 |
| Loose fine and silty sand | 1.5–1.7 | 0.7 | 28–30 |
| Medium-dense fine and silty sand | 1.7–1.9 | 0.9 | 30–35 |
| Dense to very dense fine and silty sand | 1.9–2.1 | 1.1 | 35–40 |

## A.2 FINE-GRAINED AND ORGANIC SOILS

|  | Density when drained above groundwater level (Mg/m³) | Density when submerged below groundwater level (Mg/m³) | Undrained shear strength (kN/m²) |
|---|---|---|---|
| Soft plastic clay | 1.6–1.9 | 0.6–0.9 | 20–40 |
| Firm plastic clay | 1.75–2.0 | 0.75–1.1 | 40–75 |
| Stiff plastic clay | 1.8–2.1 | 0.8–1.1 | 75–150 |
| Soft slightly plastic clay | 1.7–2.0 | 0.7–1.0 | 20–40 |
| Firm slightly plastic clay | 1.8–2.1 | 0.8–1.1 | 40–75 |
| Stiff slightly plastic clay | 2.1–2.2 | 1.1–1.2 | 75–150 |
| Stiff to very stiff clay | 2.0–2.3 | 1.0–1.3 | 150–300 |
| Organic clay | 1.4–1.7 | 0.4–0.7 | — |
| Peat | 1.05–1.4 | 0.05–0.40 | — |

## A.3 ROCKS AND OTHER MATERIALS

| Material | Density (Mg/m³) |
|---|---|
| Granite | 2.50 |
| Sandstone | 2.20 |
| Basalts and dolerites | 1.75–2.25 |
| Shale | 2.15–2.30 |
| Stiff to hard limestone | 1.90–2.30 |
| Limestone | 2.00–2.70 |
| Chalk | 0.95–2.00 |
| Broken brick | 1.10–1.75 |
| Solid brickwork | 1.60–2.10 |
| Ash and clinker | 0.65–1.00 |
| Fly ash | 1.20–1.50 |
| Loose coal | 0.80 |
| Compact stacked coal | 1.20 |
| Mass concrete | 2.20 |
| Reinforced concrete | 2.40 |
| Iron and steel | 7.20–7.85 |

Note:   Weight densities in kN/m³ are also given in EC-1-1 Annex A.

## A.4 ENGINEERING CLASSIFICATION OF CHALK[4.58]

Intact dry density scales of chalk

| Density scale | Intact dry density (Mg/m³) | Porosity n[a] | Saturation moisture content[a] (%) |
|---|---|---|---|
| Low density | <1.55 | >0.43 | >27.5 |
| Medium density | 1.55–1.70 | 0.43–0.37 | 27.5–21.8 |
| High density | 1.70–1.95 | 0.37–0.28 | 21.8–14.3 |
| Very high density | >1.95 | <0.28 | <14.3 |

[a]  Based on the specific gravity of calcite of 2.70.

Classification of chalk by discontinuity aperture

| Grade A | Discontinuities closed |
|---------|------------------------|
| Grade B | Typical discontinuity aperture <3 mm |
| Grade C | Typical discontinuity aperture >3 mm |
| Grade D | Structureless or remoulded mélange |

Subdivisions of Grades A to C chalk by discontinuity spacing

| Suffix | Typical discontinuity spacing (mm) |
|--------|-----------------------------------|
| 1 | $t > 600$ |
| 2 | $200 < t < 600$ |
| 3 | $60 < t < 200$ |
| 4 | $20 < t < 60$ |
| 5 | $t < 20$ |

Subdivisions of Grade D chalk by engineering behaviour

| Suffix | Engineering behaviour | Dominant element | Comminuted chalk matrix (%) | Coarser fragments (%) |
|--------|----------------------|------------------|------------------------------|------------------------|
| m | Fine soil | Matrix | Approx. >35 | Approx. <65 |
| c | Coarse soil | Clasts | Approx. <35 | Approx. >65 |

# Appendix B: Current British Standards and others referred to in the text

The following standards were current at the time of writing, but as the British Standards Institute (BSI) regularly reviews and updates the titles and content, readers are referred to the BSI website, www.bsigroup.com, for the most recent version.

| | |
|---|---|
| BS 4-1 | Structural steel sections (*withdrawn*) |
| BS 970 (all parts) | Specification for wrought steel (*withdrawn*) |
| BS 1377 – Parts 1–9:1990 | Methods of tests for soils for civil engineering purposes (*partially replaced*) |
| BS 2012-1:1974 | Code of practice for foundations for machinery (*reaffirmed 2010*) |
| BS 4449:2005 + A2:2009 | Steel for the reinforcement of concrete. Weldable reinforcing steel. Bar, coiled and decoiled product. Specification |
| BS 4978:2007 + A1:2011 | Visual strength grading of softwood |
| BS 5228-1:2009 | Code of practice for noise and vibration control on construction and open sites (*noise*) |
| BS 5228-2:2009 | Code of practice for noise and vibration control on construction and open sites (*vibration*) |
| BS 5268 (all parts) | Structural use of timber (*withdrawn*) |
| BS 5400 all parts | Steel, concrete and composite bridges. Codes of practice (*withdrawn*) |
| BS 5756:2007 + A1:2011 | Visual strength grading of hardwood |
| BS 5896:2012 | High-tensile steel wire and stand for the prestressing of concrete. Specification |
| BS 5930 + A2:2010 | Code of practice for site investigations (*partially replaced*) |
| BS 5950 (all parts) | Structural steelwork in buildings (*withdrawn*) |
| BS 6349-1:2000 | Maritime works. Code of practice for general criteria (*being replaced by the following*) |
| BS 6349-1-1 | Maritime works. Code of practice for planning and design of operations (*work in progress*) |
| BS 6349-1-2 | Maritime works. Code of practice for assessment of actions (*work in progress*) |
| BS 6349-1-3:2012 | Maritime works. General. Code of practice for geotechnical design |
| BS 6349-1-4:2013 | Maritime works. General. Code of practice for materials |
| BS 6349-2:2010 | Maritime works. Code of practice for the design of quay walls, jetties and dolphins (*revisions underway*) |

| | |
|---|---|
| BS 6349-4:1994 | Maritime structures. Code of practice for the design of fendering and mooring systems (*work in progress*) |
| BS 6472-1:2008 | Guide to evaluation of human exposure to vibration in buildings. Vibration sources other than blasting |
| BS 7385-2:1993 | Evaluation and measurement for vibration in buildings. Guide to damage levels from ground vibration |
| BS 8004 | Code of practice for foundations (*withdrawn*) |
| BS 8008: 1996 + A1:2008 | Safety precautions and procedures for the construction and descent of machine-bored shafts for piling and other purposes |
| BS 8081:1989 | Code of practice for ground anchorages (*partially replaced by BS EN 1537: 2000*) |
| BS 8110 | Structural use of concrete (*withdrawn*) |
| BS 8417:2011 | Preservation of wood. Code of practice |
| BS 8500-1:2006 + A1:2012 | Concrete. Complementary to BS EN 206. Method of specifying and guidance for the specifier |
| BS 8500-2:2006 + A1:2012 | Concrete. Complementary to BS EN 206. Specification for constituent materials and concrete |
| BS 10175:2011 + A1:2013 | Investigation of potentially contaminated sites. Code of practice |

The following are European Standards adopted by BSI.

| | |
|---|---|
| BS EN 197-1:2011 | Cement. Composition, specifications and common criteria for common cements |
| BS EN 206-1:2000 | Concrete. Specification, performance, production and conformity |
| BS EN 206-9:2010 | Concrete. Additional rules for self-compacting concrete |
| BS EN 338:2009 | Structural timber. Strength classes |
| BS EN 350-2:1994 | Durability of wood and wood-based products. Natural durability of solid wood. Guide to natural durability and treatability of selected wood species of importance in Europe |
| BS EN 445:2007 | Grout for prestressing tendons. Test methods |
| BS EN 446:2007 | Grout for prestressing tendons. Grouting procedures |
| BS EN 447:2007 | Grout for prestressing tendons. Basic requirements |
| BS EN 791:1995 + A1:2009 | Drill rigs. Safety (*to be replaced by BS EN 16228*) |
| BE EN 996:1995 + A3:2009 | Piling equipment. Safety requirements (*to be replaced by BS EN 16228*) |
| BS EN 1536:2010 | Execution of special geotechnical works. Bored piles |
| BS EN 1537:2000 | Execution of special geotechnical works. Ground anchors |
| BS EN 1912:2004 + A4:2010 | Structural timber. Strength classes, visual grades |
| BS EN 1990:2002 + A1:2005 | Basis of structural design. |
| BS EN 1991-1-1:2002 | Eurocode 1: Part 1-1, Actions on structures. General actions |
| BS EN 1991-1-4:2005 + A1:2010 | Eurocode 1: Part 1-4, Action on structures. General actions. Wind actions |
| BS EN 1992-1-1:2004 | Eurocode 2: Design of concrete structures, Part 1-1 General rules and rules for buildings |

| | |
|---|---|
| BS EN 1993-1-1:2005 | Eurocode 3: Design of steel structures, Part 1-1 General rules and rules for buildings |
| BS EN 1993-1-10:2005 | Eurocode 3: Design of steel structures, Part 1-10 Material toughness and through-thickness properties |
| BS EN 1993-5:2007 | Eurocode 3: Design of steel structures, Part 5 Piling |
| BS EN 1994-1:2005 | Eurocode 4: Design of composite steel and concrete structures, Part 1 General rules |
| BS EN 1995-1-1:2004 | Eurocode 5: Design of timber structures, Part 1-1 General rules |
| BS EN 1996-1:2005 | Eurocode 6: Design of masonry structures, Part 1 General rules |
| BS EN 1997-1:2004 | Eurocode 7: Geotechnical design, Part 1 General rules (corrigendum 2010) |
| NA to BS EN 1997-1:2004 | UK National Annex to Eurocode 7 Geotechnical design, Part 1 General rules |
| BS EN 1997-2:2007 | Eurocode 7: Geotechnical design, Part 2 Ground investigation and testing (*corrigendum 2010*) |
| BS EN 1998-1:2004 | Eurocode 8: Design of structures for earthquake resistance, Part 1 General rules |
| BS EN 1998-5:2004 | Eurocode 8: Design of structures for earthquake resistance, Part 5 Foundations, retaining walls and geotechnical aspects |
| BS EN 10024:1995 | Hot-rolled taper flange I-sections |
| BS EN 10025-Parts 1–6 | Hot-rolled products of structural steels |
| BS EN 10027:2005 | Designation system for steels. Steel names/numbers |
| BS EN 10080:2005 | Steel for reinforcement of concrete. Weldable reinforcing steel. General |
| pr EN 10138 | Prestressing steel (*due to be published in 2015*) |
| BS EN 10210:2006 | Hot-finished structural hollow sections |
| BS EN 10219:2006 | Cold-formed welded structural hollow sections |
| BS EN 10248:1996 | Hot-rolled sheet piling |
| BS EN 10249:1996 | Cold-formed sheet piling |
| BS EN 12063:1999 | Execution of special geotechnical works. Sheet piling |
| BS EN 12473:2000 | General principles of cathodic protection in seawater |
| BS EN 12699:2001 | Execution of special geotechnical works. Displacement piles |
| BS EN 12794:2005 | Precast concrete products. Foundation piles |
| BS EN 13369:2004 | Common rules for precast concrete products |
| BS EN 14199:2005 | Execution of special geotechnical works. Micropiles |
| BS EN 14647:2005 | Calcium aluminate cement |
| BS EN 15743:2010 | Supersulphated cement |
| BS EN 16228 Parts 2–7 | Foundation equipment (*in preparation 2014*) |

The following are International Standards Organisation standards adopted by BSI.

| | |
|---|---|
| BS EN ISO 148-1:2010 | Metallic materials. Charpy pendulum impact test. Test method |
| BS EN ISO 9001:2008 | Quality management systems. Requirements |
| BS EN ISO 17660-1:2006 | Welding of reinforcing steel |

| | |
|---|---|
| BS EN ISO 14688 – Parts 1 and 2:2002 | Geotechnical investigation and testing. Identification and classification of soil |
| BS EN ISO 14689 – Part 1:2003 | Geotechnical investigation and testing. Laboratory testing of soil |
| BS EN ISO 19901-1:2005 | Petroleum and natural gas industries. Specific requirements for offshore structures. Part 1 Metocean design and operating considerations |
| BS EN ISO 19901-4:2003 | Petroleum and natural gas industries. Specific requirements for offshore structures. Part 4 Geotechnical and foundation design considerations |
| BS EN ISO 19901-8 | Petroleum and natural gas industries. Specific requirements for offshore structures. Part 8 Marine soil investigations (*due in 2015*) |
| BS EN ISO 19902:2007 + A1:2013 | Petroleum and natural gas industries. Fixed steel offshore structures |
| BS EN ISO 22282 – Part 1:2012 | Geotechnical investigation and testing. Geohydraulic testing. General rules (*with five additional parts covering permeability testing*) |
| BS EN ISO 22475 – Parts 1–3:2011 | Geotechnical investigation and testing. Sampling methods and groundwater measurements |
| BS EN ISO 22476 – Part 1:2012 | Geotechnical investigation and testing. Field testing. Electrical cone and piezocone penetration test |
| BS EN ISO 22476 – Part 2:2005 + A1 | Geotechnical investigation and testing. Field testing. Dynamic probing |
| BS EN ISO 22476 – Part 3:2011 | Geotechnical investigation and testing. Field testing. Standard penetration test |
| BS EN ISO 22476 – Part 4:2012 | Geotechnical investigation and testing. Field testing. The Ménard pressuremeter test |
| BS EN ISO 22476 – Part 7:2013 | Geotechnical investigation and testing. Field testing. Borehole jack test |
| BS EN ISO 22476 – Part 12:2009 | Geotechnical investigation and testing. Field testing. Mechanical cone penetration test |

The following are 'Published Documents' from BSI which supplement British Standards.

| | |
|---|---|
| PD 6687-1:2010 | Background to the National Annexes to BS EN 1992-1 and 1992-3 |
| PD 6694-1:2011 | Recommendations for the design of structures subject to traffic loading to BS EN 1997-1:2004 |
| PD 6698:2009 | Recommendations for the design of structures for earthquake resistance to BS EN 1998 |

Other relevant British Standards for geotechnical works

| | |
|---|---|
| BS 6031:2009 | Code of practice for earthworks |
| BS 8006:2010 | Code of practice for strengthened/reinforced soils and other fills |
| BS 8103:1995 | Structural design of low-rise buildings |
| BS EN 1538:2010 | Execution of special geotechnical works. Diaphragm walls |
| BS EN 12715:2000 | Execution of special geotechnical works. Grouting |
| BS EN 12716:2001 | Execution of special geotechnical work. Jet grouting |

BS EN 14475:2006   Execution of special geotechnical works. Reinforced fill
BS EN 14490:2010   Execution of special geotechnical works. Soil nailing
BS EN 14679:2005   Execution of special geotechnical works. Deep mixing
BS EN 14731:2005   Execution of special geotechnical works. Ground treatment by deep vibration
BS EN 15237:2007   Execution of special geotechnical works. Vertical drainage

Relevant British Standards in preparation

BS EN ISO 17892 – Parts 1–12   Geotechnical investigation and testing. Laboratory testing of soil
BS EN ISO 22477 – Parts 1–7   Geotechnical investigation and testing. Testing of geotechnical structures

*British standards may be purchased from*
www.bsigroup.com/shop or by contacting BSI Customer Services for hard copies only: Tel +44 (0)20 8996 9001, e-mail cservices@bsigroup.com.

Current American Standards referred to in the text

ASTM D3966-07   Standard test methods for deep foundations under lateral load
ASTM D4945-08   Standard test method for high strain dynamic testing of piles
ASTM D7383-10   Standard test methods for axial compressive force pulse (rapid) testing of deep foundations

*ASTM standards may be purchased from*
ASTM International, 100 Barr Harbor Drive, PO Box C700, West Conshohocken, PA, 19428—2959, USA

AASHTO   LRFD Bridge Design Specification, 2010

*AASHTO standards may be purchased from:*
American Association of State Highway and Transport Officials, 444 North Capitol Street, NW Suite249, Washington, DC 20001. USA.

Current Australian Standard referred to in the text

AS 2159 2009   Piling. Design and installation

*Australian Standards may be purchased online from* http://www.saiglobal.com.

# Appendix C: Outline of computer software referred to in the text

There is a wide range of computer software available to the foundation designer and programs are updated and new ones produced regularly. The following summaries are indicative of the contents in the referenced programs and are for guidance only. The reader is referred to the relevant bureau for details of a particular application and relevant constitutive model.

*From Oasys Ltd, a subsidiary of Arup*
ALP
ALP represents a laterally loaded pile as a series of elastic beam elements and the soil as a series of non-linear *Winkler* independent springs acting at the nodes. The load–deflection can be modelled either as elasto-plastic behaviour (for multilayered soil) or as *p–y* curves. The program generates the deflection down the pile, together with bending moments and shear forces in the pile.

FREW
This is a program to analyse the soil–structure interaction of a flexible retaining wall and has been adapted to determine load/deflection behaviour of integral bridge abutments supported on a single row of piles. The soil is modelled by one of three methods: as an elastic solid with soil stiffness calculated from the SAFE finite element (FE) program, by using the Mindlin equations or the subgrade reaction method.

PILE
The program allows the user to determine either pile capacity or settlement analysis for a range of pile lengths and cross sections. Under-reams can be included in the capacity assessment but not for settlement. Allowable stress and limit state calculations can be performed on layered soils. Negative skin friction is treated as an *action* and not included as a *resistance*. The 2012 version includes analysis of thermal and structural properties for the design of geothermal piles.

PDISP
This program predicts vertical and horizontal displacements in the soil mass due to vertical and lateral loads, showing the likely settlement pattern beneath and beyond the loaded area. It assumes the soil is an elastic half-space and uses individual layer properties. For vertical loading, stresses in the soil mass can also be calculated.

*From Cementation Skanska Ltd*
CEMSET®
CEMSET predicts the behaviour of a pile under load using hyperbolic functions to represent the stress/strain relationship. Ten input parameters are applied to give the ultimate load,

corresponding to the asymptotic behaviour with soil resistance fully mobilised. It calculates the elastic shortening of the pile, base behaviour and pile recovery after removal of the load.

## CEMSOLVE®

CEMSOLVE is used for the back analysis of computer-controlled static load test results to determine the specific pile behaviour and soil parameters. The test must mobilise the shaft friction and a reasonable proportion of the end bearing. The load/settlement behaviour is also based on hyperbolic functions and identifies shaft and end-bearing capabilities separately.

## TIMESET®

This program models time/displacement behaviour of a pile under a constant load test to predict the final state of deformation at infinite time.

### From Ensoft Inc, Texas

## DYNAPILE

This is a program for the analysis of pile foundations under dynamic load. It computes the dynamic stiffness of single piles or pile groups for end-bearing and floating piles, based on *the consistent boundary-matrix* method. Input parameters consist of the structural and dynamic properties of the pile, layout of the pile group, soil properties, definition of excitation forces and definition of superstructure masses.

## LPILE Plus

This program analyses single piles under lateral load using $p$–$y$ curves. The program computes deflection, bending moment, shear force and soil response over the length of the pile. As an option, the components of the stiffness at the pile head can be applied to examine the soil–pile–structure interactions.

## APILE Plus5.0

The main calculation method for this program is the American Petroleum Institute (API) procedure as detailed in APIRP2A[4.15]. It is used to compute the axial capacity as a function of depth, of a driven pile in clay, sand or mixed soil profile. The offshore version also provides alternative computations for driven piles such as the Imperial College ICP method[4.41].

## GROUP (v8)

This program for pile groups will generate internally the non-linear response of the soil in terms of $t$–$z$ curves for axial loading and $p$–$y$ curves for lateral loading. For closely spaced piles, the soil–covered by introducing reduction factors for the $p$–$y$ curves for an individual pile.

### From Pile Dynamics Inc

## GRLWEAP

This is a 1D wave equation analysis program that simulates the pile response to pile driving equipment. It predicts driving stresses, hammer performance and the relationship between pile bearing capacity and net set per blow. The database has an interface with more than 800 preprogrammed hammers, diesel and hydraulic. This allows the user to investigate which hammer is best for a particular pile and soil conditions prior to mobilising and indicates the blow count needed for a given axial compressive load.

## CAPWAP® (CAse Pile Wave Analysis Program)

This program estimates the total bearing capacity of a pile and the resistance along the pile shaft and at the toe based on the wave theory approach of Smith[7.3]. The input is derived from the pile driving analyser (PDA) and completes the dynamic load testing procedure to simulate a static load test. It can be applied to driven, bored and CFA piles.

iCAP®
Based on CAPWAP, this program calculates pile capacity at the time of dynamic load testing and produces a simulated static load/settlement graph through a signal matching procedure performed during pile driving monitoring. CAPWAP is needed to produce the ultimate capacity following conclusion of the test.

*From Abaqus Inc/Abaqus UK Ltd.*
ABAQUS
ABAQUS is a general-purpose FE program with emphasis on non-linear simulations. Material models include the Mohr–Coulomb, cam-clay, and cap plasticity and jointed rock. All material models can be used in the coupled pore water flow stress analysis procedures and are available in 2D and 3D. Contact surfaces can be included to simulate the soil–pile-structure interaction.

*From Geocentrix, Ltd. UK*
REPUTE
For the design of single piles and pile groups to EC7 UK National Annex standard, using boundary element analysis, 3D loading, and linear and non-linear modelling of soil modulus variations. Elastic continuum based. Applicable to multilayered soils. ULS calculations for drained and undrained conditions.

*From University of Western Australia*
PIGLET
Uses an approximate *closed form* analysis, with interaction factors for single piles and groups. *Gibson* soil profile and variable shear modulus. Useful for spreadsheet application.

*From University of Sydney, Australia*
DEFPIG
This program calculates the deformations and load distribution within a group of piles attached to a rigid pile cap subjected to vertical, horizontal and moment loading. Piles raking in the direction of the horizontal load may be present.

*From Deltares (formerly GeoDelft), Delft, The Netherlands*
D-Pile Group
Earlier versions of this program were known as MPile; this version enables the analysis of the 3D behaviour of a single pile and a pile group, interacting via the pile cap and the soil. Modules based on the API rules and the Poulos elastic or plastic models are provided, together with options to analyse inclined piles and dynamic loading.

*From Foundation QC Pty., Victoria, Australia*
ROCKET (v3)
For design of rock socket piles in hard soils to strong rocks based on research at Monash University. Input parameters include shear strength of rock, residual friction angle and Poisson's ratio. The influence of socket roughness and asperities along the rock–pile interface is assessed to determine the rock–pile interaction, produce $t$–$z$ curves and pile top displacement.

*From Plaxis bv, Delft, The Netherlands*
PLAXIS 3D
PLAXIS 3D is a 3D FE program, developed for the analysis of geotechnical problems concerned with deformation, stability and groundwater flow. It allows for automatic generation of unstructured FE meshes for complex geotechnical structures. PLAXIS 3D Foundation

is designed for the analysis of raft foundations, piled rafts and offshore foundations. The program covers partial factors for ULS design to EC7 rules or other load-resistance factor design.

*From Fine Software, Ltd.*
GEO5Pile CPT
The program verifies the bearing capacity, shaft resistance and settlement of either an isolated pile or group of piles based on static cone penetration tests and applies the Bustamante[4.19] and Schmertmann[5.25,5.26] methods. It takes account of the installation method and type of pile and negative skin friction.

*From Civil and Structural Computer Services, MasterSeries*
RC Pile Cap
Design of pile caps to EC2 rules, based on bending or strut and tie methods, following the conventions of Whittle and Beattie[7.10].

*From Computers and Structures Inc, Berkeley*
SAFE©
Integrated design of flat slabs, ground beams and pile caps of any shape to EC2 and other LRFD codes. Includes FE analysis for complex slabs. Checks punching shear and designs strut and tie reinforcement.

*From Swiss Federal Institute of Technology, (EPFL), Lausanne*
THERMO-PILE
This software is based on the paper 'Geotechnical analysis of heat exchanger piles' by Knellwolf, Peron and Laloui from the *ASCE Journal of Geotechnical and Geoenvironmental Engineering*, Vol. 137(10) 2011. It is used to determine the extra stresses and displacements due to the temperature variations in a geothermal pile. It can also be used to calculate basic pile resistances when the temperatures are set to constant.

*From Itasca Consulting Group, Inc*
FLAC
A 2D explicit finite difference program for geotechnical analyses, particularly earth retention problems including slope stability where the slope contains pile or anchor support.

*From Geosolve, London*
WALLAP
This program provides limit equilibrium analysis of cantilevered and propped retaining walls based on EC7 partial factors and subgrade reaction analysis and 2D FEM to determine bending moments and displacements.

*From Autodesk Inc*
Autodesk BIM
The various design suites of Autodesk include Building Information Modelling with tools for documentation and visualisation, alongside 3D CAD software and the ability to coordinate design data from different file formats.

# Name index

# Subject index